W9-CRL-552

Enzymes in Action
Green Solutions for Chemical Problems

NATO Science Series

A Series presenting the results of activities sponsored by the NATO Science Committee. The Series is published by IOS Press and Kluwer Academic Publishers, in conjunction with the NATO Scientific Affairs Division.

A. Life Sciences	IOS Press
B. Physics	Kluwer Academic Publishers
C. Mathematical and Physical Sciences	Kluwer Academic Publishers
D. Behavioural and Social Sciences	Kluwer Academic Publishers
E. Applied Sciences	Kluwer Academic Publishers
F. Computer and Systems Sciences	IOS Press
1. Disarmament Technologies	Kluwer Academic Publishers
2. Environmental Security	Kluwer Academic Publishers
3. High Technology	Kluwer Academic Publishers
4. Science and Technology Policy	IOS Press
5. Computer Networking	IOS Press

NATO-PCO-DATABASE

The NATO Science Series continues the series of books published formerly in the NATO ASI Series. An electronic index to the NATO ASI Series provides full bibliographical references (with keywords and/or abstracts) to more than 50000 contributions from internatonal scientists published in all sections of the NATO ASI Series.
Access to the NATO-PCO-DATA BASE is possible via CD-ROM "NATO-PCO-DATA BASE" with user-friendly retrieval software in English, French and German (WTV GmbH and DATAWARE Technologies Inc. 1989).

The CD-ROM of the NATO ASI Series can be ordered from: PCO, Overijse, Belgium

Series 1: Disarmament Technologies – Vol. 33

Enzymes in Action
Green Solutions for Chemical Problems

edited by

Binne Zwanenburg
Department of Organic Chemistry,
NSR Institute for Molecular Structure,
Design and Synthesis,
University of Nijmegen, The Netherlands

Marian Mikołajczyk

and

Piotr Kiełbasiński
Centre of Molecular and Macromolecular Studies,
Polish Academy of Sciences,
Łódź, Poland

Kluwer Academic Publishers

Dordrecht / Boston / London

Published in cooperation with NATO Scientific Affairs Division

Chemistry Library

Proceedings of the NATO Advanced Study Institute on
Enzymes in Heteroatom Chemistry (Green Solutions for Chemical Problems)
Berg en Dal, The Netherlands
19–30 June 1999

A C.I.P. Catalogue record for this book is available from the Library of Congress.

ISBN 0-7923-6695-6

Published by Kluwer Academic Publishers,
P.O. Box 17, 3300 AA Dordrecht, The Netherlands.

Sold and distributed in North, Central and South America
by Kluwer Academic Publishers,
101 Philip Drive, Norwell, MA 02061, U.S.A.

In all other countries, sold and distributed
by Kluwer Academic Publishers,
P.O. Box 322, 3300 AH Dordrecht, The Netherlands.

Printed on acid-free paper

All Rights Reserved
© 2000 Kluwer Academic Publishers
No part of the material protected by this copyright notice may be reproduced or utilized in
any form or by any means, electronic or mechanical,including photocopying, recording or by
any information storage and retrieval system, without written permission from the copyright
owner.

Printed in the Netherlands.

TABLE OF CONTENTS

T P
248
.65
E59
E59185
2000
CHEM

Part IV: Strategic Use of Enzymes in Molecular Synthesis

PREFACE

The objective of the NATO Advanced Study Institute on enzymes in heteroatom chemistry was to discuss the current state of the art of organic enzyme chemistry, the general prospects of biocatalysis and the potential of practical application. The meeting was held in hotel Erika in Berg en Dal, near Nymegen, The Netherlands from June19-30, 1999. A group of 19 highly qualified speakers enthusiastically presented a series of 33 superb lectures, covering various aspects of enzyme chemistry. The ASI was attended by 65 scholars from 18 different countries.

In this book "Enzymes in Action" the presented chemistry is described in 22 chapters. In the section on methodologies and fundamentals of enzyme chemistry various aspects concerning the concepts of enzymes in molecular operations are covered. Heteroatom enzyme chemistry is treated in the second section. Considerable attention has been devoted to the use of enzymes in the detoxification of chemical warfare agents and the application of enzymes in solving environmental problems. In the final section the strategic use of enzymes in organic chemistry is highlighted by applications from different areas. Here the term green chemistry is appropriate, as enzyme mediated processes take place under mild environmentally benign conditions. Moreover, enzymes enable chemists to perform new chemical operations that otherwise are difficult to achieve at all.

The book gives a timely overview of a modern development in organic chemistry. It is clear that bioreagents require a different way of thinking of organic chemists. But by doing so many new avenues of exciting new chemistry open up, allowing to solve chemical problems in an elegant manner. It is clear that the impact of enzymes on organic chemistry just has started; much is to be expected in the coming years.

Prof. Dr. Binne Zwanenburg

BIO- AND CHEMO-CATALYTIC DERACEMISATION TECHNIQUES

ULRIKE T. STRAUSS AND KURT FABER*
*Institute of Organic Chemistry, University of Graz, Heinrichstraße 28,
A-8010 Graz, Austria. <kurt.faber@kfunigraz.ac.at>*

Abstract: Methods for the preparation of chiral building blocks in 100% (theoretical) chemical and optical yield from racemates are reviewed and their specific merits and limitations are discussed. This so-called 'deracemisation' is achieved by employing a bio- or chemo-catalyst or a combination of both. Four general categories of processes can be characterized: (i) Improvement of classic kinetic resolution by re-racemisation and repeated resolution, (ii) dynamic resolution based on *in-situ* racemisation of the starting material, (iii) enantioselective stereoinversion, and (iv) enantioconvergent processes.

1.   Introduction

The increased demand for chiral drugs in enantiomerically pure form, following the release of new FDA's marketing guidelines, turned the search for novel methods for the syntheses of enantiomerically pure compounds (EPC) into a major topic in contemporary organic synthesis [1]. In this context, the use of biocatalysts has found widespread application in preparative organic chemistry over the last decade [2]. From the two principles of biocatalytic reactions where chiral molecules are involved, *i.e.* (i) desymmetrization of *meso-* and prochiral compounds [3,4], and (ii) kinetic resolution of racemates [5] (Scheme 1), the latter is remarkably dominant in number of applications (~1:4) [6], which is probably due to the fact that racemates can probably more readily be synthesized than *meso-* and prochiral substrates. Despite its widespread application, kinetic resolution is impeded by several inherent disadvantages for practical applications, in particular on an industrial scale. After all, it should be kept in mind that

1

B. Zwanenburg et al. (eds.), Enzymes in Action, 1–23.
© 2000 *Kluwer Academic Publishers. Printed in the Netherlands.*

2

an ideal resolution process should provide a single enantiomeric product in 100% yield. The most obvious drawbacks are as the following:

(i) The theoretical yield of each enantiomer can never exceed the limit of 50%.

(ii) Separation of the formed product from the remaining substrate may be laborious in particular for those cases, where simple extraction or distillation fails and chromatographic methods are required [7].

(iii) In the majority of processes, only one stereoisomer is desired and there is little or no use for the other. In some rare cases, the unwanted isomer may be used through a separate synthetic pathway in an enantio-convergent fashion, but this requires additional effort put into a highly flexible synthetic strategy [8].

(iv) For kinetic reasons, the optical purity of substrate and/or product is depleted at the point, where separation of product and substrate is most desirable from a preparative point of view — *i.e.* 50% conversion [9].

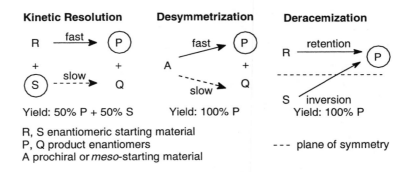

Scheme 1: Principles of kinetic resolution and desymmetrisation.

As a consequence, alternatives to kinetic resolution techniques that can deliver a single stereoisomer from a racemate are highly advantageous, such processes are generally denoted as 'deracemisation' [10] (Scheme 1). All of these techniques are dealing with a common stereochemical phenomenon, *i.e.* both substrate enantiomers have to be processed *via* two different stereochemical pathways: whereas the stereochemistry of R remains the same during its transformation to P, enantiomer S has to cross the symmetry plane (which is dividing R and S) in order to become P. As a consequence, S has to be reacted with *inversion* of configuration, whereas the stereochemistry of R is *retained* throughout the process.

In this chapter, general strategies which lead to the formation of a single enantiomeric product in 100% theoretical yield from a racemate are reviewed.

2. Improving Kinetic Resolution

2.1. RE-RACEMISATION AND REPEATED RESOLUTION

In order to avoid the loss of half of the material in kinetic resolution, it has been a common practice to racemize the unwanted isomer after separation from the desired product and to subject it again to kinetic resolution in a subsequent cycle, and so forth, until virtually all of the racemic material has been converted into a single stereoisomer [11].

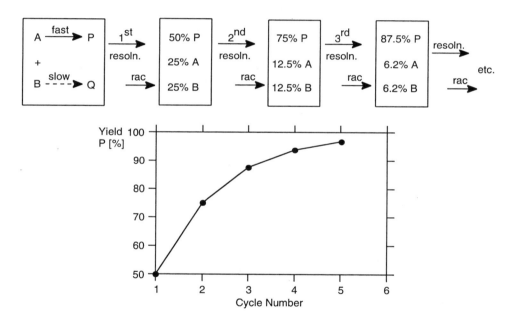

Scheme 2: Principles of deracemisation *via* repeated resolution and re-racemisation.

On a first glance, repeated resolution appears to be less than optimal and certainly lacks synthetic elegance, bearing in mind that an infinite number of cycles are theoretically required to transform all of the racemic starting material into a single stereoisomer. Upon closer examination, however, it becomes a viable option. The graph in Scheme 2 reveals that the overall yield of product P reaches a value of >95% after only five cycles, provided that both reactions - *i.e.* kinetic resolution and racemisation are essentially 'clean' without loss of material. As a consequence, racemisation is the main challenge to be met, which can be achieved by both ways, *via* chemo- or bio-catalysis. Each technique provides several merits and disadvantages [12].

Chemical racemisation can be achieved by a number of concepts (Scheme 3). In industry, acid- or base-catalysed, as well as thermal racemisation are the most widespread techniques because they are cheap processes and they are relatively easy to handle. Base-catalyzed racemisation *via* an achiral enolate-type intermediate can be applied to almost all compounds bearing an acidic hydrogen at the chiral center, and, as a consequence, it became one of the most widespread techniques used. On the contrary, acid-catalyzed racemisation is usually applicable to substances subject to keto-enol tautomerism and because of this, it is less common. Compounds showing axial chirality can be racemized *via* rotation around a σ-bond and they are viable targets for thermal racemisation. Photochemical racemisation has been employed more rarely for specific types of compounds, such as chrysanthemic acid derivatives. Chiral organohalides and nitriles are racemized *via* nucleophilic substitution catalyzed by halide and cyanide, respectively. More recently, milder methods for racemisation have been brought to attention: For instance, chiral *sec*-alcohols can be racemized *via* a transition-metal catalyzed oxidation-reduction sequence or *via* Pd^{II}-catalyzed allylic rearrangement of the corresponding acetate esters (Scheme 6). In a related fashion, interconversion of enantiomers of *sec*-amines is accomplished by Pd/C-catalyzed dehydrogenation-hydrogenation (Scheme 7).

Compounds bearing a configurationally stable center cannot be racemized directly and they have to be (chemically) converted into a configurationally unstable intermediate. For instance, amino acid amides or -esters can be racemized *via* Schiff-base derivatives of aromatic aldehydes involving the α-amino group or, alternatively, for free amino acids, mixed acid anhydrides are used. However, after acid/base-catalyzed racemisation of the respective intermediates, the starting compound has to be liberated again, which makes this technique a rather tedious multi-step procedure.

The general disadvantage of most techniques based on chemo-catalysis is the fact that they require harsh reaction conditions causing side reactions, thus leading to product degradation, which results in loss of material.

On the contrary, the most striking merit of biocatalytic racemisation is the mild reaction conditions — e.g. room temperature, atmospheric pressure and neutral pH. Under these conditions, side reactions are largely suppressed [13]. Unfortunately, Nature does not rely on racemates and, as a consequence, biochemical racemisation is a rather scarce feature, which makes racemases a small group of enzymes, that are found in certain biological niches. One of the major targets for biochemical racemisation involves stereogenic centers in carbohydrates, *i.e.* *sec*-alcohol groups. However, since both stereoisomers of these reactions represent diastereomers rather than enantiomers, 'epimerisation' would be the more correct term. Various enzymes — epimerases — are involved in the racemisation of *sec*- alcohol groups, these enzymes are usually NAD^+-dependent and of very little interest for practical applications.

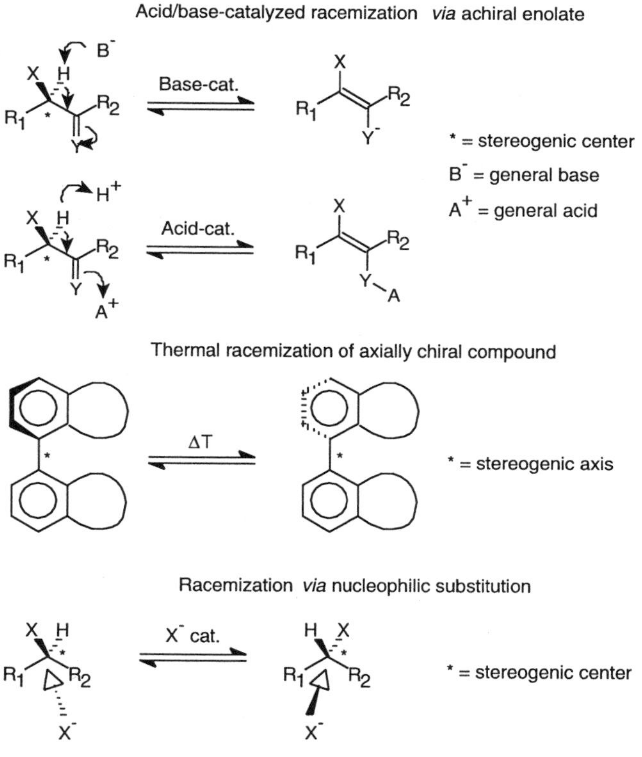

Scheme 3: Concepts of chemical racemisation.

Among non-carbohydrate 'true' racemases, base-catalysis seems to be the general scheme of biochemical racemisation, and two major groups can be classified according to their reaction mechanism [14], *viz.* racemases employing a one-base mechanism — *i.e.* the proton at the chiral center is abstracted by the same base-functionality of the enzyme, that reads it — and those employing a two-base mechanism — *i.e.* one base is capable of *abstracting* the proton at the chiral center and another (conjugated) base *puts it back on* from the opposite side, thus resembling a ping-pong mechanism.

Mandelate racemase [EC 5.1.2.2] belongs to the latter group employing a two-base mechanism [15]. The Mg^{2+}-dependent enzyme is capable of racemising a remarkably broad substrate spectrum, which opens up large possibilities for deracemisation of various kinds of α-hydroxy acids. Substrates that meet the following constraints are accepted by mandelate racemase:

(i) The α-hydroxy acid moiety is (almost strictly) required, as the only exception to this rule seems to be an α-hydroxy carboxamide functionality [16].

(ii) A π-electron system has to be present in the β,γ-position. The π-electron system can be freely varied, including the corresponding α-hydroxy-β,γ-alkenoic and α-hydroxy-α-aryl carboxylic acids. Even heteroaromatic systems are accepted at reasonable rates [17]. In general, electron-withdrawing groups attached to the β,γ-unsaturated (or aromatic) system, which help to stabilize the (anionic) transition intermediate *via* resonance, enhance the reaction rates significantly [18]. Several cofactor-independent amino acid racemases, such as proline racemase, aspartic acid racemase and glutamate racemase, also belong to the two-base type racemases. Racemases employing the one-base mechanism all display a common feature, *i.e.* they require pyridoxal phosphate as an essential cofactor for catalytic activity. Examples for these enzymes are α-amino-ε-caprolactam racemase, alanine racemase and (probably) arginine, threonine and serine racemases. Amino acid racemases are widely used in industry because of the stong commercial importance of amino acids and -derivatives.

A novel deracemisation technique based on a two-enzyme system consisting of a (i) lipase-catalyzed enantioselective acylation, followed by (ii) mandelate racemase catalyzed racemisation of the remaining non-reacted substrate enantiomer is shown in Scheme 4 [19]. Thus, in a first step, (±)-mandelic acid is subjected to lipase catalyzed *O*-acylation in an organic solvent producing (*S*)-*O*-acetyl mandelate, by leaving the (*R*)-enantiomer behind. Due to the high selectivity (E >100), the reaction comes to a standstill at 50% conversion.

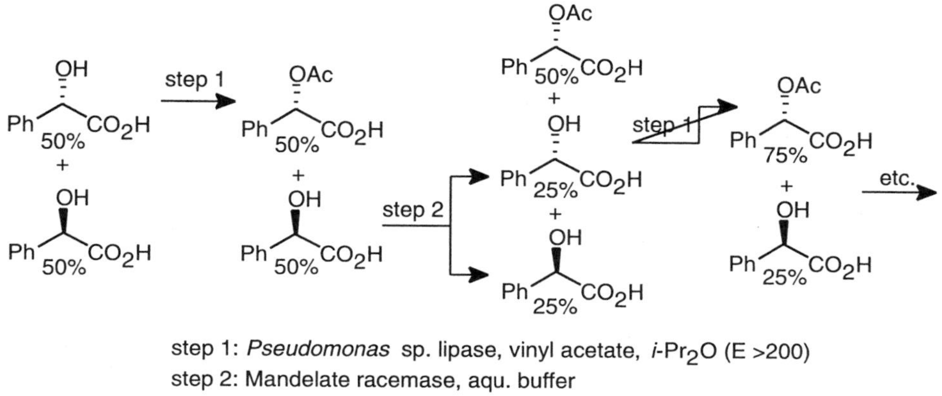

step 1: *Pseudomonas* sp. lipase, vinyl acetate, *i*-Pr$_2$O (E >200)
step 2: Mandelate racemase, aqu. buffer

Scheme 4: Deracemisation of mandelate *via* repeated resolution employing a two-enzyme process.

Then, in a subsequent step, the organic solvent is changed to aqueous buffer [20], and the remaining unreacted (*R*)-mandelic acid is racemized in the presence of (*S*)-*O*-acetyl mandelate, which is non-substrate. When this two-step process was repeated four times, (*S*)-*O*-acetyl mandelate was obtained in >80% chemical yield and >99% enantiomeric excess as the sole product. It should be emphasized that separation of the formed product from the remaining starting material is not required due to the high specificity of the racemase employed.

2.2.   DYNAMIC RESOLUTION

The disadvantages of kinetic resolution can largely be avoided by employing a so-called 'dynamic resolution' [21] (Scheme 5) Such a process comprises kinetic resolution with an additional feature, *i.e. in-situ* racemisation of the starting material, which is usually achieved using chemo-catalysis. Ultimately, all of the substrate R+S is transformed into a single product enantiomer P in 100% theoretical yield. In contrast to kinetic resolution, where the reaction slows down at 50% conversion (or comes even to a standstill, if the enantioselectivity is sufficiently high), when the fast-reacting enantiomer R is consumed and only the slow-reacting counterpart remains, substrate-racemisation ensures the continuous formation of R from S during the course of the reaction and thus avoids the depletion of R. The reaction does not come to a standstill and it therefore can be run to completion by gradually converting all racemic starting material into product P. In order to indicate the non-static behaviour of such a process, the term 'dynamic resolution' has been aptly coined. The following properties are typical for dynamic resolution processes (Scheme 5) [22]. The e.e. of the substrate (e.e.$_S$) is at its minimal at the onset of the reaction and gradually begins to increase as the faster reacting enantiomer is depleted from the reaction mixture, in particular around half-way of the reaction. On the other hand, this depletion does not occur if the substrate is constantly racemized during the resolution process and, thus, in a dynamic resolution the e.e.$_P$ is *not* a function of the conversion but remains constant throughout the reaction. Since the catalyst always faces a racemic starting material (*i.e.* [R] always equals [S]), it is understandable, that the selection of the faster reacting enantiomer from the substrate remains a simple task, as opposed to kinetic resolution where depletion of R occurs.

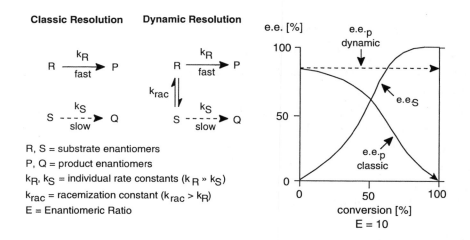

Scheme 5: Principles of kinetic and dynamic resolution.

In order to design a successful dynamic resolution process, both parallel reactions — *i.e.* kinetic resolution and in-situ racemisation — have to be carefully tuned, by taking into account the following aspects:

(i) The kinetic resolution should be irreversible in order to ensure high enantioselectivity.

(ii) The enantiomeric ratio ('E-value', E = k_R/k_S) [23] should be greater than ~20.

(iii) In order to avoid depletion of R, racemisation (k_{rac}) should be at least equal or greater than the reaction rate of the fast enantiomer (k_R).

(iv) For moderate selectivities, k_{rac} should be greater than k_R by a factor of about 10,

(v) For obvious reasons, any spontaneous reactions involving the substrate enantiomers as well as racemisation of the product should be absent.

9

(vi) Dynamic resolution is generally limited to compounds possessing one stereocenter, however, under certain circumstances, multicenter-compounds can be processed as long as the stereocenters are (stereo)chemically very similar. In such a case, the reaction is proceeding through several diastereomeric intermediates.

A common scenario for dynamic resolution processes based on enantioselective biocatalysis makes use of a combination of an enzyme-catalyzed kinetic resolution coupled with *in-situ* racemisation of the remaining substrate enantiomer through chemo-catalysis.

Four practical situations can be encountered for appropriate *in situ* racemisation processes.

(i) Compounds having a chirality center bearing an acidic proton — e.g. adjacent to an activating carbonyl group, such as an ester or ketone — usually undergo facile racemi-sation through the formation of an achiral enolate species *via* base-catalyzed proton-abstraction [24].

(ii) When such racemisation is impossible, e.g. in case of a secondary alcohol, racemisation can be achieved *via* a decomposition reaction, such as the cleavage of hemi(thio)acetals and cyanohydrins [25].

If chemo-catalysis is relying on acid or base, its combination with the biocatalyst may be very difficult due to the incompatibility of enzymes with strongly basic or acidic media. Therefore, bio-compatible *in-situ* racemisation techniques are of high value. Typical examples include:

(iii) Transition-metal catalyzed racemisation of *sec*-alcohols, based on Pd^{II}-catalyzed allylic rearrangement of allylic acetate esters [26] or Ru-catalyzed oxidation-reduction sequences [27] (Scheme 6).

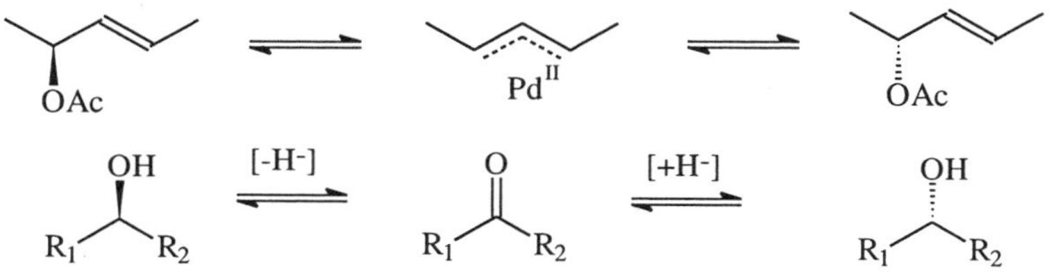

Scheme 6: Transition-metal catalyzed racemisation of *sec*-alcohols and -acetate esters.

(iv) Benzylic *sec*-amines can be racemised under bio-compatible conditions using a reversible Pd/C-catalyzed dehydrogenation-hydrogenation reaction (Scheme 7) [28].

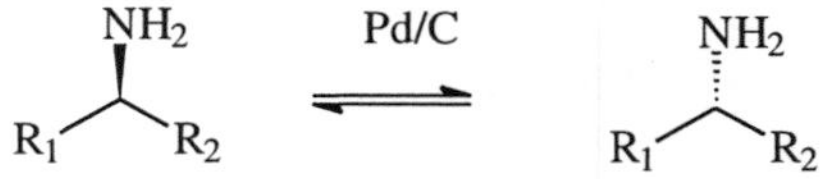

Scheme 7: Pd/C-Catalyzed racemisation of a benzylic *sec*-amine *via* dehydrogenation-hydrogenation.

All these racemisation techniques have been successfully applied for dynamic resolution. Examples are depicted in Schemes 8-10.

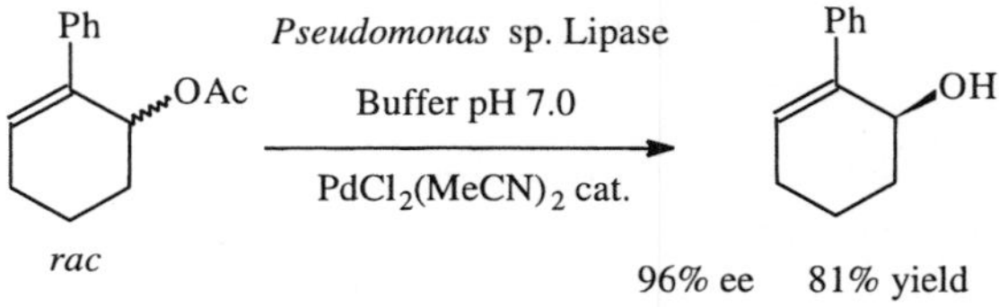

Scheme 8: Dynamic resolution of an allylic *sec*-alcohol *via* combined Pd^II- and lipase-catalysis.

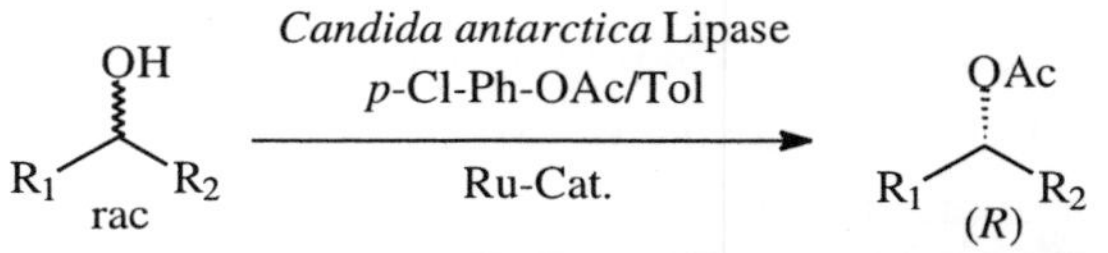

R$_1$	R$_2$	E.e. [%]	Yield [%]
Ph	Me	>99	80
Ph	Et	98	80
1-Naphtyl	Me	>99	79
cyclo-C$_6$H$_{13}$	Me	>99	79
n-C$_5$H$_{11}$	Me	>97	80
Ph-O-CH$_2$	Me	>99	88
Ph-O-CH$_2$	Cl-CH$_2$	79	68

Scheme 9: Dynamic resolution of *sec*-alcohols *via* combined Ru- and lipase-catalysis.

One approach to circumvent the incompatibility of chemical and bio-catalysts during dynamic resolution would consist in the combination of two biocatalysts, bearing in mind that enzymes are usually compatible as they generally work under the same (physiological) reaction conditions. From this point of view, the application of racemases seems to be very promising.

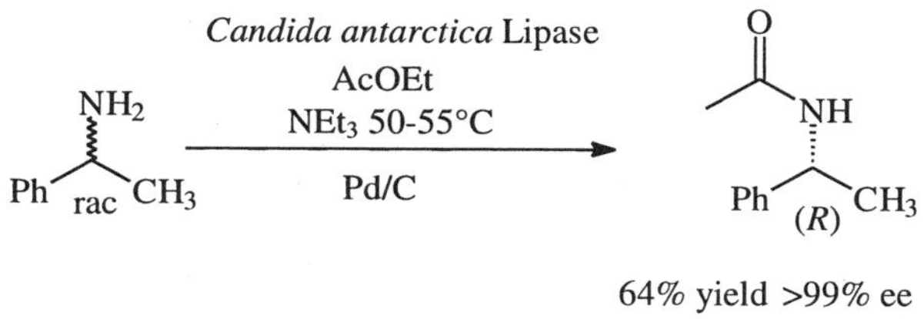

Scheme 10: Dynamic resolution of a *sec*-amine *via* Pd/C- and lipase-catalysis.

2.3. STEREO-INVERSION

The difficulty to achieve *in-situ* racemisation with compounds possessing a configurationally stable stereogenic center, such as secondary alcohols, may be overcome by employing a so-called stereo-inversion [29]. The latter may either be achieved through chemical or biocatalytic methods. .

The principle of deracemisation based on chemical stereo-inversion is outlined in Scheme 11 [30]. Thus, in a first step, kinetic resolution of a secondary alcohol was achieved using lipase-catalyzed ester hydrolysis furnishing a mixture of *sec*-alcohol and the corresponding enantiomeric ester. Without separation, the mixture was subjected to chemical inversion of the alcohol by treatment with mesyl chloride or (for large-scale reactions) with fuming nitric acid under carefully controlled reaction conditions, which gave a mixture of enantiomeric activated and non-activated esters [31]. Both compounds were hydrolyzed by strong base with concurrent *inversion* and *retention* of configuration, respectively. As a consequence, a single enantiomeric *sec*-alcohol was obtained as the sole product.

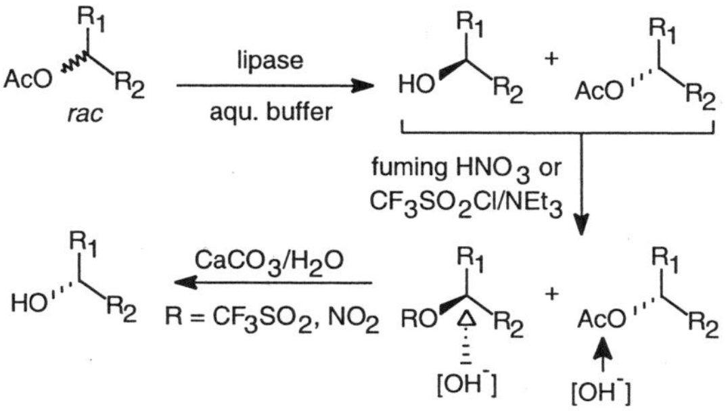

Scheme 11: Deracemisation of *sec*-alcohol based on resolution coupled with chemical stereo-inversion.

Due to the fact that in step 1 (*i.e.* kinetic resolution) the enantiomeric excess of product (e.e.p) and substrate (e.e.$_S$) are a function of the conversion, the point of stopping the lipase-catalysed reaction and the switch to the chemical inversion process has to be carefully chosen, in order to obtain a maximum e.e. of the product. This optimum can be calculated as a function of the selectivity of the reaction and is usually at or somewhat beyond a conversion of 50% [32].

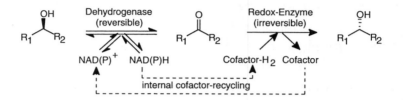

Scheme 12: Deracemisation of *sec*-alcohols based on a biocatalytic oxidation-reduction sequence.

Alternatively, stereo-inversion of *sec*-alcohols may be achieved by biocatalytic methods *via* an oxidation-reduction sequence [33] (Scheme 12). Thus, one enantiomer of a racemic mixture is selectively oxidized to the corresponding ketone under catalysis of a dehydrogenase, while the mirror-image counterpart remains unaffected. Then, the ketone is reduced again in a subsequent step by a different enzyme displaying opposite stereochemical preference. Overall, this two-step oxidation-reduction sequence constitutes formally a deracemisation process. Due to the involvement of a consecutive oxidation-reduction reaction, the net redox-balance of the process is zero and (in an ideal case) no external cofactor-recycling is neccessary since the redox-equivalents, such as NAD(P)H, may be recycled internally between both steps, e.g. by using whole-cell systems.

The success of a biocatalytic stereo-inversion *via* a redox-process is determined by the following crucial factor. For the entropy balance of the process, which is required to achieve a high optical purity of the product, at least one of both redox-reactions has to be irreversible [34].

3.   Enantioconvergent Processes

Deracemisation may be achieved through so-called enantioconvergent processes in such a way, that each of the enantiomers is converted into the same product enantiomer P *via* two independent pathways (Scheme 1). Thus, whereas enantiomer R is reacted to product P through *retention* of configuration, its counterpart S is transformed with *inversion*. In general, both reactions are conducted in a step-wise fashion. The crucial prerequisites for the proper functioning of such systems are as follows.

(i) The first step must combine excellent enantiospecificity *and* stereospecificity with regard to retention/inversion, in other words, the requirement of the chiral recognition *and* stereochemistry of the respective transformation are high and it is not surprising that these specificities are usually only achieved by enzymes.

(ii) Since the starting material for the second step (*i.e.* S) is enantiomerically enriched (or even pure), only high stereospecificity is required with respect to inversion/retention of configuration. Hence, this step may also be performed by using a chemo-catalyst.

(iii) An important factor with respect to the economy of the whole process is the compatibility of the reaction conditions of both steps. If, for instance, the conditions are incompatible, separation of product P (formed during step 1 from R) from the remaining enantiomeric starting material S is required, which is usually going in hand with loss of material. After all, it appears anachronistic, to separate materials from each other, which are to be combined at the end of the process. Thus, successful enantioconvergent processes should always be performed in a one-pot fashion.

3.1. ENANTIOCONVERGENCE USING TWO BIOCATALYSTS

To date, the only enzymes which may transform non-natural compounds with concomitant *inversion* of configuration during catalysis are (i) glycosidases [35], (ii) dehalogenases [36], (iii) sulfatases [37] and (iv) epoxide hydrolases [38]. Whereas glycosidases cannot be employed for deracemisation since their substrates are diastereomers rather than enantiomers, dehalogenases are not widely distributed in Nature and they exhibit a limited substrate tolerance. Similarly, the number of applications of sulfatases in preparative biotransformations is very limited and they are restricted to aryl-sulfatases, where no chirality is involved [39]. On the other hand, epoxide hydrolases from microbial sources, such as bacteria and fungi, have recently been shown to possess a great potential for the stereoselective hydrolysis of epoxides to furnish the corresponding vicinal diols [40]. In contrast to ester hydrolysis catalyzed by lipases, esterases or proteases, where the absolute configuration at the stereogenic center(s) always remains the same throughout the reaction, enzymatic hydrolysis of epoxides may take place *via* attack on either carbon atom of the oxirane ring and it is the structure of the substrate and of the enzyme which determine the regioselectivity of the attack [41]. This is exemplified by the following (Scheme 13). If the (*S*)-enantiomer is

14

preferentially hydrolysed from the racemate with *retention* of configuration, kinetic resolution furnishes a mixture of (S)-diol and unreacted (R)-epoxide. On the contrary, the corresponding (R)-diol is produced from the (S)-oxirane, if the enzyme acts with *inversion* of configuration. Therefore, enantioconvergent hydrolysis of epoxides should be feasible when of appropriate enzymes are available.

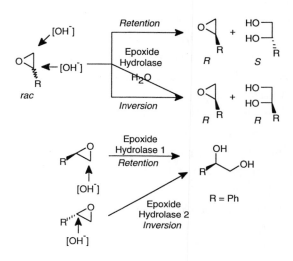

Scheme 13: Enzymatic hydrolysis of epoxides proceeding with retention or inversion of configuration.

An elegant deracemisation of (±)-styrene oxide was developed by making use of two epoxide hydrolase activities from fungal sources [42]. Whereas *Aspergillus niger* preferentially hydrolyzed the (R)-enantiomer with retention of configuration by producing (R)-phenylethan-1,2-diol (Scheme 13, epoxide hydrolase 1), *Beauveria bassiana* showed opposite enantiopreference [*i.e.* (S)] with matching opposite regioselectivity causing inversion of configuration (epoxide hydrolase 2). Combination of both biocatalysts in a single reactor led to almost complete deracemisation.

3.2.   ENANTIOCONVERGENCE   BY   COMBINED   BIO-   AND   CHEMO-
        CATALYSIS

Bacterial epoxide hydrolases have been shown to be the biocatalysts of choice for the enantioselective hydrolysis of 2,2-disubstituted oxiranes, by showing virtually absolute enantioselectivities (E >200) [43]. In this case, the reaction proved to proceed invariably with *retention* of configuration. As the existence of enzymes which attack a quarternary carbon atom with *inversion* of configuration is unlikely, deracemisation *via* a two-enzyme system as described above is impossible. However, the combination of bio- and chemo-catalysis proved to be very efficient (Scheme 14) [44].

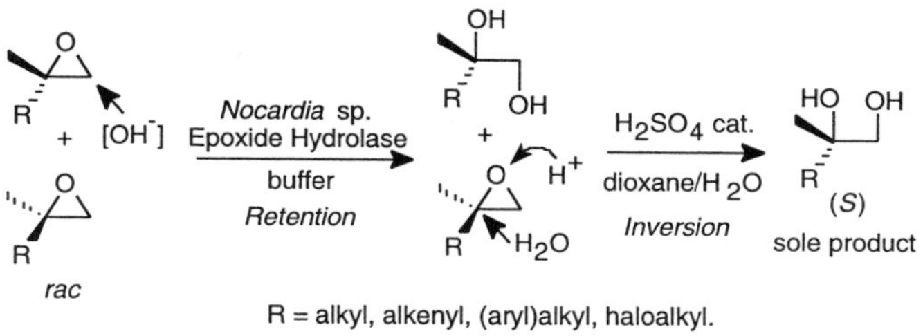

R = alkyl, alkenyl, (aryl)alkyl, haloalkyl.

Scheme 14: Enantioconvergent hydrolysis of 2,2-disubstituted epoxides through combination of bio- and chemo-catalysis.

Thus, kinetic resolution of 2,2-disubstituted epoxides using a *Nocardia* sp. epoxide hydrolase proceeded with excellent enantio- and regioselectivity by furnishing the corresponding (S)-diol and (R)-epoxide in a first step. Then, the remaining epoxide was transformed by acid-catalysis with *inversion* of configuration in a second step under carefully controlled reaction conditions to yield the corresponding (S)-diols in virtually enantiopure form and in high chemical yields (>90%). This methodology proved to be highly flexible and was also applicable to styrene-oxide type substrates [45].

3.3. ENANTIOCONVERGENCE USING A SINGLE BIOCATALYST

Processes depending on more than one catalyst are generally sensitive with respect to the tuning of both reactions and therefore enantioconvergent reactions which depend on a single catalyst would be more reliable in practice. However, the requirements for this single catalyst are extremely high, *i.e.* it has to exhibit not only high *enantioselectivity* but also show at the same time opposite *regioselectivity* for the transformation of each enantiomer in order to make the overall process enantioconvergent. As a consequence, such processes catalysed by a single (bio)catalyst are very rare (Scheme 15) [46]. For instance, an epoxide hydrolase from *Nocardia* sp. hydrolyzed both enantiomers of *cis*-2,3-disubstituted epoxides by attack at their respective (S)-oxirane carbon atom with inversion of configuration yielding the corresponding (2R,3R)-diol as the sole product in up to 92% e.e. and 85% chemical yield [47].

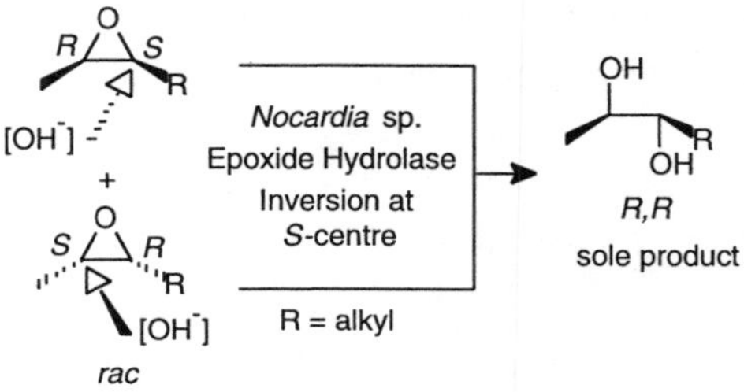

16

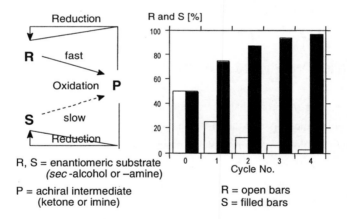

Scheme 15: Enantioconvergent hydrolysis of 2,3-disubstituted epoxides using a single biocatalyst.

4.   Deracemisation by a Cyclic Oxidation-Reduction Sequence

Deracemisation of compounds bearing a chiral *sec*-hydroxyl or *sec*-amino group can be achieved *via* a novel process consisting of a cyclic oxidation-reduction sequence [48,49]. The system consists of two independent reactions outlined in Scheme 16.

Scheme 16: Deracemisation of *sec*-alcohols and -amines by a cyclic oxidation-reduction.

First, one enantiomer of the secondary alcohol or amine (R) present in the starting racemate is selectively oxidized to yield the achiral intermediate P, *i.e.* the corresponding ketone or imine, respectively. Then, the product is chemically reduced in a non-selective fashion to yield again a racemic mixture. Both reactions alone are of limited use for the preparation of enantiopure material, since step 1 (*i.e.* a kinetic resolution by enantioselective oxidation) is limited to a 50% theoretical yield of chiral non-reacting S and achiral P, and step 2 does not show any chiral induction at all.

However, combination of both steps in a cyclic mode leads to a highly versatile deracemisation technique.

The functioning of this system is explained by the following example. If (for reasons of clarity) the selectivity of step 1 is assumed to be absolute, only R is selectively oxidized to form achiral P in 50% yield by leaving S untouched. In the second step, P is non-selectively reduced to furnish R+S in equal amounts of 25% each. Thus, the enantiomeric composition of R/S after a single cycle is now equals to 25/75. The diagram shown in Scheme 16 reveals that further cycles lead to a gradual increase of enantiomer S at the expense of R, and that the enantiomeric excess of the substrate is already well above 90% after only four cycles, in the case of absolute enantioselectivity. Overall, if the cyclic process is driven in the forward direction, enantiomer S represents the 'sink' of material in the whole system.

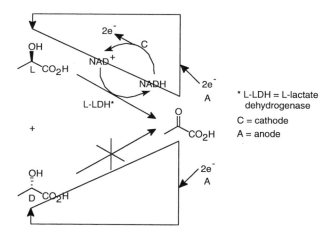

Scheme 17: Deracemisation of α-hydroxy acid *via* electro- and biochemical cyclic oxidation/reduction.

For practical applications, however, selectivities are ranging often below E-values of 100. For these cases, the enantioselectivity determines two crucial factors of the system, *i.e.* (i) the maximum obtainable e.e. at equilibrium and (ii) the number of cycles which are required to reach this value. The merits and limits of cyclic deracemisation systems have been recently described based on the underlying kinetics [48].

The practical feasibility of cyclic deracemisation based on oxidation-reduction was verified for the deracemisation of (±)-α-amino acids by combination of an amino acid oxidase coupled to NaBH₄-reduction of the corresponding intermediate imino acid [50].

18

A related approach is based on the electro- and biochemical oxidation/reduction of lactate involving the recycling of NADH [51]. Here, complete inversion of L-lactate into D-lactate was achieved in a model reactor (Scheme 17).

5. Summary

The development of methods for the preparation of chiral compounds in 100% chemical and optical yields from racemates is one of the current challenges in asymmetric synthesis. Several approaches have been described so far, which are either based on modifications of classic kinetic resolution, such as re-racemisation and repeated resolution or dynamic resolution. On the other hand, deracemisation can be achieved by transformation of enantiomers *via* enantioconvergent pathways, which is usually achieved by combining chemo- and/or biocatalysts in sequential reactions or - most elegantly - even by a single (bio)catalyst. Finally, stereo-inversion by a oxidation-reduction sequence is a feasible option. It has to be emphasized, however, that each of the above-described approachles offers a solution for only certain types of stereo-chemical problems and the corresponding substrate classes, but none of the methods can be employed in a general sense.

6. Acknowledgements

This work was performed within the Spezialforschungsbereich 'Biokatalyse' and financial support by the Fonds zur Förderung der wissenschaftlichen Forschung (project F115) and the Austrian Federal Ministry of Science and Transport is gratefully acknowledged.

7. References and Notes

1.   a) The enantiopure drug market has been estimated as US $ 18 billion worldwide. See: Stinson, S. C. (1992) Chiral drugs, *Chem. Eng. News*, Sept. 28, 46-78; b) Sheldon, R. A. (1993) *Chirotechnology*, Marcel Dekker, New York; c) Collins, A. N., Sheldrake, G. N., Crosby, J. (eds) (1992 and 1997), *Chirality in industry*, vols. I and II, Wiley, Chichester.
2.   a) Faber, K. (1997) *Biotransformations in organic chemistry*, 3rd edn., Springer, Heidelberg; b) Drauz, K., Waldmann H. (eds.) (1995) *Enzyme catalysis in organic synthesis*, vols. I and II, Verlag Chemie, Weinheim.

3. Schoffers, E., Golebiowski, A., Johnson, C. R. (1996) Enantioselective synthesis through enzymatic asymmetrisation, *Tetrahedron* **52**, 3769-3826.
4. Kagan, H. B., Fiaud, J. C. (1988) Kinetic resolution, *Topics Stereochem.* **18**, 249-330.
5. a) Horeau, A. (1975) Method for obtaining an enantiomer containing less than 0.1% of its antipode, determination of its maximum rotatory power, *Tetrahedron* **31**, 1307-1309; b) For biocatalyzed reactions: Sih, C. J., Wu, S.-H. (1989) Resolution of enantiomers via biocatalysis, *Topics Stereochem.* **19**, 63-125; c) For non-biocatalyzed reactions: Martin, V. S., Woodard, S. S., Katsuki, T., Yamada, Y., Ikeda, M., Sharpless, K. B. (1981) Kinetic resolution of racemic allylic alcohols by enantioselective epoxidation, *J. Am. Chem. Soc.* **103**, 6237-6240; d) For prebiotic reactions: Balavoine, G., Moradpour, A., Kagan, H. B. (1974) Preparation of chiral compounds with high optical purity by irradiation with circularly polarized light, a model reaction for the prebiotic generation of optical activity, *J. Am. Chem. Soc.* **96**, 5152-5158.
6. For biocatalyzed reactions, data from database Faber, ~10.000 entries, July 1999.
7. Faber, K. (1992) Chiral separation techniques using hydrolytic enzymes, *Indian J. Chem., Sect. B,* **31B**, 921-924.
8. Cotterill, I. C., Jaouhari, R., Dorman, G., Roberts, S. M., Scheinmann, F., Wakefield, B. J. (1991) Use of the two enantiomers of 7,7-dimethylbicyclo[3.2.0]hept-2-en-6-one to form complementary optically active synthons in a convergent synthesis of leukotriene-B₄, *J. Chem. Soc., Perkin Trans. 1*, 2505-2512.
9. Chen, C.-S., Fujimoto, Y., Girdaukas, G., Sih, C. J. (1982) Quantitive analysis of biochemical kinetic resolution of enatiomers, *J. Am. Chem. Soc.* **104**, 7294-7299.
10. a) Stecher, H., Faber, K. (1997) Biocatalytic deracemisation techniques: Dynamic resolutions and stereoinversions, *Synthesis*, 1-16; b) Strauss, U. T., Felfer, U., Faber, K. (1999) Biocatalytic transformation of racemates into chiral building blocks in 100% chemical yield and 100% enantiomeric excess, *Tetrahedron: Asymmetry* **10**, 107-117.
11. For a typical example see: a) Xie, Y.-C., Liu, H.-Z., Chen, J.-Y. (1998) *Candida rugosa* lipase catalysed esterification of racemic Ibuprofen with butanol: Racemisation of (R)-Ibuprofen and chemical hydrolysis of (S)-ester formed, *Biotechnol. Lett.* **20**, 455-458; b) Kamphuis, J., Boesten, W. H. J., Kaptein, B., Hermes, H. F. M., Sonke, T., Broxterman, Q. B., van den Tweel, W. J. J., Schoemaker, H. E. (1992) The production and uses of optically pure natural and unnatural amino acids, in: *Chirality in industry*, Collins, A. N., Sheldrake, G. N., Crosby, J. (eds.), Wiley, New York, pp. 187-208.
12. Ebbers, E. J., Ariaans, G. J. A., Houbiers, J. P. M., Bruggink, A., Zwanenburg, B. (1997) Controlled racemization of optically active organic compounds: Prospects for asymmetric transformations, *Tetrahedron* **53**, 9417-9476.
13. Adams, E. (1976) Catalytic aspects of enzymatic racemisation, *Adv. Enzymol. Relat. Areas Mol. Biol.* **44**, 69-138.
14. Gallo, K. A., Tanner, M. E., Knowles, J. R. (1993) Mechanism of the reaction Catalysed by glutamate racemase, *Biochemistry* **32**, 3991-3997.
15. Kenyon, G. L., Gerlt, J. A., Kozarich, J. W. (1995) Mandelate racemase: Structure-function studies of a pseudosymmetric enzyme, *Acc. Chem. Res.* **28**, 178-186.
16. Goriup, M., Strauss, U. T., Faber, K., unpublished results.
17. Felfer, U., Goriup, M., Strauss, U. T., Orru, R. V. A., Faber, K., unpublished results.
18. Hegeman, G. D., Rosenberg, E. Y., Kenyon, G. L., (1970) Mandelic acid racemase from *Pseudomonas*

putida: Purification and properties of the enzyme, *Biochemistry* **9**, 4029-4035.

19.   Strauss, U. T., Faber, K., unpublished results.

20.   Mandelate racemase was found to be inactive in various organic solvent systems: Stecher, H. Faber, K., unpublished results.

21.   a) Ward, R. S. (1995) Dynamic kinetic resolution, *Tetrahedron: Asymmetry* **6**, 1475-1490; b) Caddick, S., Jenkins, K. (1996) Dynamic resolutions in asymmetric synthesis, *Chem. Soc. Rev.* **25**, 447-456; c) Noyori, R., Tokunaga, M., Kitamura, M. (1995) Stereoselective organic synthesis via dynamic kinetic resolution, *Bull. Chem. Soc. Jpn.* **68**, 36-56.

22.   For the mathematical treatment of dynamic kinetic resolutions see: Kitamura, M., Tokunaga, M., Noyori, R. (1993) Mathematical treatment of kinetic resolution of chirally labile substrates, *J. Am. Chem. Soc.* **115**, 144-152; Kitamura, M., Tokunaga, M., Noyori, R. (1993) Quantitative expression of dynamic kinetic resolution of chirally labile enantiomers: Stereoselective hydrogenation of 2-substituted 3-oxo carboxylic esters catalysed by BINAP-Ruthenium (II) complexes, *Tetrahedron* **49**, 1853-1860.

23.   For biocatalyzed reactions, the 'binding' of the substrate enantiomers (which can be neglected with chemical catalysts) usually plays an important role in the chiral selection process and E-values of enzyme-catalyzed reactions are therefore defined through Michaelis-Menten kinetics: $E = (k_{cat}/K_M)_R/(k_{cat}/K_M)_S$.

24.   a) Um, P.-J, Drueckhammer, D. G. (1998) Dynamic enzymatic resolution of thioesters, *J. Am. Chem. Soc.* **120**, 5605-5610; b) Fülling, G., Sih, C. J. (1987) Enzymatic second-order asymmetric hydrolysis of ketorolac esters: In-situ racemisation, *J. Am. Chem. Soc.* **109**, 2845-2846.

25.   a) Inagaki, M., Hiratake, J., Nishioka, T., Oda, J. (1991) Lipase-catalyzed kinetic resolution with in-situ racemisation: One-pot synthesis of optically active cyanohydrin acetates from aldehydes, *J. Am. Chem. Soc.* **113**, 9360-9361; b) Brinksma, J., van der Deen, H., van Oeveren, A., Feringa, B. L. (1998) Enantioselective synthesis of benzylbutyrolactones from 5-hydroxyfuranone. New chiral synthons for dibenzylbutyrolactone lignans by a chemoenzymatic route, *J. Chem. Soc., Perkin Trans. 1*, 4159-4163.

26.   Allen, J. V., Williams, J. M. J. (1996) Dynamic kinetic resolution with enzyme and palladium combinations, *Tetrahedron Lett.* **37**, 1859-1862.

27.   Larsson, A. L. E., Persson, B. A., Bäckvall, J.-E. (1997) Enzymatic resolution of alcohols coupled to ruthenium-catalysed racemisation of the substrate alcohol, *Angew. Chem.* **109**, 1256-1258.

28.   Reetz, M. T., Schimossek, K. (1996) Lipase catalysed dynamic kinetic resolution of chiral amines: Use of palladium as the racemisation catalyst, *Chimia* **50**, 668-669.

29.   Carnell, A. J. (1999) Stereoinversions using microbial redox reactions, *Adv. Biochem. Eng. Biotechnol.* **63**, 57-72.

30.   a) Danda, H., Nagatomi, T., Maehara, A., Umemura, T. (1991) Preparation of optically active secondary alcohols by combination of enzymatic hydrolysis and chemical transformation, *Tetrahedron* **47**, 8701-8716; b) Lemke, K., Ballschuh, S., Kunath, A., Theil, F. (1997) An improved procedure for the lipase catalysed kinetic resolution of bicyclo[3.3.0]octane-2,6-diol – synthesis of potential C2-symmetric enantiomerically pure building blocks, *Tetrahedron: Asymmetry* **8**, 2051-2055; c) Vänttinen, E., Kanerva, L. T. (1995) Combination of the lipase catalysed resolution with the Mitsunobu esterification in one pot, *Tetrahedron: Asymmetry* **6**, 1779-1786; d) Takano, S., Suzuki, M., Ogasawara, K. (1993) Enantiocomplementary preparation of optically pure trimethylsilylethynyl-2-cyclopentenol by homochiralisation of racemic percursors: A new route to the key intermediate of

1,25-dihydroxycholecalciferol and vincamine, *Tetrahedron: Asymmetry* **4**, 1043-1046; e) Mitsuda, S., Umemura, T., Hirohara, H. (1988) Preparation of an optically pure secondary alcohol of synthetic pyrethroids using microbial lipases, *Appl. Microbiol. Biotechnol.* **29**, 310-315.

31. For small-scale reactions, Mitsunobu-conditions may be likewise employed.

32. a) Kanerva, L. T. (1996) Biocatalytic ways to optically active 2-amino-1-phenylethanols, *Acta Chem. Scand.* **50**, 234-242; b) Pedragosa-Moreau, S., Morisseau, C., Baratti, J., Zylber, J., Archelas, A., Furstoss, R. (1997) An enantioconvergent synthesis of the β-blocker (*R*)-Nifenalol using a combined chemoenzymatic approach, *Tetrahedron* **53**, 9707-9714.

33. a) Buisson, D., Azerad, R., Sanner, C., Larcheveque, M. (1992) A study of the stereocontrolled reduction of aliphatic β-ketoesters by *Geotrichum candidum*, *Biocatalysis* **5**, 249-265; b) Nakamura, K., Inoue, Y., Matsuda, T., Ohno, A. (1995) Microbial deracemisation of 1-arylethanol, *Tetrahedron Lett.* **36**, 6263-6266; c) Fantin, G., Fogagnolo, M., Giovannini, P. P., Medici, A., Pedrini, P. (1995) Combined microbial oxidation and reduction: A new approach to the high-yield synthesis of homochiral, unsaturated secondary alcohols from racemates, *Tetrahedron: Asymmetry* **6**, 3047-3053; d) Takemoto, M., Achiwa, K. (1995) The synthesis of optically active pyridyl alcohols from the corresponding racemates by *Catharanthus roseus* cell cultures, *Tetrahedron: Asymmetry* **6**, 2925-2928; e) Tsuchiya, S., Miyamoto, K., Ohta, H. (1992) highly efficient conversion of (±)-mandelic acid to its (*R*)-(-)-enantiomer by combination of enzyme mediated oxidation and reduction, *Biotechnol. Lett.* **14**, 1137-1142; f) Takahashi, E., Nakamichi, K., Furui, M. (1995) (*R*)-(-)-Mandelic acid production from racemic mandelic acids by using *Pseudomonas polycolor* IFO 3918 and *Micrococcus freudenreichii* FERM-P 13221, *J. Ferment. Bioeng.* **80**, 247-250; g) Carnell, A. J., Iacazio, G., Roberts, S. M., Willetts, A. J. (1994) Preparation of optically active cyclohexanediols and (+)-α-hydroxycycloheptanone by an enzyme catalysed stereoinversion/oxidation process, *Tetrahedron Lett.* **35**, 331-334; h) Matsumura, S., Kawai, Y., Takahashi, Y., Toshima, K. (1994) Microbial production of (2*R*,4*R*)-2,4-pentanediol by enatioselective reduction of acetylacetone and stereoinversion of 2,4-pentanediol, *Biotechnol. Lett.* **16**, 485-490; i) Shimizu, S., Hattori, S., Hata, H., Yamada, H. (1987) Stereoselective enzymatic oxidation and reduction system for the production of D-pantoyl lactone from a racemic mixture of pantoyl lactone, *Enzyme Microb. Technol.* **9**, 411-416.

34. The origin of the irreversibility of microbial/enzymatic deracemisation of *sec*-alcohols *via* an oxidation-reduction sequence is currently under investigation and the data available to date reveal a rather puzzling picture: For instance, deracemisation of various terminal (±)-1,2-diols by the yeast *Candida parapsilosis* has been claimed to operate *via* a (*R*)-specific NAD⁺-linked dehydrogenase and a (*S*)-specific NADPH-dependent reductase. Although no detailed data were given, the latter step was claimed to be irreversible: Hasegawa, J., Ogura, M., Tsuda, S., Maemoto, S., Kutsuki, H., Ohashi, T. (1990) High-yield production of optically active 1,2-diols from the corresponding racemates by microbial stereoinversion, *Agric. Biol. Chem.* **54**, 1819-1827. On the other hand, observations on the fungus *Geotrichum candidum* prove the requirement of molecular oxygen, which would suggest the involvement of an alcohol oxidase rather than an alcohol dehydrogenase: Azerad, R., Buisson, D. (1992) Stereocontrolled reduction of β-ketoesters with *Geotrichum candidum*, in: *Microbial Reagents in Organic Synthesis*, Servi, S. (ed.), NATO ASI Series C, Kluwer, Dordrecht, vol. 381, pp. 421-440.

35. Sinnott, M. L. (1990) Catalytic mechanisms of enzymatic glycosol transfer, *Chem. Rev.* **90**, 1171-1202.

36. Leisinger, T., Bader, R. (1993) Microbial dehalogenation of synthetic organohalogen compounds:

Hydrolytic dehalogenases, *Chimia* **47**, 116-121.

37. Roy, A. B. (1971) The hydrolysis of sulfate Esters, *The Enzymes* **5**, 1-19.

38. Orru, R. V. A., Archelas, A., Furstoss, R., Faber, K. (1999) Epoxide hydrolases and their synthetic applications, *Adv. Biochem. Eng. Biotechnol.* **63**, 145-167.

39. Pelsy, G., Klibanov, A. M. (1983) Preparative separation of α- and β-naphthols catalysed by immobilised sulfatase, *Biotechnol. Bioeng.* **25**, 919-928.

40. a) Faber, K., Mischitz, M., Kroutil, W. (1996) Microbial epoxide hydrolases, *Acta Chem. Scand.* **50**, 249-258; b) Archelas, A., Furstoss, R. (1997) Synthesis of enantiopure epoxides through biocatalytic approaches, *Annu. Rev. Microbiol.* **51**, 491-525; c) Archer, I. V. J. (1997) Epoxide hydrolases as asymmetric catalysts, *Tetrahedron* **53**, 15617-15662.

41. Mischitz, M., Mirtl, C., Saf, R., Faber, K. (1996) Regioselectivity of *Rhodococcus* NCIMB 11216 epoxide hydrolase: Applicability of E-values for description of enantioselectivity depends on substrate structure, *Tetrahedron: Asymmetry* **7**, 2041-2046.

42. Pedragosa-Moreau, S., Archelas, A., Furstoss, R. (1993) Enantiocomplementary epoxide hydrolases as a preparative access to both enantiomers of styrene oxide, *J. Org. Chem.* **58**, 5533-5536.

43. Mischitz, M., Kroutil, W., Wandel, U., Faber, K. (1995) Asymmetric microbial hydrolysis of epoxides, *Tetrahedron: Asymmetry* **6**, 1261-1272.

44. Orru, R. V. A., Mayer, S. F., Kroutil, W., Faber, K. (1998) Chemoenzymatic deracemisation of (±)-2,2-disubstituted oxiranes, *Tetrahedron* **54**, 859-874.

45. Pedragosa-Moreau, S., Morisseau, C., Baratti, J., Zylber, J., Archelas, A., Furstoss, R. (1997) An enantioconvergent synthesis of the β-blocker (*R*)-Nifenalol using a combined chemoenzymatic approach, *Tetrahedron* **53**, 9707-9714.

46. a) Pedragosa-Moreau, S., Archelas, A., Furstoss, R. (1996) Use of epoxide hydrolase mediated biohydrolysis as a way to enantiopure epoxides and vicinal diols: Application to substituted styrene oxides, *Tetrahedron* **52**, 4593-4606; b) Bellucci, G., Chiappe, C., Cordoni, A. (1996) Enantioconvergent transformation of racemic *cis*-β-alkyl substituted styrene oxides to (*R,R*)-*threo*-diols by microsomal epoxide hydrolase catalysed hydrolysis, *Tetrahedron: Asymmetry* **7**, 197-202.

47. Kroutil, W., Mischitz, M., Faber, K. (1997) Deracemisation of (±)-2,3-disubstituted oxiranes via biocatalytic hydrolysis using bacterial epoxide hydrolases: Kinetics of an enantioconvergent process, *J. Chem. Soc., Perkin Trans. 1*, 3629-3636.

48. Kroutil, W., Faber, K. (1998) Deracemisation of compounds possessing a *sec*-alcohol or -amino group through a cyclic oxidation-reduction sequence: A kinetic treatment, *Tetrahedron: Asymmetry* **9**, 2901-2913.

49. Free shareware programs ('Cyclo') running under Windows and Macintosh are available *via* the Internet at <http://borgc185.kfunigraz.ac.at> or directly from the authors (Kroutil, W., Kleewein, A., Faber, K. © 1998). A description how to use the program is given in the help-file which accompanies the program.

50. a) Huh, J. W., Yokoigawa, K., Esaki, N., Soda, K. (1992) Synthesis of L-proline from the racemate by coupling of enzymatic enantiospecific oxidation and chemical non-enantiospecific reduction, *J. Ferment. Bioeng.* **74**, 189-190; b) Huh, J. W., Yokoigawa, K., Esaki, N., Soda, K. (1992) Total conversion of racemic pipecolic acid into the L-enantiomer by a combination of enantiospecific oxidation with D-amino acid oxidase and reduction with sodium borohydride, *Biosci. Biotechnol. Biochem.* **56**, 2081-2082.

51. Biade, A.-E., Bourdillon, C., Laval, J.-M., Mairesse, G., Moiroux, J. (1992). Complete conversion of L-lactate into D-lactate. A general approach involving enzymatic catalysis, electrochemical oxidation of NADH and electrochemical reduction of pyruvate, *J. Am. Chem. Soc.* **114**, 893-899.

ENZYME ASSISTED ROUTES TO BIOACTIVE MOLECULES

Selective Transformations Using Lipases

P. ANDERSCH, F. FAZIO, B. HAASE, B. JAKOB AND
M.P. SCHNEIDER*
FB 9 - Bergische Universität - GH - Wuppertal
D - 42097 Wuppertal, Germany

1.   Introduction

Enzymes have emerged in recent years as highly efficient catalysts in organic synthesis. This is particularly true for ester hydrolases (esterases, lipases) many of which ideally combine the required high reaction selectivities with the synthetically so important broad substrate tolerance. Due to their ability to differentiate between (a) enantiomers, (b) enantiotopic groups attached to prochiral centers and (c) enantiotopic groups in *meso*-compounds they are highly suited for the preparation of enantiomerically pure carboxylic acids and alcohols of widely different structures. In the sections below – and based on the three different modes of substrate recognition (see chapter 3) – examples for the preparation of several classes of enantiomerically pure compounds ($\geq$ 98 % ee) are described.

2.   Lipase catalyzed reactions

Based on the generally accepted mechanism of lipase catalyzed reactions three types of transformations can be identified covering the synthetically most useful applications of these enzymes (*Figure* 1 a-c):

25

B. Zwanenburg et al. (eds.), Enzymes in Action, 25–42.
© 2000 *Kluwer Academic Publishers. Printed in the Netherlands.*

- ester hydrolysis and ester synthesis by direct esterification
- ester synthesis *via* reversible acyl transfer
- ester synthesis *via* irreversible acyl transfer.

From Figure 1 it becomes also clear that these transformations can – in principle – be employed for the synthesis of both carboxylic acids and alcohols. In aqueous media, with a large excess of water present, ester hydrolysis clearly is the dominating reaction and enzymatic hydrolyses are – next to the corresponding esterifications *via* irreversible acyltransfer – the most widely used synthetic transformations catalysed by lipases.

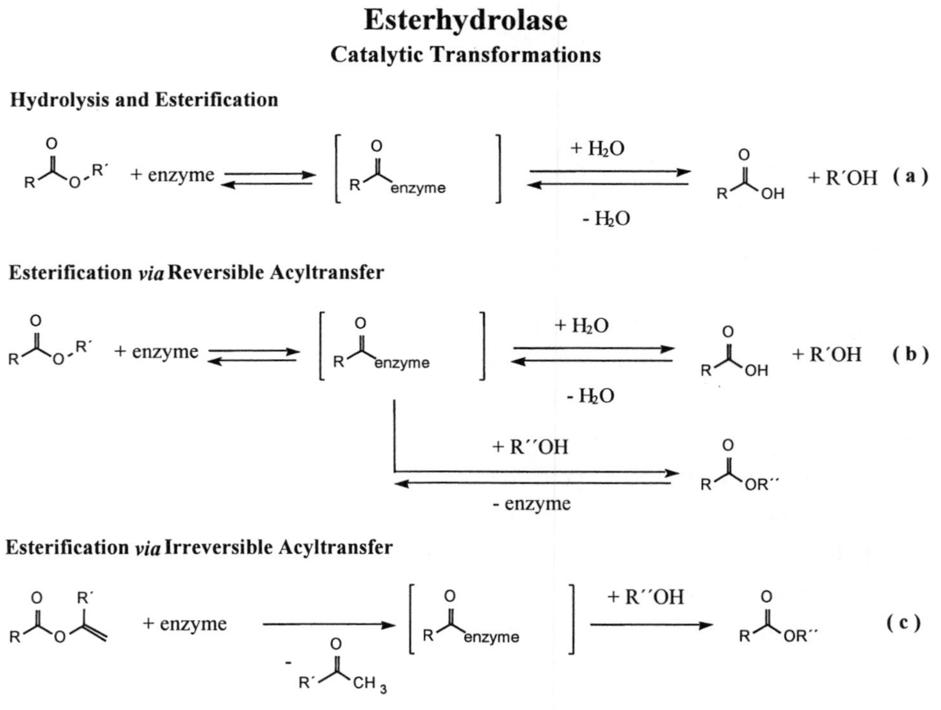

Esterhydrolase
Catalytic Transformations

Hydrolysis and Esterification

Esterification *via* Reversible Acyltransfer

Esterification *via* Irreversible Acyltransfer

Figure 1. Lipase catalyzed reactions

Esterifications - regardless of the method employed - can only be achieved successfully using low water conditions. In contrast to direct esterifications, acyl transfer reactions do not involve any water and are thus preferable especially in cases where the intermediate acyl-enzyme can be prepared irreversibly. Vinyl and isopropenyl esters of various carboxylic acids have been most widely employed for this purpose, next to anhydrides, oxime esters and vinyl carbonates. A summary of acyldonors employed for esterifications are collected in Figure 2 [1].

In the sections to follow enzymatic hydrolyses and esterifications using vinyl esters as acyl donors are used exclusively and applied to the preparation of enantiomerically pure hydroxy compounds.

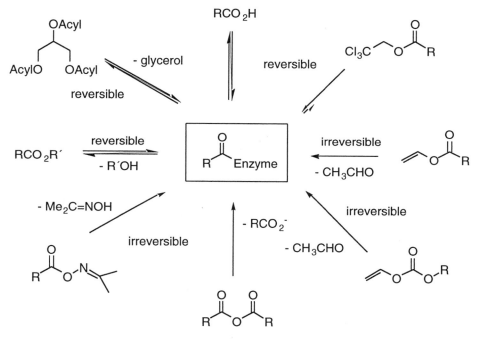

Figure 2. Acyldonors for lipase catalyzed esterifications

3.   Lipases – Modes of substrate recognition

Lipases display essentially three different modes of substrate recognition:

a.   enantiomer differentiation
b.   enantiotope differentiation in achiral precursors carrying a prochiral center
c.   enantiotope differentiation in achiral *meso*-compounds

28

The fundamental differences as applied to the preparation of enantiomerically pure hydroxy compounds are summarized schematically in Figure 3 using the hydrolytic mode of transformation.

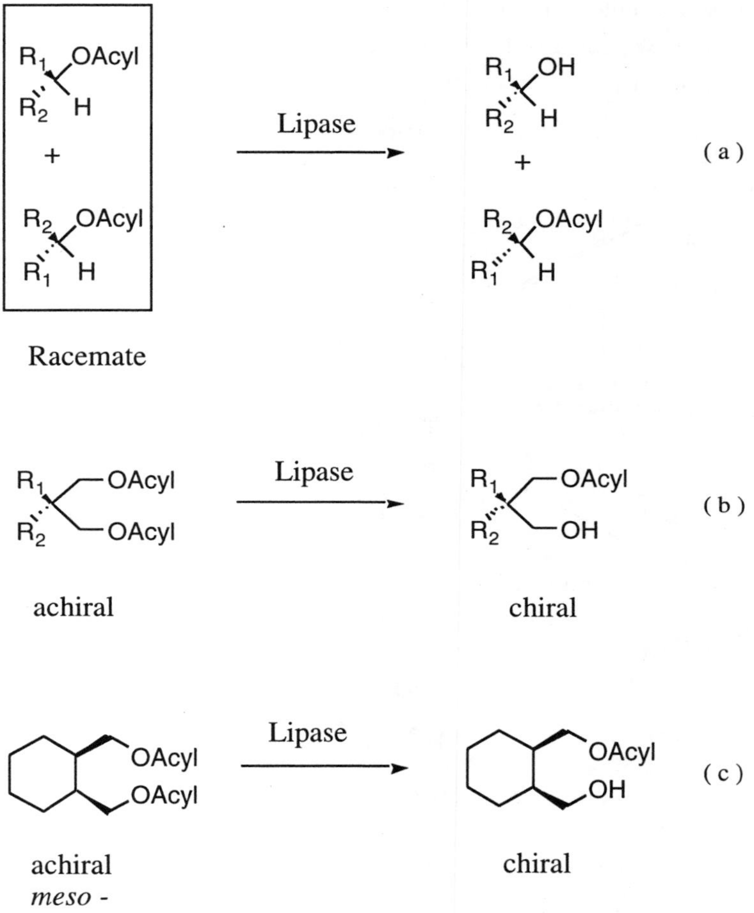

Figure 3. Lipases – modes of substrate recognition (schematic)

3.1. RESOLUTIONS (a)

Starting with racemic substrates, lipase catalyzed kinetic resolutions can lead to enantiomerically pure diastereomers which can be separated by classical techniques. The quality of such resolutions depends both on the enantioselectivity of the lipase and the conversion whereby E values of >>100 are required to obtain both enantiomers in close to enantiomerically pure form.

3.2.   ENANTIOTOPE DIFFERENTIATION (b, c)

In contrast to resolutions of racemates where – inherent to the starting material – only 50 % of one particular enantiomer can be obtained, the differentiation of enantiotopic groups in achiral substrates can lead – at least in theory – to products with 100 % optical and chemical yield.

In the examples to follow we have chosen lipase catalyzed transformations which in nearly all cases resulted in the desired high selectivities and thus produced throughout molecules in enantiomerically pure ($\geq$ 98 % ee) form.

4.   Resolutions

4.1.   INTRODUCTION

Enantiomerically pure, alkylsubstituted γ- and δ-lactones (Figure 4) are ubiquitous in nature and display a wide variety of biological activities e.g. as aroma compounds or pheromones.

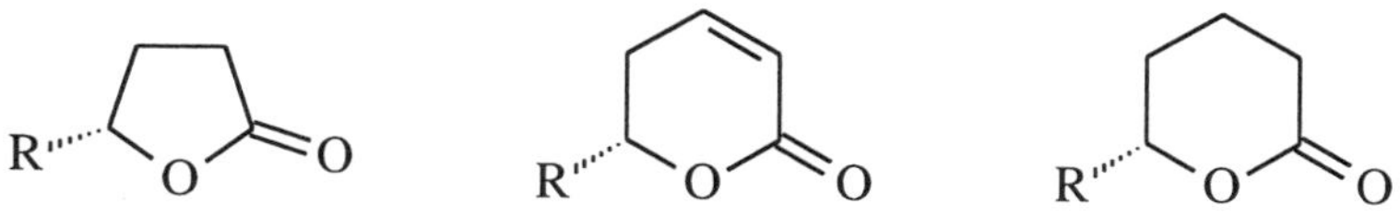

Figure 4. Enantiomerically pure, alkylsubstituted γ - and δ - lactones (only one absolute configuration is shown)

In spite of the importance of many of these molecules as aroma constituents there seems to exist no systematic study in which the relationship between organoleptic properties and their absolute configurations has been studied in detail. The reason for this probably is that enantiomerically pure γ- and δ-lactones of both absolute configurations are not easily accessible in multigram quantities. Interestingly enough – although natural products – many of these flavor lactones are isolated from natural sources frequently not optically pure but as mixtures of enantiomers with one enantiomer usually dominating.

In view of a systematic study of organoleptic properties it was our goal to synthesize both enantiomeric series of these molecules in good chemical yields and high ($\geq 99\%$ ee) enantiomeric purities.

4.2. RETROSYNTHETIC ANALYSIS

As demontrated for saturated and unsaturated 6-alkyl substituted δ-lactones, these molecules can be correlated retrosynthetically – via the corresponding δ-hydroxy-carboxylic acids – to enantiomerically pure alkyloxiranes (Figure 5).

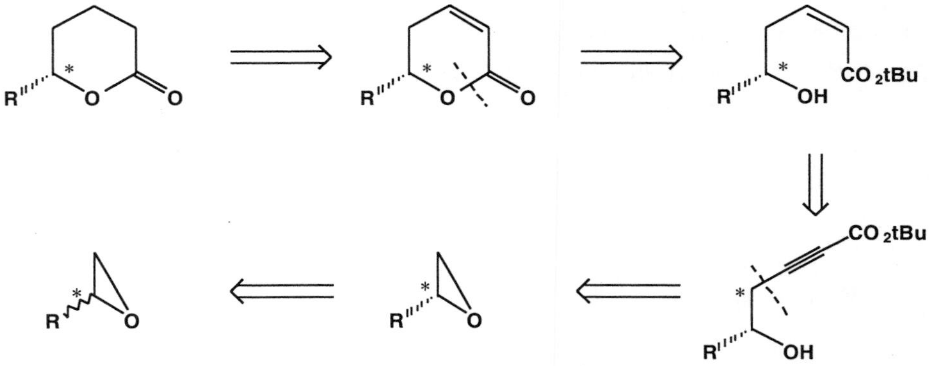

Figure 5. Retrosynthetic analysis of unsaturated and saturated δ-lactones

It can readily be envisaged that regioselective, nucleophilic ring opening of these chiral building blocks with C_3 or C_2-carbon units would lead to both series of our target molecules, i.e. γ-and δ-lactones.

4.3. ENANTIOMERICALLY PURE ALKYL OXIRANES

Based on our earlier work, concerning the development of a working model for a lipase from *Pseudomonas species* (SAM II) [2] we were able to predict and then demonstrate that the lipase-catalyzed resolutions of *t*-butyl-β-hydroxy thioethers can be achieved with very high ($\geq 98\%$ ee) enantioselectivities (Figure 6).

For this, racemic alkyloxiranes - commercially available or conveniently accessible by simple epoxidation of the corresponding 1-alkenes – were converted into the respective β-hydroxythioethers *via* regioselective ring opening using ^{t}BuSH/NaH. Treatment of the thus obtained products with chloroacetic anhydride/pyridine in the presence of *N,N*-dimethyl-4-aminopyridine (DMAP) led to the corresponding racemic chloroacetates which were then hydrolyzed in the presence of the lipase from *Pseudomonas species* (SAM II) under pH-stat conditions. These kinetic resolutions proceeded with probably the highest enantioselectivities ($E_{calc} \gg 1000$) ever recorded in our laboratory. All reactions came to a stand still after 50% conversion, the products were enantiomerically pure to the limits of detectability (GC on Cyclodex I/P). Due to the substantial differences in boiling points the resulting products – (*R*)-alcohols and (*S*)-esters – could be separated by simple distillation. Both series of products can be converted into the corresponding oxiranes by *S*-alkylation using Meerwein's salt followed by treatment with base. The very high enantiomeric purities of the resulting (*R*)- and (*S*)-alkyloxiranes were confirmed by chromatography using a recently developed method (BGIT) [3].

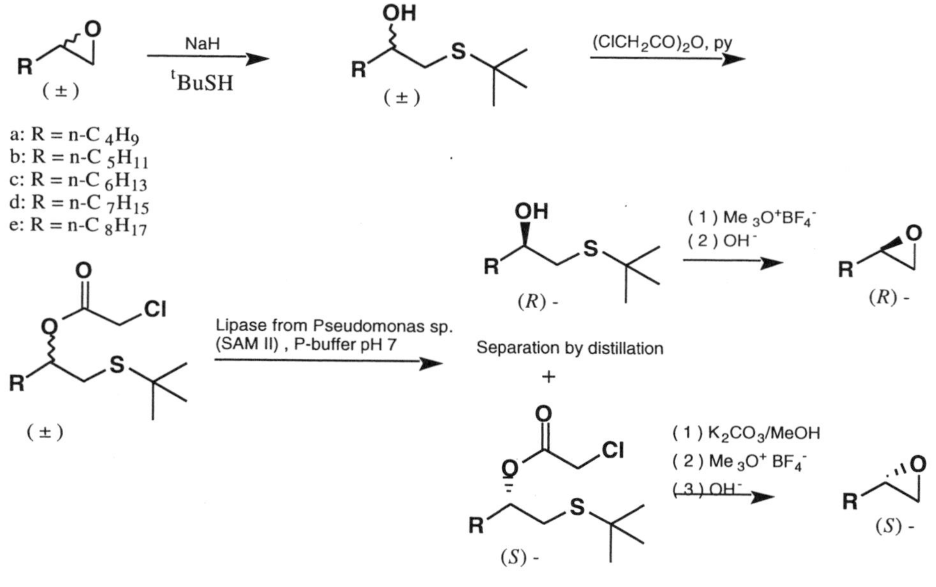

a: R = n-C$_4$H$_9$
b: R = n-C$_5$H$_{11}$
c: R = n-C$_6$H$_{13}$
d: R = n-C$_7$H$_{15}$
e: R = n-C$_8$H$_{17}$

Figure 6. Enzyme-assisted route to enantiomerically pure (*R*)- and (*S*)-alkyloxiranes

4.4.   δ-LACTONES

The thus obtained building blocks were converted into the corresponding series of unsaturated and saturated δ-lactones following the procedure outlined in Figure 7 [4], [5].

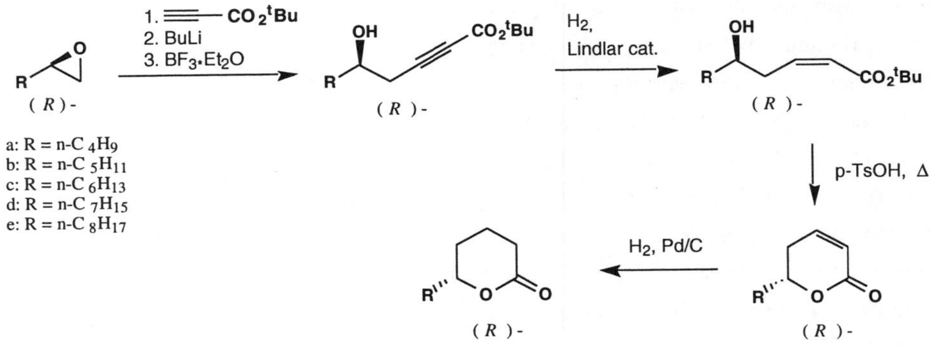

Figure 7. Synthesis of enantiomerically pure δ - lactones

Regioselective, boron trifluoride assisted ring opening of the oxirane moiety with the carbanion derived from t-butyl propiolate led to the corresponding t-butyl- (R)- and (S)-5-hydroxy-2-alkynoates. Partial hydrogenation using Lindlar catalyst proceeded quantitatively leading to the Z-configurated t-butyl- (R)- and (S)-hydroxy-2-alkenoates which were finally cyclisized to the desired unsaturated δ-lactones by treatment with p-TsOH. Their hydrogenation using Pd/C cleanly produced the saturated series of our target molecules. As confirmed by GC analysis on a chiral support (Lipodex E), the whole reaction sequence can be carried out without any measurable loss of enantiomeric purity. As a result, both enantiomeric series of these molecules are now available for studies establishing the relationship between their sensoric properties and absolute configurations.

4.5.   γ-LACTONES

Alkyl substituted γ-lactones display a wide variety of biological activities, i.e. as key aroma constituents in numerous fruits, as insect pheromones or as deterrents.

It was tempting, therefore, to use the above-mentioned enantiomerically pure alkyloxiranes also for the synthesis of this series of natural products (Figure 8) [5].

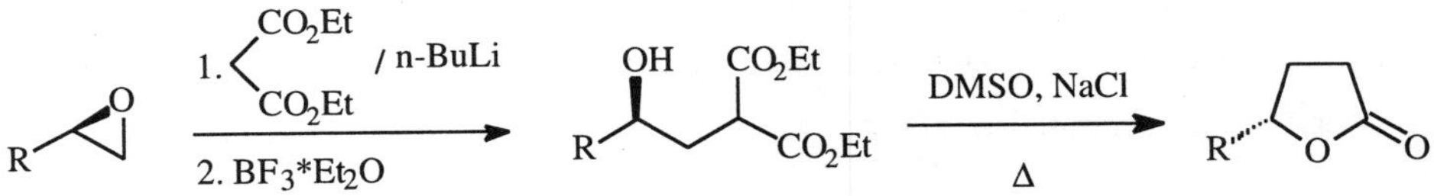

Figure 8. Synthesis of γ-lactones

Based on earlier work by Mori [6], we decided to use diethyl malonate for the introduction of the required two carbon unit.

Regioselective, boron trifluoride assisted ring opening of the oxirane function with the carbanion derived from diethyl malonate led to the corresponding malonates which are synthetic equivalents for the required (R)- and (S)-4-hydroxycarboxylic esters (precursors for our target molecules).

The conversion of these malonates – *via* ester hydrolysis, decarboxylation and ring closure – can be achieved in one step using an optimized method originally developed by Krapcho [7]. Heating in DMSO, NaCl to 150 °C for ca 6 h leads directly to the desired target molecules. The enantiomeric purities of the thus obtained series of γ-lactones are high as long as the reaction temperature is carefully controlled [5].

5. Enantiotope differentiations – Molecules of the inositolphospholipid pathway

5.1. INTRODUCTION

Based on our interest in intracellular signalling processes [8], we became increasingly engaged in the synthesis of molecules involved in the phosphatidyl inositol pathway such as enantiomerically pure structural analogues of 1,2(2,3)-*sn*-diglycerides and *myo*-inositol phosphates (see below), both types of molecules constituting important classes of second messengers.

5.2. CARBA ANALOGUES OF DI- AND TRIGLYCERIDES

Optically pure 1,2-*sn*- and 2,3-*sn*-diglycerides are notoriously unstable due to rapid acyl group migrations especially under protic conditions and at elevated temperatures causing immediate loss of optical purity (Figure 9).

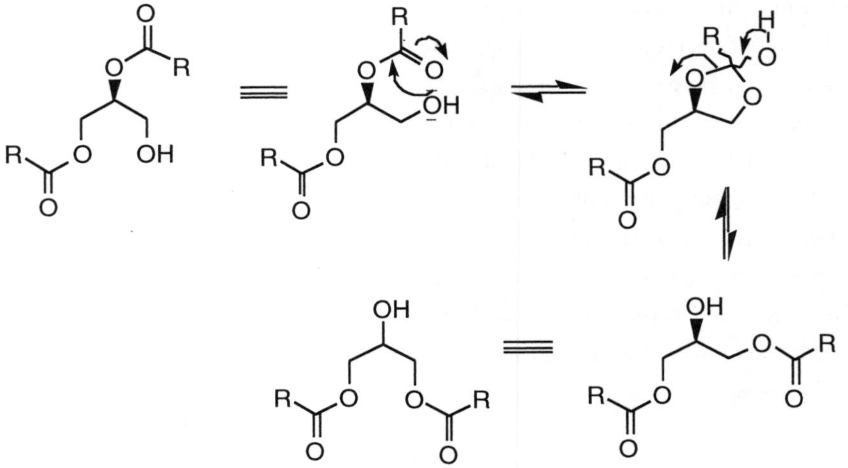

34

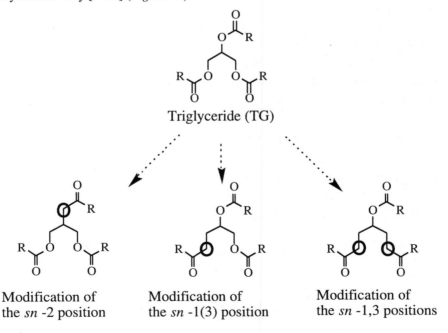

Figure 9. Acyl group migrations in 1,2(2,3)-*sn*-diglycerides

We therefore decided to explore enzyme assisted syntheses of carba-analogous triglycerides in which the sp^3-oxygens of the acyl moieties are replaced by sp^3-carbons in a systematic way [9 a-c] (*Figure* 10).

Triglyceride (TG)

Modification of
the *sn* -2 position

Modification of
the *sn* -1(3) position

Modification of
the *sn* -1,3 positions

Figure 10. Carba analogues of triglycerides as mimics of natural lipids

These structural analogues should – with the exception of hydrolytic cleavage – behave identical towards biological systems. In order to test this hypothesis first suitable synthetic routes to the corresponding C-analogues of triglycerides were developed – one of them is shown in *Figure 11* [9 c].

Conversion of the chosen acid chloride with CH_2N_2 and HBr leads *via* the corresponding diazoketones to the desired α-bromo ketone. Nucleophilic substitution with the anion of diethyl malonate leads to the required carbon backbone of the target molecule. In order to avoid intramolecular lactol formation of the ensuing 1,3-diol the carbonyl group was masked as *exo*-methylene function using a Wittig reaction before reducing the diester to the diol with $LiAlH_4$. Acylation of the diol and regeneration of the carbonyl function either by $RuCl_3/NaIO_4$ or O_3 leads to the desired target molecules.

Binding studies with lipases and lipase catalyzed hydrolyses as well as esterifications clearly demonstrated that native triglycerides and their C-analogues behave identically towards these biological systems [9 a, b]. In order to provide C-analogues of 1,2(2,3)-sn-diglycerides in optically pure form the corresponding *exo*-methylene derivatives were hydrolyzed enantioselectively or esterified under the conditions of irreversible acyl transfer in the presence of a lipase from *Pseudomonas species* (Figure 12) [9 c].

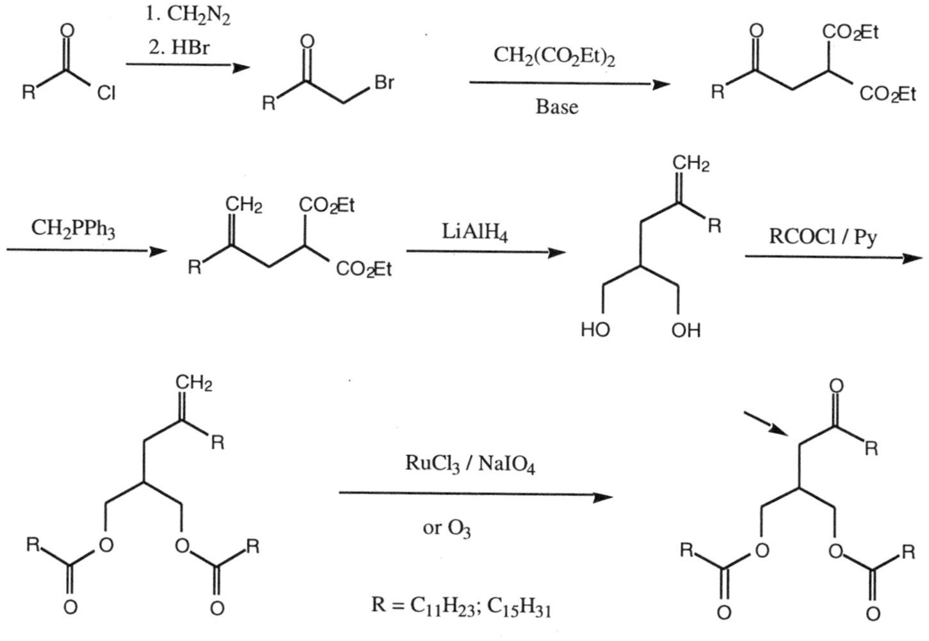

Figure 11. Synthesis of carba analogous triglycerides

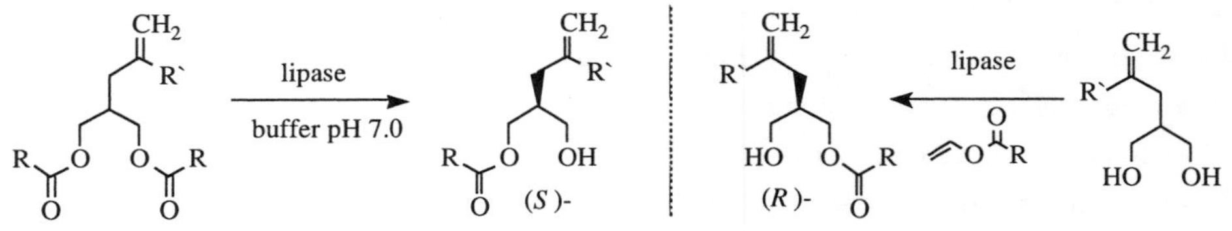

Figure 12. Carba-analogous diglycerides *via* enzymatic esterification and hydrolysis

5.3.   CARBA ANALOGUES OF PHOSPHOLIPIDS

The thus prepared C-analogues of 1,2(2,3)-*sn*-diglycerides, obtained in very high optical purities are not only interesting as potential second messengers, but can also be considered as highly useful synthetic building blocks for a new class of phospholipids including PAF-analogues and other molecules with this general backbone [9 c] (Figure 13).

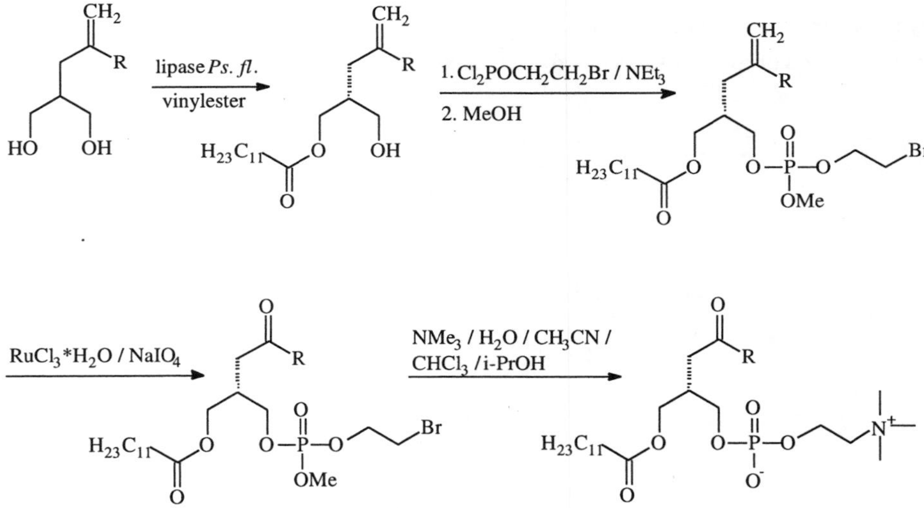

Figure 13. Carba analogues of phospholipids

5.4. *MYO*-INOSITOL PHOSPHATES

5.4.1 Introduction

Inositol phospholipids and their molecular constituents such as D-*myo*-inositol phosphates and 1,2-*sn*-diglycerides play an important role as second messengers in living cells with numerous functions as regulators and signal transducers. Unfortunately, however, frequently *myo*-inositol phosphates are only accessible in minute amounts from scarce natural sources after laborious isolation and purification procedures.

Clearly, the elucidation of their biological role would be greatly facilitated if these molecules could be made available *via* facile synthetic routes. Therefore, we explored new synthetic routes toward these molecules using enzymes for the introduction of chirality into the respective molecular backbone.

myo-Inositol itself, derived from ubiquitous and abundantly available phytic acid, is by far the most conveniently accessible and economical starting material.

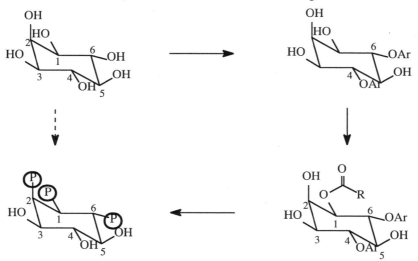

Figure 14. From *myo*-inositol to enantiomerically pure polyphosphates

It should be noted that *myo*-inositol is achiral while many *myo*-inositol phosphates are chiral and only biologically active in enantiomerically pure form. Consequently all known synthetic approaches are focussed towards the problem of converting this achiral molecule into enantiomerically pure building blocks. Thus the target was clear and well defined. For the synthesis of enantiomerically pure *myo*-inositol polyphosphates, starting from *myo*-inositol itself, a central building block was needed which could be prepared by enantioselective lipase catalyzed esterification of a suitably protected 4,6-derivatised *myo*-inositol (Figure 14).

The indicated derivatisation was required in order to obtain a *meso*-derivative with a reduced number of hydroxy groups, and to increase the solubility of the substrate in organic solvents.

5.4.2 Enantiomerically pure building blocks for myo-inositol phosphates

After screening numerous ester hydrolases (lipases) for their selectivety for this type of compounds, we found a lipoprotein lipase from the portfolio of Boehringer Mannheim which was able to convert a 4,6-protected *myo*-inositol in one step into a single enantiomer (Figure 15).

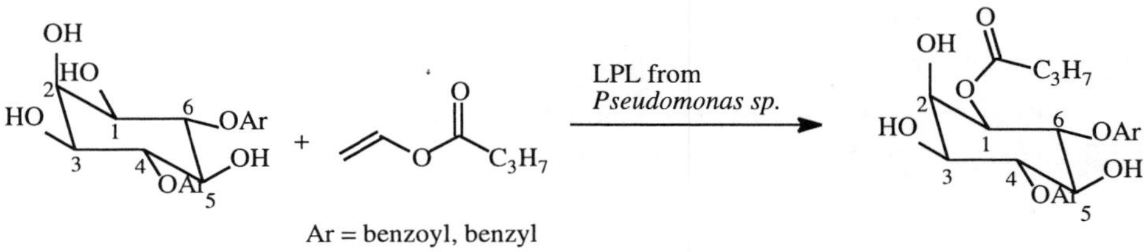

LPL from
Pseudomonas sp.

Ar = benzoyl, benzyl

Figure 15. Differentiation of enantiotopic hydroxy groups in *myo*-inositol derivatives (Lipoprotein lipase from *Pseudomonas species*, Boehringer Mannheim GmbH, Penzberg Germany)

This reaction, in which only *one* out of *four* different hydroxy groups is selectively esterified, demonstrates once more the power of enzymatic methods as applied to organic synthesis [10 a-c].

The absolute configuration of this building block was determinated unambiguously by chemical correlation [11]. After this key step, i.e. the introduction of chirality, the obtained enantiomerically pure building blocks (Ar = Bz, Bn) can be converted further into various selectively protected *myo*-inositol derivatives which can be further phophorylated leading to the corresponding *myo*-inositol phosphates (Figure 16).

The preparation of D-*myo*-inositol-1-phosphate (I-1-P) [12] and D-*myo*-inositol-1,2,6-trisphosphate (1,2,6-IP$_3$) [12], [13], which is a novel experimental drug are shown in Figures 17 and 18.

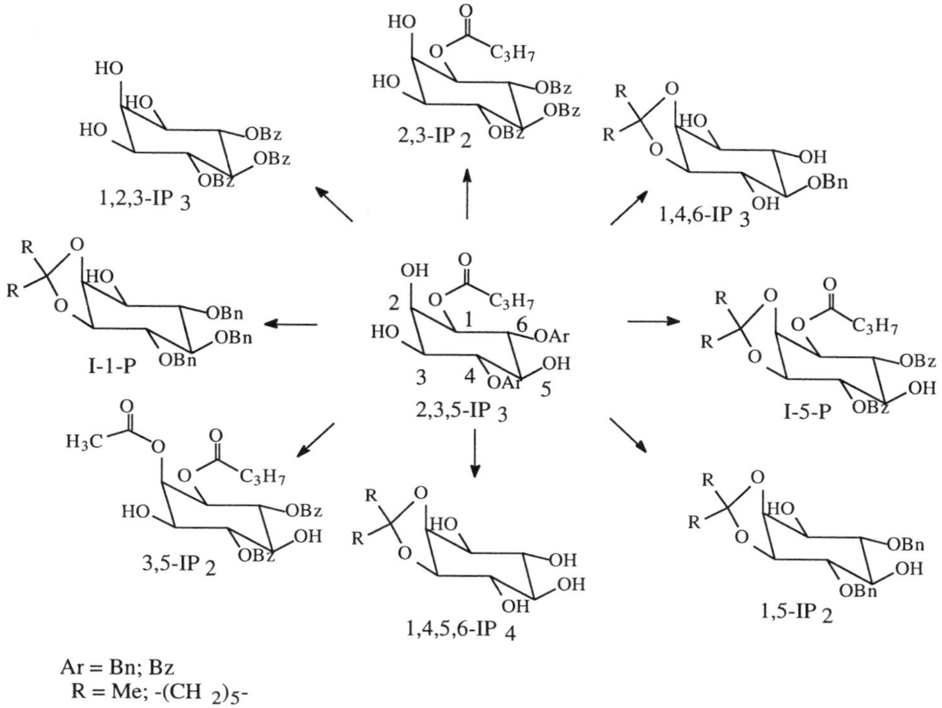

Ar = Bn; Bz
R = Me; -(CH₂)₅-

Figure 16. Selectively protected *myo*-inositol derivatives – building blocks for inositol phosphates

5.4.3 D-myo-Inositol-1-phosphate

Transketalisation with 2,2-dimethoxy propane followed by benzylation of the hydroxy group at C-5 under acidic conditions and saponification of the ester function at C-1 leads to a suitable building block for I-1-P. Phosphorylation of the free hydroxy group using *N,N*-dimethyldibenzyl phosphoamide in presence of tetrazole led to the trivalent phosphorous derivative. This in turn is oxidized to the required pentavalent phosphate using MCPBA or *t*-BuOOH. Removal of the benzyl groups by catalytic hydrogenation, addition of NaOH, followed by ion exchange chromatography leads to the chemically and optically pure *myo*-inositol-1-phosphate [12].

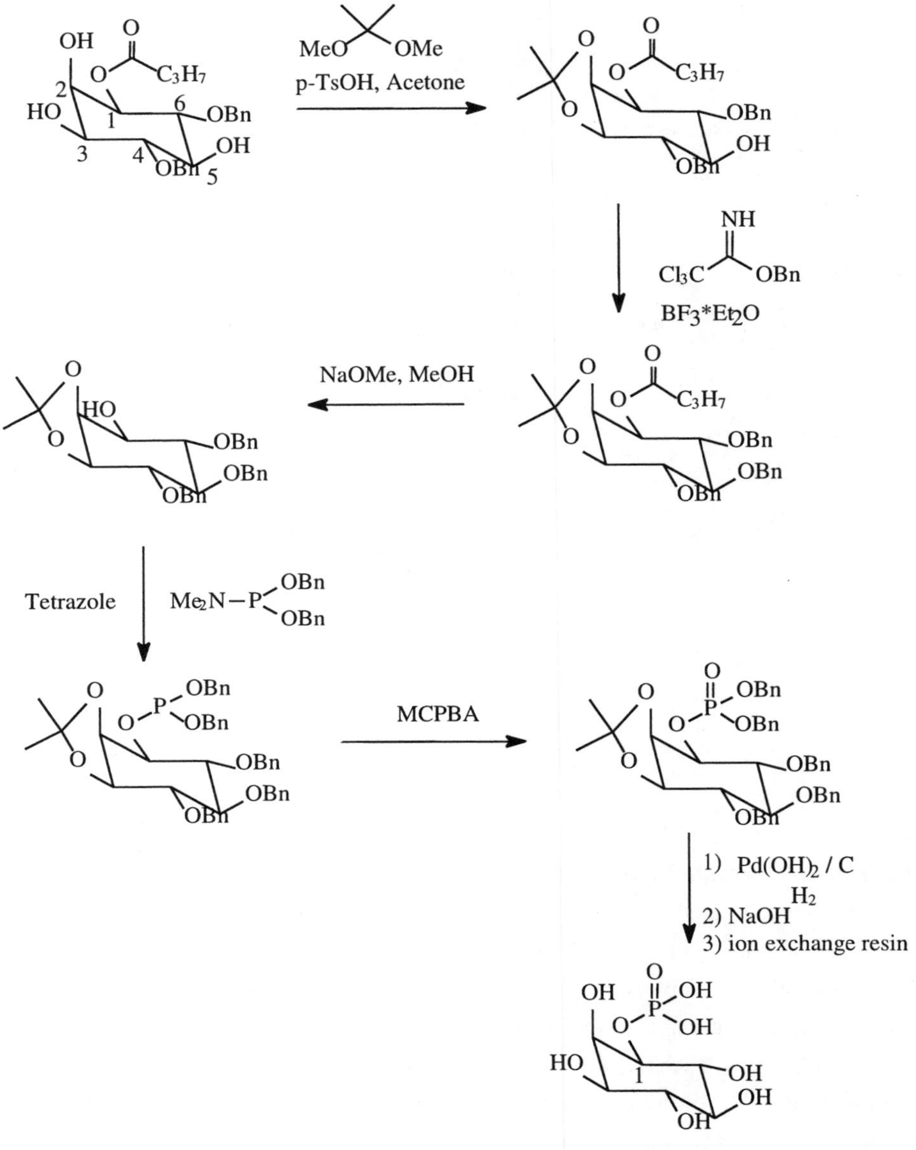

40

Figure 17. Synthesis of D-*myo*-inositol-1-phosphate (I-1-P)

5.4.4 D-myo-Inositol-1,2,6-trisphosphate

The synthesis of D-*myo*-inositol-1,2,6-trisphosphate (1,2,6-IP$_3$; PP56; α-Trinositol), which is a novel experimental drug, is outlined in Figure 18 [12], [13].

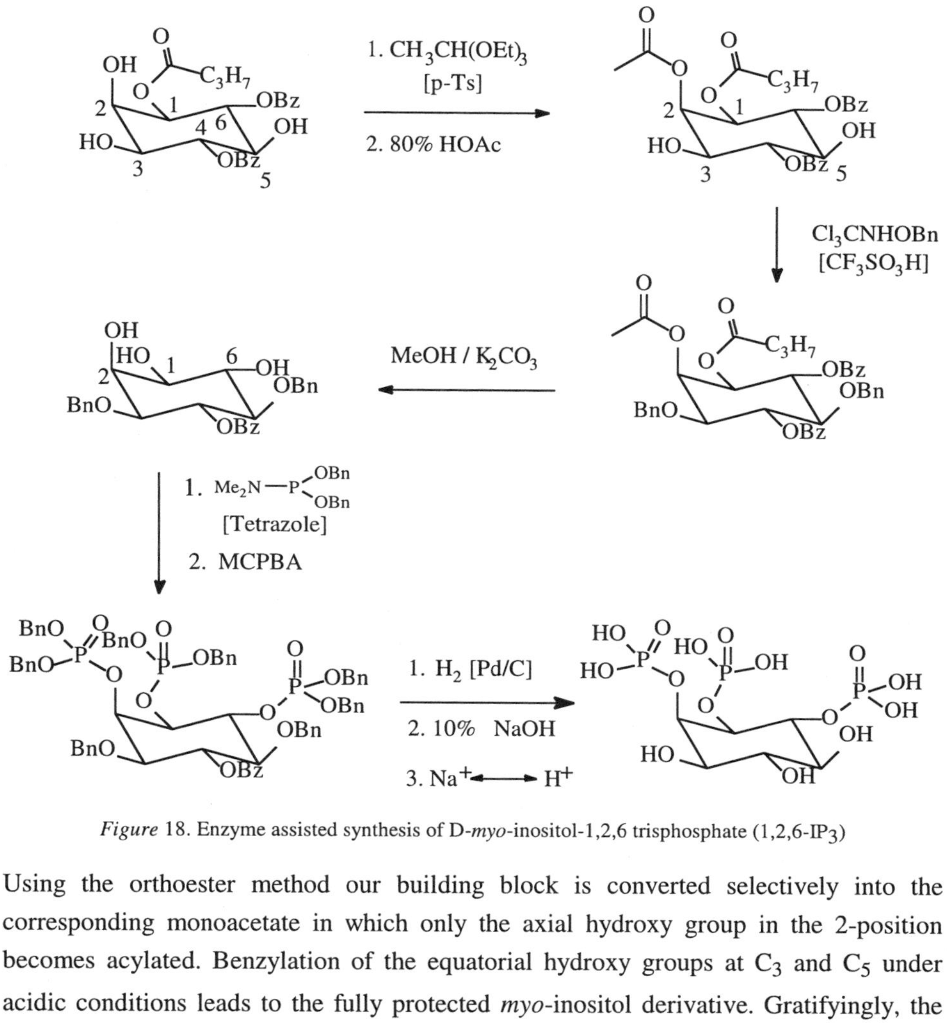

Figure 18. Enzyme assisted synthesis of D-*myo*-inositol-1,2,6 trisphosphate (1,2,6-IP$_3$)

Using the orthoester method our building block is converted selectively into the corresponding monoacetate in which only the axial hydroxy group in the 2-position becomes acylated. Benzylation of the equatorial hydroxy groups at C$_3$ and C$_5$ under acidic conditions leads to the fully protected *myo*-inositol derivative. Gratifyingly, the subsequent removal of the ester functions was highly regioselective, resulting in the formation of the free hydroxy groups in the desired positions 1, 2 and 6. It is easy understandable that in the base catalysed methanolysis the acetate and butyrate functions are removed rapidly and faster than the more stable benzoate groups. However, it was surprising that only one of the benzoate groups, the one in position 6 is removed selectively. The obtained triol can be phosphorylated as described above. Deprotection of the resulting trisphosphate ester with H$_2$/Pd-C followed by saponification (NaOH, pH 11-12) leads to 1,2,6-P$_3$ in nearly quantitative yield. All materials are obtained with very high isomeric purity as was confirmed by ion exchange chromatography [10 b].

42

6. Summary

Using the three major modes of substrate recognition as a guide line the use of ester hydrolases (lipases) for the preparation of a variety of structurally different, enantiomerically pure compounds was described. It is evident that lipases are versatile and highly useful synthetic tools in the hands of the organic chemist.

7. Acknowledgement

We thank the *Fonds der Chemischen Industrie* and the State of NRW for financial support of this work. B. J. gratefully acknowledges a stipend from the state of NRW (Nordrhein-Westfalen, Germany).

8. References

1. For a detailed discussion see: Andersch, P., Berger, M., Hermann, J., Laumen, K., Lobell, M., Seemayer, R., Waldinger, C., Schneider, M. (1997) in Ed. Dennis, A.E., Rubin, B., Ester Synthesis via Acyltransfer (Transesterification), *Methods in Enzymology*, Vol. 286, chapter 19, Academic Press, 406 - 443.

2. Laumen, K. (1987) *Ph. D. thesis*; Ader, U., Andersch, P., Berger, M., Goergens, U., Seemayer, R., Schneider, M.P. (1992) *Pure Appl. Chem.* **64**, 1165; Ader, U., Andersch, P., Berger, M., Goergens, U., Seemayer, R., Schneider, M.P. (1993) *Indian J. Chem. Sec B* **32B**, 145.

3. Lobell, M., Schneider, M.P. (1993) *J. Chromatogr.* **633**, 287.

4. Haase, B., Schneider, M.P. (1993) *Tetrahedron: Asymmetry* **4**, 1017.

5. Haase, B. (1997) *Ph. D. thesis*, Wuppertal.

6. Mori, K., Sasaki, M., Tamada, S., Suguro, T., Masuda, S. (1978) *Heterocycles* **10**, 111; Mori, K., Sasaki, M., Tamada, S., Suguro, T., Masuda, S. (1979) *Tetrahedron* **35**, 1601.

7. Krapcho, A.P., Jahngen, E.G.E.Iv., Kashdan, D.S. (1974) *Tetrahedron Lett.* **32**, 2721; Dekmezian, A.H., Kaloustian, M.K. (1979) *Synth. Commun.* **9**, 431.

8. Reitz, A.B. (1991) *Inositol Phosphates and Derivatives Synthesis, Biochemistry and Therapeutic Potential* ACS Symposium Series 463 Washington DC;Irvine, R.F. (1990) Methods in Inosite Research Raven Press New York.

9. a) Berger, M., Jakob, B., Schneider, M.P. (1994) *Bioorganic & Medicinal Chemistry* **2**, 573-588.b) Berger, M. (1993) *Ph. D. thesis*, Wuppertal.c) Jakob, B. (1999) *Ph. D. thesis*, Wuppertal

10. a) Andersch, P., Schneider, M.P. (1993) *Tetrahedron: Asymmetry* **4**, 2135. b) Andersch, P. (1995) *Ph. D. thesis*, Wuppertal. c) Ghisalba, O., Laumen, K. *Biosci. Biotech. Biochem.* **58**:(11) 2046.

11. Mercier, D., Gero, S.D. (1968) *Tetrahedron Lett.* **31**, 3459.

12. Andersch, P., Jakob, B., Schiefer, R., Schneider, M.P. (1997) in Ed. Wirtz K.W.A. *Molecular Mechanisms of Signalling and Targeting*, Springer Berlin Heidelberg, 247.

13. Andersch, P., Schneider, M.P. (1996) *Tetrahedron: Asymmetry* **7**, 349.

MOLECULAR BASIS FOR EMPIRICAL RULES THAT PREDICT THE STEREOSELECTIVITY OF HYDROLASES

ALEXANDRA N. E. WEISSFLOCH AND ROMAS J. KAZLAUSKAS
McGill University, Department of Chemistry
801 Sherbrooke St. West
Montréal, Québec H3A 2K6
Canada

Abstract: Hydrolases are the most widely used enzymes for organic synthesis, both in academic laboratories and in commercial production. Most researchers focus on a few well-studied lipases and proteases. To predict the stereoselectivity of these 'work horse' hydrolases, researchers developed empirical rules or models. These empirical rules describe the shape and other features of either a good substrate (substrate model) or of the binding site (active site model). Researchers use these rules to design new syntheses using hydrolases. Recently, x-ray crystallographers have solved the three-dimensional structures of all the workhorse lipases, but these structures are too complex to use directly in the organic synthesis design. However, by combining the empirical rules with x-ray structures, one obtains a powerful and useful tool. The empirical rules identify the most important features of the substrate binding site (usually its size and shape) and thus simplify interpretation of the x-ray structures. On the other hand, the x-ray structures provide a molecular basis for the validity of the rules, add molecular details to the rules that explain puzzling features. This review summarizes recent efforts to combine both empirical rules and three-dimensional structures to rationally-design new applications of hydrolases in organic synthesis.

1. Introduction

Organic chemists usually exploit hydrolases either for their high stereoselectivity or for their ability to act under mild conditions. For example, organic chemists often use

43

B. Zwanenburg et al. (eds.), Enzymes in Action, 43–69.
© 2000 *Kluwer Academic Publishers. Printed in the Netherlands.*

hydrolases to prepare enantiomerically pure intermediates, or to deprotect complex intermediates [1]. The ideal hydrolase for organic chemists would accept all substrates, but retain perfect stereoselectivity and regioselectivity. In practice, organic chemists chose hydrolases whose natural role is digestion, that is, breakdown of a wide range of substrates. These hydrolases have a broad substrate range, but show high selectivity toward only some substrates. Thus, a common problem is finding the most suitable hydrolase for a given application.

This review focuses on a rational approach to the stereoselectivity of hydrolases [2]. Experience has identified several particularly useful hydrolases and the types of substrates where they are most useful, Table 1. Beyond these general guidelines, researchers also developed more detailed models, or rules, for these hydrolases and these will be the main focus of this review. The aim of these rules is both to guide the choice of hydrolases and also to identify what features of the substrate are most important for high enantioselectivity. Most of the rules focus on the size and shape of the substituents; thus, these features appear to be the most important for high enantioselectivity. However, many rules also mention the polar or non-polar character of substituents, indicating that size and shape is not the only feature important for high enantioselectivity. Naemura recently reviewed empirical rules and models for hydrolases in Japanese. [3]

TABLE 1. 'Work horse' hydrolases for organic synthesis

hydrolase	abbreviation	best suited for
Candida antarctica lipase B	CAL-B	smaller secondary alcohols (e.g., MeCH(OH)R), primary amines (NH_2CHRR')
Candida rugosa lipase	CRL	large cyclic secondary alcohols, carboxylic acids
Pseudomonas cepacia lipase	PCL	secondary alcohols, primary amines (NH_2CHRR'), primary alcohols
pig liver esterase	PLE	carboxylic acids, hindered alcohols
pig pancreatic lipase	PPL	secondary alcohols, primary alcohols
subtilisin	subtilisin	carboxylic acids, esp. amino acids secondary alcohols, primary amines (NH_2CHRR')

Some rules – substrate rules - define what the substrate looks like. The simplest rules specify only relative sizes of substituents, for example, small, medium and large. The advantage and disadvantage of these rules is their simplicity. They are very easy to use and work for a wide range of substrates, but they do not provide detailed information. Other substrate rules are more detailed, but apply only to a specific type of molecule.

Other rules – active site models - define what the binding site of hydrolases looks like. These are usually two- or three-dimensional box-type models which give more details about the shape, size, hydrophobicity, and possibly the electronic character of the binding site. These models are harder to apply to a new substrate and are less general. Each hydrolase needs its own model.

In the last ten years, x-ray crystallographers have solved the structures of a number of lipases and esterases, including all the ones most commonly used for synthesis. Crystallographers had solved the structures of most synthetically useful proteases some years earlier. Researchers are now combining the molecular information in these structures with previously developed empirical rules. This combination provides a molecular basis for the empirical rules, gives molecular details to explain curious features of the rules, and provides a starting point for more detailed molecular modeling studies.

To limit the size of this review, we will discuss only those hydrolases where both empirical rules and x-ray structures are available.

2. Overview of the structures of lipases, subtilisin, and other proteases

Lipases, esterases and subtilisins are all serine hydrolases. Their catalytic machinery consists of a catalytic triad – serine, histidine, and aspartate (in a few lipases, glutamate replaces aspartate) – and the oxyanion stabilizing residues. All three of these serine hydrolases can catalyze both hydrolysis and formation of esters and amides; however, only subtilisin, the protease, is fast enough for practical use in the hydrolysis of amides. An additional difference is that subtilisin shows selectivity mainly toward the acyl portion of an ester or amide, while lipases and esterases show selectivity mainly toward the alcohol or amine portion of an ester or amide.

2.1. LIPASES AND ESTERASES FOLD ACCORDING TO THE α/β-HYDROLASE FOLD.

Even though individual lipases and esterases differ substantially in their amino acid sequence, all have similar three-dimensional structures. Lipases fold according to the α/β-hydrolase fold, which consists of a central β-sheet (eight mostly parallel strands) surrounded by α-helices on both sides [4], Figure 1. Each enzyme may have extrusions (additional sheets and helices) on the C-terminal side, especially after strands 6 and 7. For example, a large sheet extrusion at the N-terminus, two large helical extrusions after both strand 6 and 7, and several smaller extrusions increase the molecular weight of CRL to 63 kDa. On the other hand, PCL has a molecular weight of only 33 kDa because it lacks strands one and two and contains smaller helical extrusions after strands 6 and 7.

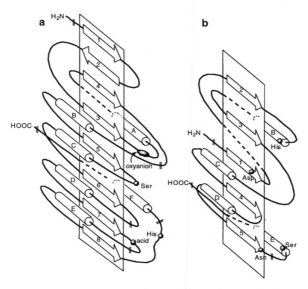

Figure 1. Schematic of protein folds. **a)** α/β hydrolase fold typical of lipases and esterases. **b)** α/β subtilase fold typical of subtilisin. Positions of the residues in the catalytic triad and in the oxyanion hole are marked. The double slashes correspond to places were there may be extrusions (additional sheets or helices) that are unique to each hydrolase. These schematic diagrams show a flat central β-sheet, but in reality, this β-sheet has the usual left-handed twist found in proteins. The degree of twist varies.

The active site is at the C-terminal edge of the β-sheet. The catalytic triad residues - serine, histidine, and aspartate (or glutamate) - are in the same order in the amino acid sequence of all esterases and lipases. In the structure, they all lie in turns after a strand that form the lipase core. The serine lies at a sharp turn after strand 5, the aspartate (or glutamate), lies in a loop following strand 7, and histidine follows strand 8. In a few lipases (e.g. pancreatic lipase), the catalytic acid lies after strand 6. The "oxyanion hole", containing the oxyanion stabilizing residues lies after strand 3. An extensive hydrogen-bonding network maintains the geometry of the catalytic triad.

These similarities in structure create a similar alcohol-binding pocket in all lipases. In particular, assembly of the catalytic machinery creates the medium-sized pocket, which will be discussed in section 3.3 below. The amino acid residue next to the catalytic histidine forms one wall of the pocket, the amino acid residue next to the catalytic serine forms the floor of the pocket and the amino acid residue next to an oxyanion-stabilizing residue forms another wall of the pocket. Note however, that the amino acid residues at these positions differ in different lipases, thus creating a larger or smaller pocket. In addition, the extrusions mentioned above can change the shape of the substrate-binding site.

2.2. SUBTILISIN FOLDS ACCORDING TO THE α/β-SUBTILASE FOLD.

The structure of subtilisin shows a different fold, the α/β subtilase fold [5] Figure 1 above. This fold consists of a sheet comprised of five parallel β-strands flanked by four α-helices, two α-helices on either side. Unlike the α/β hydrolases, where the catalytic residues lie in loops, in subtilisin the catalytic residues lie in the secondary structures. Histidine lies in the first turn of helix B, serine in the first turn of helix E, aspartate in strand 1 and asparagine, which forms part of the oxyanion hole, in the end of strand 5.

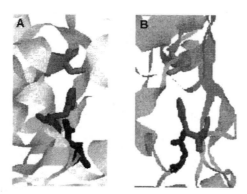

Figure 2. The active sites of subtilisins and lipases are approximate mirror images: the catalytic serine is on the opposite side of the plane formed by the imidazole ring of the catalytic histidine. **A)** The catalytic triad of subtilisin BPN' (pdb file 2st1). **B)** The catalytic triad of lipase from *Pseudomonas cepacia* (pdb file 2lip). The catalytic triad is viewed from the substrate-binding site in both cases.

Although the catalytic triad residues are the same as those found in serine esterases, the serine nucleophile lies on the opposite side of the plane formed by the imidazole ring of the histidine residue. Therefore, the three-dimensional orientations of the catalytic machinery (including the oxyanion hole) in serine esterases and serine proteases are approximately the mirror images, Figure 2. This opposite chirality is the likely reason for the opposite enantiopreference of lipases and subtilisins toward secondary alcohols, see sections 3.1 and 3.5 below.

2.3. OTHER PROTEASES

Chymotrypsin is also a serine protease, but it folds differently from both subtilisin and the lipases/esterases. The chymotrypsin fold is two antiparallel β-barrel domains, sometimes called the eukaryotic serine protease fold. The chirality of the catalytic triad is the same as that for subtilisin. Like subtilisin, chymotrypsin shows selectivity mainly for the acyl portion of an ester or amide.

There is also a third structural class of serine proteases - the carboxypeptidases. Carboxypeptidases fold according to the α/β hydrolase fold like lipases and esterases. Researchers have not yet explored the potential of carboxypeptidases for organic synthesis.

3. Selectivity of hydrolases toward secondary alcohols and primary amines

3.1. SUBSTRATE MODELS

Based on the observed enantioselectivity of lipases and subtilisins, researchers proposed rules to predict which enantiomer reacts faster [6], Figure 3. These rules are based on the size of the substituents and apply to both hydrolysis and acylation reactions. For acylation, the enantiomer shown reacts faster; for hydrolysis, the ester of enantiomer shown reacts faster.

The advantages of these rules are that they are simple to apply and work well for all secondary alcohols and primary amines. To use the rules, first, draw the substrate so that the alcohol (or ester) group points out of the page toward the reader. Second, imagine a line extending the C–O bond so that it divides the molecule in two parts. The enantiomer with the larger group on the right side of the line is the one that will react faster. Keep the flexibility of molecular structures in mind when comparing the sizes of the substituents. For instance, an alkyl chain can fold so its effective size is smaller than a substituent that cannot fold such as phenyl.

The rule for secondary alcohols applies to all lipases tested so far. These include cholesterol esterase (CE) [6a], lipase from several *Pseudomonas* species including *Ps. cepacia* (PCL) [6a], *Ps. fluorescens* (PFL) [6b], *Ps. aeruginosa* (PAL) [6c], lipase from *Rhizomucor miehei* (RML) [6d], lipase B from *Candida antarctica* (CAL-B) [6e], and porcine pancreatic lipase [6f]. The rule also works for lipase from *Candida rugosa* (CRL), but only for cyclic secondary alcohols [6a].

The different lipases differ in both their rates of reaction and enantioselectivity toward secondary alcohols. For example, CRL and CE accept large secondary alcohols such as the bicyclic alcohols, while most other lipases react only slowly. CAL-B shows the highest rates and enantioselectivities toward alcohols where the medium substituent is no larger than methyl or ethyl. The *Pseudomonas* lipases are general-purpose lipases.

This rule also applies to primary amines of the type RR'CHNH$_2$ whose shape is similar to that of secondary alcohols [7]. The lipases tested include CAL-B, PCL and PAL. Lipases are poor catalysts for the hydrolysis of amides, but these lipases efficiently catalyze the reverse reaction - acylation of amines.

Subtilisins favor the enantiomer opposite to the one favored by lipases for both secondary alcohols and primary amines of the type RR'CHNH$_2$. Thus, subtilisins and lipases are enantiocomplementary reagents. The enantioselectivity of subtilisins is usually lower than that of lipases.

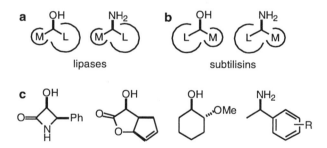

Figure 3. Empirical rules predict the fast-reacting enantiomer for lipases and subtilisins. Both secondary alcohols and amines of the type RR'CHNH$_2$ have similar structures so similar rules apply. **a)** Rules to predict the fast reacting enantiomer for lipase catalyzed reactions. M represents a medium-sized substituent such as methyl, while L represents a large substituent such as phenyl. In acylation reactions, the enantiomer shown reacts faster; in hydrolysis reactions, the ester of the enantiomer shown reacts faster. **b)** Rules to predict the fast reacting enantiomer in subtilisin catalyzed reactions. Note that lipases and subtilisins favor opposite enantiomers; thus, they are enantiocomplementary reagents. Subtilisins are usually less enantioselective than lipases. **c)** Examples of pure enantiomers prepared by lipase catalyzed reactions showing the fast-reacting enantiomer.

These empirical rulés suggest that lipases and subtilisins distinguish between enantiomeric secondary alcohols primarily by the relative sizes of the two substituents. Changing the relative sizes of the substituents might change the selectivity. Indeed, a number of researchers increased the enantioselectivity of lipase-catalyzed reactions by modifying the substrate to increase the size of the large substituent [8]. Similarly, Shimizu *et al.* reversed the enantioselectivity by converting the medium substituent into the large one [9].

However, the rules in Figure 3 cannot account for changes in selectivity that may be caused by small changes in the structure of the secondary alcohol. For example, CAL-B shows high enantioselectivity toward 3-nonanol (E > 300), but low enantioselectivity toward 1-bromo-2-octanol (E = 7.6) under the same conditions [10], Figure 4. Both an ethyl and a $-CH_2Br$ group have similar sizes, so the difference suggests that polar character of the bromine lowers the enantioselectivity. Similarly, changing the position of a methyl substituent in the aryl ring of the large substituent drastically altered the selectivity of PCL from E >100 in the para-substituted isomer to E = 27 in the ortho-substituted isomer [11].

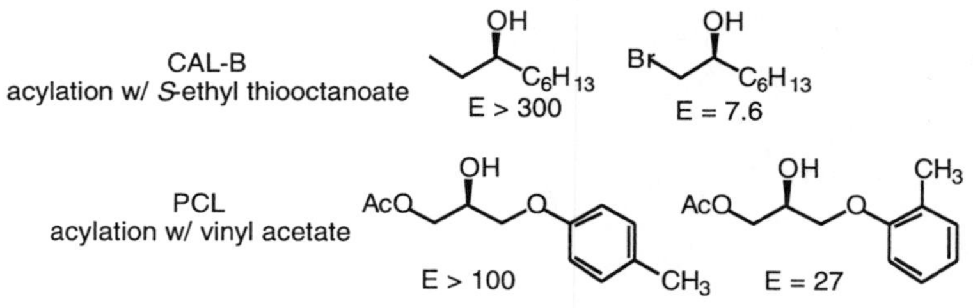

Figure 4. Small changes in structure can alter the selectivity. Nevertheless, all these examples follow the secondary alcohol rule.

To map more precisely how structure affects enantioselectivity, researchers have developed more detailed substrate models. These models apply only to substrates similar to those used to develop the model. For example, Oberhauser *et al.* developed a qualitative [12] and later a quantitative [13] model for CRL-catalyzed resolutions of bicyclic secondary alcohols, Figure 5. This model predicts the enantioselectivity for various bicyclo[2.2.1]heptanols and bicyclo[2.2.2]octanols. This model identifies regions of the substrate that tolerate substituents, as well as regions that favor positive or negative charges.

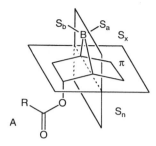

Figure 5. Model for bicyclic substrates of CRL [12]. The sites and substituents are defined as follows: **A**: reaction site (must be endo). **B**: bridge (may contain heteroatoms). S_a and S_s: *anti* and *syn* substituents (may be an ester, ether, or acetal group). S_x: *exo*-substituent (may be large). S_n: *endo*-substituent (must be very small). π: π-site (π-electron in this site enhance enantioselectivity). A later computer model gives more detailed space and electrostatic restrictions [13].

Another way of adding detail to the simple rules in Figure 3 is to define the maximum size for each substituent. For example, Exl *et al.* [14] found a maximum size of 9.2 Å for the large substituent in CRL, but a maximum size of only 7.1 Å for PCL. This finding is consistent with preference of CRL for larger alcohols, such as bicyclic alcohols.

3.2. ACTIVE SITE MODELS

A natural extension of these rules is the two-dimensional box type models, Figure 6, which attempt to define the shape of the active site. These models propose specific size limits for each substituent. The disadvantages are that they are harder to use and that each hydrolase requires separate model. Burgess and Jennings developed the first model of this type for lipases [15]. They defined the limits of a lipase from *Pseudomonas fluorescens* (lipase AK) using unsaturated secondary alcohols. Bornscheuer *et al.* assumed that the same model would also work for other *Pseudomonas* species and used this model to rationalize their measured selectivity [16]. However, when Naemura *et al.* proposed a model for lipase from *Pseudomonas fluorescens* (lipase YS) [17], it showed significant differences particularly for the medium sized substituent. One should be cautious in assuming that a model developed for one lipase will apply to another. Naemura *et al.* proposed a model for a lipase from *Alcaligenes* sp. (lipase QL) [18]. Not surprisingly, the overall shape of these models resembles the general rule in Figure 3.

To use these models, first draw the secondary alcohol using the same scale as the model. A computer-based drawing program that permits scaling of the drawing is

52

helpful. Next, fit the hydroxyl group into the catalytic site (the OH pointing out from the page). If part of a molecule goes beyond the boundaries of the model, it is not a substrate. The enantiomer with the best fit is predicted to be the major product.

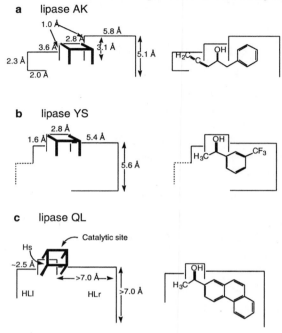

Figure 6. Two-dimensional active site models for several lipases predict which secondary alcohol reacts faster. Examples show good substrates for each enzyme. a) Model for lipase from *Pseudomonas fluorescens* (lipase AK). b) Model for another lipase from *Pseudomonas fluorescens* (lipase YS). c) Model for a lipase from *Alcaligenes* sp. (lipase QL).

The next level of detail is a three-dimensional model. Researchers overlaid energy-minimized structures of substrates and non-substrates to define the size limits and polar or nonpolar character of each pocket. Two groups independently proposed three-dimensional models for PCL [19, 20], Figure 7. Lemke *et al.* found one spherical hydrophobic pocket and another tube-like pocket. The spherical pocket accommodates bulky groups such as phenyl, phenoxymethyl, and substituted aryl derivatives, while the tube accepts only long groups such as acetoxymethyl and (*n*-hexadecanoyl-oxy)methyl. Therefore, Lemke *et al.* suggested that not just size, but also shape of the substituents determines enantioselectivity. Grabuleda *et al.* proposed a similar, but slightly smaller, model. The medium-sized pocket (H_s) was also tube-like. The differences in dimensions of the models are likely due to the different groups of substrates used to build the models.

The first three-dimensional model for hydrolases was Jones's model for pig liver esterase (PLE). This model predicts the stereoselectivity of PLE toward chiral and prochiral acids, see section 5.2 below. However, Naemura *et al.* used this model to account for PLE's enantioselectivity toward several secondary alcohols [21]. The structure of PLE is not known, but it may be an α/β-hydrolase like other esterases. Indeed, its selectivity toward secondary alcohols follows the same empirical rule as lipases. The large and small hydrophobic pockets in Jones' active site model (Figure 17 below) may correspond to the large and medium pockets of the empirical rule. Note that applying Jones' model to alcohols is not yet well tested; Jones developed the model using acids.

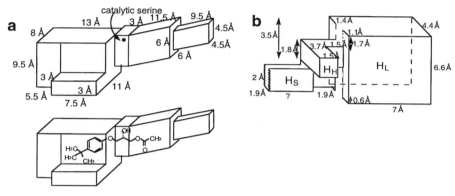

Figure 7. Three-dimensional active site models for PCL proposed after overlaying energy-minimized structures of both good and bad substrates. **a)** Lemke *et al.*'s model contains a spherical hydrophobic pocket and a tube-like hydrophobic pocket. The bottom diagram shows a good substrate within the model. **b)** Grabuleda *et al.*'s model is slightly smaller. The reacting hydroxyl group goes in the H_H site (hydrophilic), while the substituents go into the two hydrophobic sites, H_S and H_L. Both models predict the same absolute configuration. The drawing shows Grabuleda *et al.*'s model rotated by 180° about the vertical axis relative to Lemke *et al.*'s model.

3.3. STRUCTURE OF THE ALCOHOL-BINDING POCKET IN LIPASES

Cygler *et al.* [22] solved the x-ray structure of CRL containing phosphonate transition state analogs of both the fast- and slow-reacting enantiomers of a typical secondary alcohol - menthol, Figure 8. These structures show that the large substituent (isopropyl) binds in a large hydrophobic pocket formed by the side chains from three phenylalanine residues and one isoleucine residue. On the other hand, the medium substituent (methylene) binds in a smaller and less hydrophobic pocket.

54

Figure 9 compares the active sites of several lipases. Although the overall shape of the alcohol-binding site resembles the rule in Figure 3, there are also clear differences. For example, the alcohol-binding site in CE and CRL is a wide crevice consistent with the ability of these hydrolases to accept large (e.g., bicyclic) alcohols. The alcohol binding pocket in PCL narrower. Note that the medium pocket has an additional tube-like extension as suggested by the active site models in Figure 7. The alcohol-binding pocket in CAL-B is the narrowest and the stereoselectivity pocket is deeply buried. This deeply buried stereoselectivity pocket is consistent with the preference of CAL-B for alcohols where the medium substituent is no larger than methyl or ethyl.

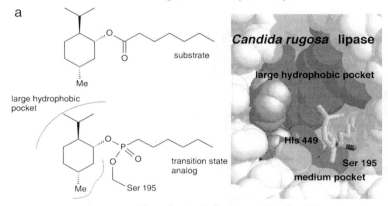

Figure 8. X-ray structure of a transition state analog covalently bound in the active site of CRL. **a**) The substrate is the fast reacting enantiomer of menthyl heptanoate. **b**) The transition state analog is a phosphonate covalently linked to CRL. Cygler *et al.* [22] prepared this derivative by reacting the corresponding phosphonyl chloride with CRL. **c**) The x-ray structure of this complex shows the isopropyl moiety in a large pocket and the methylene group in the medium pocket. This figure was created from PDB file 1lpm.

Making pictures from x-ray crystal structures, such as those in Figures 6 and 7, is not difficult. First, download the coordinates for the structures from the protein data bank at http://www.rcsb.org/pdb/. Next, display the structure using Rasmol, a free program developed by Roger Sayle at Glaxo [23]. Rasmol is available for Macintosh, PC, Unix and other platforms at http://www.umass.edu/microbio/rasmol/. After making a nice picture, it is convenient to type 'write script xxxx'. The next time you open Rasmol, you can just type 'script xxxx' and your previous picture will appear. There is a small bug in the current Mac version of Rasmol (v2.6) for exporting pictures that causes a crash. If you restart with extensions off, it crashes less frequently, but still often crashes.

3.4. MAKING MISTAKES: ORIENTATION OF THE SLOW-REACTING ENANTIOMER

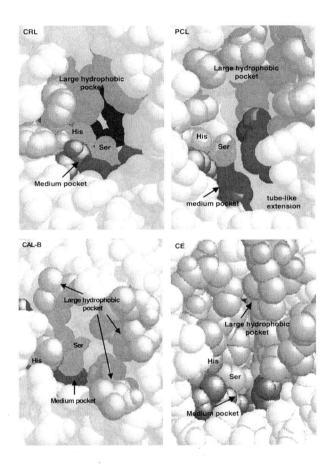

Figure 9. Proposed substrate-binding site in four synthetically useful lipases. The catalytic serine lies at the bottom of a crevice with the catalytic histidine on the left. Although the details of this crevice differ for each lipase, each crevice contains a large hydrophobic pocket (green) and medium pocket (red). This crevice is the alcohol-binding site and the two pockets resemble the empirical rule in Figure 1. The structure for CRL also shows the mouth of a tunnel (blue) that binds the acyl chain of an ester. The medium pocket in PCL contains a tube-like extension as suggested by the models in Figure 5. CAL-B shows a deeply buried medium pocket consistent with high stereoselectivity toward alcohols where the medium substituent is methyl or ethyl. CE shows a wide crevice. There is some uncertainty in the active form of CE, this is the active form suggested by Wang *et al.* [24]. Pictures were drawn starting from the following PDB files: 1lpm (CRL, [25]), 4lip (PCL, [26]), 1lbs (CAL-B, [27]), 1aql (CE, [24]).

To understand how lipases distinguish between enantiomers, one also needs to know how the slow enantiomer reacts; that is, how lipases make mistakes. The most reasonable possibilities are an H/O permutation, an M/L permutation or an M/H permutation, Figure 10. In the H/O permutation the hydrogen and alcohol oxygen substituents exchange places relative to their orientation in the fast-reacting enantiomer. In the M/L permutation, the medium and large substituents exchange places and in the M/H permutation, the medium substituent and hydrogen exchange places.

Note that all three permutations may be correct for different cases. Some lipase-substrate combinations may make mistakes by one path while others by the other path. Another possible permutation, an H/L permutation, is unlikely because it places a large substituent in the site normally occupied by hydrogen. This site is too small to accept a large substituent.

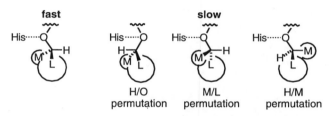

fast   **slow**

H/O permutation   M/L permutation   H/M permutation

Figure 10. Proposed catalytically productive orientations of secondary alcohols in the active site of lipases. Researchers agree that that catalytically productive orientation for the fast reacting enantiomer has the medium and large substituents in their respective pockets and the alcohol oxygen pointing toward the catalytic histidine. (The lipase orientation is assumed to be similar to that in Figure 8 and 7 above and Figure 11 below.) Three possible orientations for the slow reacting enantiomer are the H/O permutation, which exchanges the hydrogen and alcohol oxygen relative to their orientation in the fast enantiomer, the M/L permutation, which exchanges the medium and large substituents, and the H/M permutation which exchanges the hydrogen and medium substituents.

Support for the H/O permutation comes from x-ray crystal structures of CRL. A structure of a transition state analog the slow reacting enantiomer of menthol showed an orientation similar to the H/O permutation, Figure 11. In both the fast- and slow-reacting enantiomers the large and medium substituents oriented similarly, but the hydrogen and alcohol oxygen differed because of the opposite configuration of the carbinol stereocenter. For the fast reacting enantiomer, the alcohol oxygen points *toward* the histidine, but for the slow-reacting enantiomer it points *away from* the histidine. The fast-reacting enantiomer showed hydrogen bonds from the Nε–H of the catalytic histidine to both Oγ of the catalytic serine and the alcohol oxygen. Both of these hydrogen bonds are expected to speed up the reaction: the hydrogen bond to $O\gamma$ increases its nucleophilicity so that it can attack the carbonyl, while the hydrogen bond to the alcohol oxygen allows protonation of the alcohol leaving group as the tetrahedral intermediate breaks down. On the other hand, the slow-reacting enantiomer lacked the hydrogen bond between the catalytic histidine and the alcohol oxygen. The lack of this hydrogen bond may delay release of the product alcohol making it more likely that the intermediate returns to the starting ester. Molecular modeling with this structure suggested that the CRL-(1S)-menthol-transition-state-analog complex could easily distort to reform the missing hydrogen bond. Thus, this complex appears to be near to a catalytically productive complex. However, a molecular modeling study of PCL and CRL with several secondary alcohols suggested that mistakes via an M/L permutation, not the H/O permutation seen in the x-ray structure [28].

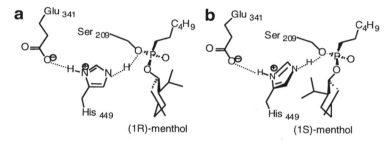

Figure 11. Schematic of the x-ray structures of transition state analogs covalently bound in the active site of CRL. a) The analog of the fast-reacting enantiomer of menthol shows hydrogen bonds from the Nϵ–H of the catalytic histidine to both $O\gamma$ of the catalytic serine and the alcohol oxygen. b) The analog of the slow-reacting enantiomer of menthol lacked the hydrogen bond between the catalytic histidine and the alcohol oxygen. The lack of this hydrogen bond likely slows down hydrolysis and makes it more likely that the intermediate returns to the starting ester.

Support for the M/L permutation a way of making mistakes comes from molecular modeling and related studies with CAL-B [27, 29, 30, 31] Researchers could not find a catalytically-productive conformation using an H/O permutation. The structure lacked a hydrogen bond between the catalytic histidine and the alcohol oxygen, similar to that seen for CRL. However, modeling suggested that this complex was quite rigid and, unlike the case for CRL, one could not restore the hydrogen bond by distorting the structure. On the other hand, upon modeling an M/L permutation, the researchers did get a catalytically productive complex with all key hydrogen bonds. Further, the researchers could predict the degree of selectivity for a number of substrates by assuming that the mistakes came from an M/L permutation. Although it is surprising that the relatively small M-binding site in CAL-B could accommodate a large substituent, the ability to predict the degree of enantioselectivity is strong support. It is reasonable that CRL with its large, open alcohol-binding site and CAL-B with its narrow, deep alcohol-binding site might make mistakes by different pathways. The H/O pathway may need a wide alcohol-binding site to permit sideways distortion to reform the key hydrogen bond.

Support for the H/M permutation comes from substrate specificity data of PCL [32], which suggested that an aromatic moiety on the substrate could be 'anchored' in the large pocket.

Kinetic measurements of PCL-, CAL-B-, and RML-catalyzed reactions of secondary alcohols show that enantiomers have similar apparent K_M values but different k_{cat} values [32, 33, 34]. Both the H/O and M/L permutations are consistent with these measurements. The H/O permutation leads to an orientation with a long or missing hydrogen bond for the slow enantiomer. Distortion required to restore this bond should lower k_{cat}, in agreement with experiments. The M/L or H/M permutations will give a nonproductive complex (similar to the H/O permutation, but nonproductive) and a productive complex (the M/L or H/M permutation). Both contribute to the observed K_M, so the value for the slow enantiomer will likely be similar to that for the fast enantiomer. However, only the productive complex reacts. The measured k_{cat} for the slow enantiomer is lower because at saturation only a small fraction of the substrate is bound productively. Thus, the M/L permutation is also consistent with similar K_Ms, but different k_{cat}s for the enantiomers. Thus, kinetic measurements do not distinguish between the three possibilities.

3.5.   EXPLAINING THE OPPOSITE ENANTIOPREFERENCE OF SUBTILISIN

Subtilisin favors the opposite enantiomer of secondary alcohol substrates as compared to lipases and esterases. The simplest explanation for this difference is that the binding site and catalytic machinery are roughly enantiomeric in the subtilisin as compared to lipases and esterases, Figure 12 and section 2 above [35].

Colombo et al. [36] modeled a transition state for acylation of 2-phenylethyl alcohol within the active site of subtilisin Carlsberg. The favored enantiomer structure contained all essential hydrogen bonds and a favorable stacking arrangement between the imidazole ring of the catalytic His and a phenyl ring, while the slow-reacting enantiomer lack some essential hydrogen bonds.

Thus, lipases and subtilisin are enantiocomplementary reagents for secondary alcohols and amines of the type $NH_2CHR_1R_2$. Subtilisin usually shows lower enantioselectivity, presumably because of the shallower binding site.

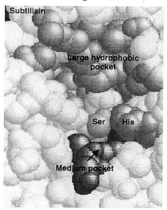

Figure 12. Substrate binding site in subtilisin Carlsberg. The labels and coloring show the amino acid residues of the catalytic triad and the residues forming the S1 binding site. This binding site is a shallow groove lined with nonpolar amino acid residues. The large hydrophobic substituent of secondary alcohols probably binds in this pocket. Note that the histidine of the catalytic triad lies to the right of the catalytic serine, while in lipases (Figure 5 above) it lies to the left of the serine. This opposite chirality of the catalytic triad in part accounts for the opposite enantiopreference of lipases and subtilisins. Coordinates are from Brookhaven protein data bank file 1sbc [37].

4.   Selectivity of hydrolases toward primary alcohols

Predicting the behavior of hydrolases toward primary alcohols is more difficult. First, most lipases show low enantioselectivity toward primary alcohols. Only lipase from *Pseudomonas cepacia* (PCL) and lipase from porcine pancreas (PPL) show moderate to high enantioselectivity toward a wide range of primary alcohols, but even for these the enantioselectivity is usually lower than toward secondary alcohols. Second, a simple rule based on the size of the substituents cannot predict the favored enantiomer for all primary alcohols. For PPL, researchers working with different substrates proposed opposite, enantiomeric rules! Of course, neither rule predicted the favored enantiomer for all substrates. For PCL, we found that a simple rule works if we exclude primary alcohols with oxygen at the stereocenter, Figure 13 [38].

A preliminary survey suggests this rule is also reliable for PPL.

60

The first puzzling feature of this empirical rule is that the rules for primary and secondary alcohols suggest a reversal of enantiopreference for PCL. In the fast-reacting enantiomer of secondary alcohols, the hydroxyl group faces toward the reader (Figure 3), but for the fast-reacting enantiomer of primary alcohols, the CH_2OH group points away from the reader (Figure 13).

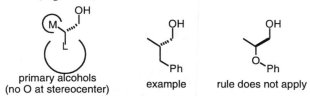

primary alcohols
(no O at stereocenter)

example

rule does not apply

Figure 13. An empirical rule to predict the fast-reacting enantiomer of primary alcohols with PCL. This rule works only for alcohols without oxygen bonded to the stereocenter. Examples of the fast-reacting enantiomers show a substrate that follows the rule and also a substrate for which the rule does not apply.

The two possible explanations for this apparent contradiction, Figure 14, differ mainly in the location of the large substituent. The first explanation suggests that the large substituent for primary alcohols bind in a similar place as for secondary alcohols. The CH_2OH group points away from the histidine to avoid disrupting the catalytic action. Support for this explanation comes mainly from molecular modeling [28], but also on the observed selectivities of substrates containing stereocenters within the L substituent [38]. One difficulty with this explanation is that increasing the difference in size between the M and L substituents does not reliably increase the enantioselectivity, while with secondary alcohols this strategy usually worked.

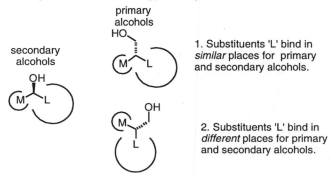

1. Substituents 'L' bind in *similar* places for primary and secondary alcohols.

2. Substituents 'L' bind in *different* places for primary and secondary alcohols.

Figure 14. Suggested relative orientations of primary and secondary alcohols in the alcohol-binding site of PCL. Suggestion 1 places the large substituent of both primary and secondary alcohols in a similar place. Suggestion 2 places the large substituent in a different place. Both suggestions assume that that the hydrogen, the M substituents, and the alcohol oxygen bind in similar places.

The second explanation suggests that the large substituent binds in a different part of the alcohol binding pocket, more specifically the region labeled 'tube-like extension' in Figure 9 or 'alternate binding pocket in Figure 15. Both labels refer to the same region. Support for this explanation also comes from modeling [39]. Modeling also suggests that mistakes occur via an M/H permutation, which explains why increasing the size of the large substituent did not reliably increase enantioselectivity.

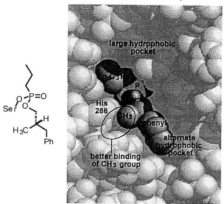

Figure 15. The active site of PCL containing a transition state analog for hydrolysis of an ester of a primary alcohol. The large substituent, benzyl in this case, does not bind in the large hydrophobic pocket, but in an alternate hydrophobic pocket.

This second explanation also suggests why primary alcohols with oxygen at the stereocenter are a special case that does not follow the rule. The 'alternate hydrophobic pocket' contains several tyrosines. One of these, Tyr29, may form a hydrogen bond to the substrate when the substrate contains an oxygen at the stereocenter. The effect of this hydrogen bond may be unpredictable; thus, the rule becomes unreliable for these substrates.

Lang *et al.* determined the x-ray crystal structure of the transition state analog shown below bound to the active site of PCL. This analog mimics hydrolysis of an ester at the primary hydroxyl group of a triacylglycerol analog [40]. This structure indeed showed one substituent bound to the 'alternate hydrophobic pocket' (HH pocket in the author's nomenclature).

5. Selectivity of hydrolases toward carboxylic acids

5.1. CANDIDA RUGOSA LIPASE AND OTHER LIPASES

Unlike most other lipases, CRL shows high enantioselectivity toward many carboxylic acids, Figure 16, and a rule can predict the enantiopreference of CRL-catalyzed reactions of carboxylic acids with a stereocenter at the α-position [41].

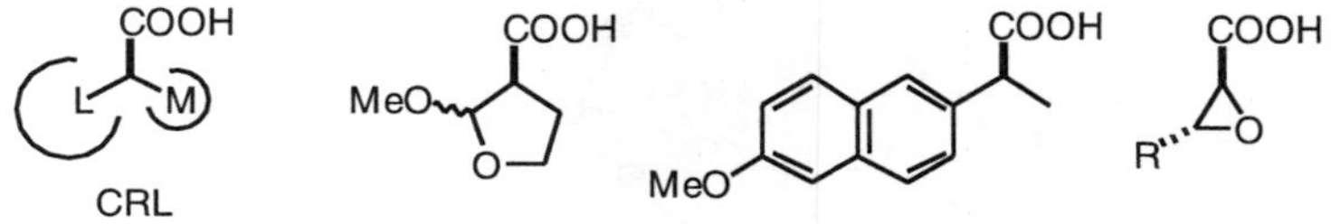

Figure 16. Empirical rule to predict the fast-reacting enantiomer of carboxylic acids with a stereocenter at the α-position. Several examples are shown.

Most lipases do not contain a special binding region for the acyl portion of the ester. The acyl portion simply binds in the large hydrophobic pocket described above in section 3.3. However, CRL is an exception - it contains a long tunnel that binds the acyl portion of an ester. This tunnel is long enough to accommodate an eighteen-carbon straight chain carboxylic acid. Longer acids fit if the side chains at the end of the tunnel rearrange. The stereocenter at the α-position of carboxylic acids lies near the mouth of this tunnel with the large substituent inside the tunnel. This mouth is shown in blue in Figure 9 above. The presence of this tunnel accounts for the high enantioselectivity of CRL toward this class of substrates. Indeed, molecular modeling supports this proposal [42]. This binding orientation also explains why the rule above often fails when the large substituent is extensively branched. An extensively branched large substituent no longer fits in the tunnel, but lies in outside the tunnel in the large hydrophobic pocket. In this orientation, modeling suggests a preference for the opposite enantiomer, in agreement with experiment. Recent site directed mutagenesis experiments on a related lipase, lipase from *Geotrichum candidum*, which also contains a tunnel, confirmed the importance of amino acid residues within the tunnel on the enantioselectivity [43].

5.2. PIG LIVER ESTERASE

Tamm's group developed a substrate model and Jones' group developed an active site model of porcine liver esterase (PLE) [44, 45]. These models predict the selectivity of

PLE toward chiral and prochiral carboxylic acids. For example, Jones' active site model explained a reversal in enantiopreference in a series of meso-dicarboxylic acid esters with varying ring size, Figure 18.

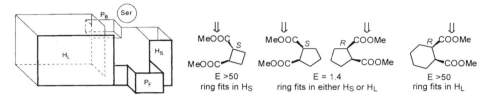

Figure 17. Active site model for PLE developed by Jones' group for predicting the selectivity of PLE toward carboxylic acids. Ser = location of ester group being hydrolyzed; H_L = large hydrophobic pocket; H_S = small hydrophobic pocket; P_F = polar front pocket; P_B = polar back pocket. The structures show how the model can explain a reversal in enantiopreference is a series of meso-dicarboxylic acids. The cyclobutane ring fits into H_S, leading to high enantioselectivity for the S-ester. Although the cyclobutane ring could also fit into H_L, hydrophobic contacts would be poor; thus, it favors H_S. The cyclopentane ring fits into either H_S or H_L, leading to low enantioselectivity. The cyclohexane ring fits only into H_L, leading to high enantioselectivity for the R-ester.

As noted above, the structure of PLE is not known, but it may be an a/b-hydrolase and the large and small hydrophobic pockets may correspond to the large and medium pockets of lipases.

5.3. CHYMOTRYPSIN AND OTHER PROTEASES

Chymotrypsin (α-CT), like other proteases, shows high enantioselectivity for the natural enantiomer of amino acids. Esters and amides containing aromatic amino acids – Phe, Tyr, and Trp – as the acyl group react most readily. Esters hydrolyze more rapidly than amides, so most preparative reactions use esters. For example, α-chymotrypsin resolves the ethyl esters of N-acetyl of Phe, Trp, and Trp with very high enantioselectivity. One exception to the high enantioselectivity rule is the moderate enantioselectivity ($E = 8$) of α-CT toward the methyl ester of N-benzoyl alanine. The lower enantioselectivity is likely due to the binding of the benzoyl group in the hydrophobic pocket in the 'wrong' enantiomer.

Researchers usually use Cohen's active site model [44] to rationalize the selectivity of α-CT toward acids, Figure 18. This model defines four sites that bind each of the four substituents at the stereocenter of a good substrate like N-acetyl-L-phenylalanine methyl ester. The strongest binding comes from the ar-site, a hydrophobic pocket, which binds aromatic or other hydrophobic groups. When the substrate has the incorrect configuration, for example, *N*-acetyl-D-phenylalanine methyl ester, it still binds to the *ar*-site, but the remaining substituents cannot orient in a productive manner. One observes nonproductive binding.

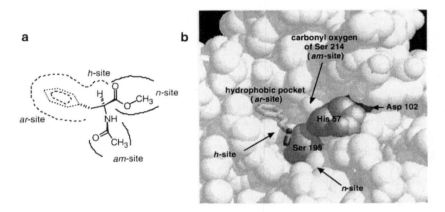

Figure 18. The active site of α-chymotrypsin. **a)** Cohen's model of α-CT [46] containing a good substrate, *N*-acetyl-L-phenylalanine methyl ester. The *ar*-site (*aromatic*-site) is a hydrophobic pocket; the *am*-site (*amide*-site) is the amide carbonyl of Ser-124, which hydrogen bonds to the amide N-H of the substrate. An OH in the substrate can also form a hydrogen bond here. Other groups such as CH₃, Cl, AcO, also fit in the *am*-site, but do not form a hydrogen bond. The *am*-site is open to the solvent and can accommodate long groups. The *h*-site (*hydrogen*-site) is a small region in the active site. It is large enough to fit an H, Cl, or OH, but not a CH₃. The *n*-site (*nucleophile*-site) binds the leaving group, an ester or amide. Like the *am*-site, the *n*-site is open-ended, that is, it is open to the solvent and can accommodate large groups. In contrast, both the *ar*- and the *h*-sites are closed; that is, they point toward the center of the protein. **b)** X-ray structure of α-chymotrypsin (spheres) contained transition state analog bound to the active site serine (pdb file 6cha [47]). The transition state analog is a 2-phenylethyl boronic acid (stick representation). Labels identify the amino acid residues of the catalytic triad and the regions that correspond to the four sites of Cohen's model.

α-CT often shows lowered stereoselectivity upon replacing the amino group of α-amino acids with an α-hydroxyl or α-*O*-acyl group. For example, α-CT showed no stereoselectivity toward mandelic acid, but high stereoselectivity toward the corresponding amino acid, phenylglycine. Presumably the OH group, but not the NH₂ group, can bind in the *h*-site. For carboxylic acids that do not resemble α-amino acids, Cohen's model is difficult to apply. Jones and Beck [48] reviewed the substrate specificity of α-CT in detail.

In recent years, researchers focus has shifted from chymotrypsin to subtilisin. Subtilisin is much less expensive because of its large-scale use as a laundry detergent additive. Although subtilisin shows similar preference for aromatic amino acids, it shows a broader substrate range. Moree *et al.* recently summarized the selectivity of subtilisin toward amino acids [49].

6. Concluding Remarks

The combination of substrate selectivity data, summarized in the form of empirical rules, with molecule structure data available from x-ray crystal structures creates a powerful tool for predicting the stereoselectivity of hydrolases. The structures often resemble the empirical rule and thus provide a molecular basis for the validity of these rules. Further, the structures add subtle molecular detail to the rules, thus explaining previously puzzling features. Molecular modeling is a powerful tool to use these structures for the next generation of explanations and predictions.

7. References

1. For example, Bornscheuer, U.T. and Kazlauskas, R.J. (1999) *Hydrolases in Organic Synthesis. Regio-and Stereoselective Biotransformations*, Wiley-VCH, Weinheim.
2. An alternative empirical approach – screening hydrolases – is also useful. For example, see Janes, L.E., Cimpoia, A., and Kazlauskas, R.J. (1999), Protease-mediated separation of cis and trans diastereomers of 2-(R,S)-benzyloxymethyl-4-(S)-carboxylic acid-1,3-dioxolane methyl ester: intermediates for the synthesis of dioxolane nucleosides, *J. Org. Chem.* **64**, in press.
3. Naemura, K. (1994) Stereoselectivity of enzymatic hydrolyses and acylations, *J. Synth. Org. Chem. Jpn.* **52**, 49-58. (in Japanese)
4. Ollis, D.L. Cheah, E., Cygler, M., Dijkstra, B., Frolow, F., Franken, S.M., Harel, M., Remington, S.J., Silman, I., Schrag, J., Sussman, J.L., Verschueren, K.H.G., and Goldman, A. (1992) The α/β-hydrolase fold, *Prot. Engineer.* **5**, 197-211.
5. Branden, C. and Tooze, J. (1991) *Introduction to Protein Structure* Garland Publishing, New York, Chapter 15.
6. a) Kazlauskas, R.J., Weissfloch, A.N.E., Rappaport, A.T., and Cuccia, L.A. (1991) A rule to predict which enantiomer of a secondary acohol reacts faster in reactions catalyzed by cholesterol esterase, lipase from Pseudomonas cepacia, and lipase from Candida rugosa, *J. Org. Chem.* **56**, 2656-2665. b) Burgess, K. and Jennings, L.D. (1991) Enantioselective esterifications of unsaturated alcohols mediated by a lipase prepared from Pseudomonas sp, *J. Am. Chem. Soc.* **113**, 6129-6139, Naemura, K., Ida, H., and Fukuda, R. (1993) Lipase YS-catalyzed enantioselective transesterification of alcohols of bicarbocyclic compounds, *Bull. Chem. Soc. Jpn.* **66**, 573-577. c) Kim, M.J. and Cho, H. (1992) Pseudomonas lipases as catalysts in organic synthesis: specificity of lipoprotein lipase, *J. Chem. Soc., Chem. Commun.* 1411-1413. d) Roberts, S. M. (1989) Use of enzymes as catalysts to promote key transformations in organic synthesis, *Philos. Trans. R. Soc. London B* **324**, 577-587. e) Johnson, C.R.

66

and Sakaguchi, H. (1992) Enantioselective transesterifications using immobilized, recombinant Candida antarctica lipase B: resolution of 2-iodo-2-cycloalken-1- ols, *Synlett* 813-816. f) Janssen, A.J.M., Klunder, A.J.H., and Zwanenburg, B. (1991) Resolution of secondary alcohols by enzyme-catalyzed transeserification in alkyl carboxylates as the solvent, *Tetrahedron* **47**, 7645-7662.

7.   Review: Kazlauskas, R.J. and Weissfloch, A.N.E. (1997) A structure-based rationalization of the enantiopreference of subtilisin toward secondary alcohols and isosteric primary amines, *J. Mol. Catal. B Enz.* **3**, 65-72.

8.   Examples: Scilimati, A., Ngooi, T.K., and Sih, C.J. (1988) Biocatalytic resolution of (±)-hydroxyalkanoic esters. A strategy for enhancing the enantiomeric specificity of lipase-catalyzed ester hydrolysis, *Tetrahedron Lett.* **29**, 4927-4930, Johnson, C.R., Golebiowski, A., McGill, T.K., and Steensma, D.H. (1991) Enantioselective synthesis of 6-cycloheptene-1,3,5-triol derivatives by enzymatic asymmetrization, *Tetrahedron Lett.* **32**, 2597-2600, Kim, M.J. and Choi, Y.K. (1992) Lipase-catalyzed enantioselective transesterification of O-trityl 1,2-diols. Practical synthesis of (R)-tritylglycidol, *J. Org. Chem.* **57**, 1605-1607, Gupta, A.K. and Kazlauskas, R.J. (1993) Substrate modification to increase the enantioselectivity of hydrolases. A route to optically-active cyclic allylic alcohols, *Tetrahedron: Asymmetry* **4**, 879-888, Adam, W., Mock-Knoblauch, C. and Saha-Möller, C.R. (1997) Kinetic resolution of hydroxy vinylsilanes by lipase-catalyzed enantioselective acetylation, *Tetrahedron: Asymmetry* **8**, 1441-1444.

9.   Shimizu, M., Kawanami, H., and Fujisawa, T. (1992) A lipase mediated asymmetric hydrolysis of 3-acyloxy-1-octynes and 3-(E)-acyloxy-1-octenes, *Chem. Lett.* 107-110.

10.   Rotticci, D., Orrenius, C., Hult, K., and Norin, T. (1997) Enantiomerically enriched bifunctional sec-alcohols prepared by Candida antarctica lipase B catalysis. Evidence of non-steric interactions, *Tetrahedron: Asymmetry* **8**, 359-362.

11.   Theil, F., Lemke, K., Ballschuh, S., Kunath, A., and Schick, H. (1995) Lipase-catalysed resolution of 3-(aryloxy)-1,2-propanediol derivatives - towards an improved active site model of Pseudomonas cepacia lipase (Amano PS), *Tetrahedron: Asymmetry* **6**, 1323-1344.

12.   Oberhauser, T., Faber, K., and Griengl, H. (1989) A substrate model for the enzymic resolution of esters of bicyclic alcohols by Candida cylindracea lipase, *Tetrahedron* **45**, 1679-1682.

13.   Faber, K., Griengl, H., Hoenig, H., and Zuegg, J. (1994) On the prediction of the enantioselectivity of Candida rugosa lipase by comparative molecular field analysis, *Biocatalysis* **9**, 227-239.

14.   Exl, C., Hoenig, H., Renner, G., Rogi-Kohlenprath, R., Seebauer, V., and Seufer-Wasserthal, P. (1992) How large are the active sites of the lipases from Candida rugosa and from Pseudomonas cepacia?, *Tetrahedron: Asymmetry* **3**, 1391-1394.

15.   Burgess, K., and Jennings, L.D. (1991) Enantioselective esterifications of unsaturated alcohols mediated by a lipase prepared from Pseudomonas sp, *J. Am. Chem. Soc.* **113**, 6129-6139.

16.   Bornscheuer, U., Herar, A., Kreye, L., Wendel, V., Capewell, A., Meyer, H.H., Scheper, T., and Kolisis, F.N. (1993) Factors affecting the lipase catalyzed transesterification reactions of 3-hydroxy esters in organic solvents, *Tetrahedron: Asymmetry* **4**, 1007-1016.

17.   Naemura, K., Fukuda, R., Murata, M., Konishi, M., Hirose, K., and Tobe, Y. (1995) Lipase-catalyzed enantioselective acylation of alcohols: a predictive active site model for lipase YS to identify which enantiomer of an alcohol reacts faster in this acylation, *Tetrahedron: Asymmetry* **6**, 2385-2394.

18.   Naemura, K., Murata, M., Tanaka, R., Yano, M., Hirose, K., and Tobe, Y. (1996) Enantioselective acylation of alcohols catalyzed by lipase QL from Alcaligenes sp.: a predictive active site model for lipase QL to identify the faster reacting enantiomer of an alcohol in this acylation, *Tetrahedron:*

Asymmetry **7**, 1581-1584.

19. Lemke, K., Lemke, M., and Theil, F. (1997) A three-dimensional predictive active site model for lipase from Pseudomonas cepacia, *J. Org. Chem.* **62**, 6268-6273.

20. Grabuleda, X., Jaime, C., and Guerrero, A. (1997) Estimation of the lipase PS (Pseudomonas cepacia) active site dimensions based on molecular mechanics calculations, *Tetrahedron: Asymmetry* **8**, 3675-3683.

21. Naemura, K., Takahashi, N., Ida, H., and Tanaka, S. (1991) Pig liver esterase-catalyzed hydrolyses of racemic diacetates of bicyclic compounds and interpretation of the enantiomeric specificity of PLE, *Chem. Lett.* 657-660.

22. Cygler, M., Grochulski, P., Kazlauskas, R.J., Schrag, J.D., Bouthillier, F., Rubin, B., Serreqi, A.N., and Gupta, A.K. (1994) Molecular basis for the chiral preference of lipases, *J. Am. Chem. Soc.* **116**, 3180-3186.

23. Sayle, R.A., and Milner-White, E.J. (1995) RASMOL: biomolecular graphics for all, *Trends Biochem. Sci.* **20**, 374-376.

24. Wang, X.Q., Wang, C.S., Tang, J., Dyda, F., and Zhang, X.J.C. (1997) The crystal structure of bovine bile salt activated lipase: insights into the bile salt activation mechanism, *Structure*, **5**, 1209-1218; For another proposal see: Chen, J.C.H., Miercke, L. J. W., Krucinski, J., Starr, J. R., Saenz, G., Wang, X.B., Spilburg, C.A., Lange, L.G., Ellsworth, J.L., and Stroud, R.M. (1998) Structure of bovine pancreatic cholesterol esterase at 1.6Å: novel structural features involved in lipase activation, *Biochemistry* **37**, 5107-5117.

25. Grochulski, P., Li, Y., Schrag, J.D., Bouthillier, F., Smith, P., Harrison, D., Rubin, B., and Cygler, M. (1993) Insights into interfacial activation from an open structure of Candida rugosa lipase, *J. Biol. Chem.*, **268**, 12843-12847.

26. Schrag, J.D., Li, Y.G., Cygler, M., Lang, D.M., Burgdorf, T., Hecht, H.J., Schmid, R., Schomburg, D., Rydel, T.J., Oliver, J.D., Strickland, L.C., Dunaway, C.M., Larson, S.B., Day, J., and McPherson, A. (1997) The open conformation of a Pseudomonas lipase, *Structure*, **5**, 187-202.

27. Uppenberg, J., Öhrner, N., Norin, M., Hult, K., Patkar, S., Waagen V., Anthonsen, T., and Jones, T.A. (1995) Crystallographic and molecular modelling studies of lipase B from Candida antarctica reveal a stereospecificity pocket for secondary alcohols, *Biochemistry* **34**, 16838-16851.

28. Zuegg, J., Hönig, H., Schrag, J.D., and Cygler, M. (1997) Selectivity of lipases: conformational analysis of suggested intermediates in ester hydrolysis of chiral primary and secondary alcohols, *J. Mol. Catal. B Enzym.* **3**, 83-98.

29. Orrenius, C., Hæffner, F., Rotticci, D., Ohrner, N., Norin, T., and Hult, K. (1998) Chiral recognition of alcohol enantiomers in acyl transfer reactions catalysed by Candida antarctica lipase B, *Biocatal. Biotransform.* **16**, 1-15.

30. Rotticci, D., Hæffner, F., Orrenius, C., Norin, T., and Hult, K. (1998) Molecular recognition of sec-alcohol enantiomers by Candida antarctica lipase B, *J. Mol. Catal. B Enzym.* **5**, 267-272.

31. Haeffner F., Norin T., and Hult K. (1998) Molecular modeling of the enantioselectivity in lipase-catalyzed transesterification reactions, *Biophys. J.* **74**, 1251-1262.

32. Nakamura, K., Kawasaki, M., and Ohno, A. (1996) Lipase-catalyzed transesterification of aryl-substituted alkanols in an organic solvent, *Bull. Chem. Soc. Jpn.* **69**, 1079-1085.

33. Nishizawa K., Ohgami Y., Matsuo N., Kisida H., and Hirohara H. (1997) Studies on hydrolysis of chiral, achiral and racemic alcohol esters with Pseudomonas cepacia lipase: mechanism of stereospecificity of the enzyme, *J. Chem. Soc., Perkin Trans. 2*, 1293-1298.

34. Ema, T., Kobayashi, J., Maeno, S., Sakai, T., and Utaka, M. (1998) Origin of the enantioselectivity of lipases explained by a stereo-sensing mechanism operative at the transition state, *Bull. Chem. Soc. Jpn.* **71**, 443-453.

35. Kazlauskas, R. J., and Weissfloch, A.N.E. (1997) A structure-based rationalization of the enantiopreference of subtilisin toward secondary alcohols and isosteric primary amines, *J. Mol. Catal. B Enz.* **3**, 65-72.

36. Colombo, G., Ottolina, G., Carrea, G., Bernardi, A., and Scolastico, C. (1998) Application of structure-based thermodynamic calculations to the rationalization of the enantioselectivity of subtilisin in organic solvents, *Tetrahedron: Asymmetry* **9**, 1205-1214.

37. Neidhart, D.J., and Petsko, G.A. (1988) The refined crystal structure of subtilisin Carlsberg at 2.5 Å resolution, *Prot. Eng.* **2**, 271-276.

38. Weissfloch, A.N.E. and Kazlauskas, R.J. (1995) Enantiopreference of lipase from Pseudomonas cepacia toward primary alcohols, *J. Org. Chem.* **60**, 6959-6969.

39. Tuomi, W.V. and Kazlauskas, R.J. (1999) Molecular basis for enantioselectivity of lipase from Pseudomonas cepacia toward primary alcohols. Modeling, kinetics and chemical modification of Tyr29 to increase or decrease enantioselectivity, *J. Org. Chem.* **64**, 2638-2647.

40. Lang, D.A., Mannesse, M.L.M., De Haas, G.H., Verheij, H.M., and Dijkstra, B.W. (1998) Structural basis of the chiral selectivity of Pseudomonas cepacia lipase, *Eur. J. Biochem.* **254**, 333-340.

41. Ahmed, S.N., Kazlauskas, R.J., Morinville, A.H., Grochulski, P., Schrag, J.D., and Cygler, M. (1994) Enantioselectivity of Candida rugosa lipase toward carboxylic acids: a predictive rule from substrate mapping and X-ray crystallography, *Biocatalysis* **9**, 209-225, Franssen, M.C.R., Jongejan, H., Kooijman, H., Spek, A.L., Camacho Mondril, N.L.F.L., Boavida dos Santos, P.M.A.C., and de Groot, A. (1996) Resolution of a tetrahdrofuran ester by Candida rugosa lipase (CRL) and an examination of CRL's stereochemical preference in organic media, *Tetrahedron: Asymmetry* **7**, 497-510.

42. Holmquist, M., Haeffner, F., Norin, T., and Hult, K. (1996) A structural basis for enantioselective inhibition of Candida rugosa lipase by long-chain aliphatic alcohols, *Prot. Sci.* **5**, 83-88; Berglund, P., Holmquist, M., and Hult, K. (1998), Reversed enantiopreference of Candida rugosa lipase supports different modes of binding enantiomers of a chiral acyl donor, *J. Mol. Catal. B Enzym.* **5**, 283-287; Botta, M., Cernia, E., Corelli, F., Manetti, F., and Soro, S. (1997) Probing the substrate specificity for lipases. 2. Kinetic and modeling studies on the molecular recognition of 2-arylpropionic esters by Candida rugosa and Rhizomucor miehei lipases, *Biochim. Biophys. Acta* **1337**, 302-310.

43. Holmquist, M. and Berglund, P. (1999) Creation of a synthetically useful lipase with higher than wild-type enantioselectivity and maintained catalytic activity, *Org. Lett.* **1**, 763-765.

44. Mohr, P., Waespe-Sarcevic, N., Tamm, C., Gawronska, K., and Gawronski, J. K. (1983), A study of stereoselective hydrolysis of symmetrical diesters with pig liver esterase, *Helv. Chim. Acta* **66**, 2501-2511.

45. Toone, E.J., Werth, M.J., and Jones, J.B. (1990) Active site model for interpreting and predicting the specificity of pig liver esterase, *J. Am. Chem. Soc.* **112**, 4946-4952; Provencher, L. and Jones, J.B. (1994) A concluding specification of the dimensions of the active site model of pig liver esterase, *J. Org. Chem.* **59**, 2729-2732.

46. Cohen, S. G. (1969), On the active site and specificity of α-chymotrypsin, *Trans. N. Y. Acad. Sci.* **31**, 705-719.

47. Tulinsky, A. and Blevins, R. A. (1987) Structure of a tetrahedral transition state complex of α-

chymotrypsin at 1.8-Å resolution, *J. Biol. Chem.* **262**, 7737-7743.

48. Jones, J. B. and Beck, J. F. (1976) Asymmetric syntheses and resolutions using enzymes, In *Applications of Biochemical Systems in Organic Chemistry*, Jones, J. B., Sih, C. J., Perlman, D., Eds. Techniques in Chemistry Series Vol. X, Wiley: New York, pp 107-401.

49. Moree, W.J., Sears, P., Kawashiro, K., Witte, K., and Wong, C.H. (1997) Exploitation of subtilisin BPN' as catalyst for the synthesis of peptides containing noncoded amino acids, peptide mimetics and peptide conjugates, *J. Am. Chem. Soc.* **119**, 3942-3947.

FUNCTIONAL GROUP TRANSFORMATIONS MEDIATED BY WHOLE CELLS AND STRATEGIES FOR THE EFFICIENT SYNTHESIS OF OPTICALLY PURE CHIRAL INTERMEDIATES

NICHOLAS J. TURNER

Department of Chemistry, Centre for Protein Technology, University of Edinburgh, King's Buildings, West Mains Road, Edinburgh EH9 3JJ, UK.

1.   Functional Group Transformations Mediated by Whole Cells

The use of whole cells, as opposed to isolated enzymes, has proven to be extremely effective for certain types of biotransformations, especially cofactor dependent processes such as oxidation and reduction. We have investigated whole cell reactions for effecting i) nitro group reduction ii) carbonyl reduction iii) C=C reduction and iv) nitrile hydrolysis. For each of these transformations the following topics have been addressed; substrate range, type of selectivity displayed *e.g.* regio, stereo, chemo-, mechanistic aspects of the reactions, applications to the synthesis of biologically active molecules and lastly scale-up issues [1].

1.1.   NITRO GROUP REDUCTION

The metabolism of nitro containing compounds has been well studied and involves a series of different pathways including reduction of the nitro group to the corresponding amine. This reaction is well documented and can be carried out using the yeast *Saccharomyces cerevisiae* (Scheme 1). In general for nitroarene substrates **1** to undergo reduction they need to possess an additional electron withdrawing group for activation. The electron withdrawing substituent can be a second nitro group in which case issues of regioselectivity arise (see below).

B. Zwanenburg et al. (eds.), Enzymes in Action, 71–94.
© 2000 *Kluwer Academic Publishers. Printed in the Netherlands.*

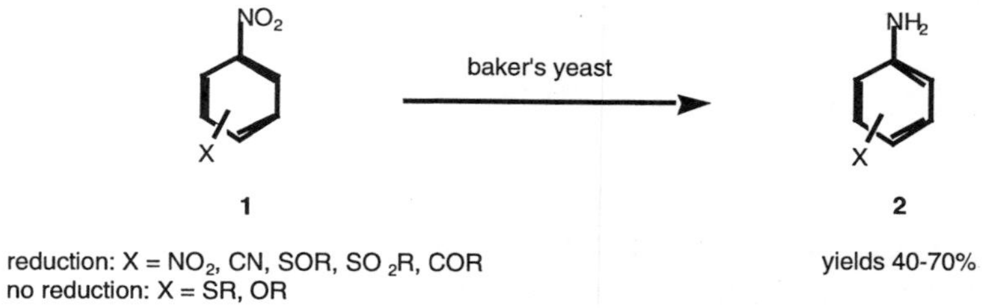

reduction: X = NO$_2$, CN, SOR, SO$_2$R, COR
no reduction: X = SR, OR

yields 40-70%

Scheme 1: Reduction of nitroarenes using *Saccharomyces cerevisiae*.

In addition to a knowledge of the range of substrates able to be reduced, information has recently be gained concerning the mechanism of the reduction as well as the nature of the enzymes involved in the reaction [2]. Using 2-nitrobenzonitrile **3** as substrate we observed that the major product was 2-aminobenzamide **4** (56%) and that only a small amount (5%) of the expected 2-aminobenzonitrile **5** was obtained (Scheme 2). The result was rationalised by proposing an initial reduction of the nitro group to the hydroxylamine **6** which cyclises to give a iminobenzisoxazolidine **7** which subsequently undergoes further reduction to the amine **4**.

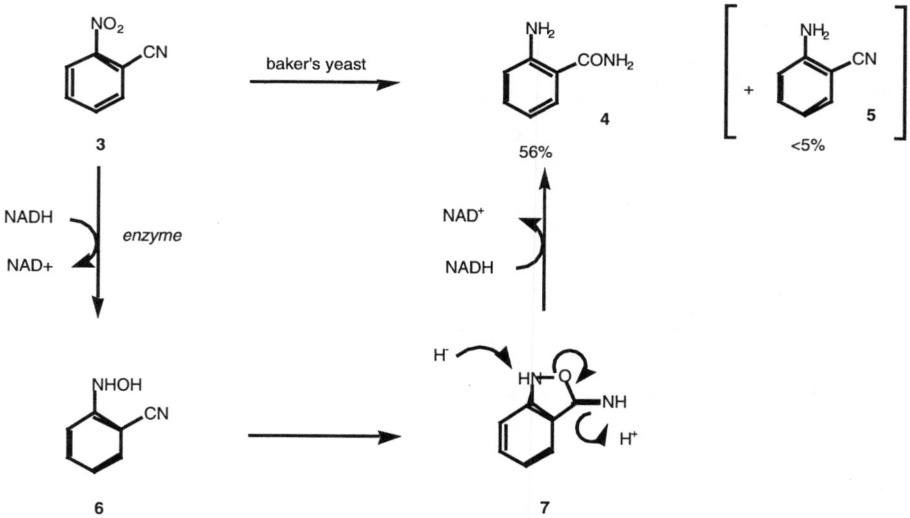

Scheme 2: Proposed mechanism of reduction of 2-nitrobenzonitrile **3** by *S. cerevisiae*.

We have subsequently exploited this mechanism for the conversion of wide range of *ortho*-cyanonitroarenes **8** to the corresponding *ortho*-aminobenzamides **9** [3] (Scheme 3).

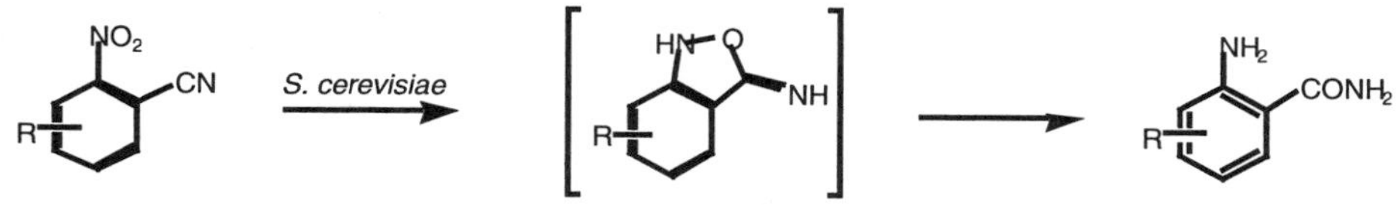

R	yield/%
H	79
3-CN	75
4-CN	--
5-CN	84
6-CN	70
4-Cl	69
5-CF$_3$	84

Scheme 3: Reduction of substituted 2-nitrobenzonitriles by *S. cerevisiae*.

We have also examined the regioselective reduction of non-symmetrical dinitroarenes. This class of substrates reacts readily in view of the activating effect of the second nitro group. In some cases high levels of regioselectivity can be achieved and we have attempted to rationalise the selectivity of the basis of an interplay between steric and electronic effects (Scheme 4). For example, reduction of 2,4-dinitroanisole **10** yields mainly 2-amino-4-nitroanisole **11** (80%) compared with the regioisomeric product **12** (15%).

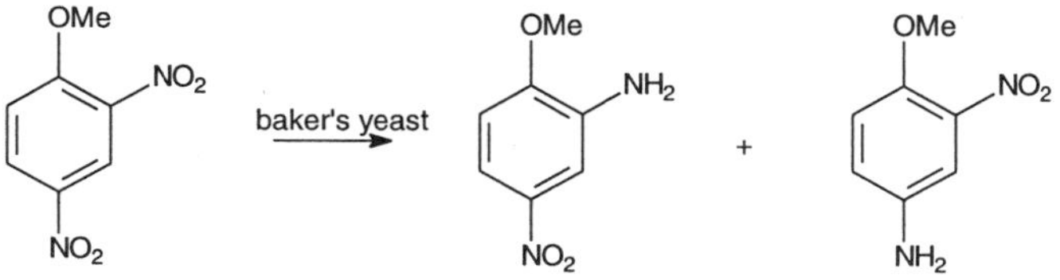

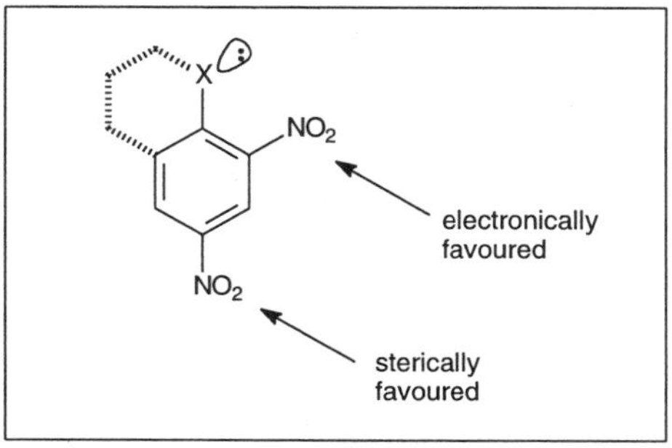

Scheme 4: Regioselective reduction of dinitroarenes by *S. cerevisiae*.

In order to shed light on the enzymes involved in nitroarene reduction by *S. cerevisiae* we have recently purified 3 enzymes that are able to catalyse the reduction of nitroaromatic substrates. Using 1,4-dinitrobenzene as the substrate we identified three NADPH dependent enzymes that could catalyse the reduction, namely NADPH dehydrogenase 2, thioredoxin reductase and dihydrolipoamide dehydrogenase (Figure 1).

NRase 1: NADPH Dehydrogenase 2 (Old Yellow Enzyme)	EC 1.6.99.1	44886 Da
NRase 2: Thioredoxin Reductase	EC 1.6.4.5	34107 Da
NRase 3: Dihydrolipoamide Dehydrogenase	EC 1.8.1.4	51558 Da

Figure 1: Purification of nitroreductases (NRases) from *S. cerevisiae*.

1.2. YEAST REDUCTION OF β-KETO ESTERS

There are many reports in the literature describing the use of *S. cerevisiae* for the reduction of carbonyl groups including isolated ketones, 1,2- and 1,3-diketones and β-ketoesters. We have extended the β-ketoester range of substrates by using racemic δ-hydroxy-β-keto esters as substrates with the intention of carrying out a simultaneous kinetic resolution and diasteroeselective reduction of the carbonyl group. The aim of this project was to devise a rapid entry into the δ-lactone moiety of the cholesterol lowering agent mevinolin **13**. These compounds exert their effect by inhibiting the enzyme HMG CoA reductase, a key enzyme in the biosynthesis of cholesterol. The required ester **15** was easily prepared by alkylation of the dianion derived from ethylacetoacetate **14** with cinnamaldehyde **16**. When ester **15** was subjected to yeast catalysed reduction the diol **17** was obtained as a single diastereoisomer in 87% e.e. together with the recovered alcohol **15** which had an e.e. of 50%.

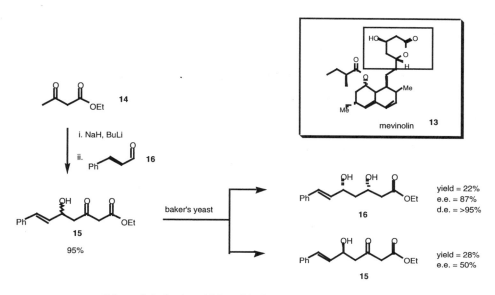

Scheme 5: Reduction of β-keto-δ-hydroxy esters using *S. cerevisiae*.

Hydrolysis of the ethyl ester of **17** followed by cyclisation of the hydroxy acid yielded the lactone **18**. Unfortunately the lactone turned out to be of the opposite absolute configuration to that required for the correct enantiomer of the mevinolin lactone, as determined by optical rotation measurements. Despite varying the conditions for the yeast reduction we were unable to switch the enantioselectivity to provide the correct stereoisomer (Scheme 6).

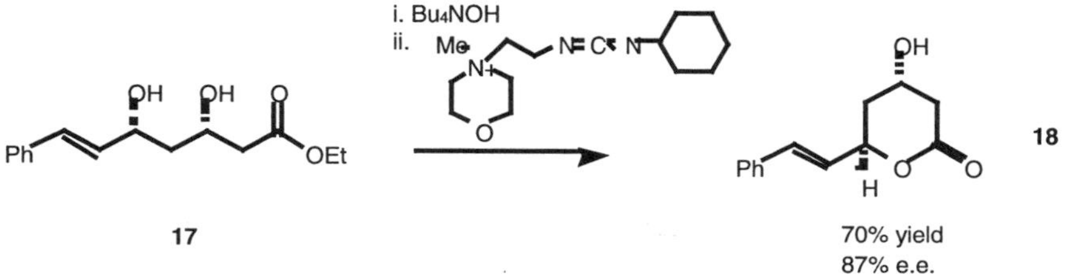

Scheme 6: Conversion of carboxylic acid **16** to lactone **18**.

In a parallel approach we examined the kinetic resolution of ester **19** using pig liver esterase (PLE). Starting from the alcohol **15** it proved possible to prepare both the protected *syn* and *anti*-diols using diastereoselective reductions. Treatment of protected syn-diol **19** with PLE yielded the acid **21** in 77% e.e and in this case we established that the acid product was of the required absolute configuration (Scheme 7).

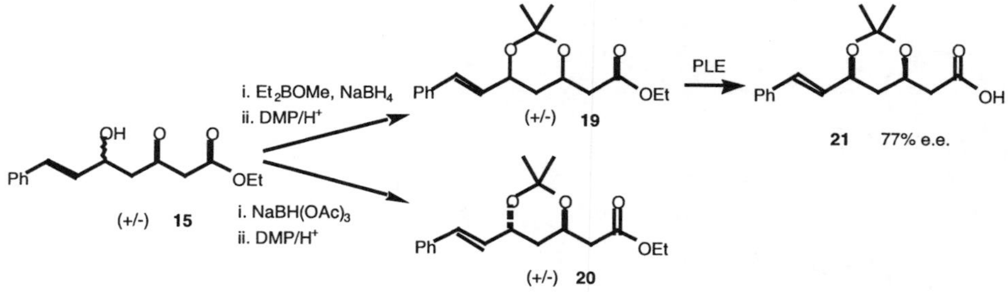

Scheme 7: Kinetic resolution of ester **19** using pig liver esterase.

1.3. CARBON-CARBON DOUBLE BOND REDUCTION

Yet another reaction known to be catalysed by yeasts and related organisms is reduction. However, whereas reduction of carbonyl groups has been examined in some detail, microbial carbon-carbon double bond reduction is relatively under explored. Enoate reductases, which catalyse the conjugate reduction of α,β-unsaturated carbonyl groups have been documented and used in synthesis. Other types of double bond are also known to undergo reduction and the general observation is that some activation of the double bond is required, usually by the introduction of electron withdrawing groups.. A group from SmithKline Beecham recently reported the reduction of unsaturated thiazolidinediones *e.g.* **22** using the red yeast *Rhodotorula rubra*. The product **23** is a member of a new class compounds for treatment of non-insulin dependent diabetes. The SB group examined the feasibility of developing a large scale process for C=C bond reduction in the synthesis of the anti-diabetic compound BRL 49653 **23** [4].

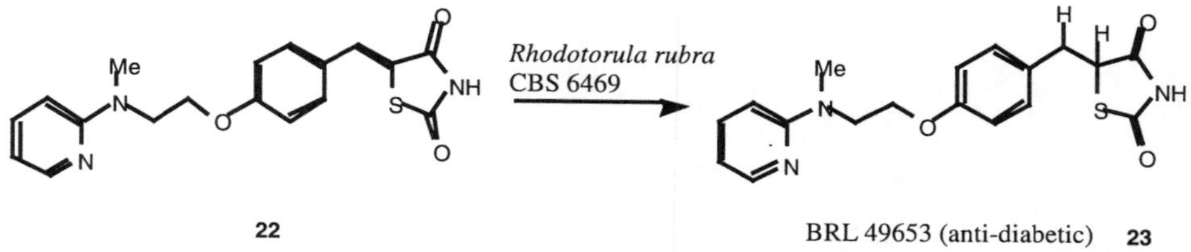

Scheme 8: Reduction of unsaturated thiazolidinediones using *Rhodotorula rubra*.

We have investigated this reaction in some detail, particularly with respect to the substrate specificity and mechanism of reduction. The reaction is known to produce optically enriched products which racemise rapidly. By challenging the yeast with a wide range of substrates we established that only minor changes of the thiazolidine dione ring could be tolerated. In order to investigate the mechanism of reduction we examined both the reduction of deuterated substrate **24a** in H$_2$O and secondly the reduction of protiated substrate **24b** in D$_2$O. In both cases we obtained products **25** that contained the deuterium atom distributed in a 1:1 ratio between the two diastereotopic sites.

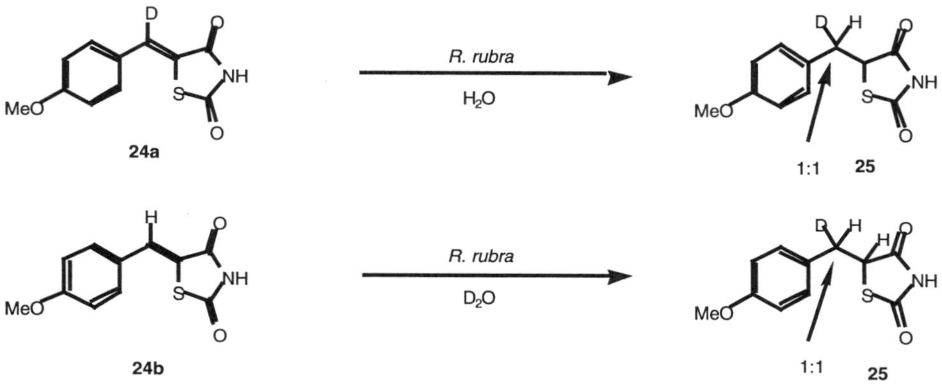

Scheme 9: Reduction of thiazolidinediones using *Rhodotorula rubra*.

The incorporation of deuterium from D$_2$O (via D$^+$) strongly suggests that the mechanism of reduction involves initial protonation of the double bond of **26** to give **27** followed by reduction of the sulfenium ion to yield the saturated product **28**. The alternative route of initial conjugate reduction to give **29** followed by enantioselective protonation would not result in incorporation of deuterium from D$_2$O at the β-position.

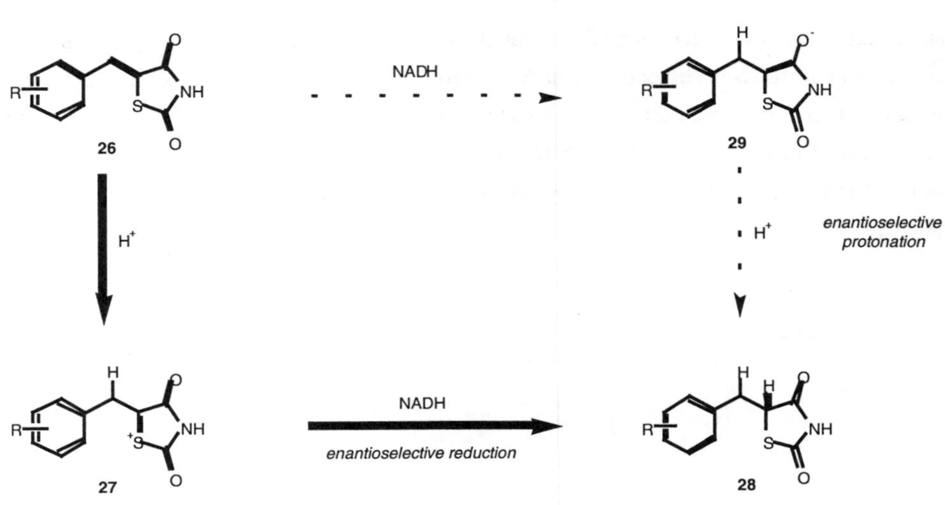

Scheme 10: Mechanism of reduction of thiazolidinediones using *Rhodotorula rubra*.

1.4.   NITRILE HYDROLYSIS

Nitrile groups can be hydrolysed to amides (*hydratase*) and thereafter carboxylic acids (*amidase*) or alternatively in a direct manner without the intermediate amide (*nitrilase*) [5]. Nitrile hydratases have found commercial application (*e.g.* Nitto process for conversion of acrylonitrile to acrylamide) and recently have been the focus of attention in terms of their substrate specificity and possible selectivity. A major advantage of using both nitrile hydratases and nitrilases as catalysts, compared to chemical hydrolysis, is that both of these enzymes catalyse the hydrolysis of nitriles under mild conditions allowing for the presence of acid and base sensitive functionality in the molecule [6] (Scheme 11).

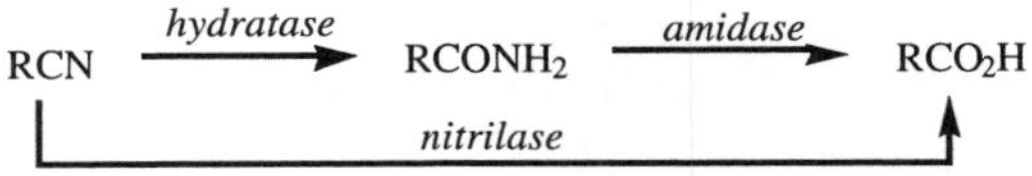

Scheme 11: Enzymes that hydrolyse nitriles.

Through a thorough investigation of the substrate specificity of organisms known to contain nitrile hydratase enzymes we were able to establish that certain classes of nitrile containing compounds were indeed effective substrates for hydrolysis [7]. For example, we discovered that a range of 2-arylpropionitriles were hydrolysed to the corresponding arylpropionic acids including the non steroidal anti-inflammatory compound (S)-ibuprofen **30.**

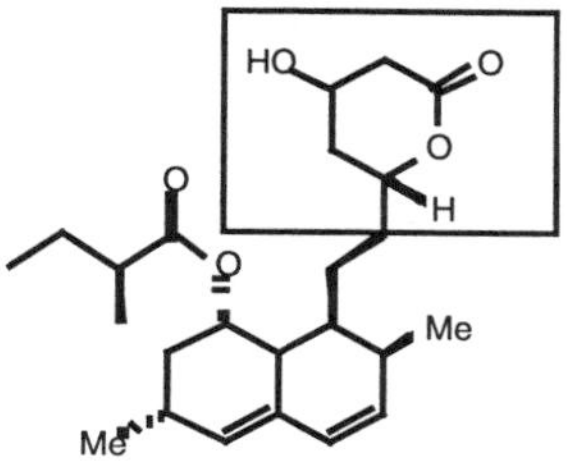

mevinolin **13**

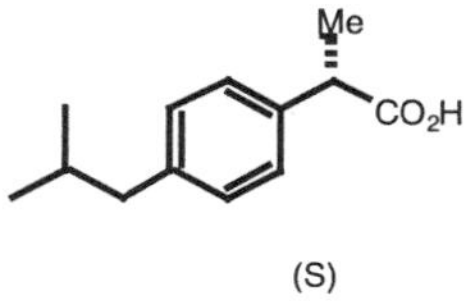

(S)-ibuprofen **30**

Figure 2: Targets using nitrile hydrolase methodology.

Another class of substrate we investigated were protected 3-hydroxyglutaronitriles **32**. A range of protected 3-hydroxyglutaronitriles **32** were easily prepared by treatment of epicholorohydrin **31** with two equivalents of potassium cyanide followed by reaction with the appropriate protecting group. Hydrolysis with either *Rhodococcus* sp. or *Brevibacterium* sp., both of which contain nitrile hydratase enzymes, resulted in the production of the corresponding cyano acids **33** in a wide range of yields and enantiomeric excesses [9,10] (Scheme 12). The substrates are initially hydrolysed to the nitrile-amide by a pro-(S)-selective hydratase followed by non-selective conversion ot the acid catalysed by an amidase. Interestingly the e.e. of the unproteced derivative proved to be higher than when an acetyl group was present.

80

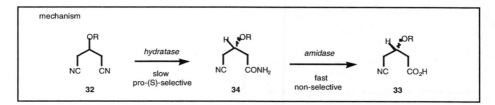

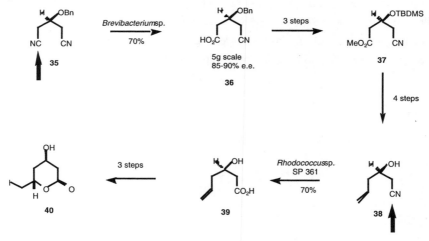

Scheme 12: Hydrolysis of substituted glutaronitriles using *Rhodococcus* sp.

Since the benzyl protected dinitrile **35** was transformed to the (S)-cyano acid **36** in good yield and e.e., we selected this intermediate for further development and were able to develop a synthesis of the δ-lactone moiety **40** of the mevinic acids **13** (see above). In this sequence it is noteworthy that nitrile hydratases are used twice, once for the initial asymmetric hydrolysis and secondly for the mild hydrolysis of the nitrile group of **38** during its conversion to **39** [11] (Scheme 13).

Scheme 13: Synthesis of a key δ-lactone intermediate **40** of the mevinic acids using nitrile hydratase enzymes.

2. Strategies for the Efficient Synthesis of Optically Pure Chiral Intermediates

There is an increasing need to devise new methods for the synthesis of enantiomerically pure chiral compounds that avoid kinetic resolution of racemates [12]. In order to overcome the limitation of 50% yield associated with the resolution approach, three basic strategies have been adopted that still utilise racemates as the starting materials i) dynamic kinetic resolution ii) deracemisation and iii) enantioconvergent resolution.

2.1. DYNAMIC KINETIC RESOLUTION (DKR)

Dynamic kinetic resolution is the most popular approach and involves a kinetic resolution coupled with *in situ* racemisation of the starting material such that the unreactive enantiomer is converted to the reactive one. Many examples of this reaction have been reported including the lipase catalysed DKR of 5(4*H*)-oxazolones **41** as an entry into enantiomerically pure protected α-amino acids **42** (Scheme 14) [13].

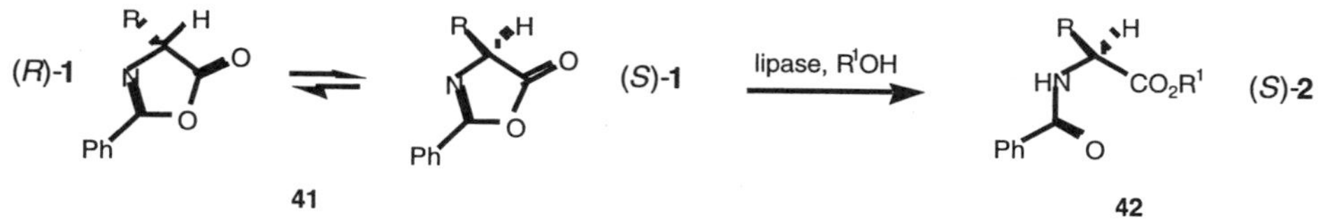

Scheme 14: Lipase catalysed dynamic kinetic resolution of 5(4*H*)-oxazolones **42**.

Our interest in the above reaction arose from a desire to develop a new method for the synthesis of optically pure L-(S)-tert-leucine **43** [14]. L-tert-leucine is an unnatural α-amino acid that presents special challenges for synthesis in optically pure form in view of the sterically demanding tert-butyl group. This tert-butyl group plays an important role in the use of L-tert-leucine, and derivatives, as a chiral auxiliary in asymmetric synthesis and catalysis (Figure 3).

L-(*S*)-*tert*-leucine **43**

tert-LeuPhos

Evans oxazolidinone
J. Org. Chem., 1993, **58**, 3569

Bisoxazoline
ligands

Chiral building block
Synthesis 1993, 137

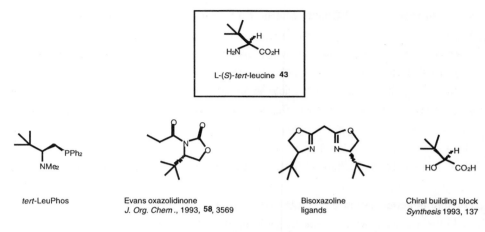

Figure 3: L-(S)-tert-leucine and its applications as a chiral auxiliary.

L-tert-leucine is also found as a component of a series of peptidomimetics that are inhibitors of matrix metalloproteinases (MMI's) [15]. This class of MMI's is currently in clinical trials for treatment of arthritis and cancer (e.g. **44** in Figure 4).

44

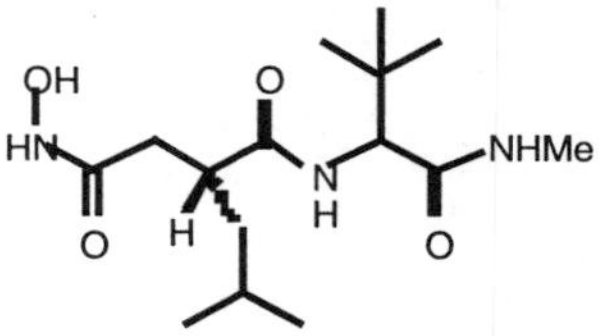

Figure 4: Peptidomimetics that contain L-(S)-tert-leucine.

The approach that we developed is shown in Figure 19 and involves conversion of racemic tert-leucine **43** to the corresponding oxazolones followed by treatment with Lipozyme (*Mucor miehei* lipase) in toluene containing BuOH and triethylamine. Despite the slow rate of the reaction (5 days) the product N-benzoyl-L-tert-leucine butyl ester **46** was obtained in 94% yield and 99.5% e.e indicative of a highly efficient dynamic kinetic resolution [16]. The enantioselectivity of the reaction was found to be highly dependent on the alcohol used and also on the presence of triethylamine for reasons that are discussed below.

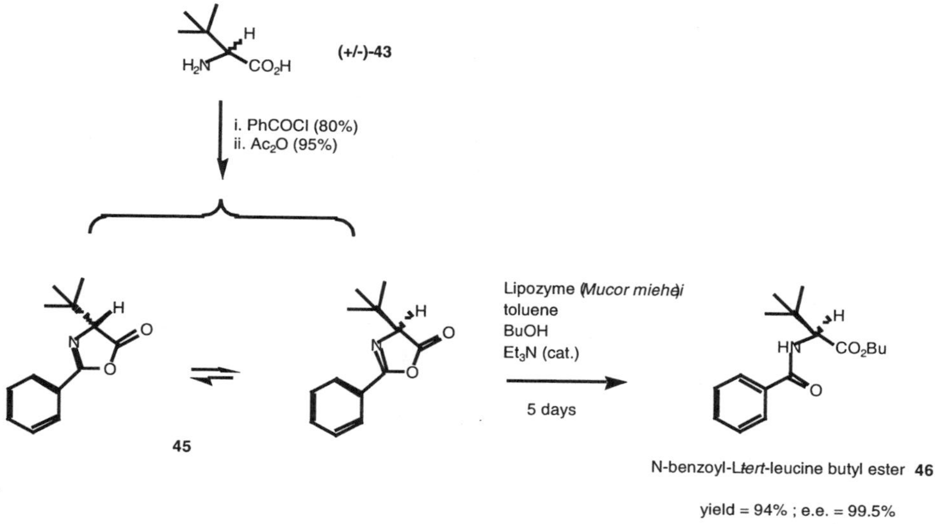

(+/-)-43

i. PhCOCl (80%)
ii. Ac$_2$O (95%)

Lipozyme (Mucor miehei)
toluene
BuOH
Et$_3$N (cat.)

5 days

45

N-benzoyl-L-tert-leucine butyl ester 46

yield = 94% ; e.e. = 99.5%

Scheme 15: DKR of 4-tert-butyl substituted oxazolone.

In collaboration with Chiroscience (now Chirotech) in Cambridge, UK, we were able to scale up the reaction to a concentration of $200gl^{-1}$. By the addition of a higher ratio of catalyst to substrate the reaction time was shortened from 5 days to 24 hours [17] (Scheme 16).

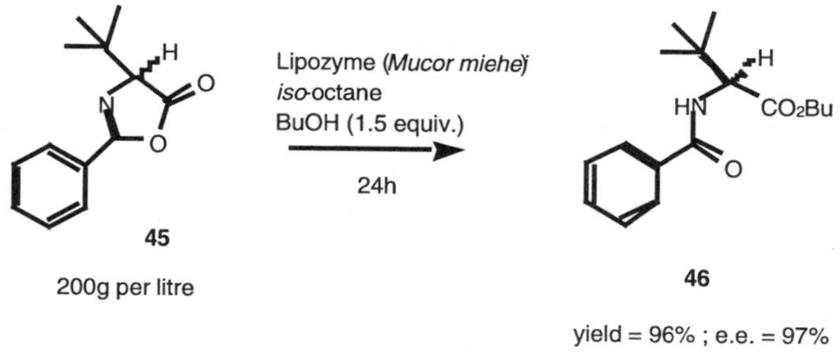

Lipozyme (Mucor miehei)
iso-octane
BuOH (1.5 equiv.)

24h

45

200g per litre

46

yield = 96% ; e.e. = 97%

Scheme 16: Scale-up of oxazolone biotransformation.

In order to complete the synthesis of L-tert-leucine we needed to remove the protecting groups and at this stage some problems were encountered. Treatment of enantiomerically pure protected amino acid 46 with 6M HCl led to partial racemisation of the product 43 (e.e. 73%) presumably due to competing formation of the oxazolone 45 which rapidly racemises. Formation of oxazolone 45 is favoured due to relief of steric compression at the α-amino acid stereogenic centre. By carrying out a two step deprotection procedure involving initial ester hydrolysis using alcalase (protease from

84

Bacillus licheniformis) it proved possible to overcome this problem of partial racemisation and obtain enantiomerically pure product (Scheme 17).

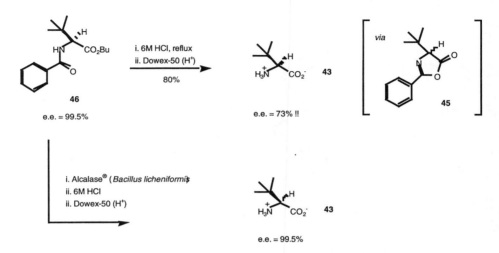

Scheme 17: Deprotection of N-benzoyl-L-tert-leucine butyl ester.

In addition to tert-leucine we examined a series of other amino acid derived oxazolones. In general the enantioselectivity of Lipozyme towards these other substrates proved lower although Novozyme (*Candida antarctica* lipase B) proved more effective. We have also probed in detail the mechanism of these reactions by examining the effect on the reaction of a number of parameters including solvent, water activity and the presence and absence of a catalytic amount of triethylamine. After a series of careful studies we have developed optimised conditions that are applicable to the synthesis of a range of amino acids (Scheme 18 and Scheme 19).

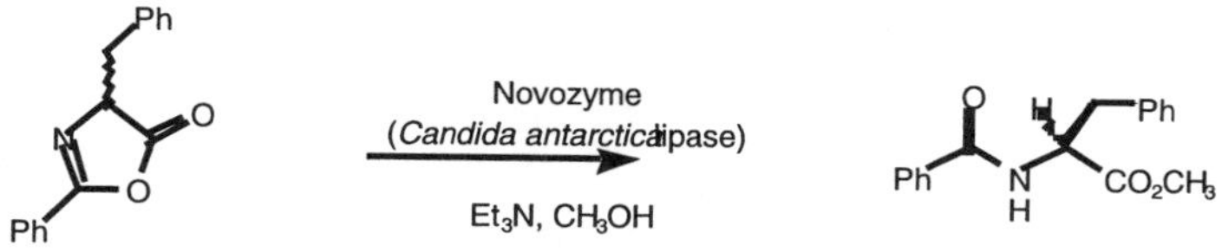

Solvent	Et₃N		No Et₃N	
	Yield/%	e.e./%	Yield/%	e.e./%
DCM	78	89	65	75
CHCl₃	66	75	63	83
THF	64	95	71	**97**
Et₂O	87	97	90	58
tBuOMe	90	96	91	34
DIPE	86	96	90	33
toluene	82	94	82	71
CH₃CN	44	97	88	**98**

Scheme 18: Effect of solvent on biotransformation.

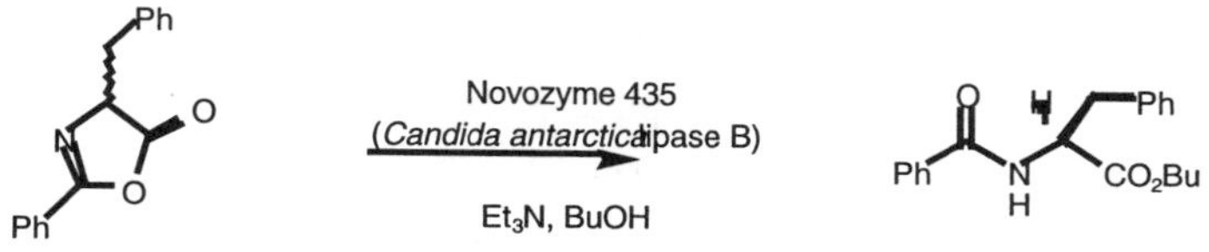

Solvent	a_w	No Et₃N		Et₃N	
		initial rate	e.e./%	initial rate	e.e./%
hexane	0.01	26	85	30	90
hexane	0.69	4	55	20	87
hexane	0.97	1.5	30	18	80
toluene	0.01	15	85	27	93
toluene	0.22	3	61	17	95
CH₃CN	0.01	15	99	10	97
CH₃CN	0.10	NR	--	5	90

Scheme 19: Effect of triethylamine on biotransformation.

During the work with L-tert-leucine we observed that both the particular alcohol (nucleophile) used and also the presence of triethylamine had a dramatic effect on the enantioselectivity of the lipase catalysed ring-opening reaction. This effect can best be seen by comparing the results of reactions involving BuOH/Et3N (99.5% e.e.) with MeOH/no Et$_3$N (33.5% e.e.). Although it is reasonable to expect a variation of e.e. with reaction conditions this data indicated a substantial change in enzyme selectivity (Scheme 20) [18, 19].

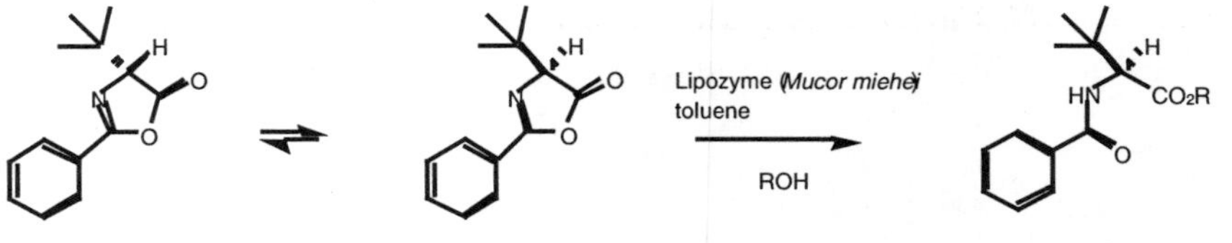

ROH	e.e./%	
	Et$_3$N	no Et$_3$N
MeOH	80.1	38.6
EtOH	97.1	68.0
BuOH	99.5	79.9

Scheme 20: Effect of alcohol and triethylamine on biotransformation.

We have rationalised the requirement for E$_t3$N on the basis that under the reaction conditions there is a small amount of hydrolysis of the substrate to the carboxylic acid **49** Scheme 21). This hydrolysis is unavoidable since even if great efforts are taken to remove water from the system there is always a small amount of tightly bound water that remains. The production of carboxylic acid under essentially anhydrous conditions appears to effect the enzyme enantioselectivity in ways which are not fully understood. What is clear is that addition of triethylamine reverse this effect presumably by formation of an ion pair **50** with the acid. This process is clearly solvent dependent as supported by the data shown in Scheme 19.

Finally, we prepared a series of more complex oxazolones that contained two stereogenic centres, one of which is epimerisable and the other of which is not [20]. When a solution of **51a/b** (7% d.e.) in petrol:DCM 100:1 was heated a precipitation occurred yielding crystals of one of the two diastereoisomers **51b** (99% d.e.) in 70% yield. In this process, generally termed 'crystallisation induced dynamic resolution' the inherent chirality of the molecule determines which diastereomer preferentially precipitates from solution (Scheme 22). It was also possible to establish that the minor diastereoisomer in solution is the one that preferentially precipitates from solution, an

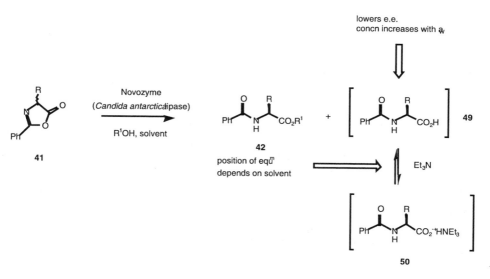

Scheme 21: Explanation of requirement for triethylamine.

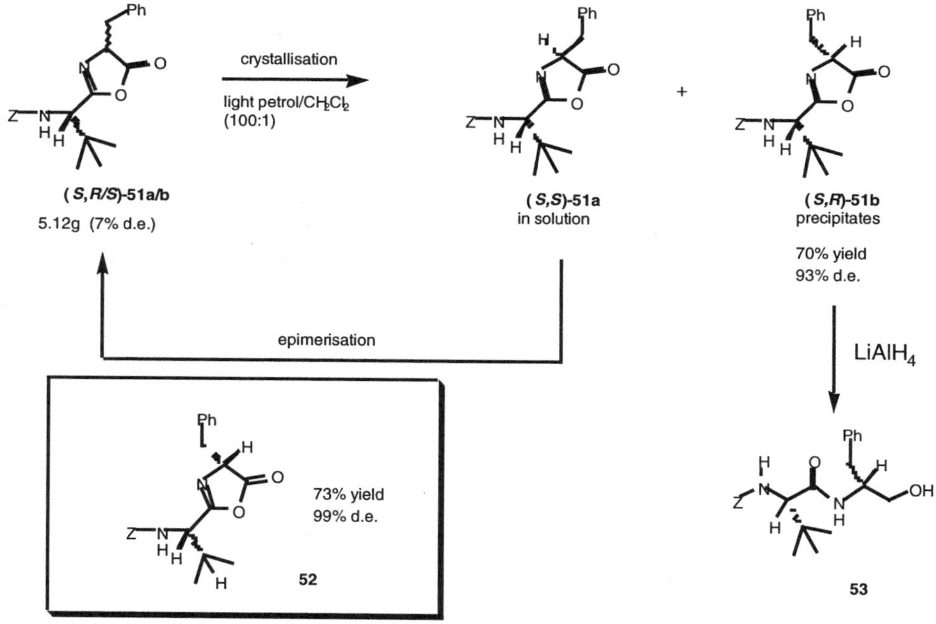

Scheme 22: Crystallisation induced dynamic resolution of oxazolones.

observation that is in accord with the van't Hoff-Dimroth rule. Reductive ring-opening of oxazolone **51b** with lithium aluminium hydride yielded the optically and diastereomerically pure peptide alcohol **53**.

2.2. DERACEMISATION (DR)

Deracemisation involves the conversion of a racemate to the single enantiomer form *via* a stereoinversion process. This second order asymmetric transformation can be achieved either by using two enzymes with opposite enantioselectivity (*e.g.* deracemisation of secondary alcohols using alcohol dehydrogenases [21]) or by using an enzyme in combination with a chemical reagent (*e.g.* amino acid oxidase and a reducing agent for the deracemisation of amino acids [22, 23]). An example of the former process is given in Scheme 23 in which racemic 2-arylethanols were deracemised using *Geotrichum candidum*.

mechanism

* (S)-isomer is consumed first --> acetophenone.

* (R)-isomer is produced by reduction of acetophenone

* Use of (S)-isomer results in conversion to (R)-isomer

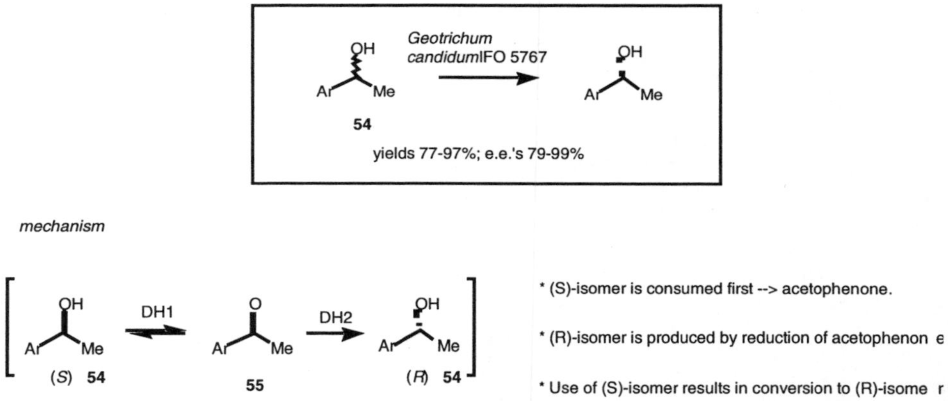

Scheme 23: Deracemisation of secondary alcohols using *Geotrichum candidum*.

In 1992 Soda *et al.*, [22] reported a method for the deracemisation of DL-proline **56** (n = 0) and DL-pipecolic acid **56** (n = 1) using D-amino acid oxidase from porcine kidney in combination with sodium borohydride. The key to the reaction is the reduction, *in situ*, of the intermediate imino acid **57**. Although the reaction consumes large quantities of borohydride, owing to the spontaneous reaction of this reagent with water at pH 7, the overall result is a quantitative conversion of D/L-proline to the optically pure L-isomer (Scheme 24). Remarkably the enzyme is able to operate at a reaction pH of 10.5.

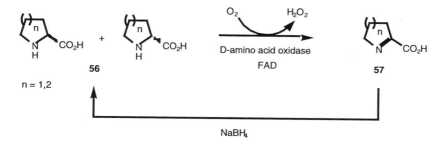

Scheme 24: Deracemisation of proline and pipecolic acid using D-amino acid oxidase and sodium borohydride.

We have recently investigated this reaction in some detail [24], particularly with respect to the range of D- and L-amino acid oxidases that can be used and also seeking alternatives to sodium borohydride as a reducing agent. An important commercially available D-amino acid oxidase is the one from *Trigonopsis variabilis* and this enzyme is currently marketed by Boehringer Mannheim in Germany. This oxidase is available in immobilised form and is used in the process for the conversion of cephalosporin C to 7-aminocephalosporanic acid. Treatment of cephalosporin C **58** with the D-amino acid oxidase converts the amino acid of the side-chain to the α-keto acid which undergoes oxidative decarboxylation to the carboxylic acid **59**. This modified side-chain is then a substrate for an amidase from *E.coli* which catalyses removal to give 7-ACA (Scheme 25).

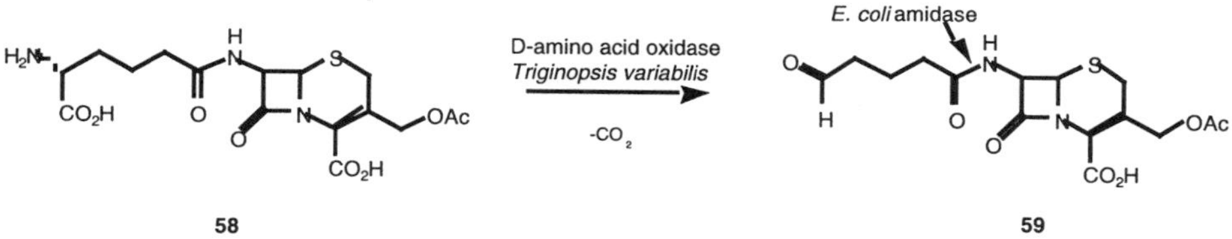

Scheme 25: D-amino acid oxidase from *Triginopsis variabilis*.

We have used the *Trigonopsis variabilis* D-amino acid oxidase to convert DL-piperazine-2-carboxylic acid **60** to the enantiomerically pure L-isomer in 86% yield and 99% e.e. L-piperazine-2-carboxylic acid is a component of the Merck anti-HIV protease inhibitor Crixivan **61**. In an improvement on the original Soda procedure, sodium borohydride was replaced by the more stable sodium cyanoborohydride enabling the number of equivalents of reducing agent to be drastically reduced to 5-10 (Scheme 26).

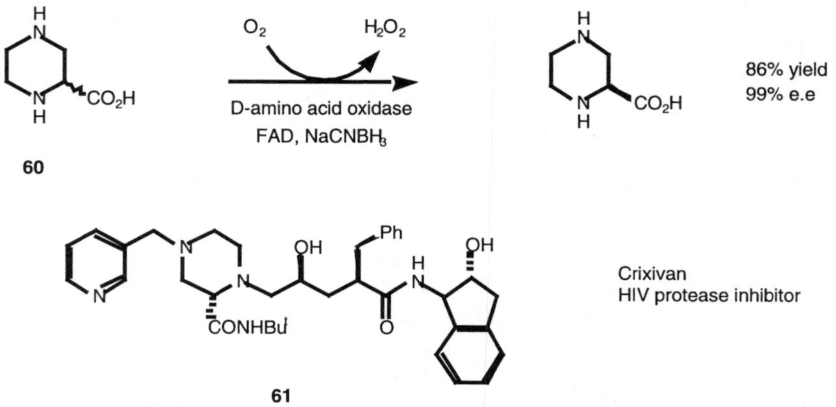

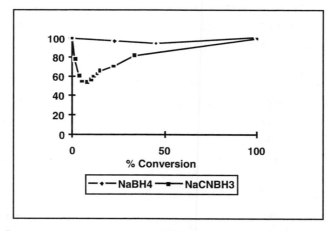

Scheme 26: Deracemisation of D/L-piperazine-2-carboxylic acid, a component of Crixivan.

Comparison of sodium borohydride with cyanoborohydride reveals a difference in the kinetics of the process. Since cyanoborohydride is less reactive, reduction of the imine becomes rate limiting whereas with the more reactive borohydride the enzyme oxidation is the slowest step (Figure 5).

Figure 5: Kinetics of reduction of sodium borohydride compared to cyanoborohydride.

We were especially interested to see whether the deracemisation methodology could be extended to include acyclic, as well as cyclic amino acids. The major concern here was that the corresponding imino acids **63**, resulting from the oxidation, would undergo spontaneous hydrolysis faster than reduction back to the amino acid precursor. Remarkably, acyclic amino **62** acids work well giving yields of 80-90% and excellent e.e.'s. Some loss of material does indeed occur due to competing hydrolysis of the imine **63** to the keto acid **64** (Scheme 27).

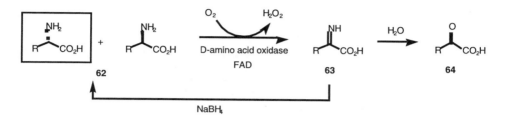

Scheme 27: Deracemisation of acyclic amino acids.

As an example of the use of acyclic amino acids we have looked at the deracemisation of DL-leucine **65** (Scheme 28).

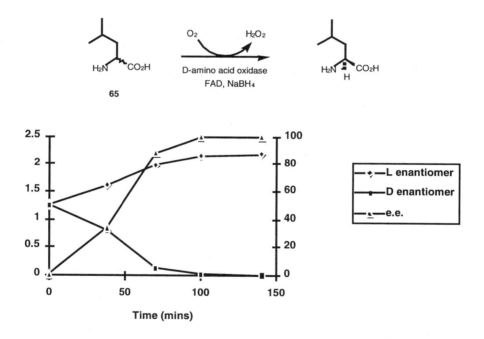

Scheme 28: Deracemisation of DL-leucine **65**.

92

2.3.   ENANTIOCONVERGENT RESOLUTION (ER)

For the special case of epoxides a third type of total conversion of a racemate is possible [25]. This transformation, termed enantioconvergent resolution, involves initial epoxide hydrolase catalysed kinetic resolution of a racemic epoxide **66** to give a single enantiomer of the diol **67** and unreacted epoxide. The remaining epoxide can then be converted to the same enantiomer of the diol by treatment with acid which stereoselectively reacts at the more hindered end. It is also possible to effect both steps using a single epoxide hydrolase (Scheme 29).

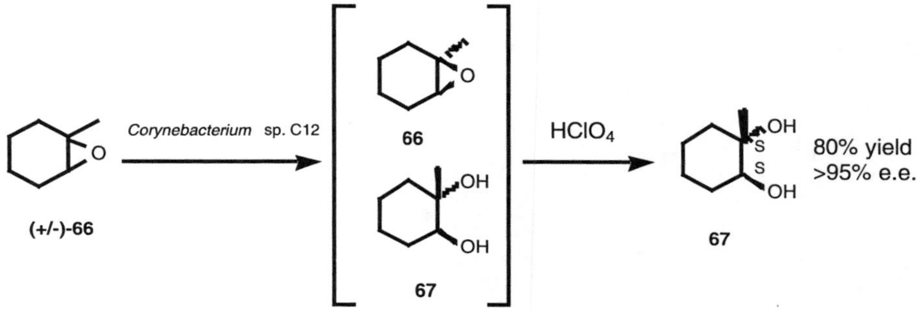

Scheme 29: Enantioconvergent resolution of epoxides using an epoxide hydrolase.

3.   Acknowledgements

The author gratefully acknowledges the enthusiastic and hard-working contributions of the co-workers involved in these projects including Francois-Rene Alexandre, Marina Alexeeva, Tim Beard, Josie Blackie, Stuart Brown, Alexis Carstairs, Dean Johnson, Sam Maddrell, John McGuinn, Marie-Claire Parker, Shagufta Rafiq, Thijs Stock, and James Winterman. We are also indebted for funding to the Biotechnology and Biological Sciences Research Council (BBSRC) and to the following companies who have generously provided support; AstraZeneca, Chirotech, GlaxoWellcome, Pfizer, SmithKline Beecham, Ultrafine Chemicals and NSC Technologies.

4. References

1. Turner, N.J. (**1997**) *Current Organic Chemistry*, **1**, 21-36.
2. Davey, C.L., Powell, L.W., Turner, N.J. and Wells, A. (1994) *Tetrahedron Lett.* **35**, 7867-7870.
3. Blackie, J.A., Turner, N.J. and Wells, A.S. (1997) *Tetrahedron Lett.* **38**, 3043-3046.
4. Cantello, B.C.C., Eggleston, D.S., Haigh, D., Haltiwanger, R.C., Heath, C.M., Hindley, R.M., Jennings, K.R., Sime J.T. and Woroniecki, S.R. (1994) *J.Chem.Soc., Perkin Trans. 1*, 3319.
5. Crosby, J., Moilliet, J., Parratt, J.S. and Turner, N.J. (1994) *J. Chem. Soc., Perkin Trans. 1*, 1679-1687.
6. Hönicke-Schmidt, P. and Schneider, M.P. (1990) *J. Chem. Soc., Chem. Commun.*, 648; Klempier, N., de Raadt, A., Faber K. and Griengl, H. (1991) *Tetrahedron Lett.*, **32**, 341; de Raadt, A., Klempier, N., Faber, K. and Griengl, H. (1992) *J. Chem Soc., Perkin 1*, 137; de Raadt, A., Griengl, H., Klempier, N. and Stütz, A.E. (1993) *J. Org. Chem.* **58**, 3179.
7. Cohen, M.A., Sawden, J. and Turner, N.J. (1990) *Tetrahedron Lett.* **31**, 7223.
8. Beard, T., Cohen, M.A., Crosby, J., Moilliet, J.S., Parratt, J.S. and Turner, N.J. (1993) *Tetrahedron Asymmetry* **4**, 1085; Cohen, M.A., Parratt, J.S. and Turner, N.J. (1992) *Tetrahedron Asymmetry* **3**, 1543.
9. Crosby, J., Parratt, J.S. and Turner, N.J. (1992) *Tetrahedron Asymmetry*, **3**, 1547; Kerridge, A., Parratt, J.S., Roberts, S.M., Theil, F., Turner, N.J. and Willetts, A.J. (1994) *Bioorg. Med. Chem. Lett.* **2**, 447-455.
10. Kakeya, H., Sakai, N., Sugai, T. and Ohta, H. (1991) *Tetrahedron Lett.* **32**, 1343; Kakai, H., Sakai, N., Sugai, T. and Ohta, H. (1991) *Agric. Biol. Chem.* **55**, 1877; Kakeya, H., Sakai, N., Sano, A., Yokoyama, M., Sugai, T. and Ohta, H. (1991) *Chem. Lett.*, 1823.
11. Maddrell, S.J., Turner, N.J., Kerridge, A., Willetts, A. J. and Crosby, J. (1996) *Tetrahedron Lett.* **37**, 6001-6004.
12. For recent reviews see: Ward, R.S. (1995) *Tetrahedron Asymmetry* **6**, 1475; Noyori, R., Tokunuga, M. and Kitamura, M. *Bull. Chem. Soc. Jpn.* **68**, 36.
13. Bevinakatti, H.S., Newadkar, R.V. and Banerji, A.A. (1990) *J. Chem. Soc., Chem. Commun.*, 1091; Gu, R.-L., Lee, I.S. and Sih, C.J. (1992) *Tetrahedron Lett.* **33**, 1953; Crich, J., Brieva, R., Marquart, P., Gu, R.-L., Flemming, S. and Sih, C.J. (1993) *J. Org. Chem.* **58**, 3252.
14. For a review on the synthesis and applications of *tert*-leucine see: Bommarius, A.S., Schwarm, M., Stingl, K., Kottenhahn, M., Huthmacher, K. and Drauz, K. (1995) *Tetrahedron Asymmetry* **6**, 2851.
15. Brown, P.A., Johnson, W.H. and Lawton, G. (Hofmann-La Roche) (1992) EP Appl., 0497 192.
16. Turner, N.J., Winterman, J.R., McCague, R., Parratt, J.S. and Taylor, S.J.C. (1995) *Tetrahedron Lett.* **36**, 1113.
17. Turner, N.J., Winterman, J.R. and McCague, R. (Chiroscience) (1995) PCT Intl. Appl., WO 9512573.
18. Stock, H.T. and Turner, N.J. (1996) *Tetrahedron Lett.* **37**, 6575-6578.
19. Parker, M.-C., Brown, S.A., Robertson, L. and Turner, N.J. (1998) *J. Chem. Soc., Chem. Commun.*, 2247.
20. Yagnik, A.T., Littlechild, J.A. and Turner, N.J. (1997) *J. Computer-Aided Molecular Design* **11**, 256-264.
21. Nakamura, K., Inoue, Y., Matsuda, T. and Ohno, A. *Tetrahedron Lett.* **36**, 6263.
22. Huh, J. W., Yokoigawa, K., Esaki, N. and Soda, K. (1992) *Biosci. Biotech. Biochem.* **56**, 2081; Huh, J.W., Yokoigawa, K., Esaki, N. and Soda, K. (1992) *J. Ferment. Bioeng.* **74**, 189.

23. For a recent review and discussion of the kinetics of deracemisation reactions see: Strauss, U.T., Felfer, U. and Faber, K. (1999) *Tetrahedron Asymmetry* **10**, 107.
24. Beard, T.M. and Turner, N.J. unpublished results.
25. Archer, I.V.J., Leak, D.J. and Widdowson, D.A. (1996) *Tetrahedron Lett.* **37**, 8819.

RECENT ADVANCES ON BIOREDUCTIONS MEDIATED BY BAKER'S YEAST AND OTHER MICROORGANISMS

E. SANTANIELLO*

Dipartimento di Scienze Precliniche LITA Vialba, Università di Milano, Via G.B. Grassi,74 - 20157 Milano, Italy

P. FERRABOSCHI, A. MANZOCCHI

Dipartimento di Chimica e Biochimica Medica, Università di Milano, Via Saldini 50-20133 Milano, Italy

Abstract: The bioreductions mediated by BY and other microorganisms are considered mainly from a preparative point of view. The use of other microorganisms is reported when BY is either unable to perform the reaction or if this proceeds with moderate enantioselectivity or when the opposite stereochemical outcome of the reaction should be obtained. The biotransformation of nitrogen-containing groups is especially addressed. The bioreductions of trinitrotoluene is presented as an example of the bioremediation of explosives.

1. Introduction

Microorganisms may at present be regarded as biotechnological reagents that can be used in several chemical processes of special interest to organic synthesis [1]. Although the applications of whole-cell biocatalysts in oxido-reduction reactions and other areas are in continuous development [2], for an organic chemist the use of these biocatalysts requires a training in microbiology and expertise in the related techniques or collaboration with microbiologists. An exception is baker's yeast (**Saccharomyces cerevisiae**) that has gained a great popularity as a versatile microbial reagent among organic chemists. Baker's yeast (BY) applications have been summarized in two reviews

B. Zwanenburg et al. (eds.), Enzymes in Action, 95–115.
© 2000 *Kluwer Academic Publishers. Printed in the Netherlands.*

which appeared a few years ago [3], followed by others like the recent overview on the synthetic capabilities of BY [4]. Reviews on synthetic applications of BY, including the large-scale production of alcohols in an enantiomerically pure form, have recently appeared [5]. We will highlight the recent advances on microbial bioreductions starting from our review [6] in order to furnish an up-to-date of applications of this special biocatalytic approach.

2. Baker's yeast-mediated Bioreductions: General Considerations

BY is available in any food market and no special equipment or skill in microbiology are required to set up the fermenting conditions that allow the biotransformation. The bioreductions with BY are experimentally simple and the protocol generally used for

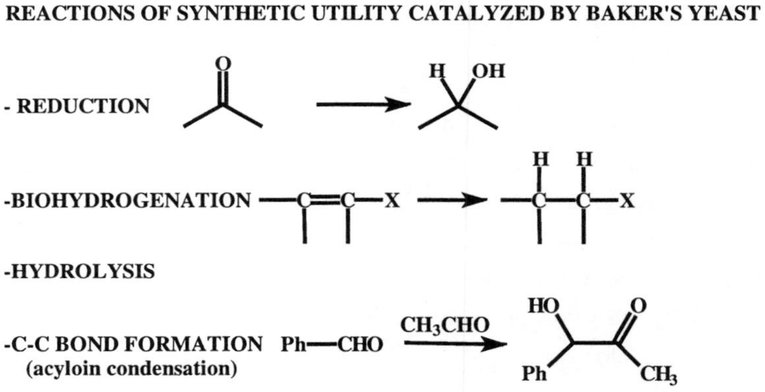

REACTIONS OF SYNTHETIC UTILITY CATALYZED BY BAKER'S YEAST

Scheme 1

BY relies mainly upon the fermenting conditions that use glucose as energy source in tap water or sometimes in buffer solutions. Addition of the substrates as such or in solution ensures dispersion of the organic compound that can be biotransformed while mechanical stirring at ambient temperature. Use of detergents and emulsifiers may, in some instance, prevent the problems of poisoning living cells and alter the course of the biotransformation. If the substrate does not cross the yeast cell membrane sometimes it can be necessary to disrupt the cells by chemical or physical methods and in this case the cofactors and oxidoreductases responsible for the bioreduction are still capable to carry out the desired biotransformations. In spite of the drawbacks of the whole-cell biocatalysis, such as low substrate concentration, high biomass ratio, problems of product recovery, etc., the application of baker's yeast-mediated reactions has been recognized as a useful technique in synthetic organic chemistry. Bioreductions of carbonyl compounds are the most exploited reactions of the BY's biochemical machinery capability, but also other biotransformations (Scheme 1) such as hydrolysis or a C-C forming bond reaction (acyloin condensation) can be accomplished [4]. The recently reported [7] rearrangement of an allyl chloride is a good example of an unusual reaction that can be mediated by BY (Scheme 2).

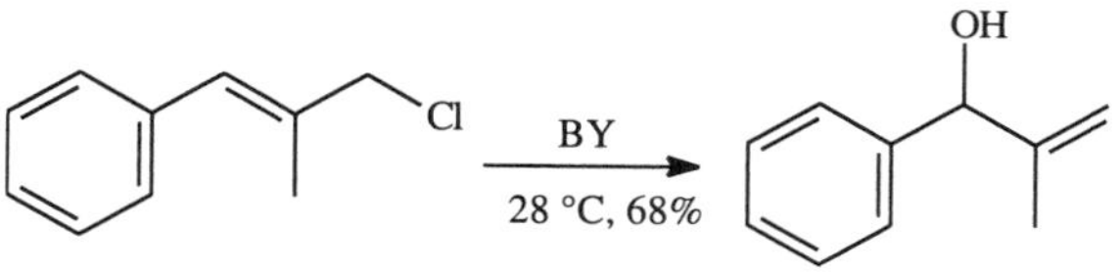

Scheme 2

One of the most peculiar reactions catalyzed by BY certainly is the biohydrogenation of activated double bonds and several examples are already available [4]. We have recently reported that the biohydrogenation process can also be applied to the transformation of 2-substituted allyl alcohols to give the corresponding 2-methyl-1-alkanols with excellent enantiomeric excess (ee) [8,9], as shown in Scheme 3.

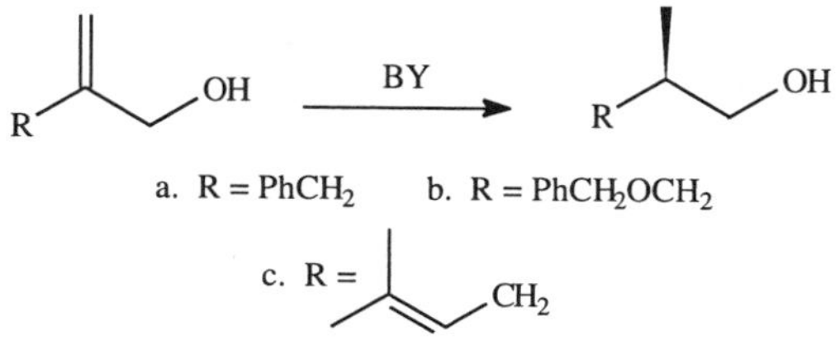

a. R = PhCH$_2$ b. R = PhCH$_2$OCH$_2$

c. R =

Scheme 3

3. Reduction of Carbonyl Compounds

3.1. THE STEREOCHEMICAL COURSE OF THE BIOREDUCTION

Detailed procedures for the bioreduction of a few carbonyl compounds have been also reported in *Organic Syntheses* [10]. A BY-catalyzed biotransformation may be brought about by a single enzyme and stereomerically pure compounds may be formed. The classical substrates for microorganism-mediated bioreductions are carbonyl compounds and the product configuration may follow a general rule originally postulated by Prelog some years ago for the bioreduction catalyzed by the *Curvularia falcata* oxidoreductase [11]. Provided that the size of the carbonyl substituents are relatively different (S, small

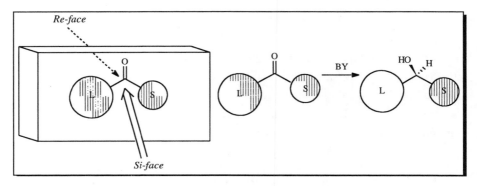

Scheme 4

size; L, large size), the attack of the incoming hydrogen usually occurs from the *re*-face in order to produce the (S) stereoisomer (Scheme 4). Although for many substrates BY seems to follow Prelog's rule, it should be reminded that the yeast contains several different oxidoreductases [12] that may catalyze the reduction of the same substrate. If this occurs with opposite stereochemical outcome, the products of the biotransformation may be obtained with low to moderate ee. Many microorganisms may carry out an "anti Prelog" bioreduction when the stereoisomer formed has the configuration opposite to that expected according to Prelog rule. Several methods have been reported to control the stereochemical course of BY-mediated bioreductions, including the immobilization of the yeast, the addition of inhibitors of undesired enzymatic activities and the use of organic solvents [13]. Chemical modifications of the substrate may lead to a compound whose structure better corresponds to the stereochemical demand of the oxidoreductase that is supposed to catalyze the bioreduction. An early example of this approach was the BY reduction of acetophenones [14] that proceeded in 20% yield and 69% ee. 4-Iodo and 4-phenylsulfonylacetophenone were reduced to the corresponding (S)-alcohols in 30 and 84% yield, the ee being considerably higher (96 and 92%, respectively). In this case, the L substituent became larger by substitution of the phenyl ring and this enhanced the difference between the S and L substituents size, so that, according to Prelog's rule, a high ee could be reached (Scheme 5).

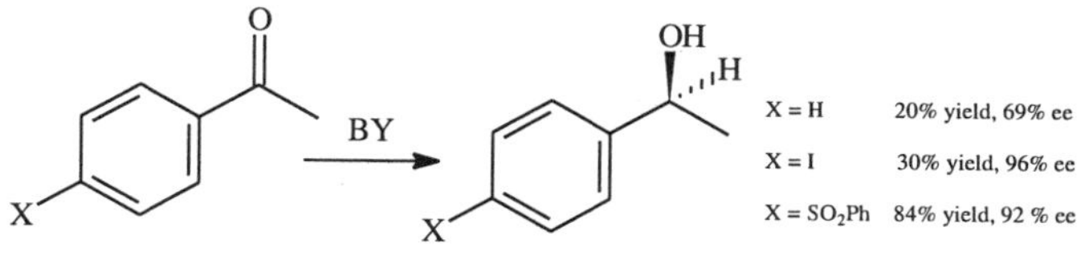

X = H	20% yield, 69% ee
X = I	30% yield, 96% ee
X = SO$_2$Ph	84% yield, 92 % ee

Scheme 5

3.2. REDUCTION OF DIFFERENT SERIES OF KETONES

BY reduction of ketones to enantiomerically enriched alcohols, reported at the beginning of the century [15] is still actively investigated. As BY, many microorganisms can perform the same kind of bioreduction [16], whereas other microorganisms can carry out an "anti-Prelog" reduction [17] as shown in Scheme 6.

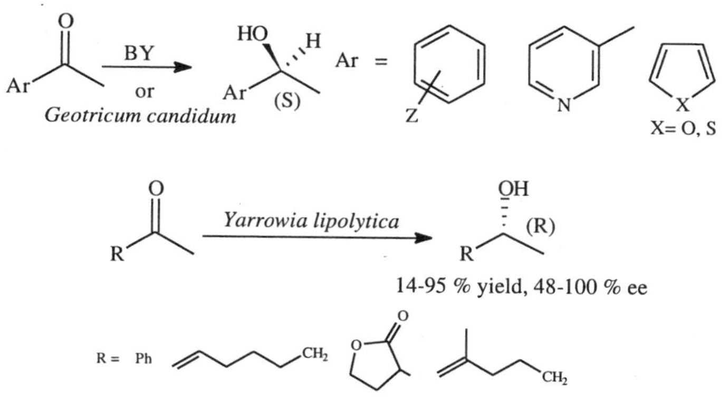

14-95 % yield, 48-100 % ee

Scheme 6

Among many other examples of reduction of aromatic ketones, we report here (Scheme 7) the case of heterocyclic ketones such as 1-(1,2,4-triazine-5-yl)alkanones to the corresponding (S)-alcohols [18] and a bulky substrate as phenyl pyridyl ketones that can be reduced to the corresponding R or S alcohols [19] with baker's yeast, free or entrapped on alginate, with variable ee (up to 96%).

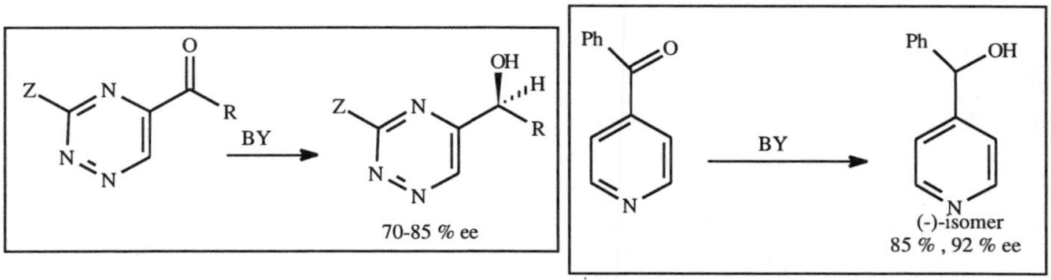

70-85 % ee

(-)-isomer
85 % , 92 % ee

Scheme 7

Fermenting BY is able to reduce an extremely large number of carbonyl compounds without being effected by other groups present in the structure. In Scheme 8 only a few examples have been selected within the vast number of substrates subjected BY bioreductions for the preparation of stereomerically pure compounds. In many instances other microorganisms may afford the same (with higher ee than BY) or the opposite stereoisomer. These structures may contain, among many others, silicon [20] or sulfur [21] groups, esters as in 2- and 3-oxoesters [22,23]. In contrast to keto esters, ketoamides are not reduced with BY, whereas the fungus *Mortierella isabellina* is able to transform β–oxoamides into enantiomerically pure hydroxy amides [24, 25].

We report here in some details the reduction of ketoacetals with BY and other yeasts. (Figure 9). The corresponding (S)-hydroxy derivatives were obtained with moderate enantioselectivity with BY, whereas enantiomerically pure (R)- and (S)-products could be prepared using a few selected microorganisms [26].

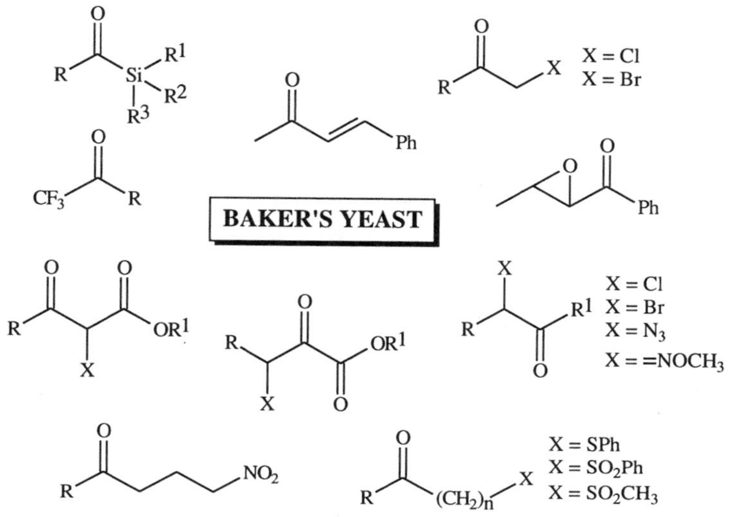

Scheme 8

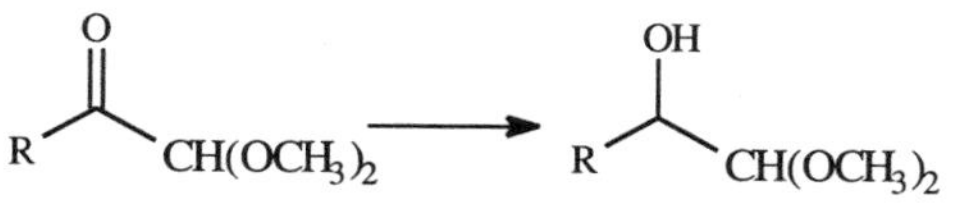

	R = Ph		R = C₄H₉	
	Yields (%, 48h)	ee (%)[a]	Yields (%, 48h)	ee (%)
Baker' yeast	45	46	26	60 (S)
Hansenula glucozyma	100	60	100	94 (S)
Hansenula minuta	100	96	50	20 (R)
Kloeckera saturna	30	90	46	92 (S)
Kluyveromyces marxianus	53	92	60	100 (R)

[a] Prevalent S configuration with all the strains.

Scheme 9

3.3.   REDUCTION OF CYCLIC CARBONYL COMPOUNDS

The current literature reports a remarkably high number of cycloalkanones that can be stereoselectively reduced to the corresponding alcohols. A few structures have been collected in Figure 1 including the most recent examples of functionalized cycloalkanones [27], cyclic ketoesters [28], or bicyclic compounds [29]. Many products of the above bioreduction are optically pure and have been used as chiral intermediates in asymmetric syntheses of natural products or biologically active compounds.

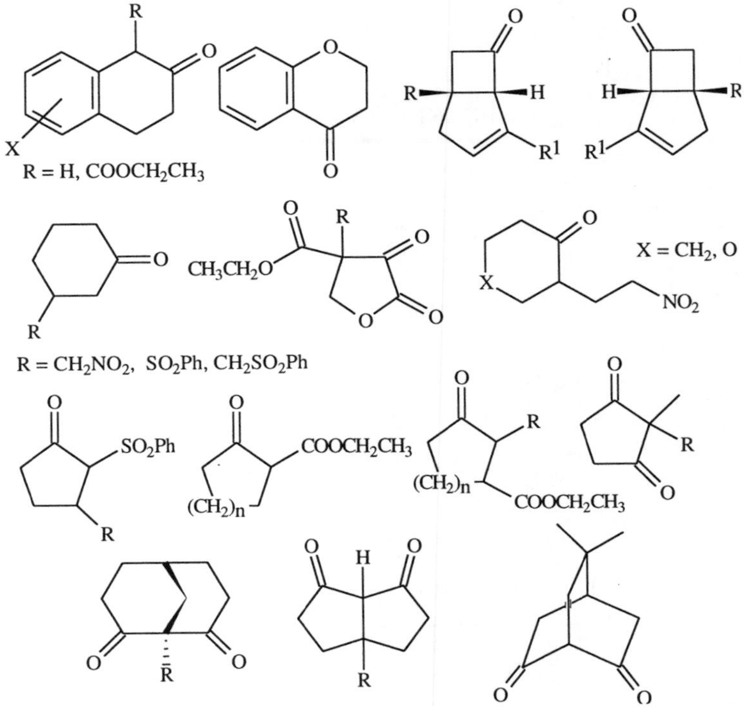

Figure 1

As previously reported for cyclic β–ketoamides [25], BY was unable to reduce an oxo lactam required for the synthesis of a chiral intermediate for the synthesis of an antibacterial quinolonic compound (Scheme 10). After screening of several strains of microorganisms it was found that a *Phaeocreopsis sp.* could accomplish the bioreduction with virtually complete enantioselectivity [30].

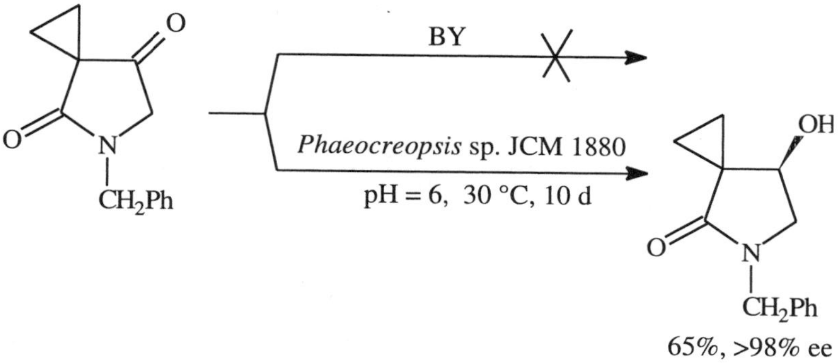

·103

Scheme 10

Several other oxolactams can be reduced by BY and in Scheme 11 we report the interesting case of the bioreduction of a benzothiazepinone that affords the enantiomerically pure intermediate for the synthesis of the drug Diltiazem [31].

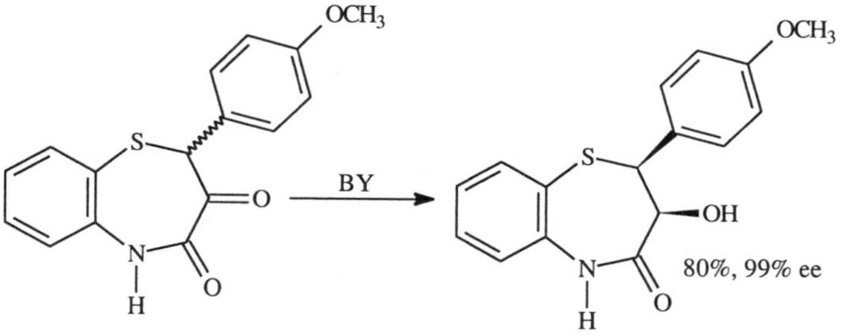

Scheme 11

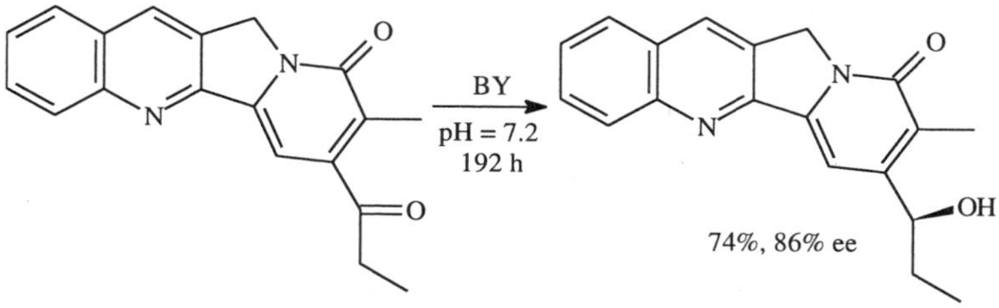

Scheme 12

In Scheme 12 it is shown that a carbonyl group in another complex heterocyclic structure, such as the so called mappicine ketone (in turn chemically prepared from camptothecin) can be enantioselectively reduced with BY [32].

3.4. CHEMO AND STEREOSELECTIVE REDUCTION OF CARBONYL COMPOUNDS: THE CASE OF DIKETONES

Diketones are excellent substrates for the biocatalytic process of microbial reduction and the products may be either hydroxy ketones or diols, depending on the relative position of the carbonyl groups and the microorganism chosen for the bioreduction. In the case of α-diketones the bioreduction with BY was not selective, but the products were enantiomerically pure. In Scheme 13 the results of the bioreduction with other microorganisms are also reported [33]. In the case of the BY-mediated reduction of 1-phenyl-2,3-propanedione (Scheme 14) a highly regio and enantioselective reduction was obtained only in the presence of methyl vinyl ketone as additive [34].

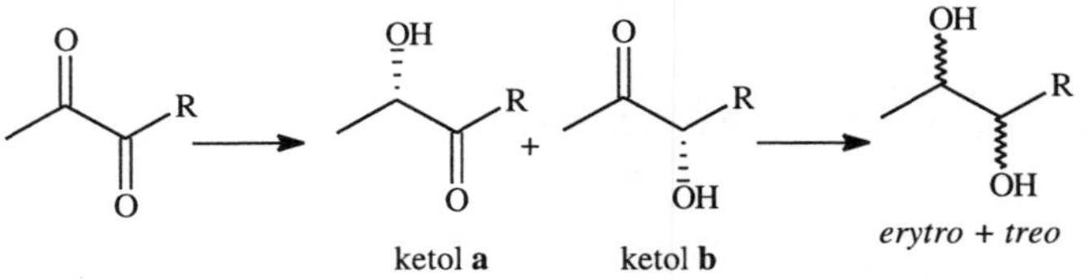

ketol a ketol b

α-Diketones	R $=C_2H_5$			R $= C_6H_5$		
Reaction products	Ketol **a**	Ketol **b**	Diols	Ketol **a**	Ketol **b**	Diols
Baker's yeast	17	14	69	70	-	- (30% ketone)
Aspergillus niger	4	12	84	-	50	50
Geotrichum candidum	25	15	60	60	-	17 (23% ketone)
Rhodotorula rubra	42	10	48	62	-	13 (25% ketone)

Scheme 13

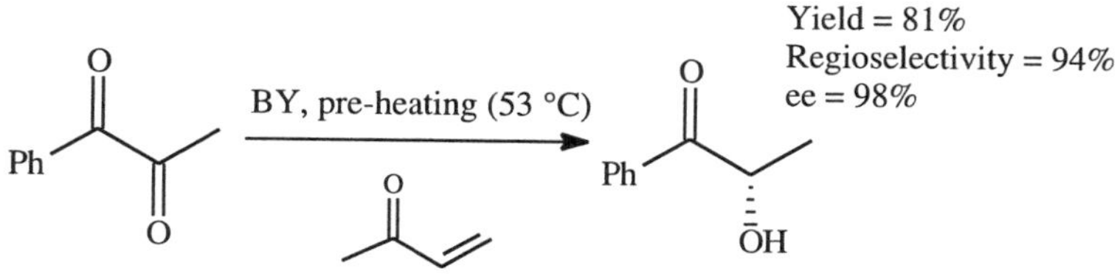

Scheme 14

A special case of chemo, regio and stereoselective bioreduction mediated by baker's yeast is the synthesis of the steroidal hydroxy diketone trimegestone (Scheme 15), a new progestinomimetic compound developed for the treatment of postmenopausal diseases. The bioreduction is the key-step that can be performed at an industrial scale. This constitutes one of the first industrial application of BY-mediated reduction of a carbonyl group [35].

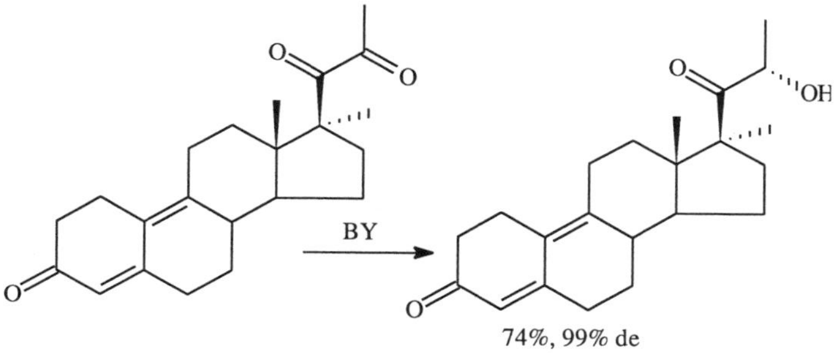

Scheme 15

The well known BY-mediated bioreduction of the simplest β-diketone, i.e. acetyl-acetone, proceeds to the corresponding (S)-hydroxy ketone and in the presence of other yeasts the (2R,4R)-diol is formed according to an "anti-Prelog" mechanism of reduction [36]. γ-Diketones behave differently from the just-mentioned β-diketones with respect to the selectivity of the bioreduction, due to the different chemical properties of the non-enolizable carbonyl groups. 2,5-Hexanedione has been already reported to be a good substrate for the BY-mediated bioreduction and, by recent improvements in the experimental protocol, stereomerically pure (2S,5S)-hexanediol can be obtained [37]. It has been also reported that the enantiomeric diastereomer can be prepared using *Pichia farinosa* [38] as a biocatalyst (Scheme 16).

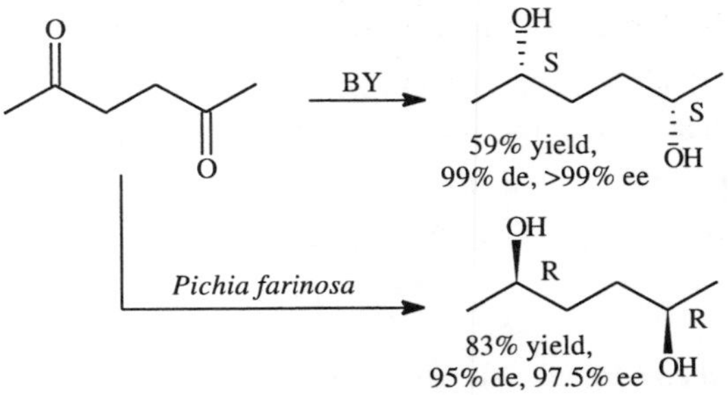

Scheme 16

Various aromatic symmetric diacetyls can be substrate for BY [39] and have been reduced to nearly pure (S)-monoalcohols (Scheme 17).

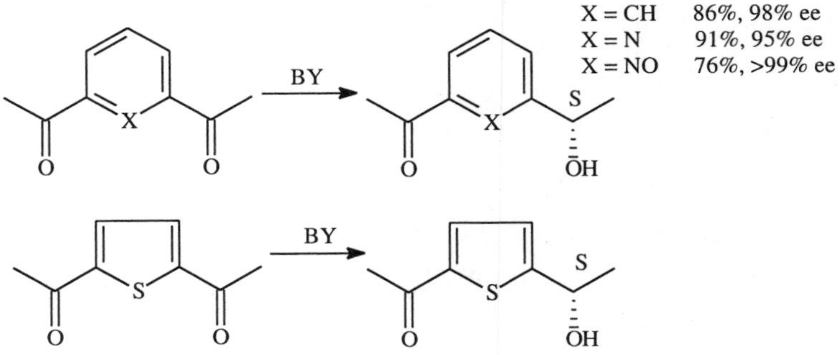

Scheme 17

3.5. STEREOCHEMICAL CONTROL OF THE BIOREDUCTION OF CARBONYL COMPOUNDS

Several methods are available to control the stereochemical outcome of microbial reductions. A classical example is the case of the BY reduction of ethyl 4-chloroaceto-acetate that affords the (S)-hydroxy ester with modest ee, whereas the octyl ester is reduced to the enantiomerically pure (R)-hydroxy ester [40]. On the other hand, the reduction of ethyl 4-chloroacetoacetate by different microorganisms furnished the enantiomerically pure (S)-isomer or the (R)-product, that was required for the synthesis of L-carnitine [41]. Another interesting case of stereochemical control of the bioreduction by substrate modification was observed in the case of hydroxy ketones. These compounds are efficiently reduced to the corresponding diols with BY in an apparently "anti-Prelog" manner as exemplified by the bioreduction of hydroxyacetone to (R)-1,2-propanediol [10a]. It was shown that 1-benzyl hydroxyacetone was reduced with BY to the (S)-alcohol [42] and that phenacyl alcohol and its acetate were reduced to the corresponding products with opposite configuration [43] as shown in Scheme 18.

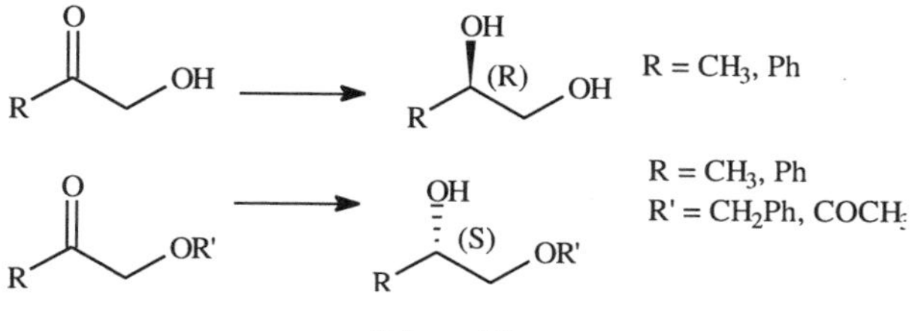

Scheme 18

When 3-phenyl-2-oxo-1-propanol, its acetate and methyl ether were subjected to BY reduction (Figure 20), (R)-diol (95% ee), and the (R)-methyl ether (72% ee) were formed. The (S)-acetate (72% ee) confirmed that only this modification reversed the stereochemical outcome of the process [44].

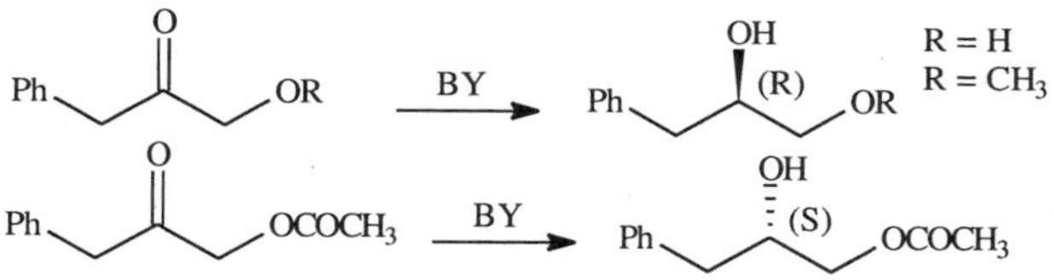

Scheme 19

Other examples of baker's yeast-mediated bioreductions of hydroxy ketones or the corresponding esters can be found in the recent literature, confirming the just-reported results and the sterochemical outcome outlined above [45].

4. Bioreduction of Sulfur Compounds

In 1874 the French chemist M. Dumas had already observed that powdered sulfur can be reduced to hydrogen sulfide by fermenting BY [46] and a few decades later the reduction of thioacetaldehyde to ethanethiol with BY has been reported [47]. Other sulfur-containing compounds undergo different type of transformations when added to fermenting BY [4]. Recently, it has been reported that the BY-mediated bioreduction of thiooxoesters and thioketones may proceed satisfactory in terms of chemical and optical yields [48] as shown in Scheme 20.

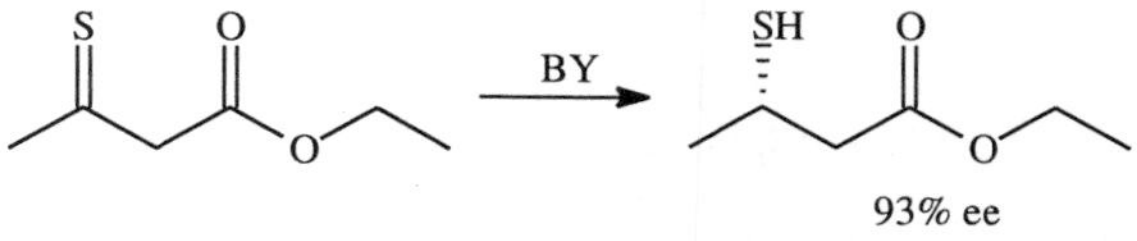

Scheme 20

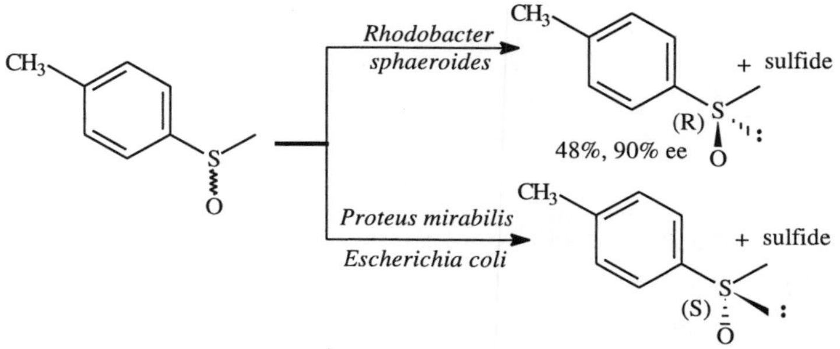

Scheme 21

Other microorganisms can perform the chemically difficult reduction of aryl sulfoxides to the corresponding sulfides and the reaction leaves enantiomerically pure unreacted sulfoxide (Scheme 21). Depending on the micrrganism used, both enantiomeric sulfoxides can be prepared, but the enantiomerically pure (S)-sulfoxide could be obtained mainly with the purified dimethyl sulfoxide reductase [49].

5. Reduction of Nitrogen Containing Groups

5.1. REDUCTION OF AZIDO ARENES

Recently, the enormous potential of BY to carry out the biotransformation of reducible nitrogen-containing functional groups has been expanded to the reduction of azidoarenes (Scheme 22) under non-fermenting conditions [50]. The reported results are even more surprising in the light of the fact that under bioreductive conditions the azido moiety of alkyl keto azides remains uneffected and only alkyl azido alcohols are formed [51].

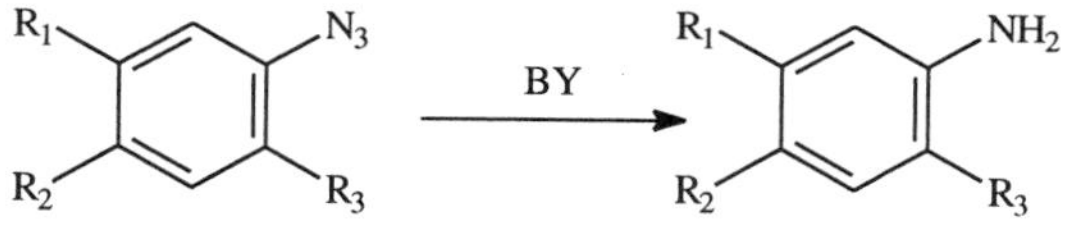

$R_1 = R_2 = R_3 = H$ $R_1 = R_3 = H; R_2 = CH_3$
$R_1 = R_3 = H; R_2 = Cl$ $R_1 = R_2 = H; R_3 = COOH$

83-92%

Scheme 22

The biocatalytic reduction of aryl azides by means of BY has been applied to azidoarenes of more complicated structure for the chemoenzymatic synthesis of pyrrolo [2,1-c][1,4]benzodiazepine antibiotics [52]. In this example the reduction product cyclized with a proximal carbomethoxy or aldehyde function to afford the heterocyclic system (Scheme 23).

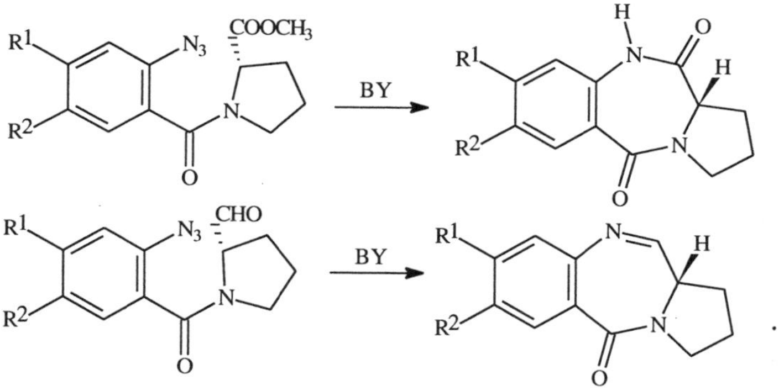

Scheme 23

It has also been reported that the BY mediated bioreduction of the azidoarene that is the key step for a stereo- and chemoselective synthesis of 4β-aminopodophyllotoxin [53].

5.2.   REDUCTION OF NITRO AND NITROSO ARENES

The capability of BY to accomplish the bioreduction of aromatic nitro groups has been reported more than eighty years ago [54], but only recently this reaction has attracted renewed attention in a study [55] dealing with the reduction of a series of substituted nitrobenzenes (Scheme 24). Interestingly, in the case of acyl nitrobenzenes the BY reduction furnished either nitro alcohols or amino ketones without giving any amino alcohol. A series of reports about the BY-mediated bioreduction of nitroarenes in basic aqueous medium and ethanol or methanol at temperature ranging from 70 to 85 °C have been critically revised [56]. The above harsh conditions destroy BY implying that simply a source of glucose remains, that is able to carry out the chemical reductions in basic conditions. On the other hand, it has been demonstrated that under classical fermenting conditions (pH 5.5-6.0, 30 °C), no reduction of nitrobenzene or azoxybenzene was observed with BY. Also a claimed BY-catalyzed 1,3-dipolar cycloaddition reaction of nitrile oxides has been revised in the light of more carefully experimental conditions and appropriate controls [57].

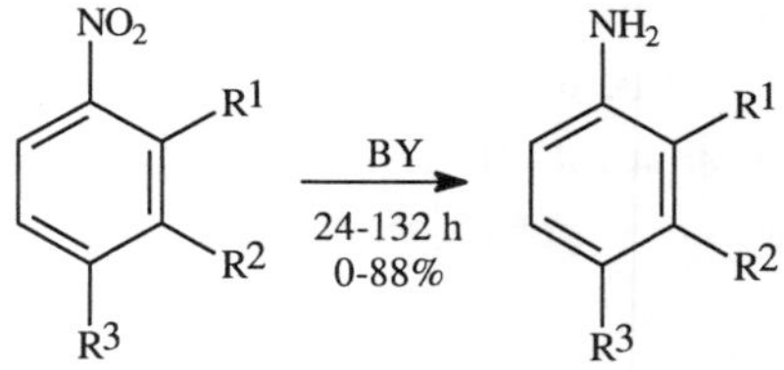

$R^1, R^2, R^3 = H, CH_3, COCH_3, COOCH_2CH_3, CF_3, CN, NH_2,$
$NHCOCH_3, NO_2, OH, OCH_3, Br$

Scheme 24

5.3. BIOREDUCTION OF TRINITROTOLUENE AS AN EXAMPLE OF BIOREMEDIATION OF EXPLOSIVES

The previously reported work on the BY-mediated bioreduction of nitroarenes [56] has been preceeded by a study on the reduction of a wide range of substituted dinitroarenes [58]. The results revealed that products were nitroanilines and that highly regioselective reactions can be carried out. The above work is representative of the extremely important topic of the biodegradation of nitroaromatic compounds, that are used as synthetic intermediates, dyes, pesticides, and finally explosives. TNT disposal and remediation has become a significant enviromental concern, due to the high production levels (1000 tons per year in the 1980s) and to the toxicity connected with the compound. Several microbial systems are capable of transforming or degrading nitroaromatic compounds and, in the case of the explosive 2,4,6-trinitrotoluene (TNT), a few are able to reduce the compound to aminotoluenes (Scheme 25), whereas others can degrade the molecule to small aliphatic acids [59].

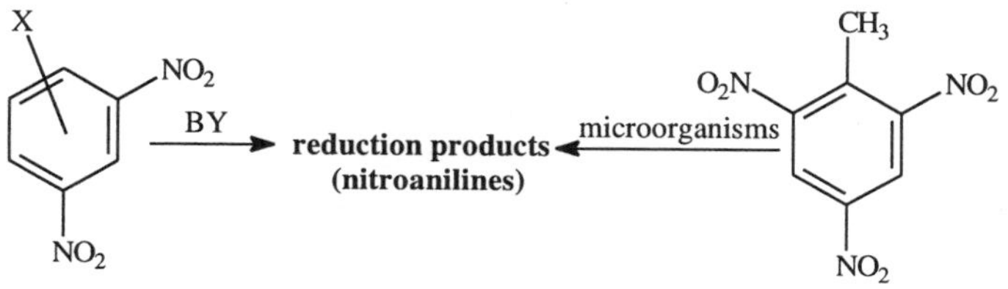

Scheme 25

6. Final Remarks

BY-mediated bioreductions are still actively investigated and many synthetic applications of the stereomerically pure products have been reported in the current literature. One of the main reason for the success of this biocatalytic approach is due to the ease to perform biotransformations and to the numerous methods available to control the stereochemical outcome of the reactions. Relatively unusual reactions such as biohydrogenations, rearrangements, acyloin condensations and a few others mediated by this friendly and easy to use microorganism offer a relatively unexplored area of investigations. A nice recent example of an unprecedented BY-catalyzed reduction-cyclization (Scheme 26) is the synthesis of substituted isoxazoles from (Z)-3-alkyl-3-nitro-2-phenylpropenenitriles [60].

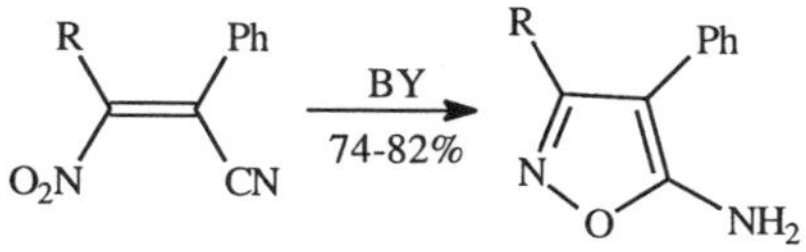

Scheme 26

7. Acknowledgements

The financial support from Consiglio Nazionale per le Ricerche (CNR, *Target Project on Biotechnology*) is gratefully acknowledged.

112

8. References

1. S. Servi, Ed. (1992) *Microbial Reagents in Organic Synthesis*, Kluwer Academic Publishers, Dordrecht.
2. Holland, H.L. (1998) Microbial trasformations, *Current Opinion in Chemical Biology* **2**, 77-84.
3. (a) Servi, S. (1990) Baker's yeast as a reagent in organic synthesis, *Synthesis* 1-25. (b) Csuk, R. and Glänzer, B.I. (1991) Baker's yeast mediated transformations in organic chemistry, *Chem. Rev.* **91**, 49-97.
4. D'Arrigo, P., Pedrocchi-Fantoni, G. and Servi, S. (1997) Old and new synthetic capacities of baker's yeast, *Adv. Appl. Microbiol.* **44**, 81-123.
5. (a) Pereira, R.S. (1998) The use of baker's yeast in the generation of asymmetric centers to produce chiral drugs and other compounds, *Crit. Rev. Biotechnol.* **18**, 25-83. (b) Kometani,T., Yoshii, H. and Matsuno, R. (1996) Large-scale production of chiral alcohols with bakers' yeast, *J. Mol. Catal. B: Enzym.* **1**, 45-52.
6. Santaniello, E., Ferraboschi, P., Grisenti, P. and Manzocchi, A. (1992) The biocatalytic approach to the preparation of enantiomerically pure chiral building blocks, *Chem. Rev.* **92**, 1071-1140.
7. Takeshita, M., Yaguchi, R. and Unuma, Y. (1995) Enzymatic synthesis of (1R, 2S)- and (1S, 2R)-2-methyl-2,3-epoxy-1-phenylpropanols, *Heterocycles* **40**, 967-974.
8. Ferraboschi, P., Casati, S. and Santaniello, E. (1994) Baker's yeast-mediated hydrogenation of 2-substituted allyl alcohols: a biocatalytic route to a new highly enantioselective synthesis of (R)-2-methyl alkanols, *Tetrahedron: Asymmetry* **5**, 19-20.
9. Ferraboschi, P., Reza Elahi, S., Verza, E., Meroni Rivolta, F. and Santaniello, E. (1996) Baker's yeast mediated biohydrogenation of 2-substituted allyl alcohols: synthesis of enantiomerically pure (2S)-3-benzyloxy-2-methyl-1-propanol, *Synlett* 1176-1178.
10. (a) Levene, P.A. and Walti, A. (1943) *l*-Propylene glycol, *Organic Syntheses Coll. Vol. II*, 545-547. (b) Seebach, D., Sutter, M.A., Weber, R.H. and Züger, M.F. (1985) Yeast reduction of ethyl acetoacetate: (S)-(+)-ethyl 3-hydroxybutanoate, *Organic Syntheses* **63**, 1-9. (c) Mori, K. and Mori, H. (1990) Yeast reduction of 2,2-dimethylcyclohexane-1,3-dione: (S)- (+)-3-hydroxy-2,2-dimethylcyclohexanone, *ibid.* **68**, 56-63.
11. Prelog, V. (1964) Specification of the stereospecificity of some oxidoreductases by diamond lattice sections, *Pure Appl. Chem.* **9**, 119-130.
12. Ward, O.P. and Young, C.S. (1990) Reductive biotransformations of organic compounds by cells or enzymes of yeast, *Enzyme Microb. Technol.* **12**, 482-493.
13. Nakamura, K. (1992) Stereochemical control in microbial reduction, in S. Servi (ed.); *Microbial Reagents in Organic Synthesis*, Kluwer Academic Publishers, Dordrecht, pp.388-398.
14. Nakamura, K., Ushio, K., Oka, S. and Ohno, A. (1984) Stereochemical control in yeast reduction, *Tetrahedron Lett.* **25**, 3979-3982.
15. Neuberg, C. and Nord, F.F. (1919) Die phytochemische reduktion der ketone. Biochemische darstellung optisch-aktiver sekundärer alkohole, *Ber.* **52**, 2237-2248.
16. Fogagnolo, M., Giovannini, P.P., Guerrini, A., Medici, A., Pedrini, P. and Colombi, N. (1998) Homochiral (*R*)- and (*S*)-1-heteroaryl- and 1-aryl-2-propanols via microbial redox, *Tetrahedron: Asymmetry* **9**, 2317-2327.
17. Fantin, G., Fogagnolo, M., Giovannini, P.P., Medici, A., Pedrini, P., Gardini, F. and Lanciotti, R. (1996) Anti-Prelog microbial reduction of prochiral carbonyl compounds, *Tetrahedron* **52**, 3547-3552.

18. Rykowsky, A., Lipinska, T., Guzik, E., Adamiuk, M. and Olender, E. (1997) 1,2,4 - Triazines in organic synthesis. 6. Enantioselective reduction of 5-acyl-1,2,4-triazines and their oximes by baker's yeast, *Pol. J. Chem.* **71**, 69-76.

19. Takemoto, M. and Achiwa, K. (1994) Synthesis of optically active α-phenylpyridylmethanols with baker's yeast, *Chem. Pharm. Bull.* **42**, 802-805.

20. Huber, P., Bratovanov, S., Bienz, S., Syldatk, C. and Pietzsch, M. (1996) Chiral silicon groups as auxiliaries for enantioselective synthesis: access to optically active silanes by biotransformation and the enantiospecific preparation of (*R*)-(+)-1-phenylethanol, *Tetrahedron: Asymmetry* **7**, 69-78.

21. Maguire, A.R. and Lowney, D.G. (1997) Asymmetric reduction of 1-methylsulfonylalkan-2-ones with baker's yeast, *J. Chem. Soc., Perkin I*, 235-238.

22. Dao, D.H., Okamura, M., Akasaka, T., Kawai, Y., Hida, K. and Ohno, A. (1998) Stereochemical control in microbial reduction. Part 31: reduction of alkyl 2-oxo-4-arylbutyrates by baker's yeast under selected reaction conditions, *Tetrahedron: Asymmetry* **9**, 2725-2737.

23. Eh, M. and Kalesse, M. (1995) Remarkable kinetic resolution of chiral β-keto esters by baker's yeast reduction, *Synlett*, 837-838.

24. Quirós, M., Rebolledo, F., Liz, R. and Gotor, V. (1997) Enantioselective reduction of β-keto amides by the fungus *Mortierella isabellina*, *Tetrahedron: Asymmetry* **8**, 3035-3038.

25. Quirós, M., Rebolledo, F. and Gotor, V. (1999) Bioreduction of 2-oxocyclopentanecarboxamides: syntheses of optically active 2-aminomethyl- and 2-aminocyclopentanols, *Tetrahedron: Asymmetry* **10**, 473-486.

26. Ferraboschi, P., Santaniello, E., Tingoli, M., Aragozzini, F. and Molinari, F. (1993) Microbial reduction of 2-keto acetals as a biocatalytic approach to the enantioselective synthesis of optically active 2-hydroxy acetals, *Tetrahedron: Asymmetry* **4**, 1931-1940.

27. Tanikaga, R., Obata, Y. and Kawamoto, K.-i. (1997) Baker's yeast mediated reduction of cyclohexanones containing a nitro or a sulfonyl group at C-3, *Tetrahedron: Asymmetry* **8**, 3101-3106.

28. Danchet, S., Bigot, C., Buisson, D. and Azerad, R. (1997) Dynamic kinetic resolution in the microbial reduction of α-monosubstituted β-oxoesters: the reduction of 2-carbethoxy-cycloheptanone and 2-carbethoxy-cyclooctanone, *Tetrahedron: Asymmetry* **8**, 1735-1739.

29. Fantin, G., Fogagnolo, M., Medici, A., Pedrini, P., Marotta, E., Monti, M and Righi, P. (1996) Microbial reduction of methyl-substituted bicyclo[3.2.0]hept-3-en-6-ones : a screening to homochiral *endo*- and *exo*-alcohols, *Tetrahedron: Asymmetry* **7**, 277-282.

30. Satoh, K., Imura, A., Miyadera, A., Kanai, K. and Yukimoto, Y. (1998) An efficient synthesis of a key intermediate of DU-6859a *via* asymmetric microbial reduction, *Chem. Pharm. Bull.* **46**, 587-590.

31. Kometani, T., Sakai, Y., Matsumae, H., Shibatani, T. and Matsuno, R. (1997) Production of (2*S*,3*S*)-2,3-dihydro-3-hydroxy-2-(4-methoxyphenyl)-1,5-benzothiazepin-4(5*H*)-one, a key intermediate for Diltiazem synthesis, by bakers' yeast-mediated reduction, *J. Ferment. Bioeng.* **84**, 195-199.

32. Das, B., Madhusudhan, P. and Kashinatham, A. (1998) The first conversion of camptothecin to (*S*)-mappicine by an efficient chemoenzymatic method, *Bioorg. Med. Chem. Lett.* **8**, 1403-1406.

33. Boutoute, P., Mousset, G. and Veschambre, H. (1998) Regioselective or enantiogenic electrochemical and microbial reductions of 1,2-diketones, *New J. Chem.*, 247-251.

34. Nakamura, K.,Kondo, S.-i., Kawai, Y., Hida, K., Kitano, K. and Ohno, A. (1996) Enantio- and regioselective- reduction of α-diketones by baker's yeast, *Tetrahedron: Asymmetry* **7**, 409-412.

35. Crocq, V., Masson, C., Winter,J., Richard, C., Lemaitre, G., Lenay, J., Vivat, M., Buendia, J, and Prat, D. (1997) Synthesis of Trimegestone: the first industrial application of bakers' yeast mediated reduction of a ketone, *Org. Process. Res. Dev.* **1**, 2-13.

36. (a) Fauve, A., Veschambre, H. (1988) Microbiological reduction of acyclic β-diketones, *J. Org. Chem.* **53**, 5215-5219. (b) Matsumura, S., Kawai, Y., Takahashi, Y. and Toshima, K. (1994) Microbial production of (2R,4R)-2,4-pentanediol by enantioselective reduction of acetylacetone and stereoinversion of 2,4-pentanediol, *Biotechnol. Lett.* **16**, 485-490.

37. Ikeda, H., Sato, E., Sugai, T. and Ohta, H. (1996) Yeast-mediated synthesis of optically-active diols with C-2-symmetry and (R)-4-pentanolide, *Tetrahedron* **52**, 8113-8127.

38. Otten, S., Fröhlich, R. and Haufe, G. (1998) Synthesis and structural characterization of enantiopure (2R,5R)-(+)-2,5-dimethylthiolane, *Tetrahedron: Asymmetry* **9**, 189-191.

39. Uchiyama, M., Katoh, N., Mimura, R., Yokota, N., Shimogaichi, Y., Shimazaki, M. and Ohta, A. (1997) Highly enantioselective reduction of symmetrical diacetylaromatics with baker's yeast, *Tetrahedron: Asymmetry* **8**, 3467-3474.

40. Zhou, B.-n., Gopalan, A.S., VanMiddlesworth, F., Shieh, W.-R. and Sih, C.J. (1983) Stereochemical control of yeast reductions. 1. Asymmetric synthesis of L-carnitine, *J. Am. Chem. Soc.* **105**, 5925-5926.

41. Aragozzini, F., Valenti, M., Santaniello, E., Ferraboschi, P. and Grisenti, P. (1992) Biocatalytic enantioselective preparations of (R)- and (S)- ethyl 4-chloro-3-hydroxybutanoate, a useful chiral synthon, *Biocatalysis* **5**, 325-332.

42. Manzocchi, A., Fiecchi, A. and Santaniello, E. (1987) Stereochemically controlled bakers'-yeast-mediated reductions: synthesis of (S)-(+)-1,2-propanediol and (S)-(−)-1,3-butanediol, 1-benzyl ethers, *Synthesis*, 1007-1009.

43. Manzocchi, A., Fiecchi, A. and Santaniello, E. (1988) Stereochemical control of bakers' yeast mediated reduction of a protected 2-hydroxy ketone, *J. Org. Chem.* **53**, 4405-4407.

44. Ferraboschi, P., Grisenti, P., Manzocchi, A. and Santaniello, E. (1994) Baker's yeast-mediated reduction of α-hydroxy ketones and derivatives: the steric course of the biotransformation, *Tetrahedron* **35**, 10539-10548.

45. (a) Utaka, M., Ito, H., Mizumoto, T. and Tsuboi, S. (1995) Regio- and enantioselective synthesis of (S)-1-acetoxy-2-hydroxy-4-alkanones by use of bakers' yeast reduction of 1-acetoxy-2,4-alkanediones, *Tetrahedron: Asymmetry* **6**, 685-686. (b) Egri, G., Kolbert, A., Bálint, J., Fogassy, E., Novák, L. and Poppe, L. (1998) Baker's yeast mediated stereoselective biotransformation of 1-acexy-3-aryloxypropan-2-ones, *ibid.* **9**, 271-283.

46. Dumas, M. (1874) Recherches sur la fermentation alcoolique *Ann. Chim. Phys.* **3**, 59-108.

47. Neuberg, C. and Nord, F. F. (1914) Phytochemische Bildung von Äthylmercaptan, *Chem. Ber.* **47**, 2264-2271.

48. Nielsen, J. K. and Madsen, J. Ø. (1994) Stereoselective reduction of thiocarbonyl compounds with baker's yeast, *Tetrahedron: Asymmetry* **5**, 403-410.

49. (a) Abo, M., Okubo, A. and Yamazaki, S. (1997) Preparative asymmetric deoxygenation of alkyl aryl sulfoxides by *Rhodobacter sphaeroides* f.sp. *denitrificans*, *Tetrahedron: Asymmetry* **8**, 345-348. (b) Hanlon, S. P., Graham, D. L., Hogan, P. J., Holt, R. A., Reeve, C. V., Shaw, A. L., McEwan, A. G. (1998) Asymmetric reduction of racemic sulfoxides by dimethyl sulfoxide reductases from *Rhodobacter capsulatus*, *Escherichia coli* and *Proteus* species, *Microbiology* **144**, 2247-2253.

50. Baruah, M., Boruah, A., Prajapati, D. and Sandhu, J.S. (1996) Bakers' yeast mediated chemoselective reduction of azidoarenes, *Synlett* 1193-1194.

51. Besse, P., Veschambre, H., Chênevert, R., Dickman, M. (1994) Chemoenzymatic synthesis of chiral β-azidoalcohols. Application to the preparation of chiral aziridines and aminoalcohols, *Tetrahedron: Asymmetry* **5**, 1727-1744.

52. Kamal, A., Damayanthi, Y., Reddy, B.S.N., Lakminarayana, B. and Reddy, B.S.P. (1997) Novel biocatalytic reduction of aryl azides: chemoenzymatic synthesis of pyrrolo[2,1-c][1,4]benzodiazepine antibiotics, *Chem. Commun.*, 1015-1516.

53. Kamal, A., Laxminarayana, B. and Gayarti,N.L. (1997) Stereo and chemoselective enzymatic reduction of azido functionality: synthesis of 4-β-aminopodophyllotoxin congeners by baker's yeast, *Tetrahedron Lett.* **38**, 6871-6874.

54. Neuberg, C. and Welde, E. (1914) Phytochemical reductions. II. Transformation of aliphatic nitro compounds into amino compounds, *Biochem. Z.* **60**, 470-476.

55. Takeshita, M., Yoshida, S., Kiya, R., Higuchi, N. and Kobayashi, Y. (1989) Reduction of aromatic nitro compounds with baker's yeast, *Chem. Pharm. Bull.* **37**, 615-617.

56. Blackie, J.A., Turner, N.J. and Wells, A.S. (1997) Concerning the baker's yeast (*Saccharomyces cerevisiae*) mediated reduction of nitroarenes and other N-O containing functional groups, *Tetrahedron Lett.* **38**, 3043-3046

57. Easton, C.J., Huges, M.M., Tiekink, E.R.T., Savage, G.P. and Simpson, G.W. (1995) Aryl nitrile oxide cycloaddition reactions in the presence of baker's yeast and β-cyclodextrin, *Tetrahedron Lett.* **36**, 629-632.

58. Davey, C.L., Powell, L.W., Turner, N.J. and Wells, A. (1994) Regioselective reduction of substituted dinitroarenes using baker's yeast, *Tetrahedron Lett.* **35**, 7867-7870.

59. (a) Spain, J.C. (1995) Biodegradation of nitroaromatic compounds, *Ann. Rev. Microbiol.* **49**, 523-555. (b) Kalafut, T., Wales, M.E., Raspogi, V.K., Naumova, R.P., Zaripova, S.K. and Wild, J.R. (1998) Biotranformation patterns of 2,4,6-trinitrotoluene by aerobic bacteria, *Curr. Microbiol.* **36**, 45-54.

60. Navarro-Ocaña, A., Jiménez-Estrada, M., González-Paredes, M.B. and Bárzana, E. (1996) Synthesis of substituted isoxazoles from (Z)-3-alkyl-3-nitro-2-phenylpropenenitriles using baker's yeast, *Synlett* 695-696.

ENZYMATIC AMINOLYSIS AND AMMONOLYSIS REACTIONS

VICENTE GOTOR

*Departamento de Química Orgánica e Inorgánica, Facultad de Química,
Universidad de Oviedo, Avenida Julián de Clavería, 8; 33006-Oviedo,
Spain*

Abstract: Aminolysis and ammonolyis reactions of esters catalysed by enzymes, especially lipases, are very efficient methods for the preparation of amides in an enantio- and chemoselective manner. In addition, this methodology is very useful for the resolution of amines and esters of high added value, and for the chemoenzymatic synthesis of products of biological and industrial interest. The enzymatic alkoxycarbonylation of racemic amines yields optically active carbamates.

Keywords: Enzymes, lipases, amines, esters, enantioselective, chemoselective, amides, carbamates, aminolysis, acylation, alkoxycarbonylation, resolution.

1. Introduction

Nowadays, for many organic chemists, biotransformations is an area of great interest for the preparation of products which are difficult to obtain by conventional chemical methods. Enzymes, especially lipases, have been widely used for the preparation of optically active esters, lactones, carboxylic acids and alcohols through the corresponding asymmetric transesterification and esterificacion reactions [1]. Over the last few years, the enzymatic acylation of nitrogen organic compounds has emerged as a useful process in organic synthesis [2]. Similarly, the enzymatic alkoxycarbonylation of alcohols and amines can be of utility for the preparation of chiral carbonates and carbamates. Recently, some examples of enzymatic alkoxycarbonylation reactions have also been reported [3].

B. Zwanenburg et al. (eds.), Enzymes in Action, 117–132.
© 2000 *Kluwer Academic Publishers. Printed in the Netherlands.*

It is now well established that the lipase-catalyzed aminolysis and ammonolysis of esters is an useful reaction in organic synthesis. This method is a complement or alternative to other more usual syntheses such as the enzymatic hydrolysis, esterification and transesterification processes. This methodology has the advantage of its irreversibility because of the low protease activity of the lipase. Moreover, the formation of an amide bond is always an operation of synthetic utility due, in some cases, to the difficulty in obtaining these products under mild conditions or from difunctional starting compounds.

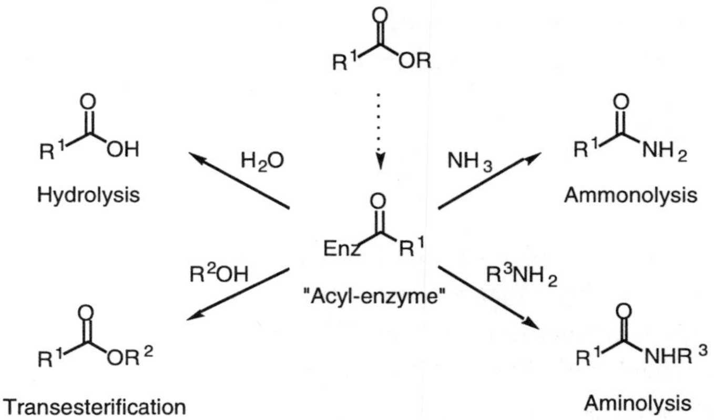

Scheme 1

In this chapter, we describe the utility of hydrolases (mainly lipases) in enzymatic aminolysis and ammonolysis reactions. Thus, in the main part of this review we discuss the possibilities of these processes for the preparation of amides and the resolution of esters and amines, as well as certain other of their applications. The second part will deal briefly with the alkoxycarbonylation of amines for the synthesis of carbamates.

2. Results and discussion

2.1. ENZYMATIC ACYLATION OF AMINES AND AMMONIA

2.1.1. Enzymatic Preparation of Amides

For an enzymatic transesterification reaction, one of the most important variables for a successful process is the choice of an adequate acyl donor. In the last few years, several strategies have been developed to avoid the reversibility of these reactions so as to achieve a more effective process. Activated esters [4], oxime esters [5], vinyl esters [6], anhydrides [7], thioesters [8], and recently 1-ethoxyvinyl esters [9], have been described as good acyl donors for enzymatic transesterification. However, the enzymatic acylation of amines is not possible with these reagents because the chemical reaction competes with the enzymatic process due to the higher nucleophilicity of amines compared with alcohols. For this reason, in many cases non-activated esters can be used as acyl agents in enzymatic aminolysis and ammonolysis reactions.

Lipases are the most efficient hydrolases to catalyse the acylation of amines and ammonia, and only in some cases proteases have been used for the resolution of certain primary amines [10]. Several lipases have proven their usefulness for the preparation of amides from non-activated esters [11], and the enzymatic amidation reaction is an efficient procedure for the synthesis of fatty acid amides [12]. From the data reported in the literature, the lipase of *Candida antarctica* (CAL) is the most efficient biocatalyst for the enzymatic aminolysis and ammonolysis processes.

Acrylic or propargylic esters are less reactive towards nucleophiles, such as amines or ammonia, than the corresponding saturated ester. Moreover, the normal reaction is the formation of the corresponding Michael adducts. We have reported the chemoselective formation of propargyl and acrylic amides with amines and ammonia [13]. Thus, *N*-unsubstituted acrylic and propiolic amides are obtained in high yields by reaction of the corresponding α,β-unsaturated esters with ammonia in the presence of CAL.

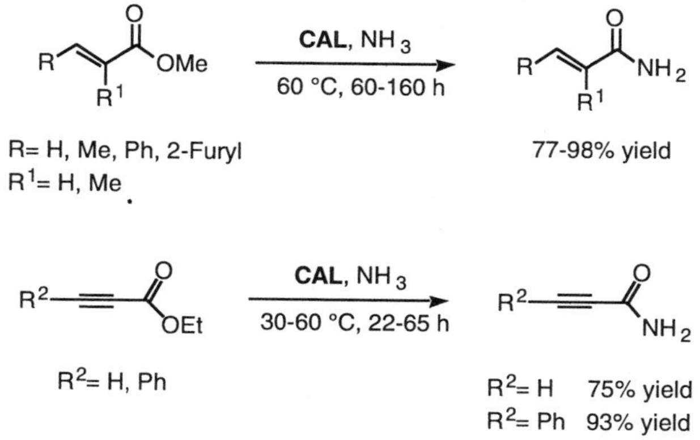

R= H, Me, Ph, 2-Furyl
R^1= H, Me

77-98% yield

R^2= H, Ph

R^2= H   75% yield
R^2= Ph   93% yield

Scheme 2

Also of interest is the enzymatic reaction of diethyl maleate and fumarate with ammonia and amines. When ethyl fumarate is used in the presence of CAL, the corresponding *trans*-amido ester is formed in good yields. However, the maleate gives the same amido ester as the fumarate. The formation of these compounds can be rationalized by a Michael/retro-Michael type isomerization of diethyl maleate to fumarate, preceeding the enzymatic reaction. In consequence, maleates are not adequate substrates for the enzyme [14].

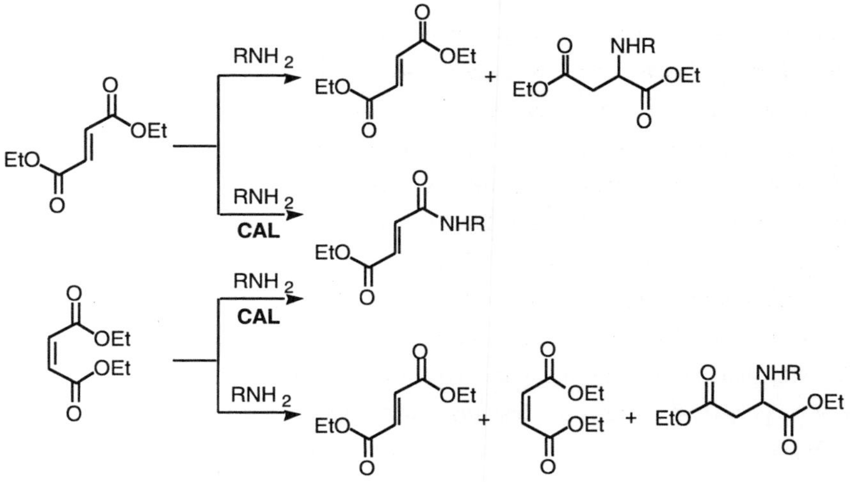

Scheme 3

Other interesting processes with difunctional esters are, for example, the formation of α-hydroxy amides from the corresponding benzyl esters using *Pseudomonas cepacia* lipase (PSL) as the biocatalyst [15], and the preparation of chiral amides from 2-chloro-propionic esters catalyzed by *Candida cylindracea* lipase (CCL) [16]. Of special relevance is the preparation of β-hydroxy amides [17] and β-keto amides [18] from 3-hydroxy and 3-keto esters. Both processes are catalyzed by CAL and in the case of 3-hydroxy amides the (*R*) and (*S*) enantiomers can be prepared in very high enantiomeric excesses. This reaction is of utility for the resolution of racemic ethyl 3-hydroxybutyrate and for the preparation of chiral 1,3-aminoalcohols.

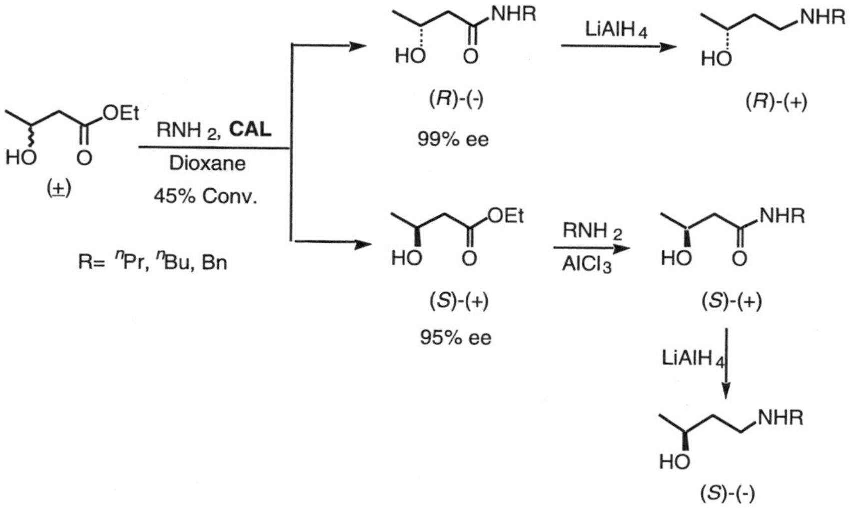

Scheme 4

The reaction of β-keto esters with ammonia and amines in the presence of CAL produces the corresponding 3-oxo amides in excellent yields. These compounds are difficult to obtain by other methods because in the absence of the biocatalyst the formation of enamino esters takes place. Conversely, if racemic amines are used as starting materials, the enzyme has preference towards the (R) isomer of the amine, and β-keto amides with (R) configuration can be prepared in high enantiomeric excesses.

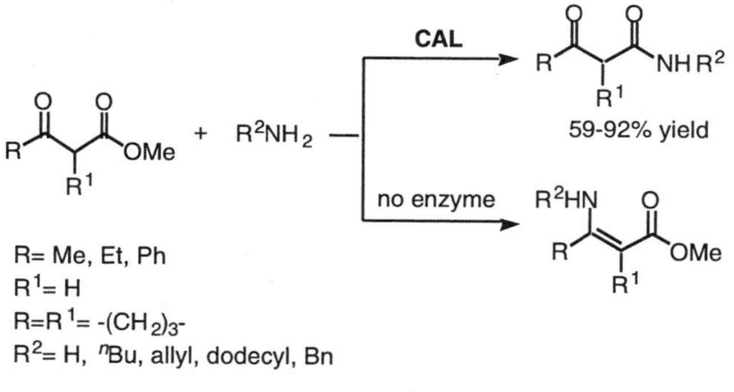

Scheme 5

2.1.2. Resolution of Esters

As we have already mentioned, the enzymatic aminolysis and ammonolysis reaction can be an alternative to the hydrolysis or transesterification reaction for the resolution of esters. For example, the ammonolysis of ibuprofen 2-chloroethyl ester is more enantioselective than the corresponding hydrolysis [19]. Recently, Zwanenburg et al.

have reported an enzymatic resolution of *N*-alkylated azetidine esters by CAL mediated ammonolysis [20].

The kinetic resolution of racemic ethyl 4-chloro-3-hydroxybutanoate has been carried out by an enzymatic ammonolysis reaction using again CAL as biocatalyst [21]. The reaction rate changes can be explained in terms of the solubility of ammonia in the different solvents used in this process. Thus, good reaction rates were attained with hydrogen-bond acceptor solvents, although in some of them the enantioselectivity is low. The highest enantiomeric ratio (E= 44, c= 48%) was achieved when 1,4-dioxane was used as the solvent. In all cases the enzyme reacts faster with the (*S*) isomer of the ester. The obtained (*S*)-amide is a precursor for the preparation of compounds of medicinal interest.

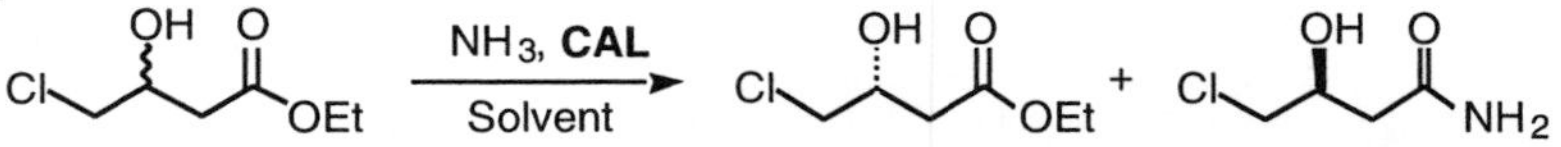

Solvent	t (h)	c (%)	ee (S)	ee (P)	E
1,4-Dioxane	4.75	48	82	89	44
THF	3.5	30	40	92	36
^{t}BuOH	0.75	51	79	75	17
CH$_2$Cl$_2$	4.75	13	14	95	45

Scheme 6

β-Amino esters are attractive and versatile substrates because they are able to behave as nucleophiles (amino group) as well as acyl donors (ester function) in aminolysis processes. We have carried out the resolution of racemic ethyl 3-aminobutyrate by acylation and amidation with CAL as biocatalyst [22]. However, the best results are achieved by acetylation of the ester and hydrolysis of their Cbz-derivative.

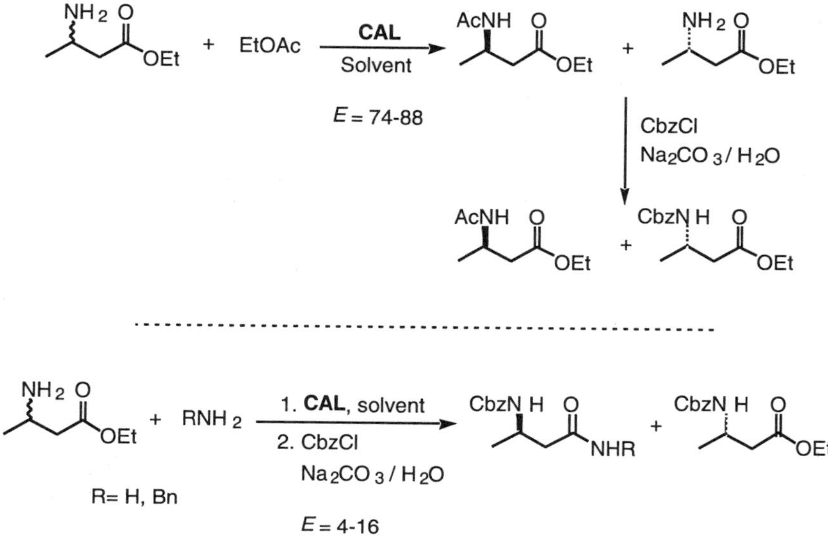

Scheme 7

We have reported the first example of the asymmetrization of prochiral esters by an aminolysis and ammonolysis reaction. The reaction of dimethyl 3-hydroxyglutarate with ammonia or amines and CAL as the biocatalyst gives the corresponding enantiomerically pure amido esters. In all solvents tested, the nucleophile reacts with the pro-(R) ester group, and the (S) amido ester is always obtained [29].

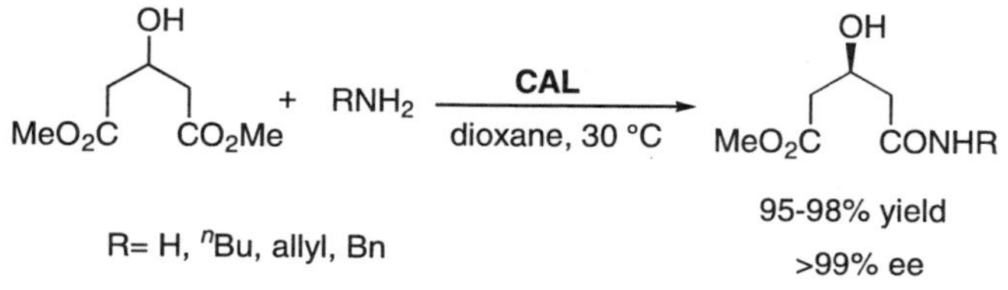

Scheme 8

2.1.3. Resolution of Amines

Proteases have scarcely been used for the resolution of amines. However, lipases themselves have shown to be more versatile for this type of process, and in the last few years, several groups have reported the resolution of different primary amines by means of lipases. Of the lipases tested, CAL again is the most active to obtain a good resolution. Of interest is the dynamic kinetic resolution of phenylethylamine in the presence of Pd, which promotes the racemization of the remaining isomer of the amine

[24]. The resolution of primary amines and their corresponding alcohols and thiols has also been investigated [25]. In this report, it is possible to note the different behaviour of these three classes of compounds in the enzymatic acylation reaction. This methodology is of great utility for the resolution of certain amines of limited stability, such as hetero-arylamines. Recently, we have demonstrated a very simple procedure for the resolution of 1-(heteroaryl)-ethylamines using ethyl acetate as solvent and acyl donor [26].

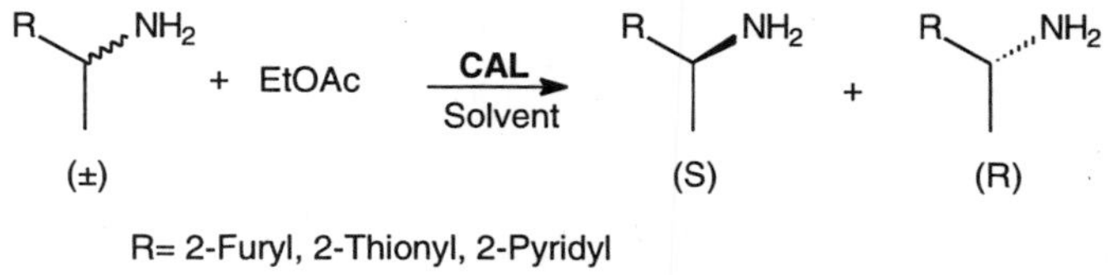

R= 2-Furyl, 2-Thionyl, 2-Pyridyl

Scheme 9

The enzymatic acylation of secondary amines is a process that presents more difficulty than the acylation of primary amines, and accordingly only a few examples have been described in the literature. The best results are obtained with cyclic secondary amines [27]. Surprisingly, the kinetic resolution of propanolol has been described by a lipase-catalysed N-acylation of a secondary amino group. The best resolution takes place with isopropenyl acetate in ether and CCL as catalyst [28].

Racemic compounds possessing more than one functional group can also undergo kinetic resolution by coupling of two sequential enzyme-catalysed processes. In this manner material can be obtained with acceptable enantiomeric purity. We have used this strategy for the resolution of trans-cyclohexane-1,2-diamine using dimethyl malonate as acyl donor and CAL as the catalyst [29]. Enantiopure bisamido ester is obtainable when equivalent amounts of diamine and diester, and a reaction time of 7 h, are employed to effect the resolution (Scheme 10). Under these conditions, the diamine is obtained with 83% ee. However, if a 1,2 equivalent of diester is used, the unreacted diamine can be obtained enaniomerically pure and the bisamido ester with 97% ee [29].

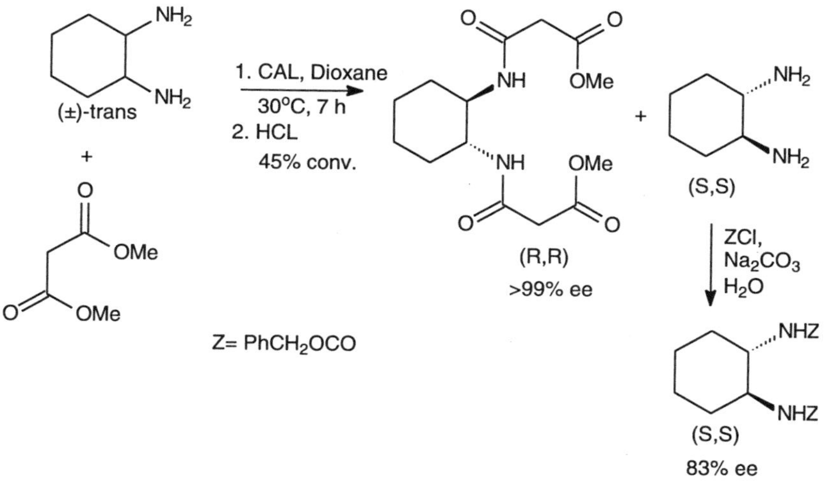

Scheme 10

Some examples of double resolution catalysed by lipases have been reported with moderate to low diasteroselectivities and conversions.

We have investigated a double enantioselective enzymatic aminolysis reaction of racemic and prochiral 3-hydroxy esters with racemic amines [30]. A CAL catalysed reaction between ethyl (±)–4-chloro-3-hydroxybutanoate and racemic amines has been carried out in dioxane at 30 °C. We have tested this reaction with three different amines: phenylethylamine, 2-furylethylamine and 2-heptylamine. In general, very good diasteroselectivities and enantioselectivies are obtained, although they depend on the starting amine. In all cases, the lipase is selective towards the (S) isomer of the ester and the (R) isomer of the amine.

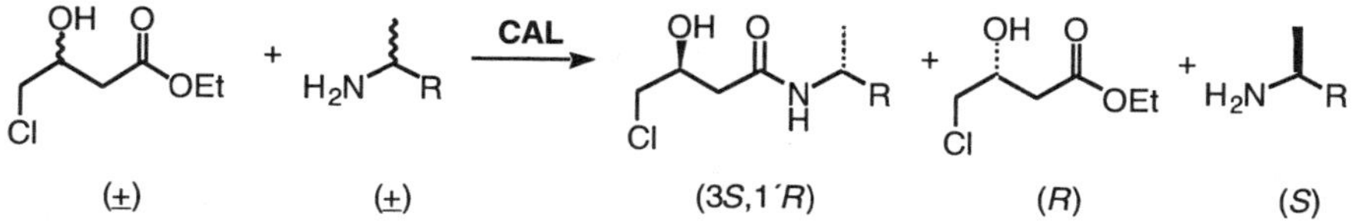

Scheme 11

The asymmetrization-resolution reaction of diethyl 3-hydroxyglutarate and phenyl-ethylamine or 2-furylethylamine takes place under similar conditions as employed for 4-chloro-3-hydroxybutanoate. In this case a 2:1 molar ratio ester/amine is used to allow the 100% conversion of the ester. CAL is totally enantioselective towards the (R) isomer of the amines, and the correponding (3S,1R') amido ester is obtained with very high diastereoselectivity and enantioselectivity.

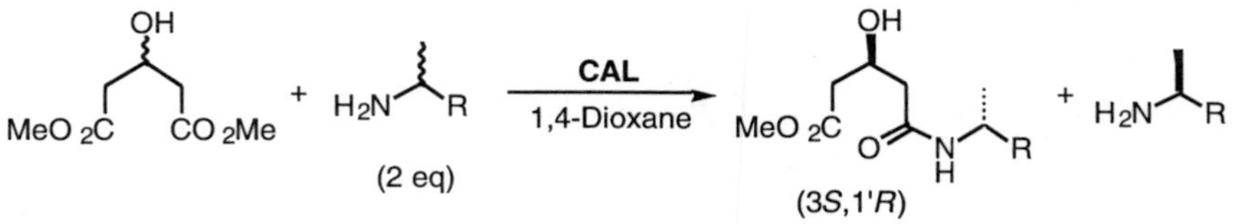

126

			Stereoisomeric Ratio of Amide (%)			
R	c (%)	ee$_{amine}$ (%)	(3S,1'R)	(3R,1'S)	(3R,1'R)	(3S,1'S)
Ph	86	75	98	---	2	---
2-Furyl	85	72	96	---	4	---

Scheme 12

2.1.4. Applications

Three examples will be described which demonstrate the utility of the enzymatic acylation of amines and ammonia.

Synthesis of Polyamines and Macrocycles. We have used the amido esters obtained in the enzymatic aminolysis and ammonolysis of diesters for the preparation of different types of diamines. Treatment of the amido esters obtained by enzymatic aminolysis with ammonia in methanol, and subsequent reduction of the resulting tetramide with BH$_3$· THF, afford the corresponding tetramines in very high yields [32]. For example, spermine can be prepared from the amido ester which is obtained by enzymatic aminolysis of dimethyl maleate and 1,4-butanediamine. These resulting polyamines are excellent starting materials for the preparation of different kinds of chiral macrocycles, compounds which are not previously described (Scheme 13).

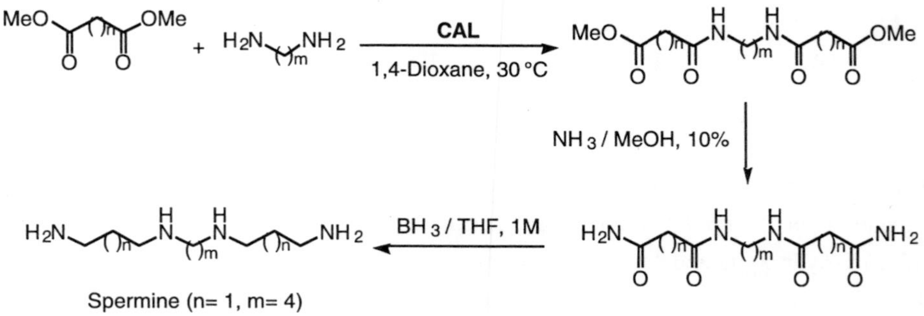

Scheme 13

The chiral amido esters and *trans*-1,2-cyclohexanediamine are good starting materials for the preparation of optically active tetraazamacrocycles and hexaazamacrocycles with C_2 symmetry. Such compounds have received much attention in recent years as chiral ligands in asymmetric synthesis, and as organic receptor molecules capable of binding anionic species. Following the Richman-Akins procedure, we have prepared two optically active hexaazamacrocycles [33], as well as tetraazamacrocycles for a C_2 symmetrical cyclam derivative (Scheme 14) [34].

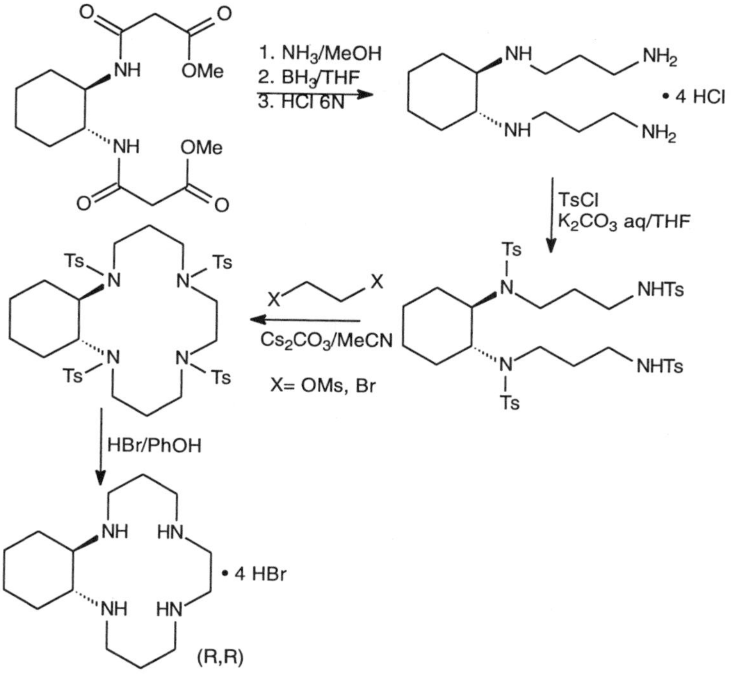

Scheme 14

Preparation of (R)-GABOB. The synthetic utility of the amido esters obtained in the asymmetrization of diethyl 3-hydroxyglutarate has been demonstrated in the preparation of biologically active (R)-4-amino-3-hydroxybutanoic acid, (R)-GABOB. The compound prepared by the corresponding enzymatic ammonolysis reaction, is acetylated and subjected to the Hoffman rearrangement with $Hg(OAc)_2$, NBS, and BnOH to give the corresponding benzylcarbamate. By subsequent hydrolysis and hydrogenolysis, this then results in (R)-GABOB (Scheme 15) [23].

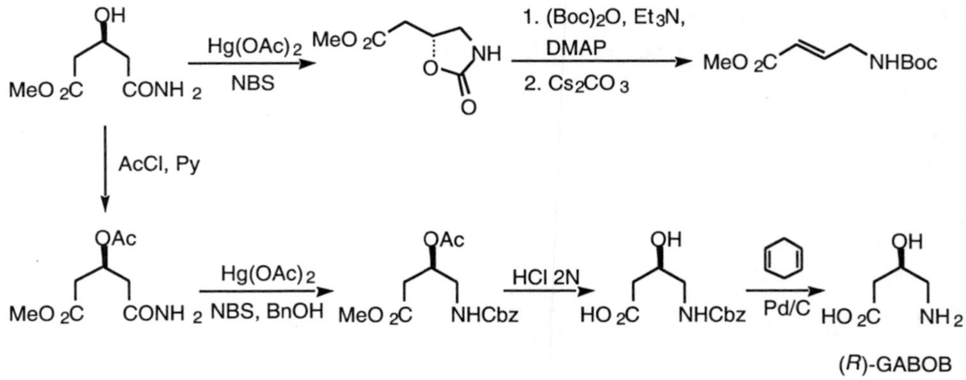

Scheme 15

Synthesis of Pyrrolidine Derivatives. The enantiopure 4-chloro-3-hydroxy amides prepared by the double enzymatic aminolysis resolution mentioned above, are very good starting materials for the synthesis of optically active 3-hydroxy and 2-oxopyrrolidines, both of which containing two stereogenic centers. Treatment of these amides (3S,1'R) with BH$_3$· THF gives the corresponding pyrrolidinol (3S,1'R) in very high yields, and high purity in most cases. No racemization takes place in this reaction. For the conversion of the starting amides into the corresponding γ-lactams, the hydroxyl group had to be protected in order to avoid side reactions. The cyclisation of silylated amides was performed in THF using NaH as the base (Scheme 16).

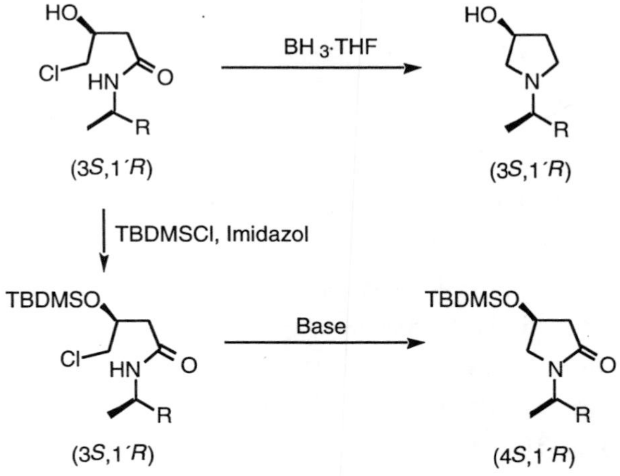

Scheme 16

2.2. ENZYMATIC ALKOXYCARBONYLATION OF AMINES

The alkoxycarbonylation of amines yields carbamates, which are of a great value in some fields, especially in medicinal chemistry. Although the acylation of alcohols and amines is now well established, a few examples of enzymatic carbonation reactions have been reported. We have investigated the alkoxycarbonation of amines because this process also permits the protection of the amino group under mild conditions.

To carry out the carbonylation of amines, several alkoxycarbonylation agents have been tested. Alkyl carbonates are much less reactive than the corresponding esters in enzymatic processes, although vinyl alkyl carbonates prove to be very useful alkoxy-carbonylation agents for alcohols and amines (Scheme 17).

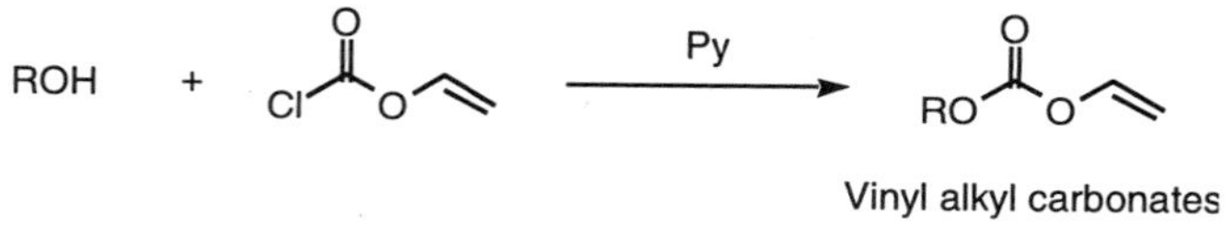

Vinyl alkyl carbonates

Scheme 17

CAL is again the most active lipase for this kind of reaction [34]. Chiral carbamates can be prepared by the enzymatic alkoxycarbonylation of vinyl carbonates and racemic amines [35]. The enantioselectivity achieved depends on the nature of the substrate and the solvent. In all cases CAL has preference for the (R) isomer. When employing racemic vinyl alkyl carbonates and primary amines, the corresponding (S)-carbamates are prepared with higher enantioselectivity in shorter reaction times (Scheme 18) [36].

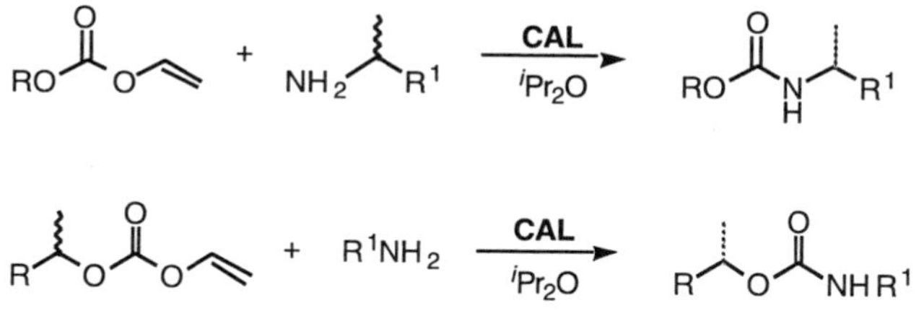

Scheme 18

The double enantioselective lipase-catalysed alkoxycarbonylation of racemic amines with racemic vinyl carbonates has been carried out using CAL, to yield carbamates with two stereogenic centers (Scheme 19) [37].

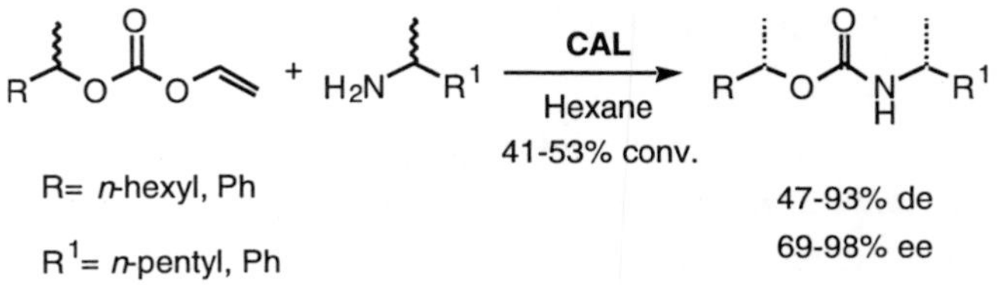

130

R= n-hexyl, Ph

R^1= n-pentyl, Ph

47-93% de

69-98% ee

Scheme 19

3. Conclusions

In view of the large number of enzymes yet to be evaluated for their utility in synthesis, and the possibilities that new methods in biotechnology are continuously offering, biotransformations will enable chemists to discover new ways for the preparation of compounds of relevance, which are difficult to obtain by chemical procedures. The results described in this chapter concerning the enzymatic aminolysis and ammonolysis reactions, substantiate the enormous synthetic potential predicted by different groups over the last decade about the use of enzymes in non-aqueous media.

4. Acknowledgements

I gratefully acknowledge the work of my co-workers, which forms the basis of this chapter. Financial support has been provided by CICYT (projets BIO95-0687 and BIO98-0770), and Universidad de Oviedo.

5. References

1.   Faber, K. (1997) *Biotransformation in Organic Chemistry* 3rd ed.; Springer-Verlag: Berlin.

2.   Gotor, V. (1992) Enzymatic Aminolysis, Hydrazinolysis and Oximolysis Reactions, *Microbial Reagents in Organic Synthesis*; in Servi, S. Ed.; Kluver Academic.Publishers, Dordrecht, pp. 199-208.

3.   Ferrero, M., Fernández, S., and Gotor, V. (1997) Selective Alkoxycarbonylation of A-Ring Precursors of Vitamin D Using Enzymes in Organic Solvents. Chemoenzymatic Synthesis of 1α,25-Dihydroxyvitamin D$_3$ C-5-A-Ring Carbamate Derivatives, *J. Org. Chem.* **62**, 4358-4363.

4.   Kircher, G., Scollar, M.D., and Klibanov, A.M. (1985) Resolution of Racemic Mixtures via Lipase Catalysis in Organic-Solvents, *J.Am. Chem. Soc.***107**, 7072-7076.

5. Ghogare, A., and Kumar, G.S. (1989) Oxime Esters as Novel Irreversible Acyl Transfer Agents for Lipase Catalysis in Organic Solvents, *J. Chem. Soc. Chem. Commun.*, 1533-1535.

6. Deguil-Castaing, M., De Jeso, B., Drouillard S., and Maillard, B. (1987) Enzymatic Reactions in Organic Synthesis: Ester Interchange of Vinyl Esters, *Tetrahedron Lett.* **28**, 953-954.

7. Bianchi, D., Cesti, P., and Battistel, E. (1988) Anhydrides as Acylating Agents in Lipase-Catalysed Stereoselective Esterification of Racemic Alcohols, *J.Org.Chem.* **53**, 5531-5534.

8. Frykman, H., Öhrner, N., Norin, T., and Hult, K. (1993) S-Ethyl Tiooctanoate as Acyl Donor in Lipase Catalysed Resolution of Secondary Alcohols, *Tetrahedron Lett.* **34**, 1367-1370.

9. Kita, Y., Takebe, Y., Murata K., Naka, T., and Akai, S. (1996) 1-Ethoxyvinyl Acetate as a Novel, Highly Reactive, and Reliable Acyl Donor for Enzymatic Resolution of Alcohols, *Tetrahedron Lett.* **37**, 7369-7372.

10. a) Kitaguchi, H., Fitzpatrick, P.A., Huber, J.E., and Klibanov, A.M. (1989) Enzymatic Resolution of Racemic Amines: Crucial Role of the Solvent, *J. Am. Chem. Soc.* **111**, 3094-3095. b) Gutman, A.L., Meyer, E., Kalerin E., Polyak, F., and Sterling, J. (1992) Enzymatic Resolution of Racemic Amines in a Continous Reactor in Organic Solvents, *Biotechnol.Bioeng* **40**, 760-767.

11. Castro, M.S., and Sinisterra, J.V. (1998) Lipase-Catalyzed Synthesis of Chiral Amides. A Systematic Study of the Variables that Control the Synthesis, *Tetrahedron* **54**, 2877-2892.

12. Kabata, K., Kawamura, M., Toyoshima, M., Tamura, Y., Ogawa, S., and Watanabe, T. (1998) *Biotechnology Lett.***20**, 451-454.

13. a) Puertas, S., Brieva, R., Rebolledo, F., and Gotor, V. (1993) Lipase Catalyzed Aminolysis of Ethyl Propiolate and Acrylic Esters. Synthesis of Chiral Acrylamides, *Tetrahedron* **49**, 4007-4014. b) Sanchez, V., Rebolledo, F., and Gotor, V. (1994) Higly Enzymatic Ammonolysis of α,β-Unsaturated Esters, *Synlett*, 529-530.

14. Quirós, M., Astorga, C., Rebolledo, F., and Gotor, V. (1995) Enzymatic Selective Transformation of Diethyl Fumarate, *Tetrahedron* **51**, 7715-7720.

15. Adamczyk, M., Grote, J., and Rege, S. (1997) Stereoselective *Pseudomonas cepacia* Lipase Mediated Synthesis of α-Hydroxyamides, *Tetrahedron: Asymmetry* **8**, 2509-2512.

16. Quirós, M., Sanchez, V., Brieva, R., Rebolledo, F., and Gotor, V. (1993) Lipase-Catalyzed Synthesis of Optically Active Amides in Organic Media, *Tetrahedron: Asymmetry* **4** , 1105-1112.

17. Garcia, M.J., Rebolledo, F., and Gotor, V. (1993) Practically Enzymatic Route to Optically Active 3-Hydroxyamides. Synthesis of 1,3-Aminoalcohols, *Tetrahedron: Asymmetry* **4**, 2199-2210.

18. Garcia, M.J., Rebolledo, F., and Gotor, V. (1994) Lipase-Catalyzed Aminolysis and Ammonolysis of β-Ketoesters. Synthesis of Optically Active β-Ketoamides, *Tetrahedron* **50**, 6935-6940.

19. De Zoote, M.C., Kock-van Malen, A.C., van Rantwijk, F., and Sheldon, R.A. (1993) Ester Ammonolysis a New Enzymatic Reaction, *J. Chem. Soc.Chem. Commun.*, 1831-1832.

20. Starmans, W.A.J., Doppen, R.G., Thijs, L., and Zwanenburg B. (1998) Enzymatic Resolution of Methyl *N*-Alkyl Azetidine-2-Carboxylates by *Candida antarctica* Lipase-Mediated Ammoniolysis, *Tetrahedron: Asymmetry* **9**, 429-435.

21. Garcia-Urdiales, E., Rebolledo, F., and Gotor, V. (1999) Enzymatic Ammonolysis of Ethyl (±)-4-Chloro-3-hydroxybutanoate. Chemoenzymatic Synthesis of Both Enantiomers of Pyrrolidin-3-ol and 5-(Chloromethyl)-1,3-Oxazolidin-2-one, *Tetrahedron: Asymmetry* **10**, 721-726.

22. Sánchez, V.M., Rebolledo, F., and Gotor, V. (1997) *Candida antarctica* Lipase Catalyzed Resolution of Ethyl (±)-3-Aminobutyrate, *Tetrahedron: Asymmetry* **8**, 37-40.

23. Puertas, S., Rebolledo, F, and Gotor, V. (1996) Enantioselective Enzymatic Aminolysis and Ammonolysis of Dimethyl 3-Hydroxyglutarate. Synthesis of (R)-4-Amino-3-hydroxybutanoic Acid, *J. Org. Chem.* **61**, 6024-6027.

24. Reetz, M.T., and Schimosseck, K. (1996) Lipase-Catalyzed Dinamic Kinetic Resolution of Chiral Amines: Use of Palladium as the Racemization Catalyst, *Chimia* **50**, 668-669.

25. Öhrner, N, Orrenius, Ch., Mattson, A., Norin, T. and Hult, K. (1996) Kinetic Resoilution of Amines and Thiol Analogues of Secondary Alcohols Catalized by the *Candida antarctica* Lipase B, *Enzyme Microbiol. Technol.* **19**, 328-331.

26. Iglesias, L.E., Sánchez, V.M., Rebolledo, F., and Gotor, V. (1997) *Candida anatrctica* B Lipase Catalysed Resolution of (±)-1-(Heteroaryl)-ethylamines, *Tetrahedron: Asymmetry* **8**, 2675-2677.

27. Djeghaba, Z., Delauze, H., Maillard, B, and De Jéso B. (1995) Les Enzymes en Synthese Organique IX: Acylation Enzymatica des Amines Secondaires Cycliques, *Bull Soc. Chim. Bel.* **104**, 161-164.

28. Chiou, T.W, Chang, C.C., Lai, C.T., and Tai, D.F., (1997) Kinetic Resolution of Propanolool by a Lipase-Catlyzed N-Acetilation, *Bioorg. Med. Chem. Lett.* **7**, 433-436.

29. Alfonso, I., Astorga, C., Rebolledo, F., and Gotor, V. (1996) Sequential Biocatalytic Resolution of (±)-*trans*-Cyclohexane-1,2-Diamine. Chemoenzymatic Synthesis of an Optically Active Polyamine, *J. Chem. Soc. Chem. Commun.*, 2471-2472.

30. Sánchez, V. M., Rebolledo, F., and Gotor, V. (1999) *Candida antarctica* Lipase Catalyzed Doubly Enantioselective Aminolysis Reaction. Chemoenzymatic Synthesis of 3-Hydroxypyrrolidines and 4-(Silyloxy)-2-oxopyrrolydines with Two Stereogenic Centers, *J. Org. Chem.* **64**, 1464-1470.

31. Astorga, C., Rebolledo, F. and Gotor, V. (1994) Enzymatic Aminolysis of Non-activated Diesters with Diamines (1994), *J. Chem. Soc. Perkin Trans 1*, 829-832.

32. Alfonso, I. Rebolledo, F. and Gotor, V. (1999) Chemoenzymatic Synthesis of two Optically Active Hexaazaacrocycles, *Tetrahedron: Asymmetry*, **10**, 367-374.

33. Alfonso, I., Astorga, C., Rebolledo, F., and Gotor, V. (1999) Optically Active Tetraazamacrocycles Analogous to Cyclam, *Tetrahedron: Asymmetry*, in press.

34. Pozo, M., Pulido, R., and Gotor,V. (1992) Vinyl Carbonates as Novel Alkoxycarbonylation Reagents in Enzymatic Synthesis of Carbonates, *Tetrahedron* **48**, 6477-6484.

35. Pozo, M. and Gotor, V. (1993) Chiral Carbamates through an Enzymatic Alkoxycarbonylation Reaction, *Tetrahedron* **49**, 4321-4326.

36. Pozo, M., and Gotor, V. (1993) Kinetic Resolution of Vinyl Carbonates through a Lipase Mediated Synthesis of their Carbonates and Carbamates, *Tetrahedron* **49**, 10725-10732.

37. Pozo, M., and Gotor, V. (1995) Double Enantioselective Enzymic Synthesis of Carbonates and Urethanes, *Tetrahedron: Asymmetry* **6**, 2797-2802.

ENANTIOSELECTIVE OXIDATIONS CATALYZED BY PEROXIDASES AND MONOOXYGENASES

S. COLONNA[a], N. GAGGERO[a], C. RICHELMI[a], P. PASTA[b]

[a] *Istituto di Chimica Organica, Facoltà di Farmacia, Università degli Studi di Milano, Via Venezian 21, I 20133 Milano, Italy*

[b] *Istituto di Biocatalisi e Riconoscimento Molecolare CNR, Via Mario Bianco 9, I 20131 Milano, Italy*

Abstract. Enantiomerically pure sulfoxides are excellent chiral auxiliaries for asymmetric synthesis and in the preparation of several enantiopure biologically active compounds. For these reasons we have explored biocatalytic approaches for the synthesis of sulfoxides based on the use of heme peroxidases and flavin monooxygenases, such as chloroperoxidase (CPO) and cyclohexanone monooxygenase (CMO), respectively. By using isolated enzymes or whole-cell biotransformations, we have prepared alkyl aryl sulfoxides, 1,3-dithioacetal-1-oxides and dialkyl sulfoxides in high enantiomeric excess (ee). Recent experiments from our and other laboratories have shown that chloroperoxidase is also able to catalyze a broad spectrum of stereoselective epoxidation reactions. The substrate repertoire includes substituted styrenes, straight chain aliphatic and cyclic *cis*-olefins. The enzyme is also able to perform benzylic hydroxylation and hydroxylation of alkynes with high ee.

1. Introduction

The current interest in catalytic oxidative transformations has been stimulated by two major fundamental issues. The first one is the need for the replacement of methods which use stoichiometric amounts of heavy metal salts by catalytic alternatives that use hydrogen peroxide or dioxygen as oxidant. The second major demand is the request of chemical reactions occurring with high chemo, regio or enantioselectivity in order to

133

B. Zwanenburg et al. (eds.), Enzymes in Action, 133–160.
© 2000 *Kluwer Academic Publishers. Printed in the Netherlands.*

134

improve chemical yield and avoid enantiomeric ballast. Oxidoreductases are enzymes suitable for achieving these two goals.

There are two major classes of proteins which oxidize organic substrates, namely monooxygenases and peroxidases [1]. Monooxygenases, which mediate the introduction of an oxygen atom into an unactivated substrate, require stoichiometric amounts of expensive cosubstrates such as NADH, that must be regenerated for economic reasons in large scale applications. Furthermore, they are relatively unstable and difficult to isolate. For these reasons monooxygenases are often used in whole-cell systems, but the occurrence of competing reactions often lowers the enantioselectivity and the chemical yield of the oxidation reaction. Peroxidases on the other hand can perform selective oxidations using clean oxidants such as hydrogen peroxide or organic peroxides, without the need of expensive cosubstrates. Moreover, most peroxidases are relatively stable extracellular enzymes, that catalyze a variety of reactions for a large range of substrates. From synthetic point of view the most interesting reactions catalyzed by peroxidases are the oxygen-transfer reactions, as in the case of monooxygenases. Peroxidases and monooxygenases can be subdivided in two types, those depending upon heme or upon a flavin molecule for oxygen activation and transfer. Heme peroxidases and monooxygenases share many structural and mechanistic properties. Indeed, much of the chemistry of all these hemoproteins can be rationalized through two common intermediates called compound I and compound II, which are one electron or two electrons above the ferric oxidation state, respectively. In this paper we shall confine ourselves to horseradish peroxidase (HRP) and chloroperoxidase (CPO) as heme enzymes, and to cyclohexanone monooxygenase (CMO) as flavin monooxygenase.

Peroxidases are ubiquitously found in animals, plants and microorganisms. Ferric protoporphyrin is the prosthetic group and imidazole the fifth axial iron ligand. Our understanding of structure–function relationship is largely based on HRP. The peroxidase catalytic activity is an one-electron oxidation of substrates, such as phenols and aromatic amines (Scheme 1).

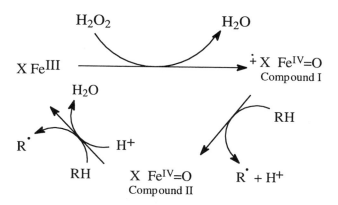

Scheme 1: Catalytic cycle of peroxidases (X = porphyrin).

In the catalytic cycle hydrogen peroxide binds to the ferric ion and its fast reduction to water is coupled to two-electron oxidation of the enzyme. This generates the compound I intermediate, which is reduced by the substrate molecule, with the formation of compound II as the second intermediate. The radical produced in the reaction generally evolves nonenzymatically to nonradical products by pathways which are characteristic of each substrate (coupling, dismutation, etc.) [2]. Of the two electrons required for the peroxide reduction one comes from Fe(III), while the other in HRP comes from the porphyrin, producing a porphyrin radical cation. In the native enzyme iron(III) is present in the high spin state. A basic amino acid residue (histidine) abstracts a proton from the hydroperoxide whilst stabilization of the developing negative charge on the peroxide by a positively charged amino acid (arginine) would assist the heterolytic cleavage of the oxygen-oxygen bond with formation of compound I.

According to the crystal structure of HRP isozyme C (HRPC) [3], the heme group is sandwiched between the distal and proximal domains of the enzyme, and there are two Ca^{++} binding sites. The key catalytic residues for a series of peroxidases, namely Arg38, Phe41 and His42, are highly conserved and have similar disposition. The distal pocket thus confirms the expectations for the proposed mechanism for compound I formation. The substrate access channel of HRPC is an example of a common feature shared among peroxidases, namely a solvent accessible or exposed heme edge close to heme meso C20 and heme methyl C18, and a protected heme group.

2.   Sulfoxidation Reactions

Enantiomerically pure sulfoxides are important chiral synthons for the stereoselective synthesis of natural products and are among the most powerful stereodirecting groups in carbon-carbon bond formation [4]. Current interest in chiral sulfoxides also reflects the existence of products with biological properties determined by the configuration of the sulfinyl group [5]. Some materials such as liquid crystals are based on a chiral sulfoxide structure. The utility of the sulfinyl group in the synthesis of drugs is exemplified by the preparation of (S)-sulforaphane, a naturally occurring methyl sulfoxide, which stimulates the production of anticarcinogenic protective enzymes [6].

We have been the first to show that HRP is able to catalyze the enantioselective oxidation of alkyl aryl sulfides to sulfoxides with hydrogen peroxide [7]. Substituted methyl phenyl sulfides were chosen as model substrates in view of the optical stability of the corresponding sulfoxides to pyramidal inversion. In order to verify the generality of this type of reaction, we also investigated the behaviour of benzyl methyl and methyl-2-pyridyl sulfide. Asymmetric induction took place only with methyl phenyl and methyl *para*-substituted phenyl sulfides, the ee being in the range of 30-68%. In the cases examined the prevailing sulfoxide had the (S) absolute configuration. It should be pointed out that the enzyme-catalyzed reaction is in competition with the spontaneous oxidation by hydrogen peroxide at a rate dependent on the concentration and nature of the substrate. The substantial enantioselectivity, together with oxygen labelling studies and sulfide binding studies, indicates that aryl methyl sulfides can approach the heme active site and makes possible an oxygen transfer from compound I.

Later the enantioselectivity of the HRP oxidation of aryl alkyl sulfides has been substantially increased by Ortiz de Montellano by molecular engineering [8]. The Phe41/Leu mutant of HRP shows a higher turnover and enantioselectivity with respect to native HRP. This is particularly evident in the case of phenyl cyclopropyl sulfide that affords the corresponding sulfoxide with F41L HRP and native HRP having 94% and 7% ee, respectively. Sulfoxidation rates for both native and modified HRP correlate well with the σ^+ constants of the substituents and the ρ values are very similar. The negative values of ρ indicate that a similar degree of positive charge is developed in the transition state. More important is the fact that the substitution of Phe by the smaller Leu favours the access of the substrate to the ferryl oxygen responsible for the peroxygenase activity, thus transforming native HRP into a nearly stereospecific sulfoxidation catalyst. The increased peroxygenase activity of the mutant has been confirmed in the epoxidation of styrene and β-methyl styrene, a reaction not catalyzed by native HRP [8].

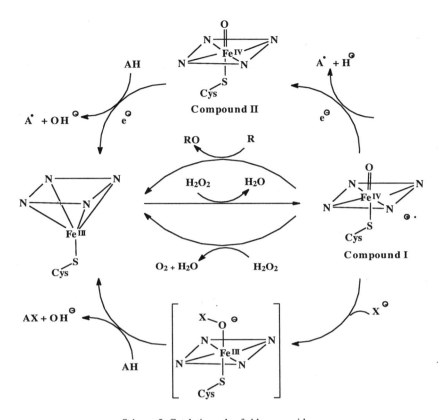

Scheme 2: Catalytic cycle of chloroperoxidase.

Haloperoxidases are grouped according to the range of halide ions they can oxidize in the presence of hydrogen peroxide. In particular chloroperoxidase (CPO) from the marine fungus *Caldariomyces fumago*, first isolated by Hager in 1961, is a very versatile catalyst, that not only performs halogenations (Cl, Br, I) of organic substrates, but also functions as classical peroxidase and catalase. It is unique among the peroxidases because it contains a cystein axial ligand of the heme instead of the classical imidazole ligand. This enzyme is a heme peroxidase/cytochrome P-450 functional hybrid. The crystal structure shows that CPO folds into a novel tertiary structure dominated by eight helical segments [9]. Hydrogen bonds involve the distal site water molecule, Glu183 (the proposed acid-base catalyst) and His. The proposed substrate binding site pocket is bracketed by Phe103 and Phe186. Cys29 provides the thiolate heme ligand and the heme pocket is polar as predicted, but the acid-base catalyst is not histidine, as in the other peroxidases, but Glu183, that is positioned directly adjacent to the peroxide binding site, thus favouring the peroxide bond cleavage with formation of compound I.

The catalytic cycle of CPO [1] is represented in Scheme 2. The addition of hydrogen peroxide to the high spin ferric iron gives compound I. In the peroxidase like activity, compound I takes an electron from the substrate and is transformed in to compound II, that reacts with a second molecule of substrate and regenerates the native enzyme. In the catalase pathway compound I reacts with a second molecule of hydrogen peroxide to give dioxygen and native enzyme. In the halogenation, compound I reacts with halide ions to give an iron hypohalide which acts as a source of halogen. The details of the halogenation reaction are still matter of debate. In the monooxygenase pathway the ferryl oxygen is transferred to the substrate to give the oxidized product and the native enzyme.

In contrast with HRP, the iron heme in CPO is much more accessible to the substrates such as organic sulfides and alkenes. We have been able to show that CPO is a very efficient catalyst for the oxidation of organic sulfides to sulfoxides [10]. The study of a large series of alkyl aryl and heterocyclic sulfides has shown that electronic and particularly steric factors dramatically affect the outcome of the reaction. Excellent yields and very high ee have been obtained in the sulfoxidation of a large number of sulfides structurally related to phenyl methyl sulfide (Table 1). Binding experiments show that these substrates better fit the topology of the active site of CPO.

Both the ee and the chemical yield are decreased by enlarging the size of the alkyl substituent at the sulfur atom or by introducing the substituent in the *ortho* position of the aromatic ring.

In order to ascertain the origin of the enantioselectivity in the sulfoxidation we have investigated the binding of alkyl aryl sulfides to CPO using optical, CD and NMR spectroscopy [11]. The iron-proton distances for CPO-sulfide and HRP-sulfide complexes, derived from relaxation time measurements, show that in general the substrates are located slightly closer to the iron heme in CPO complexes than in HRP complexes. We have also measured the spectral titration of native CPO with several substrate molecules and proved the formation of enzyme-sulfide complexes with an 1:1 stoichiometry. Increasing the size of the sulfur alkyl substituent reduces the affinity for the enzyme. Also the change in the substitution pattern of the aromatic nucleus from *para* to *ortho*, and an increase in the electron-withdrawing power of the phenyl ring substituent leads to a decrease in the affinity for the enzyme. It is significant that the trend in affinity of the sulfides for CPO parallels the enantioselectivity pattern observed in the CPO catalyzed oxidation of these substrates, the more immobilized substrates giving the higher enantioselectivity. The linear correlation between the binding constants of *para*-substituted sulfides to CPO and the Taft σ constants and the negative slope of the resulting plot, suggest that these substrates act as donors in donor-acceptor complexes involving some residues of the active site. The polar residues in the CPO distal region are candidates for acting as acceptor groups. In all cases the prevailing enantiomer in the sulfoxide formed has the (R) absolute configuration.

TABLE 1. Aryl alkyl sulfides oxidation catalyzed by CPO

Sulfide	Oxidant	Yield (%)	Ee (%)	Control reaction Yield (%)
p-CH$_3$-C$_6$H$_4$-S-CH$_3$	H$_2$O$_2$	98	91	18
	t-BuOOH	80	70	24
o-CH$_3$-C$_6$H$_4$-S-CH$_3$	H$_2$O$_2$	27	33	13
	t-BuOOH	56	43	33
p-CH$_3$O-C$_6$H$_4$-S-CH$_3$	H$_2$O$_2$	72	90	16
	t-BuOOH	70	61	31
o-CH$_3$O-C$_6$H$_4$-S-CH$_3$	H$_2$O$_2$	24	27	23
	t-BuOOH	30	37	18
C$_6$H$_5$-S-CH$_3$	H$_2$O$_2$	100	98	30
	t-BuOOH	90	80	65
p-Cl-C$_6$H$_4$-S-CH$_3$	H$_2$O$_2$	77	90	13
	t-BuOOH	60	70	20
o-Cl-C$_6$H$_4$-S-CH$_3$	H$_2$O$_2$	33	85	13
	t-BuOOH	17	45	7
p-NO$_2$-C$_6$H$_4$-S-CH$_3$	H$_2$O$_2$	10	80	2
	t-BuOOH	16	80	2
p-CH$_3$-C$_6$H$_4$-S-C$_2$H$_5$	H$_2$O$_2$	50	68	15
	t-BuOOH	50	68	20
p-CH$_3$-C$_6$H$_4$-S-n-C$_3$H$_7$	H$_2$O$_2$	53	5	22
	t-BuOOH	30	5	28
p-CH$_3$-CO-NH-C$_6$H$_4$-S-CH$_3$	H$_2$O$_2$	86	67	23
	t-BuOOH	86	70	33
C$_6$H$_5$-CH$_2$-CH$_3$	H$_2$O$_2$	100	90	33
	t-BuOOH	73	55	50
2-pyridyl-S-CH$_3$	H$_2$O$_2$	100	99	26
	t-BuOOH	61	89	16
p-F-C$_6$H$_4$-S-CH$_3$	H$_2$O$_2$	100	97	27
	t-BuOOH	90	70	29

The sulfoxidation reaction catalyzed by CPO is likely to proceed through a direct oxygen transfer from compound I or a rebound mechanism involving the formation of compound II and a sulfenium radical cation, followed by recombination of these two

140

species within the same cage of solvent as indicated by kinetic, spectroscopic and ^{18}O labelling studies (Scheme 3).

$$Fe^{3+} + H_2O_2 \xrightarrow{K_1} (Fe=O)^{3+} + H_2O \quad (1)$$

$$(Fe=O)^{3+} + S \xrightarrow{K_2} Fe^{3+} + SO \quad (2)$$

$$Fe^{3+} + H_2O_2 \xrightarrow{K_1} (Fe=O)^{3+} + H_2O \quad (1)$$

$$(Fe=O)^{3+} + S \xrightarrow{K_3} (Fe=O)^{2+}\ S^{+\cdot} \quad (3)$$

$$(Fe=O)^{2+}\ S^{+\cdot} \xrightarrow{K_4} Fe^{3+} + SO \quad (4)$$

Scheme 3: Mechanisms for oxygen-transfer in reactions catalyzed by CPO.

The synthetic utility of CPO as enantioselective catalyst for the sulfoxidation reaction has been recently expanded by our group [14]. A wider application of methodologies based on the sulfinyl function is limited by the relative paucity of general one-step procedures for obtaining enantiomerically pure dialkyl sulfoxides. We have found that CPO affords dialkyl sulfoxides with high ee (Table 2).

TABLE 2. Dialkyl sulfides sulfoxidation catalyzed by CPO

Sulfide	Conversion (%)	Ee (%)	Absolute Configuration
Cyclopentyl methyl sulfide	≥ 98	≥ 98	R
Cyclohexyl methyl sulfide	85	85	R
Allyl methyl sulfide	≥ 98	≥ 98	R
Pentyl methyl sulfide	75	≥ 98	R
Octyl methyl sulfide	40	54	R
iso-Propyl methyl sulfide	≥ 98	≥ 98	R
tert-Butyl methyl sulfide	80	85	R
tert-Butyl ethyl sulfide	30	35	R
Bis(thiomethyl)methane	75	≥ 98	R

Cyclopentyl methyl sulfide, isopropyl methyl sulfide and bis(thiomethyl) methane are excellent substrates for CPO in terms of chemical yield and ee. The increase in the size of the cycloalkyl group bound to the sulfur (cyclohexyl instead of cyclopentyl) leads to an appreciable decrease in the chemical and optical yield. The importance of steric factors is confirmed by the results obtained with methyl sulfides with different alkyl chains. An increase in the branching of the alkyl chain decreases the chemical yield and the enantioselection. These results can be rationalized by taking into account the higher contribution of the spontaneous oxidation in these substrates. As already observed for alkyl aryl sulfides, the oxidation of dialkyl sulfides affords predominantly or exclusively the corresponding (R)-sulfoxide.

In spite of the remarkable synthetic potentiality of CPO, commercial processes based on this enzyme have not developed yet. One of the reasons of this is that hydrogen peroxide (or other peroxides) rapidly inactivates the enzyme by oxidation of the porphyrin ring. Furthermore, the enantioselective oxidation of various substrates is in competition with their spontaneous oxidation by hydrogen peroxide, which reduces the enantiomeric purity of the products.

This year Van der Velde *et al.* have found that CPO can selectively oxidize methyl phenyl sulfide to (R)-methyl phenyl sulfoxide in the presence of dihydroxyfumaric acid (DHFA) and dioxygen [15]. They have hypothesized that oxidation occurred through a mechanism involving the initial formation of hydrogen peroxide via autooxidation of DHFA.

We have reinvestigated the CPO-catalyzed oxidation of phenyl methyl sulfide with dioxygen in the presence of either DHFA or ascorbic acid as external reductants [16]. The reaction was carried out in a membrane reactor, which allowed the reuse of the enzyme for several conversion cycles. Table 3 shows the results obtained using ascorbic acid/dioxygen as oxidizing system. Almost complete conversion up to the seventh cycle with high optical purity was observed. Similar results were found in the case of DHFA. This is a consequence of the high enzyme stability of CPO with the two systems and of the absence of the aspecific oxidation of the substrate.

Concerning the mechanism, the involvement of hydrogen peroxide via autooxidation of DHFA is not completely convincing. First, as reported by Van der Velde *et al.* [15], catalase had no effect on the reaction and, second, the high stability of CPO with DHFA/dioxygen is not compatible with the presence of consistent concentrations of hydrogen peroxide. We have found that upon reaction of CPO with excess ascorbic acid in an anaerobic optical cell, the enzyme is reduced to the Fe(II) state. This reduced form reacts rapidly with dioxygen to fully regenerate the native Fe(III) state. The capacity of the enzyme to undergo reactions with dioxygen in the presence of reducing agents raises the possibility that it may perform oxygen-transfer reactions to exogenous substrates according to a true monooxygenase pathway as in the case of cytochrome P-450.

TABLE 3. *p*-Tolyl-methyl-sulfide oxidation catalyzed by CPO with external reductant

Cycle no.	Conversion (%)	e.e. (%)
1	92	98
2	91	98
3	90	98
4	87	98
5	88	98
6	89	98
7	85	98
8	75	98

The synthetic utility of CPO as catalyst has been extended by Allenmark and coworkers to the oxidation of a series of rigid aromatic sulfides [17]. The almost planar 1-thiaindane gave quantitative yield of the (*R*)-1-oxide in 99% ee. The more sterically hindered homologue 1-thiatetrahydronaphtalene (1-thiachroman) also afforded the corresponding sulfoxide in high ee, but the turnover rate was lower. Similar results were observed with 1-thiachroman-4-one. A racemic sulfoxide was obtained with the symmetric disulfide 1,3-benzodithiol, indicating an equal accessibility of both sulfur atoms on the ferryl oxo complex.

In conclusion the differences in reactivity and enantioselectivity of the peroxidases can be explained by the differences in the environment of the active site. Heme peroxidases in which the heme iron is less accessible give a lower selectivity and reactivity. Indeed several other peroxidases catalyze the enantioselective sulfoxidation of alkyl aryl sulfides. These include microperoxidase-11, lactoperoxidase and cytochrome c peroxidase. However, their turnover number, as well as enantioselectivity are much lower than those observed with CPO.

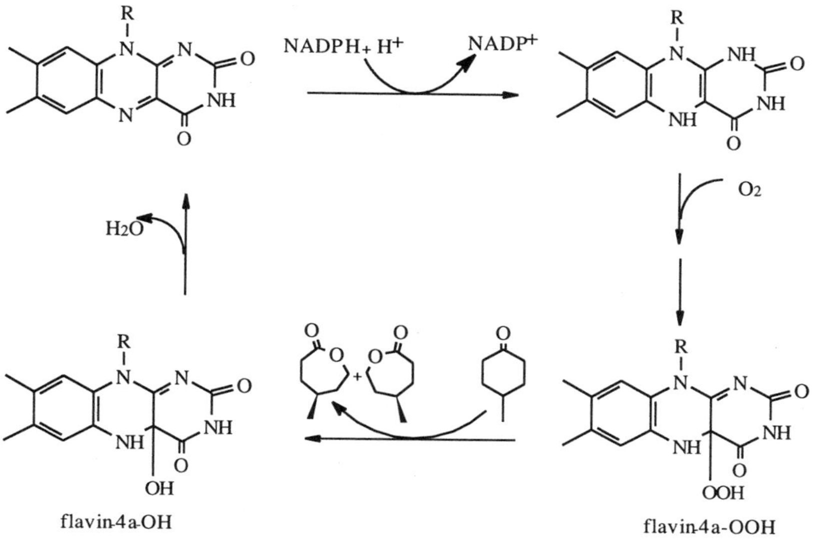

Scheme 4: Catalytic cycle of cyclohexanone monooxygenase.

The biosulfoxidation reaction is also catalyzed by cyclohexanone monooxygenase from *Acinetobacter calcoaceticus* (CMO). CMO is a flavoenzyme of about 60,000 daltons, active as a monomer, which contains one firmly, but noncovalently bound FAD unit per monomer. It has a large potential application in the manufacture of organic fine chemicals and in organic syntheses, based on the Baeyer-Villiger reaction. The only reagents consumed are dioxygen, a reductant and a reacting ketone which is transformed enantioselectively into the corresponding ester with formation of water. According to the proposed mechanism the 4a-hydroperoxyflavin intermediate acts as a nucleophile at the carbonyl carbon. Intramolecular elimination of water from the 4a-hydroperoxyflavin generates FAD for another catalytic cycle [18] (Scheme 4).

Walsh and co-workers described the synthesis of (*S*)-ethyl-*p*-tolyl-sulfoxide (64% ee) using CMO [19]. However, their investigation was not extended to other sulfides.

Our interest was to study the stereochemistry of oxidation at sulfur, catalyzed by cyclohexanone monooxygenase, using various alkyl aryl sulfides (Table 4) [20].

TABLE 4. Alkyl aryl sulfides sulfoxidation catalyzed by CMO

	Sulfide	Yield (%)	Ee (%)	Absolute configuration	
1	C_6H_4-S-CH_3	88	99	**R**	
2	p-F-C_6H_4-S-CH_3	91	92	**R**	
3	o-CH_3-C_6H_4.S-CH_3	90	87	**R**	
4	2-pyridyl-S-CH_3	86	87		R
5	p-C_2H_5O-C_6H_4-S-CH_3	92	59		R
6	C_6H_4-CH_2-S-CH_3	97	54		R
7	o-CH_3 -C_6H_4-S-CH_3	81	51		R
8	C_6H_4S-C_2H_5	86	47		R
9	m-CH_3-C_6H_4-S-CH_3	90	40		R
10	o-Cl-C_6H_4.S-CH_3	35	32		R
11	C_6H_4-S-i-C_3H_7	93	3		S
12	p-CH_3-C_6H_4-S-CH_3	94	37		S
13	p-Cl-C_6H_4-S-CH_3	78	51		S
14	p-CH_3O-C_6H_4-S-CH_3	89	51		S
15	p-CH_3-C_6H_4-S-i-C_3H_7	99	86		S
16	p-CH_3-C_6H_4-S-C_2H_5	89	89	S	
17	p-F-C_6H_4-S-C_2H_5	96	93	S	
18	t-C_4H_9-S-CH_3	98	99	**R**	·
19	C_4H_9-S-S-C_4H_9	85	32	ND^b	
20	CH_3-S-S-C_3H_7	92	64; 34^a	ND	

[a]For the two regioisomeric thiosulfinates. [b]ND, not determined.

The oxidation of sulfides by the enzyme (reaction 1) was coupled to a second enzymatic reaction to regenerate NADPH; therefore only catalytic quantities of NADPH were required. The regenerating system used was either glucose-6-phosphate and glucose-6-phosphate dehydrogenase (G6PDH) (reaction 2) or L-malate and malic enzyme (reaction 3):

R-S-R^I + NADPH + O_2 + H^+ + CMO $\rightarrow$ R-SO-R^I + $NADP^+$ + H_2O (reaction 1)

D-glucose-6P + $NADP^+$ + G6PDH $\rightarrow$ D-gluconate-6P + NADPH + H^+ (reaction 2)

L-malate + $NADP^+$ + malic enzyme $\rightarrow$ pyruvate + CO_2 + NADPH + H^+ (reaction 3)

The increase in size of the alkyl chain increased the initial oxidation rates of alkyl aryl sulfides. The benzyl group was more activating than the phenyl group, but the activation of the latter was augmented by introducing a substituent in the aromatic ring. With regard to the stereoselectivity of the enzymatic reaction, the data indicate that it is highly dependent on substrate structure. Thus, for alkyl aryl sulfoxides the optical purity of the products ranges from 99% ee with the (R) configuration for methylphenylsulfoxide, to

93% ee with the (*S*) configuration for ethyl*p*-fluorophenylsulfoxide. The enzyme showed very high enantioselectivity (99% ee) for *tert*-butylmethylsulfide. The enantiomeric excess of the sulfoxides did not change appreciably with the progress of the reaction. The oxidation of the sulfoxides to the corresponding sulfones was very slow and could not be exploited for kinetic resolution purposes. Interestingly, similar results in terms of enantioselectivity were obtained using either crude or purified CMO, so the high sensitivity of the cyclohexanone monooxygenase to the structural variations of the substrate is an intrinsic property of a single enzyme. Similar results were obtained with functionalized sulfides [21] and benzyl alkyl sulfides [22]. The use of macromolecular NADPH in a membrane reactor increased the efficiency of coenzyme recycling, a critical step in this kind of biotransformation. Poly(ethylene glycol)-NADPH was used and coenzyme regeneration was carried out with the propan-2-ol-alcohol dehydrogenase system. Both CMO and alcohol dehydrogenase from *Thermoanaerobium brockii* (TBADH) maintained high activities with the macromolecular coenzyme [23].

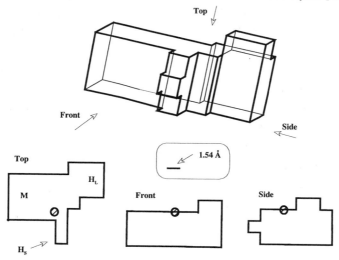

Figure 1: Active site model of CMO.

We have proposed an active site model of the enzyme to explain the stereoselectivity of sulfoxidation and to predict the absolute configuration of the products (Fig. 1) [24].

The versatility of CMO is further exemplified by its ability to promote enantioselective oxidation of 1,3-dithioacetals [25]. This is a major finding since 1,3-dithioacetal monosulfoxides serve as chiral acyl anion equivalents. In particular, 2-acyl-2-alkyl-1,3-dithiane monosulfoxide is an extremely effective compound for imparting stereocontrol in enolate alkylations and aminations, Mannich reactions, organometallic additions, heterocyclic cycloadditions and so on. In spite of their synthetic utility, the preparation of these chirons remains difficult. The oxidation of 1,3-dithiane and of its 2-alkyl derivatives using the Kagan and Modena procedure (the Sharpless modified procedure) leads to the monosulfoxides with poor optical purities (30%). We have found that the CMO-catalyzed oxidation of 1,3-dithiane, 1,3-dithiolane and bis(methylthio)-methane gives enantiomerically pure (R)-monosulfoxides in chemical yields ranging from 81 to 94%. Starting from racemic 1,3-dithiane monosulfoxide the enzyme was able to oxidize the (S)-enantiomer to the corresponding monosulfone faster than the (R)-enantiomer, the enantiomeric ratio E value being 20. As a consequence, enantiomerically pure (R)-1,3-dithiane monosulfoxide was obtained as a result of both asymmetric synthesis and kinetic resolution. The same behaviour was observed for bis(methiltio)methane, whereas only asymmetric synthesis was operating in the case of 1,3-dithiolane since the V_S/V_R value was as high as 49.

The kinetic parameters for the CMO-catalyzed oxidation of dithioacetals and racemic dithioacetal monosulfoxides are in agreement with the preference of CMO for the dithiane over the monosulfoxide. This is confirmed by the fact that no 1,3-dithiane monosulfone was formed as long as 1,3-dithiane was present in the reaction mixture.

The main limitation of the exploitation of CMO as catalyst for the biosulfoxi-dations stems from the use of the isolated enzyme necessitating the recycling of the expensive NADPH coenzyme. Although improvement was gained by us by using the immobilized coenzyme with a membrane reactor, this technology is still unsuitable for large scale applications. We have found a much more practical solution to this problem by using whole cell of *Acinetobacter calcoaceticus* [26]. As an example 1,3-dithiane is efficiently oxidized to the corresponding (R)-sulfoxide. The ee of the sulfoxide stayed constant (83%) as long as the starting sulfide remained in the bioconversion medium, and increased up to 99% afterwards, in agreement with the results previously obtained with the isolated enzyme. In a preparative scale-up experiment 1 g of dithiane led to 860 mg (76%) of isolated (R)-1,3-dithiane oxide with 98% ee. It should be noted that up to now the best chemical preparation of 1,3-dithiane-1-oxide necessitated the use of a three-step procedure (overall yield 40% with the modified Sharpless procedure).

For the CMO promoted oxidation of dialiphatic sulfides high enantioselectivity has been observed by us for cycloalkyl methyl sulfoxides and alkyl methyl sulfoxides with a linear or branched alkyl chain of up to four carbon atoms. In all cases the absolute configuration of the prevailing enantiomer is (R) [14]. A longer alkyl chain not only causes a drastic drop in the chemical and optical yield, but also changes the stereochemical course of the reaction, since the prevailing enantiomer has now the (S)

absolute configuration. A comparison of the results obtained in the oxidation of dialkyl sulfides with CPO and CMO leads to the following conclusions: (i) both enzymes exhibit high enantioselectivity in the oxidation of cycloalkyl methyl and alkyl methyl sulfides with limited steric requirements; (ii) the two enzymatic systems are enantiocomplementary for *n*-pentylmethylsulfide and for *n*-octylmethylsulfide; (iii) chloroperoxidase is more convenient than cyclohexanone monooxygenase since it is commercially available, uses hydrogen peroxide as oxidant and does not require the regeneration of the cofactor.

3.   Epoxidation Reactions

Optically active epoxides are very useful chiral synthons as they can undergo facile stereospecific ring-opening to form bifunctional compounds. They are very important as key intermediates in the production of bioactive chiral compounds or as final products with biological activity (for example the drug Diltiazem, used as calcium antagonist).

Several chiral porphyrin complexes have been reported to catalyze the epoxidation reaction using iodosoarenes as oxidants [27]. Recently chiral salen manganese compounds have been used succesfully for the enantioselective epoxidation reaction of conjugated olefins by Jacobsen and coworkers [28]. In this manner epoxychromans have been prepared in enantiomerically pure form as intermediates for potassium channel activators with promising antihypertensive activity.

Furthermore, asymmetric epoxidation is of fundamental importance in biological processes. Xenobiotics containing an olefinic function are biotransformed in mammals to epoxides by the action of cytochrome P-450 dependent monooxygenases [29]. These epoxides exhibited high electrophilic reactivity and unless promptly detoxified, they can produce toxic, mutagenic and carcinogenic effects due to the formation of covalent adducts at the nucleophilic sites of DNA. Epoxide detoxification occurs mainly by two enzymatic processes, namely by glutathione transferase or by epoxide hydrolase.

To retain the chirality present in the enantiomerically pure epoxide the most useful reactions are stereocontrolled nucleophilic additions and reductions. There are three general approaches to biological epoxide formation. The direct stereospecific formation of epoxides whereby only a single enantiomer is obtained, is the most attractive route. It can be accomplished by direct oxygen insertion. Another possibility is the formation of an α,β–halohydrin, catalyzed by a haloperoxidase, which may be

148

chemically or enzymatically converted into an epoxide. A process was developed for the production of propylene oxide with *in situ* generation of hydrogen peroxide by glucose oxidase, however the process has not been commercialized. An important drawback of the halohydrin method is that the products are racemic. The formation of α,β–halohydrin is not stereospecific, since the reaction involves the initial formation of hypohalous acid, followed by chemical reaction with olefins [30]. It should be noted that some haloperoxidases have a wide substrate range. Enantiomerically enriched epoxides may be also prepared by enzymatic resolution via epoxide hydrolase.

The direct oxygen insertion is most important. Cytochrome P-450 and the non-heme iron monooxygenases such as ω-hydroxylase and methane monooxygenase (MMO) serve mainly as hydroxylating enzymes. However, in the absence of steric constraints olefin epoxidation is kinetically more favourable than hydroxylation. For example, ω-hydroxylase produces predominantly epoxyoctane from octene [31] and MMO produces propene oxide from propene. Internal alkenes are also epoxidized, whereas the alkane hydroxylase from *Pseudomonas oleovorans* forms epoxides from terminal alkenes only. It should be stressed that this type of monooxygenase produce a very "electrophilic" oxygen intermediate which could attack a substrate at a number of sites. Bacterial monoxygenases, such as *Nocardia* sp., can oxidize alkenes to epoxides with an ee up to 99%. They are very effective also in the epoxidation of allyl-aryl ethers, which are valuable intermediates in the synthesis of beta-blockers. In the context of internal olefin epoxidation it is relevant that some bacteria which degrade α–pinene, form pinene epoxides as the first catabolic product. With limonene as substrate, only the cyclic epoxide is formed, whereas no products are formed from cyclohexene.

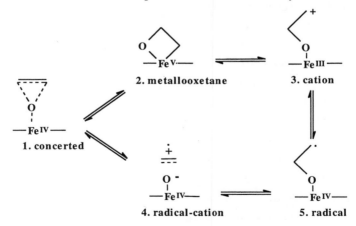

Scheme 5: Mechanisms of the epoxidation reactions catalyzed by CPO.

Although P-450$_{cam}$ is generally considered to be a camphor specific enzyme, it has been shown to oxidize organic compounds unrelated to camphor. Styrene and *cis*- and *trans*-β-methyl-styrene are catalytically oxidized to the corresponding epoxides. Benzaldehyde is also formed, due to the oxidation of the olefin by hydrogen peroxide by uncoupled turnover of the enzyme. Oxidation of styrenes by P-450$_{cam}$ is less efficient than that of the natural substrate camphor. The large degree of uncoupled reaction depends on the loose binding of styrene in the active site, a feature that is correctly modelled by molecular dynamic calculations [32]. Even simple aliphatic alkenes are epoxidized by cytochrome P-450 isozymes of liver microsomes from different species. The configuration of the prevailing enantiomer is species dependent. Enantioselectivity differences of aliphatic alkene epoxidation by human liver microsomes of four individuals are negligible [33].

Possible mechanisms for alkene oxidation are shown in Scheme 5. The various intermediates are interconvertible via electron-transfer processes. The following four possibilities can be considered.

1. Concerted mechanism. The oxidation proceeds in one, concerted step.

2. Oxygen transfer via a radical cation. In this mechanism initial one-electron transfer is followed by oxygen transfer from the iron(IV) oxo species to the radical cation intermediate (oxygen rebound). This intermediate has been trapped for a synthetic iron (IV) oxoporphyrin radical cation-catalyzed epoxidation of styrene [34].

3. The formation of a cationic, or radical intermediate via direct electrophilic attack of the iron(V) oxo species on the electron rich olefin or via further reaction of the radical cation. If the intermediate is a cation and the adjacent carbon atom has an hydrogen atom a rearrangement occurs with hydride transfer from the neighbouring carbon to the cationic carbon. This may account for the already mentioned formation of phenylacetaldehyde.

4. Oxygen transfer via metalloxetane intermediate. This intermediate is unlikely because of steric restrictions.

In 1987 Ortiz de Montellano found that CPO catalyzes the oxidation of styrene to styrene oxide [35] with hydrogen peroxide as oxidant. It was shown by studies with ^{18}O-enriched hydrogen peroxide that the epoxide oxygen is derived quantitatively from the peroxide. The epoxidation of *trans*- deuterated styrene proceeds without detectable loss of stereochemistry. The epoxidation by CPO, according to Ortiz de Montellano, involves a ferryl oxygen transfer similar to that proposed for cytochrome P-450. The phenylacetaldehyde produced as a coproduct, does not arise from decomposition of

150

styrene oxide and is probably generated by 1,2 rearrangement of a hydrogen cation formally generated by heterolytic opening of the epoxide ring (Scheme 6).

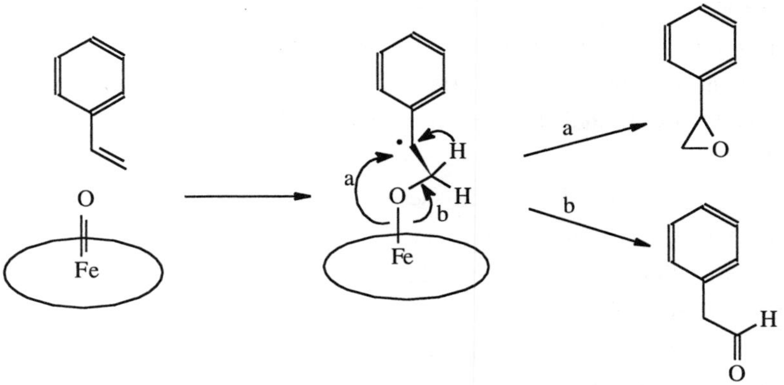

O

Fe

a

O

Fe

H

H

a

b

O

H

O

Scheme 6: Mechanism of stereoselective epoxidation of styrene as proposed by Ortiz de Montellano

Epoxidation of alkenes by chloroperoxidase has been investigated also by Geigert and co-workers [36]. Reactions with various alkenes were examined. Although no attempt was made to quantify the yield in these reactions, the most reactive alkene was styrene. The presence of a halide or a conjugated double bond does not prevent epoxidation, however the presence of an allylic alcohol or carbonyl does.

Hydrogen peroxide dependent oxidation of 1,3-butadiene by CPO produces both butadiene monooxide and crotonaldehyde [37]. The formation of crotonaldehyde but not of methyl vinyl ketone indicates regioselectivity in oxygen transfer from hemoprotein to 1,3-butadiene. The intermediate formed may undergo either ring closure to form butadiene monooxide or a hydrogen shift to form 3-butanal which tautomerizes to produce crotonaldehyde.

We wanted to ascertain whether CPO is able to transfer enantioselectively an oxygen atom to olefins and to investigate the stereochemical course of the reaction. As substrates we chose styrene, o-chloro-styrene, m-chloro-styrene and p-nitro-styrene (Table 5). Preliminary experiments indicated that t-BuOOH is the oxidant of choice instead of hydrogen peroxide because of the higher chemical yield. The ee were in the range of 28-68% and the prevailing epoxide had the (R) absolute configuration [38]. This means that the ferryl oxygen attacks the double bond from the side of the smaller substituent, as already found in the sulfoxidation reaction. As proposed by Ortiz de Montellano [35], the stereoselective epoxidation reactions are mediated by ferryl oxygen transfer to the olefin to give a metal-stabilized free radical, that may evolve either to epoxide or to phenylacetaldehyde, which is the coproduct, via hydrogen migration (Scheme 6). In about the same time as we did, Hager and Jacobsen have studied the CPO-catalyzed epoxidation reaction [39]. They have used disubstituted alkenes and hydrogen peroxide as oxidant. Compared to cytochrome P-450, CPO can be considered a better candidate as practical epoxidation catalyst since it uses hydrogen peroxide instead of dioxygen and NADH. Particularly good results were obtained with *cis*-substituted alkenes bearing alkyl substituents, whereas *trans*-olefins were found very unreactive. The enzyme tolerated branching in the alkyl substituents, although enantioselectivities were dependent on branching. Certain trisubstituted alkenes were also accepted by CPO with moderate to good enantioselectivity. Suicidal formation of *N*-alkyl porphyrins occurred during the epoxidation of terminal alkenes (and alkynes) (Table 6) [40].

TABLE 5. Epoxidation of styrene derivatives catalyzed by CPO

Substrate	Yield (%)	Ee (%)	Absolute Configuration
Styrene	23	49	*R*
p-Chlorostyrene	35	66	*R*
m-Chlorostyrene	34	62	*R*
o-Chlorostyrene	3	64	*R*
p-Bromostyrene	30	68	*R*
p-Nitrostyrene	5	28	*R*

CPO is an unusual peroxidase since it catalyzes benzylic hydroxylation, thus resembling cytochrome P-450 [41]. The reaction occurs with high enantioselectivity. On passing from ethyl- to propylbenzene opposite stereochemistry is observed in the formation of the corresponding alcohol. The mechanism of benzylic hydroxylation may be either concerted or proceeding via hydrogen abstraction, followed by oxygen rebound. The utilization of *trans*-2-phenyl-1-methylcyclopropane as radical probe is in favour of a concerted process since no rearranged alcohol is formed in contrast with, for instance, the *Pseudomonas oleovorans* monooxygenase (POM)-catalyzed epoxidation.

TABLE 6. Epoxidation of olefins catalyzed by CPO

Olefin	Product yield, ee	Byproducts yield	Turnover number
	yield 100%, ee 95%		2100
	yield 40%, ee 95%	50%	840
	yield 3%	35%	63
	yield 2%, ee 10%		34
	yield 23%, ee 95%		1700
	yield 40%, ee 49%	CHO 24%, COOH 5%	840

As mentioned above, the epoxidation of *trans*-alkenes catalyzed by CPO gives very low yield of epoxide [42]. The main reaction is allylic oxidation with formation of allylic alcohol, that can be further oxidized to the corresponding aldehyde. Allylic hydroxylation is also observed with alkenes having the double bond far removed from the chain terminus (*cis*-3-alkenes). Terminal alkenes deactivate the enzyme [42]. As a consequence linear long chain aliphatic 1-alkenes are poor substrates for epoxidation. The turnover number for terminal alkenes can be increased by introducing an alkyl chain (Table 7). Indeed epoxidation of 2-methylalkenes showed a dramatic increase both in turnover and enantioselectivity. Further, while the epoxidation of allyl propionate with CPO leads rapidly to the formation of an inactive green enzyme derivative, no detection of such species was found during the epoxidation of methallyl propionate, at the same time the ee increased from 24 to 94%. With α-ethyl-styrene the catalytic turnover declined, but a high enantioselectivity was observed. The results show that the products cannot be fit to a single stereochemical pattern, since configurations are inverted for three of the five matched pairs. The most salient regularity in the series is that only methyl substituted alkenes show the same faceselectivity.

TABLE 7. Epoxidation of olefins catalyzed by CPO

Substrate	Major Product	Ee (%)	Yield (%)	Turnovers
		49	89	900
		89	55	440
		81	1	26
		37	7	77
		70	41	3400
		46	1	28
		89	22	1700
		24	12	5
		94	34	4200
		10	2	34
		95	23	1700
		n.d.	4	25

(R)-(-)-mevalonolactone is an important intermediate in biosynthetic pathways leading to sterols, terpenes, etc. It has been obtained by Hager through a multistep synthesis featuring enantioselective epoxidation by CPO [43]. The enzyme is quite sensitive to hydrogen peroxide, loosing activity in excess reagent. For this reason Hager decided to use *tert*-butyl hydroperoxide as oxidant in the epoxidation reaction, since the reaction runs smoothly even when a large excess of this oxidant is present. Especially for larger scale reactions, the use of *tert*-BuOOH appears efficacious. The system examined was a series of ω–bromo-2-alkenes as a function of chain length [44]. In all cases the predominant enantiomer had the (*R*) configuration, except for 3-bromo-2-metylpropene-1-oxide which was predominantly (*S*) simply because of a priority switch.

The crystal stucture of CPO reveals, as we have already seen, a catalytic heme functionality immersed in a proteinaceous pocket. Entry into this active site is likely to be increasingly inhibited for substrates with greater steric requirements. The kinetic parameters V_{max} and K_M are in line with faster reaction rates for shorter chain substrates, while binding is somewhat enhanced for larger alkyl chain alkenes.

The effect of chain length on enantioselectivity has been investigated recently by Hager in the epoxidation of functionalized alkenes, with *tert*-butyl hydroperoxide as terminal oxidant [45]. These functionalized olefins are effectively converted into epoxides with excellent enantioselectivity (ee's up to 95%). As the carbon chain lengthens the ee remains high, while the yield progressively decreases as expected. When a terminal bromine atom is present the enantioselectivity is good, although the yields are moderate. The unsaturated carboxylic ester methallyl propionate is an excellent substrate, but the optical yield and especially the chemical yield decreases by adding a carbon atom. In all the products the carbon atom adjacent to the terminal methyl group is of the (*R*) absolute configuration. Again the presence of the methyl group is essential for the selectivity and the efficiency of the epoxidation reaction. This study shows that CPO can tolerate a variety of functional groups.

Chiral propargylic alcohols are important building blocks for the enantioselective synthesis of complex molecules, in particular biologically active compounds. There are two general strategies for the asymmetric synthesis of propargylic alcohols: the enantioselective alkynylation of aldehydes and the stoichiometric or catalytic reduction of acetylenic ketones. However, as recently found by Hager [46] the most direct procedure for the preparation of propargylic alcohols with high ee is the asymmetric hydroxylation of alkynes catalyzed by CPO. This method works with a variety of unfunctionalized and functionalized alkynes yielding chiral propargylic alcohols. Both hydrogen peroxide and *tert*-butyl hydroperoxide can serve as terminal oxidant in the reaction. In the case of unfunctionalized acetylenes, the results record the effect of the carbon chain length on the enantioselectivity and yield of the reaction. When the smallest alkyne serves as a substrate, both the ee and yield of propargylic alcohol are very moderate. However, the enantioselectivity and yield dramatically increase by adding one carbon atom. CPO proved to be progressively less stereo-selective towards alkynes possessing additional carbons and the chemical yield also decreases. Obviously steric requirements of the active site of the enzyme limit the entry of larger substrates and the size restrictions are similar to those noted in the CPO epoxidation of *cis*-alkenes. Very surprisingly, when an electron-withdrawing group (acetoxy or bromo) is attached to the methyl group remote of the prochiral propargylic carbon, the enantioselectivity and yield show a remarkable enhancement.

The mechanism of the propargylic hydroxylation catalyzed by CPO is ambiguous. However, since radical intermediates can be stabilized by electron-withdrawing groups, it is tempting to speculate that propargylic hydroxylations proceed via a free-radical intermediate. This hypothesis differs from the mechanism proposed for the CPO benzylic hydroxylation reaction.

Hydrogen peroxide as terminal oxidant is preferred over *tert*-butyl hydro-peroxide that produces a significant amount of α,β-ynone arising from oxidation of the propargylic alcohol. In all cases the chiral product has produced the (R) configuration. These results are similar to the enantioselective epoxidation of alkenes by CPO. Modifications of the active site of CPO by random and/or site-directed mutagenesis would increase, in principle, the substrate specificity and improve the selectivity of the propargylic hydroxylations.

Native HRP does not catalyze the epoxidation of styrene and of *cis*-β-methyl styrene and only traces of the corresponding epoxides are formed in the reaction with hydrogen peroxide. However, decreasing the volume of the amino acid at position 41 by replacing phenylalanine by leucine allows the epoxidation of styrene, but not of *cis*-β-methyl-styrene, whereas the mutant containing threonine oxidizes both substrates [8]. Styrene is oxidized to styrene oxide, phenylacetaldehyde and benzaldehyde, the latter by an unknown mechanism. The oxidation of *cis*-β-methyl styrene by the F41T mutant yields *cis*-β-methyl-styrene oxide, *trans*-β-methyl-styrene oxide, 1-phenyl-2-propanone. This epoxidation therefore proceeds with partial loss of the substrate stereochemistry and formation of a rearranged ketone. The ee of the *cis*-epoxide formed is greater than 99%, whereas the *trans*-epoxide has only 10% ee. ^{18}O labelling studies clearly indicate that the epoxidation with retention of stereochemistry occurs exclusively by the ferryl oxygen transfer mechanism. On the contrary, the epoxidation with inversion of stereochemistry involves two reaction mechanisms, one leading to the incorporation of ferryl oxygen and the other not. Surprisingly *trans*-β-methyl-styrene is oxidized not only by F41T HRP, but also by native HRP. It appears that F41T HRP increases not only the ee, but also the fraction of the reaction that incorporates peroxide oxygen in the epoxide. These results clearly substantiate that the epoxidation of all three substrates, but more so the epoxidation of *trans*-β-methyl-styrene, is mediated by at least two mechanisms, one of which involves a "P450-like" ferryl oxygen transfer and the other, obscure, with incorporation of molecular oxygen [8]. The similarity in styrene K_M and V_{max} values for the two mutants suggests that both open the active site to a similar extent, but this is contradicted by the lower stereospecificity of the F41T mutant and by the observation that cis-β-methyl styrene is only oxidized by the F41T mutant. The inference that the cavity adjacent to the ferryl species is less restricted in the F41T than in F41L mutant is supported by the finding that F41T HRP is less stereoselective in the sulfoxidation of thioethers than F41L.

4.   Miscellaneous

In the doctoral thesis of Dr. Seelbach "Chloroperoxidase - An industrial catalyst ?" (Bonn, 1997) it has been shown that indeed with a fed-batch reactor and the use of a hydrogen peroxide-stat the oxidation of indole to oxindole occurs with a turnover higher than 860,000 [47]. Substituted oxindoles are interesting compounds due to their biological properties, for instance as anti-inflammatory drugs. Direct chemical oxidation of indole to 2-oxindole is difficult as oxidation occurs at the more electron rich 3-position. The turnover number is still high for the sulfoxidation of phenyl-methyl sulfide (150,000) and drops substantially in the epoxidation of cis-2-heptene (7,400).

When compared with other enzymatic and microbial epoxidation methods, the CPO-based system appears to accept a broader range of substrates and to effect epoxidation with much higher enantioselectivity. Perhaps most important, this method is highly complementary to the existing asymmetric epoxidations involving either synthetic or biological catalysts, that show very poor enantioselectivity with most of the substrates which give good results with CPO. The lack of functional expression in hosts other than *Caldariomyces fumago* has not deterred Hager from the goal of engineering this enzyme. The development of a *Caldariomyces fumago* integrative vector system has allowed his group to create random libraries, although of relative small size, from which interesting chloroperoxidase variants have been isolated [48], that resist inactivation by terminal alkenes and hydrogen peroxide.

5. Acknowledgements

This work was supported by CNR, MURST. (Ministero dell'Università e della Ricerca Scientifica e Tecnologica), Programmi di Ricerca Scientifica di Interesse Nazionale "Processi efficienti per l'ossidazione controllata di composti organici" and the Biotechnology Programme of the European Commission (BIO 4-CT98-0267).

6. References

1. Dawson, J.H. (1988) Probing structure-function relations in heme-containing oxygenases and peroxidases, *Science* **240**, 433-439.
2. Marnett, L.J. and Kennedy, T.A. (1995) Comparison of the peroxidase activity of hemoproteins and cytochrome P450, in Ortiz de Montellano, P.R. (ed.), *Cytochrome P450. Structure, Mechanism, and Biochemistry*, Plenum Press, New York, pp. 49-80.
3. Gajhede, M., Schuller, D.J., Enriksen, A., Smith, A.T., and Poulos, T.L. (1997) Crystal structure of horseradish peroxidase C at 2.15 Å resolution, *Nature Structural Biology* **4**, 1032-1038.
4. Mikolajczyk, M., Drabowicz, J., and Kielbasinski, P. (1997) *Chiral Sulfur Reagents; Applications in Asymmetric and Stereoselective Synthesis*, CRC Press, Boca Raton.
5. Carreño, M.C. (1995) Applications of sulfoxides to asymmetric synthesis of biologically active compounds, *Chem. Rev.* **95**, 1717-1760.
6. Whitesell, J.K. and Wong, M.-S. (1994) Asymmetric synthesis of chiral sulfinate esters and sulfoxides. Synthesis of sulforaphane, *J. Org. Chem.* **59**, 597-601.
7. Colonna, S., Gaggero, N., Carrea, G., and Pasta, P. (1992) Horseradish peroxidase catalysed sulfoxidation is enantioselective, *J. Chem. Soc., Chem. Commun.*, 357-358.
8. Ozaki, S. and Ortiz de Montellano, P.R. (1995) Molecular engineering of horseradish peroxidase: thioeter sulfoxidation and styrene epoxidation by Phe-41 leucine and threonine mutants, *J. Am. Chem. Soc.* **117**, 7056-7064.
9. Sundaramoorthy, M., Terner, J., and Poulos, T.L. (1995) The crystal structure of chloroperoxidase: a heme peroxidase-cytochrome P450 functional hybrid, *Structure* **3**, 1367-1377.
10. (a) Colonna, S., Gaggero, N., Manfredi, A., Casella, L., Gullotti, M., Carrea, G., and Pasta, P. (1990) Enantioselective oxidations of sulfides catalyzed by chloroperoxidase, *Biochemistry* **29**, 10465-10468. (b) Colonna, S., Gaggero, N., Casella, L., Carrea, G., and Pasta, P. (1992) Chloroperoxidase and hydrogen peroxide: an efficient system for enzymatic enantioselective sulfoxidation, *Tetrahedron: Asymmetry* **3**, 95-106.
11. Casella, L., Gullotti, M., Ghezzi, R., Poli, S., Beringhelli, T., Colonna, S., and Carrea, G. (1992) Mechanism of enantioselective oxygenation of sulfides catalyzed by chloroperoxidase and horseradish peroxidase. Spectral studies and characterization of enzyme-substrate complexes, *Biochemistry* **31**, 9451-9459.
12. Fu, H., Kondo, H., Ichikawa, Y., Look, G.C., and Wong, C.-H. (1992) Chloroperoxidase-catalyzed asymmetric synthesis: enantioselective reactions of chiral hydroperoxides with sulfides and bromohydration of glycals, *J. Org. Chem.* **57**, 7265-7270.

158

13. van Deurzen, M.P.J., Remkes, I.J., van Rantwijk, F., and Sheldon, R.A. (1997) Chloroperoxidase catalyzed oxidations in *t*-butyl alcohol/water mixtures, *J. Mol. Catal. A: Chem.* **117**, 329-337.

14. Colonna, S., Gaggero, N., Carrea, G., and Pasta, P. (1997) A new enzymatic enantioselective synthesis of dialkyl sulfoxides catalysed by monooxygenases, *J. Chem. Soc., Chem. Commun.*, 439-440.

15. van de Velde, F., van Rantwijk, F., and Sheldon, R.A. (1999) Selective oxidations with molecular oxygen, catalyzed by chloroperoxidase in the presence of a reductant, *J. Mol. Catal. B: Enz.* **6**, 453-461.

16. Pasta, P., Carrea, G., Monzani, E., Gaggero, N., and Colonna, S. (1999) Chloroperoxidase catalyzed enantioselective oxidation of methyl phenyl sulfide with dihydroxyfumaric acid/oxygen or ascorbic acid/oxygen as oxidant, *Biotechnol. and Bioengineering* **62**, 489-493.

17. Allenmark, S.G. and Andersson, M.A. (1996) Chloroperoxidase-catalyzed asymmetric synthesis of a series of aromatic cyclic sulfoxides, *Tetrahedron: Asymmetry* **7**, 1089-1094.

18. Walsh, C.T. and Chen, Y.-C.J. (1988) Enzymic Baeyer-Villiger oxidations by flavin-dependent monooxygenases, *Angew. Chem. Int. Ed. Engl.* **27**, 333-343.

19. Light, D.R., Waxman, D.J., and Walsh, C. (1982) Studies on the chirality of sulfoxidation catalyzed by bacterial flavoenzyme cyclohexanone monooxygenase and hog liver flavin adenine dinucleotide containing monooxygenase, *Biochemistry* **21**, 2490-2498.

20. Carrea, G., Redigolo, B., Riva, S., Colonna, S., Gaggero, N., Battistel, E., and Bianchi, D. (1992) Effects of substrate structure on the enantioselectivity and stereochemical course of sulfoxidation catalyzed by cyclohexanone monooxygenase, *Tetrahedron: Asymmetry* **3**, 1063-1068.

21. Secundo, F., Carrea, G., Dallavalle, S., and Franzosi, G. (1993) Asymmetric oxidation of sulfides by cyclohexanone monooxygenase, *Tetrahedron: Asymmetry* **4**, 1981-1982.

22. Pasta, P., Carrea, G., Holland, H.L., and Dallavalle, S. (1995) Synthesis of chiral benzyl alkyl sulfoxides by cyclohexanone monooxygenase from *Acinetobacter* NCIB 9871, *Tetrahedron: Asymmetry* **6**, 933-936.

23. Secundo, F., Carrea, G., Riva, S., Battistel, E., and Bianchi, D. (1993) Cyclohexanone monooxygenase catalyzed oxidation of methyl phenyl sulfide and cyclohexanone with macromolecular NADP in a membrane reactor, *Biotechnol. Letters* **15**, 865-870.

24. Ottolina, G., Pasta, P., Carrea, G., Colonna, S., Dallavalle, S., and Holland, H.L. (1995) A predictive active site model for the cyclohexanone monooxygenase catalyzed oxidation of sulfides to chiral sulfoxides, *Tetrahedron: Asymmetry* **6**, 1375-1386.

25. Colonna, S., Gaggero, N., Bertinotti, A., Carrea, G., Pasta, P., and Bernardi, A. (1995) Enantioselective oxidation of 1,3-dithioacetals catalysed by cyclohexanone monooxygenase, *J. Chem. Soc., Chem. Commun.*, 1123-1124.

26. Alphand, V., Gaggero, N., Colonna, S., Pasta, P., and Furstoss, R. (1997) Microbiological transformations 36: preparative scale synthesis of chiral thioacetal and thioketal sulfoxides using whole-cell biotransformations, *Tetrahedron* **53**, 9965-9706.

27. Naruta, Y., Tani, F., Ishihara, N., and Maruyama, K. (1991) Catalytic and asymmetric epoxidation of olefins with iron complexes of "twin-coronet" porphyrins. A mechanistic insight into the chiral induction of styrene derivatives, *J. Am. Chem. Soc.* **113**, 6865-6872.

28. Jacobsen, E.N., Zhang, W., Muci, A.R., Ecker, J.R., and Deng, L. (1991) Highly enantioselective epoxidation catalysts derived from 1,2-diaminocyclohexane, *J. Am. Chem. Soc.* **113**, 7063-7064.

29. Ortiz de Montellano, P.R. (1995) Oxygen activation and reactivity, in Ortiz de Montellano, P.R. (ed.), *Cytochrome P450. Structure, Mechanism, and Biochemistry*, Plenum Press, New York, pp. 245-303.

30. Neidleman, S.L. and Geigert, J. (1992) *Biohalogenation: Principles, Basic Roles, and Applications*, Ellis Horwood, Chichester.

31. May, S.W., Schwartz, R.D. Abbott, B.J., and Zaborsky, O.R. (1975) Structural effects on the reactivity of substrates and inhibitors in the epoxidation system of *Pseudomonas oleovorans*, *Biochim. Biophys., Acta* **403**, 245-255.

32. Fruetel, J.A., Collins, J.R., Camper, D.L., Loew, G.H., and Ortiz de Montellano, P.R. (1992) Calculated and experimental absolute stereochemistry of the styrene and β-methylstyrene epoxides formed by cytochrome P450$_{cam}$, *J. Am. Chem. Soc.* **114**, 6987-6993.

33. Wistuba, D., Nowotny, H.-P., Träger, O., and Schurig, V. (1989) Cytochrome *P*-450-catalyzed asymmetric epoxidation of simple prochiral and chiral aliphatic alkenes: species dependence and effect of enzyme induction on enantioselective oxirane formation, *Chirality* **1**, 127-136.

34. Gross, Z. and Nimri, S. (1995) Seeing the long-sought intermediate in the reaction of oxoiron(IV) porphyrin cation radicals with olefins, *J. Am. Chem. Soc.* **117**, 8021-8022.

35. Ortiz de Montellano, P.R., Choe, Ortiz de Montellano, P.R., Choe, Y.S., DePillis, G., and Catalano, C.E. (1987) Structure-mechanism relationships in hemoproteins, *J. Biol. Chem.* **262**, 11641-11646.

36. Geigert, J., Lee, T.D., Dalietos, D.J., Hirano, D.S., and Neidleman, S.L. (1986) Epoxidation of alkenes by chloroperoxidase catalysis, *Biochem. Biophys. Res. Commun.* **136**, 778-782.

37. Elfarra, A.A., Duescher, R.J., and Pasch, C.M. (1991) Mechanisms of 1,3-butadiene oxidations to butadiene monoxide and crotonaldehyde by mouse liver microsomes and chloroperoxidase, *Arch. Biochem. Biophys.* **286**, 244-251.

38. Colonna, S:, Gaggero, N., Casella, L., Carrea, G., and Pasta, P. (1993) Enantioselective epoxidation of styrene derivatives by chloroperoxidase catalysis, *Tetrahedron: Asymmetry* **4**, 1325-1330.

39. Allain, E.J., Hager, L.P., Deng, L., and Jacobsen, E.N. (1993) Highly enantioselective epoxidation of disubstituted alkenes with hydrogen peroxide catalyzed by chloroperoxidase, *J. Am. Chem. Soc.* **115**, 4415-4416.

40. Dexter, A.F. and Hager, L.P. (1995) Transient heme N-alkylation of chloroperoxidase by terminal alkenes and alkynes, *J. Am. Chem. Soc.* **117**, 817-818.

41. Zaks, A. and Dodds, D.R. (1995) Chloroperoxidase-catalyzed asymmetric oxidations: substrate specificity and mechanistic study, *J. Am. Chem. Soc.* **117**, 10419-10424.

42. Dexter, A.F., Lakner, F.J., Campbell, R.A., and Hager, L.P. (1995) Highly enantioselective epoxidation of 1,1-disubstituted alkenes catalyzed by chloroperoxidase, *J. Am. Chem. Soc.* **117**, 6412-6413.

43. Lakner, F.J. and Hager, L.P. (1996) Chloroperoxidase as enantioselective epoxidation catalyst: an efficient synthesis of (*R*)-(-)-mevalonolactone, *J. Org. Chem.* **61**, 3923-3925.

44. Lakner, F.J., Cain, K.P., and Hager, L.P. (1997) Enantioselective epoxidation of ω-Bromo-2-methyl-1-alkenes catalyzed by chloroperoxidase. Effect of chain length on selectivity and efficiency, *J. Am. Chem. Soc.* **119**, 443-444.

45. Hu, S. and Hager, L.P. (1999) Asymmetric epoxidation of functionalized cis-olefins catalyzed by chloroperoxidase, *Tetrahedron Letters* **40**, 1641-1644.

46. Hu, S. and Hager, L.P. (1999) Highly enantioselective propargylic hydroxylations catalyzed by chloroperoxidase, *J. Am. Chem. Soc.* **121**, 872-873.

47. Seelbach, K., van Deurzen, M.P.J., van Rantwijk, F., Sheldon, R.A., and Kragl, U. (1997) Improvement of the total turnover number and space-time yield for chloroperoxidase catalyzed oxidation, *Biotechnol. Bioengineering* **55**, 283-288.

48.   Schmidt-Dannert, C. and Arnold, F.H. (1999) Directed evolution of industrial enzymes, *TIBTECH* **17**, 135-136.

HYDROLYTIC ENZYMES IN THE SYNTHESIS OF NON-RACEMIC HETEROORGANIC COMPOUNDS WITH A STEREOGENIC CENTRE ON THE HETEROATOM

PIOTR KIEŁBASIŃSKI AND MARIAN MIKOŁAJCZYK
Centre of Molecular and Macromolecular Studies,
Polish Academy of Sciences, 90-363 Łódź, Sienkiewicza 112,
Poland

Abstract: Commonly available hydrolytic enzymes are capable of recognizing heteroatom stereogenic centers. A review of the application of these enzymes in the preparation of optically active phosphorus, sulfur, silicon and germanium compounds is presented.

1. Introduction

Application of enzymes in organic chemistry, particularly in the synthesis of optically active compounds both from racemic and prochiral substrates, is rapidly becoming a powerful methodology [1,2]. A great advantage of biocatalysts is that there is an enzyme-catalyzed process equivalent to almost every type of organic reaction. Moreover, enzymes display three major types of selectivity: chemoselectivity, regioselectivity and diastereo- or enantioselectivity. Although it is generally true that an enzyme displays its highest catalytic activity in water, quite recently some noteworthy rules for conducting biotransformations in organic media have been delineated [3]. Therefore, biocatalysis offers an attractive option for the development of new economical and ecologically acceptable processes.

The chemical literature records an enormously great number of examples of enzyme-promoted syntheses of optically active compounds in which carbon is the centre of chirality. Scheme 1 shows one of the most common approaches utilizing hydrolytic enzymes. The stereogenic or prostereogenic centre is here recognized and stereo-

161

B. Zwanenburg et al. (eds.), Enzymes in Action, 161–191.
© 2000 Kluwer Academic Publishers. Printed in the Netherlands.

selectively bound in the active site of the enzyme, but is not involved in the chemical transformations which take place at a more or less remote reacting site. In this case, it is an ester group which is either formed or cleaved. The stereogenic (prostereogenic) centre may be located either in the acid or in the alcohol portion of the ester molecule.

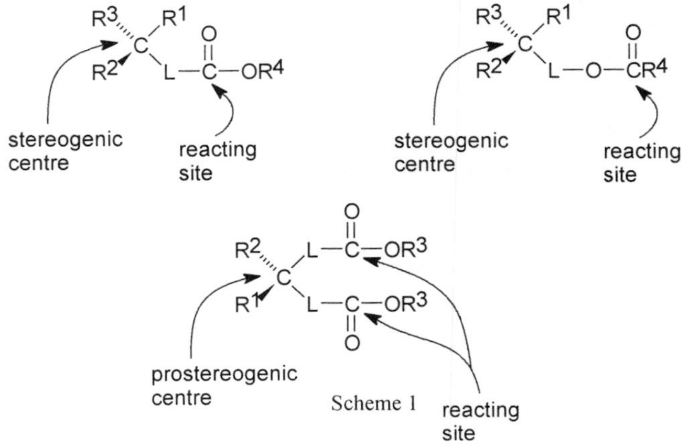

$$R^3\!\!\!\!\diagdown\!\!\!\!\underset{R^2}{\overset{R^1}{C}}\!\!-\!\!L\!-\!\overset{\overset{O}{\|}}{C}\!-\!OR^4$$

stereogenic centre   reacting site

$$R^3\!\!\!\!\diagdown\!\!\!\!\underset{R^2}{\overset{R^1}{C}}\!\!-\!\!L\!-\!O\!-\!\overset{\overset{O}{\|}}{C}R^4$$

stereogenic centre   reacting site

$$R^2\!\!\!\!\diagdown\!\!\!\!\underset{R^1}{\overset{}{C}}\!\!\diagup\!\!\!\!\begin{smallmatrix}L\!-\!\overset{\overset{O}{\|}}{C}\!-\!OR^3\\[2pt]L\!-\!\underset{\underset{O}{\|}}{C}\!-\!OR^3\end{smallmatrix}$$

prostereogenic centre   Scheme 1   reacting site

However, in contrast to carbon compounds, the corresponding enzymatic syntheses of chiral heteroorganic compounds are much more scarce. In principle, they may be accomplished according to two strategies. In the first one the chemical transformation takes place directly at the stereogenic or prostereogenic heteroatomic centre which is at the same time stereoselectively bound with the enzyme active site (Scheme 2). Such an approach was utilized in the enzyme-catalyzed enantioselective oxidation of unsymmetrical sulfides to sulfoxides [4], bacterial reduction of racemic sulfoxides [5] and pepsin-catalyzed hydrolysis of unsymmetrical sulfites [6]. In the case of phosphorus derivatives, the phosphotriesterase promoted hydrolysis of racemic phosphates [7] and phosphorothionates [8] belongs under this methodology.

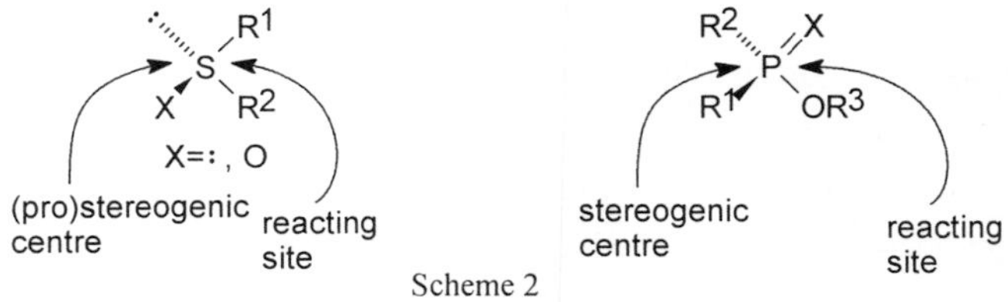

$$\underset{X}{\overset{R^1}{\underset{R^2}{S}}}\qquad X=:,O$$

(pro)stereogenic centre   reacting site

$$R^2\!\!\!\!\diagdown\!\!\!\!\underset{R^1}{\overset{X}{P}}\!\!\diagup\!\!\!\!OR^3$$

stereogenic centre   reacting site

Scheme 2

The second approach shown in Scheme 3 resembles the situation depicted in Scheme 1, the only difference being the fact that the heteroatom and not carbon is a stereogenic centre. It is worth noting that this approach to the synthesis of optically active heteroorganic compounds envisages the use of common hydrolytic enzymes which catalyze the formation or cleavage of the ester moiety.

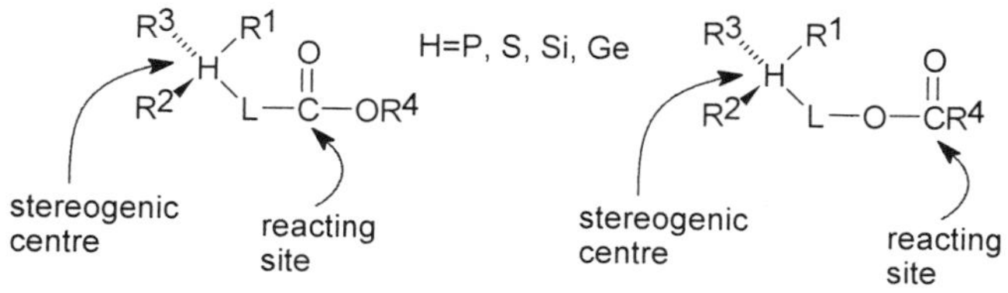

Scheme 3

This methodology has been a subject of continuously growing interest [9]. The aim of the present account is to summarize the results obtained in the author's laboratory on the hydrolytic enzyme-mediated synthesis of optically active heteroatom compounds as well as to present the achievements in this field published by other authors.

2. Synthesis of Optically Active Organophosphorus Compounds

Phosphorus compounds play a significant role in various areas of current research such as organic synthesis, biochemistry and catalysis. Especially interesting are chiral, non-racemic compounds containing the stereogenic phosphorus atom because of their application in asymmetric synthesis and catalysis. Among various classes of chiral phosphorus compounds, chiral phosphine oxides and phosphines are of particular importance [10]. The first of them are used as chiral auxiliaries in stoichiometric reactions and are the best precursors of chiral phosphines which, in turn, are widely used as chiral ligands in the transition metal catalysts. Therefore, efficient and general methods for the synthesis of chiral phosphine oxides and related compounds are a permanent subject for research of organic chemists. The successful application of enzymatic methods in the preparation of optically active carbon compounds prompted us to check whether this methodology could be applied to organophosphorus compounds having the stereogenic phosphoryl group.

2.1. KINETIC RESOLUTION OF PHOSPHORYLACETATES

In the first set of experiments the enzymatic hydrolysis of racemic phosphinylacetates **1a-e** in the presence of porcine liver esterase (PLE) was investigated [11,12]. The reaction was carried out until ca. 50% conversion and, after quenching the reaction mixture and work-up, both the unreacted esters **1*** and the corresponding acids **2** were isolated in good yields and with high enantioselectivity (Scheme 4, Table 1). The only exception was methyl *t*-butylphenylphosphinylacetate **1e** which was unreactive under the reaction conditions applied. Due to the fact that the acids **2** were difficult to obtain in a pure state, they were reesterified to the methyl esters **1**.

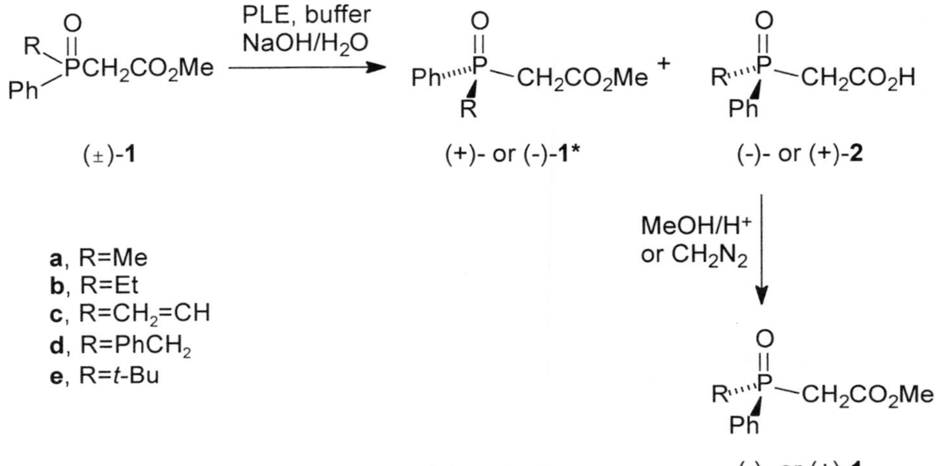
PLE, buffer
NaOH/H2O
(±)-1
(+)- or (-)-1*
(-)- or (+)-2
MeOH/H+
or CH2N2
a, R=Me
b, R=Et
c, R=CH2=CH
d, R=PhCH2
e, R=t-Bu
(-)- or (+)-1

Scheme 4

TABLE 1. Enzymatic hydrolysis of phosphoryl acetates **1**.

Substrate	Ester 1*				Acid 2			
	Yield (%)	$[\alpha]_D$ (MeOH)	ee (%)	Abs. conf.	Yield (%)	$[\alpha]_D$ (MeOH) ($[\alpha]_D$,MeOH)[a]	ee (%)	Abs. conf.
1a	50	+22.0	82	*R*	42	- 22.2[a]	82	*S*
1b	45	+9.7	>96	*R*	41[a]	- 8.2[a]	81	*S*
1c	40	- 54	~100	*S*	22	+54.5	-	*R*
1d	46	- 23.3	80	*R*	43	+16.9 (+21.7)[a]	~79	*S*
1f	40	- 16.1	~95	*S*	44	+9.1 (+10.8)[a]	~64	*R*
1g	46	- 11.3	~67	*S*	40	+13.8 (+11.8)[a]	~71	*R*
1h	69	- 8.7	~48	*S*	30[a]	+13.0 (+9.1)[a]	~50	*R*
1i	50	- 5.8	~26	*S*	47[a]	+3.0[a]	~15	*R*
1j	54	- 3.0	~30	*S*	33[a]	+4.0[a]	~25	*R*
1k	50	+8.5	~38	-	34[a]	- 10.0[a]	~42	-
1l	40	+1.9	-	-	60[a]	- 1.4[a]	-	-
1m	66	- 3.0[b]	~20	-	22	+8.7[b]	52	-
1n	20	- 21.7	~90	-	58	+5.8	25	-

a) After reesterification; b) In $CHCl_3$

The enantiomeric excess values of both enantiomeric methyl esters **1*** and **1** obtained in this way were determined by ^{1}H NMR spectra of their complexes with (S)-(+)-t-butylphenylphosphinothioic acid as a chiral solvating agent [13].

Since phosphinylacetates **1** are a new class of chiral, non-racemic compounds and have not been prepared before, it was desirable to establish their absolute configuration. To this end, the unreacted esters (+)-**1b*** and (-)-**1d*** recovered from the enzymatic hydrolysis were subjected to decarboxylation [11] to give the well known (S)-(-)-ethylmethylphenylphosphine oxide [14] and (S)-(-)-benzylmethylphenylphosphine oxide [14], respectively. As the bonds around chiral phosphorus are not broken in this reaction, the absolute configuration R was assigned to (+)-**1b** and (-)-**1d**. In the case of (-)-**1a**, the chemical correlation involved dimethylation of the methylene carbon atom and decarboxylation [12] to give (S)-(-)-isopropylmethylphenylphosphine oxide [15]. For similar reasons as mentioned above, the absolute configuration S was assigned to (-)-**1a**. In turn, hydrogenation of the acetate (-)-**1c*** to (R)-(+)-**1b** allowed to ascribe the S-chirality to the former.

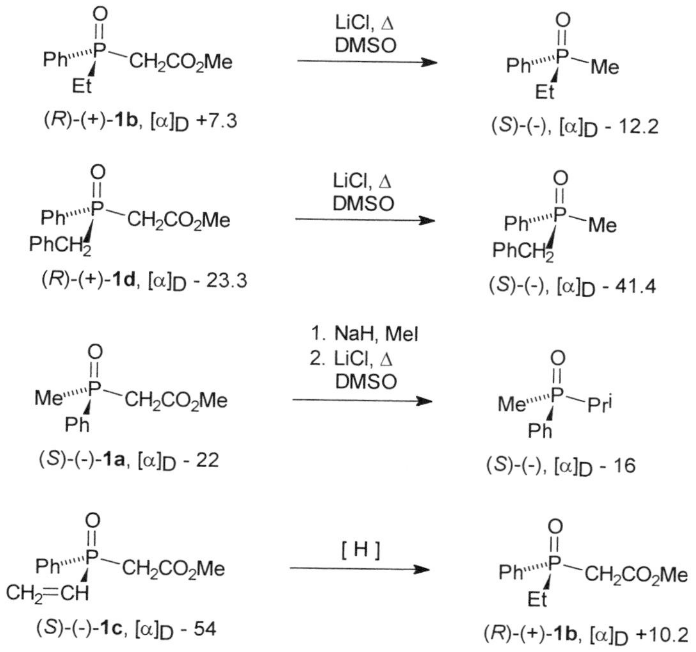

Scheme 5

Determination of the absolute configuration of the enzymatic hydrolysis products revealed that in all cases the enantiomers of **1** of the same spatial structure shown below are preferably recognized by PLE and hydrolyzed faster.

$$R^{''''} \overset{\overset{\displaystyle O}{\|}}{\underset{Ph}{P}} - CH_2CO_2Me$$

This observation prompted us to extend our kinetic resolution studies to the phosphonyl acetates **1f-j** and structurally related compounds **1k-n**.

$$\underset{Ph}{\overset{R}{\diagdown}} \overset{\overset{\displaystyle O}{\|}}{P} - CH_2CO_2Me \qquad \underset{R^1}{\overset{R^2}{\diagdown}} \overset{\overset{\displaystyle O}{\|}}{P} - CH_2CO_2Me$$

$$(\pm)\text{-}\mathbf{1} \qquad\qquad\qquad (\pm)\text{-}\mathbf{1}$$

f, R=MeO	**k**, R^1=Et, R^2=MeO
g, R=EtO	**l**, R^1=Et, R^2=EtO
h, R=*n*-PrO	**m**, R^1=PhO, R^2=EtO
i, R=*i*-PrO	**n**, R^1=Et$_2$N, R^2=MeO
j, R=*n*-BuO	

As expected, the PLE-catalyzed hydrolysis of these two series of phosphoryl acetates was also occuring in an enantioselective way and afforded the optically active products **1*** and **2** in good yields (Table 1) [16]. However, their enantiomeric excess values were generally lower, an exception being acetates **1f** and **1n** (ee over 90%). We were also able to assign the absolute configuration to the esters **1*f-j** and acids **2f-j** on the basis of the literature data (for **1f** and **2f**) [17], chemical correlation, CD spectra and X-ray analysis. Thus, all the levorotatory esters **1*f-j** have the *S* configuration and the dextrorotatory ones *R*, with the latter being preferentially hydrolyzed by PLE. Moreover, comparison of the absolute configuration of the enantiomers of phosphinylacetates **1a-e** and phosphonylacetates **1f-j**, that undergo faster enzymatic hydrolysis, clearly shows that they have the same three-dimensional shape.

$$R^{''''} \overset{\overset{\displaystyle O}{\|}}{\underset{Ph}{P}} - CH_2CO_2Me \qquad \begin{array}{l} R=Me, Et, CH_2=CH, PhCH_2 \\ MeO, EtO, \textit{n}\text{-PrO}, \textit{i}\text{-PrO}, \textit{n}\text{-BuO} \end{array}$$

2.2. APPLICATION OF THE JONES ACTIVE-SITE MODEL OF PIG LIVER ESTERASE (PLE) TO PHOSPHORYL SUBSTRATES

An interesting observation on the uniform steric course of the PLE-catalyzed hydrolysis of the phosphoryl acetates **1** discussed above encouraged us to try to identify the molecular interactions that determine enantiorecognition. In the absence of an X-ray structure of PLE we turned our attention to the empirical model of the active-site of PLE proposed by Jones and based on his extensive studies on the PLE-catalyzed reactions of C-chiral and prochiral substrates [18,19]. This model, which is constructed in a cubic-space form, is comprised of five binding loci. The important binding regions for specificity determination are two hydrophobic pockets, large H_L and small H_S, and two polar pockets P_F and P_B. The fifth catalytically essential region is the serine locus. Though this model had been originally developed for C-chiral substrates, we have applied it to our P-chiral substrates [12,16]. Thus, in the case of a series of the phosphoryl acetates **1a-j**, the faster hydrolyzed enantiomers should be accommodated in the PLE active site as follows. The methoxycarbonyl group should be located within the locus of serine, the phenyl group in H_L, the alkyl or alkoxyl group in H_S and the phosphoryl group in P_B (Fig. 1a). The location of the three former groups is obvious. There is, however, some uncertainty as far as the location of the phosphoryl group is concerned since it may alternatively be placed in the front polar pocket P_F (Fig. 1b). This would lead to the preferential hydrolysis of the opposite enantiomers of **1a-j**. This is, however, not the case. According to Jones predictions, the P_F pocket is mostly used to bind the second nonhydrolyzed ester function of diester substrates, but it can also accept nonpolar groups. On the other hand, the P_B site interacts well, among others, with carbonyl functions, but is too polar to accept hydrophobic moieties. Therefore, the strongly polar phosphoryl group (P=O) should be preferentially accommodated in P_B hence favoring the model shown in Fig. 1a, which is in full agreement with the experiment.

168

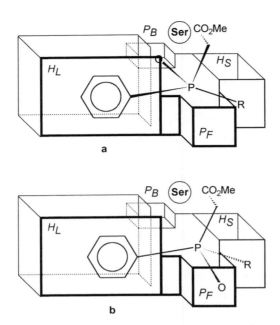

Figure 1: Binding orientation of **1** in the active-site of PLE
(**a**) of the enantiomer preferentially hydrolyzed
(**b**) of the enantiomer recovered

2.3. KINETIC RESOLUTION OF HYDROXYMETHYLPHOSPHONATES AND PHOSPHINATES

α-Hydroxyalkanephosphonates have recently attracted attention because of their interesting biological activity. Chemical synthesis of P-chiral α-hydroxyphosphonates and structurally related compounds involves the Abramov and Pudovik reactions of carbonyl compounds with the appropriate chiral H-phosphonates [20]. However, the main limitation of this approach is the poor availability of the latter reaction partners. Encouraged by successful enzymatic resolution of phosphoryl acetates we decided to apply enzymatic methods for the preparation of this class of optically active phosphorus compounds. In fact, we have chosen a lipase-mediated acetylation of hydroxymethyl-phosphonates and phosphinates **3** and a reverse hydrolysis of the corresponding O-acetyl derivatives **4** (Scheme 6) [21].

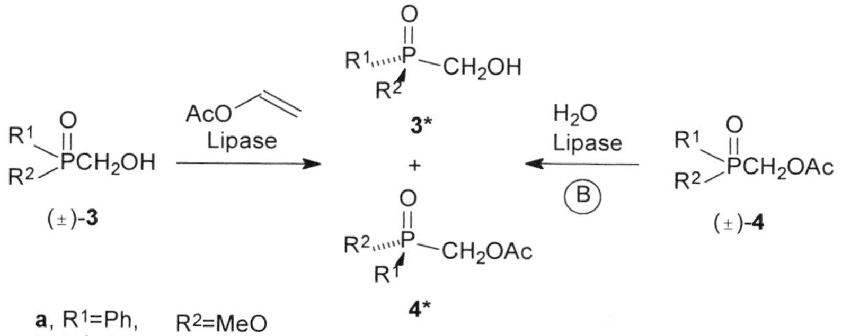

a, R^1=Ph,   R^2=MeO
b, R^1=Ph,   R^2=EtO
c, R^1=Ph,   R^2=i-PrO
d, R^1=i-PrO, R^2=Me
e, R^1=i-PrO, R^2=Et
f, R^1=i-PrO, R^2=MeO

Scheme 6

TABLE 2. Lipase-promoted resolution of racemic **3** and **4**.

Substrate	Lipase[b]	Proced.	3*				4*			
			Yield (%)	[α]$_D$ (MeOH)	ee (%)	Abs. conf.	Yield (%)	[α]$_D$ (MeOH)	ee (%)	Abs. conf.
a	PFL	A	44	- 24.7	80	R	39	+53.5	89	S
a	AM	A	42	- 29.7	92	R	44	+51.4	86	S
b	PFL	A	42	- 18.5	58	R	53	+35.5	47	S
b	AM	A	30	- 17.2	54	R	68	+14.9	21	S
c	PFL	A	37	- 46.6	80	R	46	+27.1	21	S
d	PFL	A	46	- 15.6	85	n.d.[a]	54	+13.2	85	n.d.[a]
				- 18.4[c]	>95[c,d]			+15.5[c]	>95[c,d]	
e	PFL	A	41	- 13.5	86	n.d.[a]	59	+13.3	64	n.d.[a]
				- 14.5[c]	>95[c,d]			+18.4	>95[c,d]	
f	PFL	B	55	+0.5	16	S	45	- 0.93	34	R
				+3.0[C]	>92[c,d]			- 2.3[c,d]	>92[d]	

a) n.d. - not determined; b) PFL: Lipase from *Pseudomonas fluorescens* in *i*-Pr$_2$O or *t*-BuOMe, AM: AMANO PS Lipase in CH$_2$Cl$_2$;

c) After repeated hydrolysis of the enantiomerically enriched substrates; d) None of the other enantiomer was detected; e) Neat.

The racemic **1a-e** were acetylated using vinyl acetate in the presence of lipases of *Pseudomonas cepacia*. The reaction was carried out under kinetic resolution conditions, i.e. it was stopped at ca. 50% conversion. The unreacted substrate **3*** and the acetylated product **4*** were separated by column chromatography. In the case of hydroxymethyl-phosphonate **3f**, hydrolysis of its O-acetyl derivative **4f** proved to be more efficient. However, **3d*** and **4d***, **3e*** and **4e*** as well as **3f*** and **4f*** with the ee values over 90%

could be prepared after a threefold repetitive acetylation or hydrolysis of the enantiomerically enriched substrates. The results obtained are summarized in Table 2. Determination of the absolute configuration of the optically acitve **3** and **4** revealed that within the series of structurally similar substrates, enantiomers of the same tetrahedral configuration, shown in Scheme 6 are preferentially recognized by the lipases used.

The resolved hydroxymethylphosphinates **3d** and **3e** were used for the synthesis of enantiomeric forms of a new class of herbicides **5** exhibiting activity against a variety of grassy weeds [22].

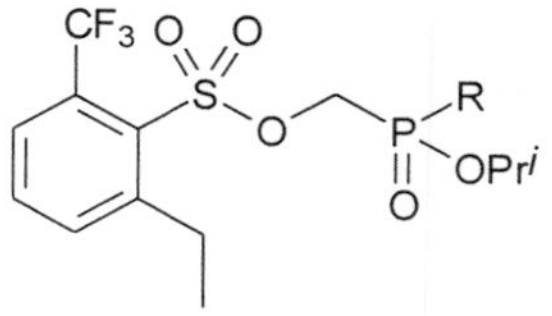

5

R=Me, Et

Biological testing of the enantiomeric phosphosulfonate herbicides **5** demonstrated that, in each case, the herbicidal activity was attributed to the (+)-enantiomer and that the (+)-enantiomer is more active than the racemate.

2.4. ENZYMATIC SYNTHESIS OF MISCELLANEOUS P-CHIRAL COMPOUNDS

Serreqi and Kazlauskas reported that the P-chiral aromatic phosphines and phosphine oxides **6** and **8** may be enzymatically resolved using the *ortho*-acetoxy group as the resolution handle [23]. A screening of commercially available hydrolases showed that two of them i.e. cholesterol esterase (CE) and lipase from *Candida rugosa* catalyzed hydrolysis of the acetates **6** and **8** with good enantioselectivity (Scheme 7, Table 3). It is interesting to point out that in this case the phosphorus stereocentre is separated by four bonds from the reacting carbonyl group.

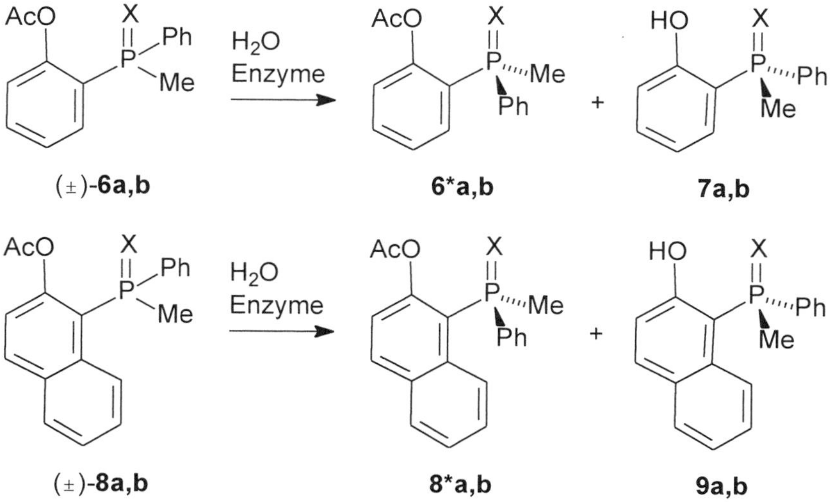

Scheme 7

TABLE 3. Enzymatic hydrolysis of acetates **6** and **8**.

Substrate	Enzyme	Product			Acetate*	
		No	ee (%)	Abs. conf.		ee (%)
6a	CE	**7a**	53	*R*	**6a***	49
6a	CRL	**7a**	19	*R*	**6a***	12
6b	CE	**7b**	49	*S*	**6b***	33
8a	CE	**9a**	89	*R*	**8a***	61
8a	CRL	**9a**	95	*R*	**8a***	69
8b	CE	**9b**	43	*S*	**8b***	44
8b	CRL	**9b**	15	*S*	**8b***	11

Acetylation using lipases from *Pseudomonas cepacia* was applied by Baba *et al.*, for kinetic resolution of the racemic hydroperoxycyclophosphamide **10** which is a metabolite of the well known anticancer drug cyclophosphamide [24]. However, this kinetic resolution was not very efficient and gave the products with the ee values not exceeding 34%.

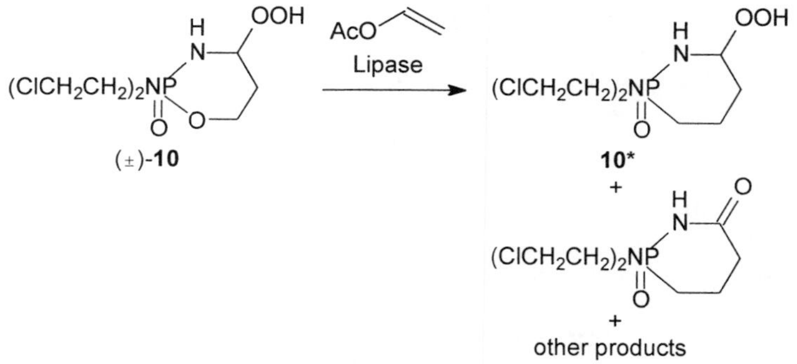

Scheme 8

In contrast to many successful examples of enzymatic kinetic resolution of racemic, P-chiral compounds discussed above, the use of prochiral phosphorus substrates in the enzymatic synthesis attracted little attention so far. Our preliminary experiments using prochiral phosphinyldiacetates **11** showed that their PLE-catalyzed hydrolysis proceeded slowly and with very low enantioselectivity (ee up to 10%) [25].

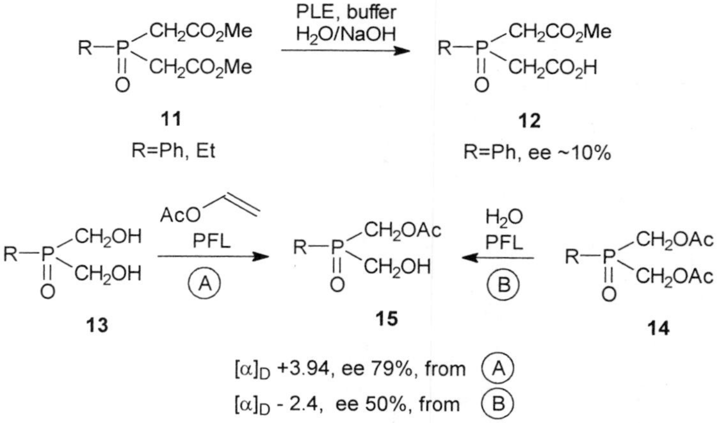

$[\alpha]_D$ +3.94, ee 79%, from (A)

$[\alpha]_D$ - 2.4, ee 50%, from (B)

Scheme 9

Better results in terms of enantioselectivity were obtained with (bis-hydroxymethyl)-phenylphosphine oxide **13** and (bis-acetoxymethyl)phenylphosphine oxide **14**. Acetylation of the former and hydrolysis of the latter in the presence of PFL afforded the product **15** with the ee value 79 and 50%, respectively [25].

3.   Synthesis of Optically Active Sulfur Compounds

Chiral, non-racemic sulfur compounds are widely used in asymmetric and stereoselective synthesis [26]. Their application as chiral synthons is mainly due to the high asymmetric induction exerted by S-chiral groups which makes the transfer of chirality from sulfur to other centres possible. In this way, many reactions may be efficiently stereocontrolled by sulfur auxiliaries which later are easily removable under mild conditions by reductive or eliminative methods.

3.1.   OPTICALLY ACTIVE SULFOXIDES

Among ca. forty different classes of chiral sulfur compounds sulfoxides **16** are of particular importance and they have become a class of commonly used organic reagents.

16

Although a number of approaches to their synthesis has been developed [27], a search for new, efficient and highly stereoselective methods continues. The enzymatic approach described in this section, which makes use of commonly available hydrolytic enzymes, enables the synthesis of a large diversity of chiral, non-racemic sulfoxides bearing an ester group. Additionally, the sulfoxide moiety is located either in the acid or in the alcohol portion of the ester.

3.1.1 Acyclic Sulfoxides

The first successful attempt was the kinetic resolution of racemic sulfinylacetates (**17**, n=1) and β-sulfinylpropionates (**17**, n=2) using bacteria (also in the form of its acetone dried cells). Ohta *et al.* [28] managed to obtain the recovered esters in good yields and with ee's of 90-97%, but were unable to isolate the acidic products of the reaction, thus losing the reacting enantiomer in each case. The latter problem was solved by Burgess *et al.* [29,30], who employed a crude lipase preparation from *Pseudomonas* sp. K-10. In this case, both unreacted esters and acids were isolated in good to high yields and with almost full stereoselectivity. Noteworthy, also transesterification mediated by the same enzyme and performed under similar conditions gave both separable esters with ee's up to 95% [30]. However, the above transformations proved to be limited to the substrates with one or two methylene groups between the sulfinyl and ester groups. The substrates having a longer linker did not undergo enzyme-promoted hydrolysis/transesterification at all (Scheme 10).

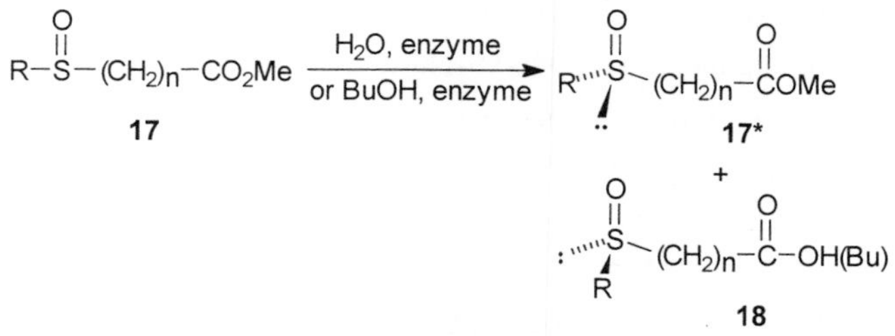

Corynebacterium equi. IFO 3730, 30°C
R=Ph, p-Tol, p-ClC₆H₄; n=1, 2, ee (ester **17***); 90-97%
Pseudomonas sp. K-10 (Amano), 25°C
Hydrolysis: pH 7.5, H₂O/toluene,
R=Ph, p-Tol, p-ClC₆H₄, p-O₂NC₆H₄, p-MeOC₆H₄, 2-Naph, n-Bu, c-C₆H₁₁;
n=1, 2; ee (ester **17***) >95%, acid **18** 80 - >95%
Transesterifcation: in n-BuOH,
R=p-ClC₆H₄, 2-Naph; n=1, 2; ee: 85 - >95%

Scheme 10

Naso *et al.* [31] subjected a series of 3-arylsulfinylpropenoates **19** to enzymatic hydrolysis and found that those having **Z** configuration are better substrates for enzymes. Moreover, they also found that the *S* enantiomers of the substrates are preferentially hydrolyzed in the presence of *Candida cylindracea* lipase (CCL), while the *R* ones in the presence of α-chymotrypsin (α-CT). Enantioselectivity of the latter reaction is considerably enhanced by the addition of t-butyl methyl ether to the buffer solution (Scheme 11).

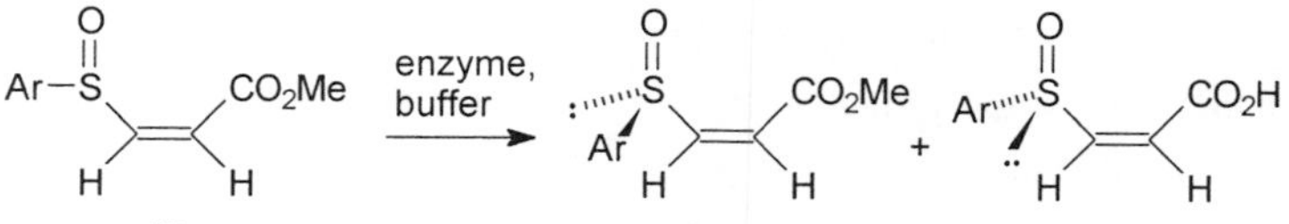

CCL: ee (ester *R*-**19***) 85%, (acid *S*-**20**) not reported
α-CT/t-BuOMe: ee (ester *S*-**19***) 91%, (acid *R*-**20**) 65%

Scheme 11

In turn, Allenmark and Andersson [32] screened 2-(alkylsulfinyl)benzoates **21** and found that the enantioselectivity of their hydrolysis, assisted by the lipase from *Candida* *rugosa* (CRL), is very high and that the enzyme shows a strong kinetic preference for the *S*-(-) enantiomers (Scheme 12).

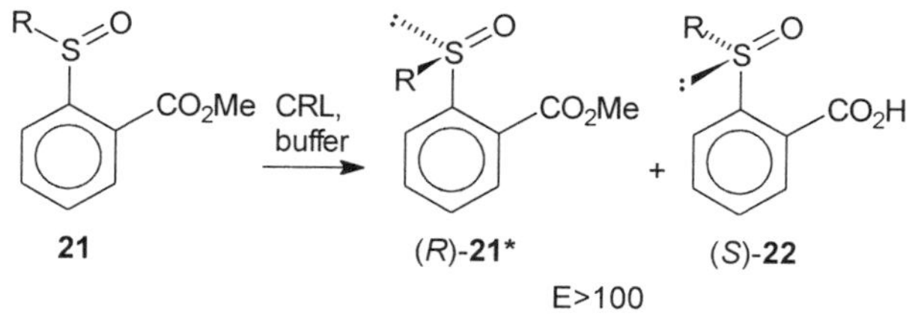

Scheme 12

All the reactions presented above were performed under the kinetic resolution conditions, with all the inherent disadvantages of this approach, such as a necessity of product/substrate separation and the 50% - yield limit of each enantiomer. Therefore, we decided to check whether desymmetrization of an appropriate prochiral substrate would occur during its enzyme-promoted transformations. As a substrate, dimethyl sulfinyldicarboxylate **23** was chosen. Among several enzymes tested, two of them, i.e. Porcine liver esterase (PLE) and α-chymotrypsin (α-CT), proved to be useful and gave enantiomerically enriched monoester **24** [33]. Interestingly, the two enzymes exhibited opposite enantioselectivity which enabled us to obtain each enantiomer of **24**. We faced certain difficulties in isolating the product as it proved to be very well soluble in water and insoluble in any organic solvent, except methanol. However, the difficulties were successfully overcome by lyophilizing the reaction solution followed by extraction of the residue with methanol and chromatographic purification. Repeated crystallization of the enantiomerically enriched samples from acetone afforded pure (-) and (+)-enantiomers which allowed their X-ray analysis [34] and determination of their absolute configuration as *S*-(+) and *R*-(-).

Soon thereafter a publication of Tamai *et al.* [35] came out which fully confirmed our results. However, these authors did not try to isolate the enantiomerically enriched products **24** but transformed them *in situ* into the corresponding mixed alkyl phenacyl diesters **25**, which were also used for X-ray crystallographic analysis (Scheme 13).

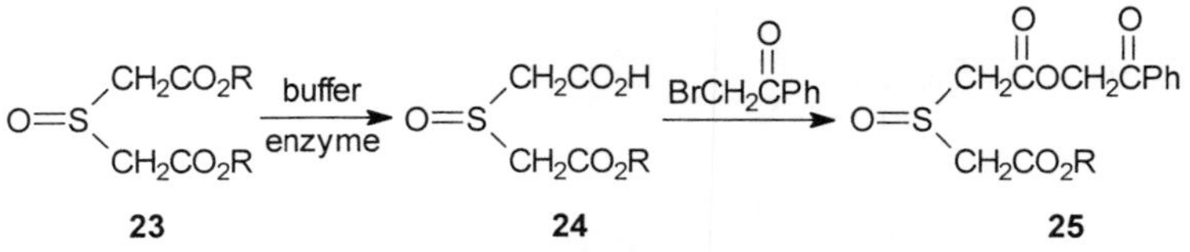

$$O=S\overset{CH_2CO_2R}{\underset{CH_2CO_2R}{\diagdown}} \xrightarrow[\text{enzyme}]{\text{buffer}} O=S\overset{CH_2CO_2H}{\underset{CH_2CO_2R}{\diagdown}} \xrightarrow{\text{BrCH}_2\overset{O}{\overset{\|}{C}}\text{Ph}} O=S\overset{CH_2COCH_2CPh}{\underset{CH_2CO_2R}{\diagdown}}$$

	23		24		25

R	Enzyme	Yield (%)	$[\alpha]_D$ (MeOH)	ee (%)	Abs. conf.	Ref.
Me	PLE	70[a]	+15.8[a]	79	*S*	33
Me	PLE	74[b]	+45.2[b]	86	*S*	35
Me	α-CT	63[a]	- 18.3[a]	92	*R*	33
Me	PPL	86[b]	- 50.0[b]	91	*R*	35
Et	PLE	74[b]	+40.1[b]	82	*S*	35
Et	PPL	92[b]	- 37.6[b]	76	*R*	35

Scheme 13

As PLE was the first and main enzyme used both by us and the Japanese authors in the above transformation, we became interested to identify the molecular interactions that determine enantiorecognition. Therefore, we used again, as in the case of phosphoroacetates described in Section 2.2, the empirical Jones' model for the PLE active site [18,19]. According to this model, our prochiral substrate 23 should be located in the PLE active site as follows [36]. The ester group that undergoes hydrolysis should be placed within the spherical locus of the catalytically active serine function, the second, unreacting ester group - in the front polar pocket (P_F) and the sulfinyl oxygen atom in the back polar pocket (P_B). The lone electron pair should be treated here as the smallest substituent and located in the small hydrophobic pocket (H_S, Fig. 2a). However, as the lone electron pair is known to possess a highly polar character, an alternative location, namely in the front polar pocket (P_F), together with the non-reacting ester group, seems more likely (Fig. 2b). Both modes of location lead to the same conclusion that the (*S*) enantiomers of 24 should be preferentially produced, which is in agreement with the experiment.

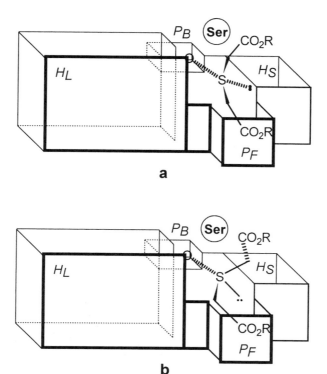

Figure 2: Preferred binding orientations of sulfinyldiacetates **23** in the PLE active site leading to (+)-(S)-**24**.

To check whether the same model can be applied to predict the absolute configuration of the preferentially hydrolyzed enantiomers of racemic sulfoxides, we performed the PLE-catalyzed hydrolysis of a series of racemic sulfinylacetates **26** under the kinetic resolution conditions [36], and obtained both the unreacted esters **26*** and acids **27** in enantiomerically enriched forms. Since the absolute configurations of the products were known from the literature [28,30,37], we were able to easily determine the enantioselectivity of this reaction, which was identical within the series of substrates investigated (Scheme 14).

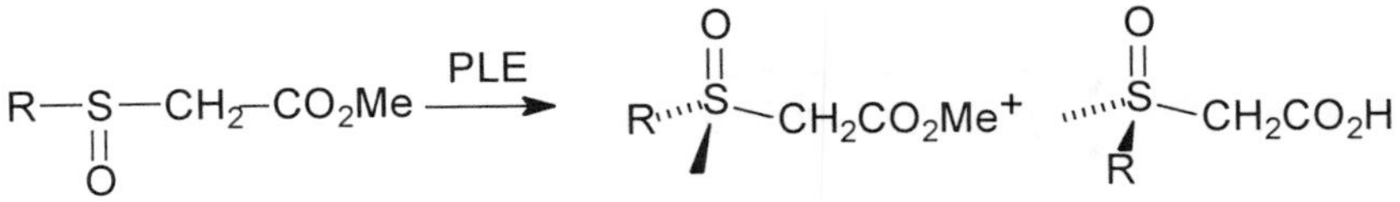

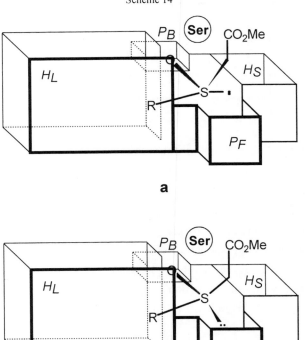

$$R-\underset{\underset{O}{\overset{\overset{}{|}}{S}}}{}-CH_2-CO_2Me \xrightarrow{PLE}$$

(±)-**26**

(R)-**26***   (S)-**27**

R	(R)-26*			(S)-27			E
	Yield (%)	[α]$_D$ (CHCl$_3$)	ee (%)	Yield (%)	[α]$_D$ (CHCl$_3$)	ee (%)	
t-Bu	53	+48	48	38	-19.1	38	3.5
Ph	52	+37	21	40	-58	34	2.5
p-Tol	32	+145	80	58	-82.8	46	6.1

Scheme 14

a

b

Figure 3: Binding orientations in the PLE active site of the preferably hydrolyzed (S)-enantiomers of **26**.

According to the above model, the enantiomers of **26** that undergo faster hydrolysis require the following location in the PLE active site. The ester group should be located within the serine locus, the sulfinyl oxygen atom in the back polar pocket (P_B) and the large organic group in the large hydrophobic pocket (H_L). The remaining lone electron pair (again treated here as the smallest substituent) should be located either in the small hydrophobic pocket (H_S, Fig. 3a) or rather in the front polar pocket (P_F, Fig. 3b). In both cases the model predicts preferential accommodation of the same enantiomer, i.e. *S*, which is in full agreement with the results obtained [36].

All the successful enzyme-promoted transformations discussed so far were performed using substrates containing the stereogenic sulfinyl moiety in the acid portion of the ester. However, there are also several examples of the use of hydrolytic enzymes in transformations of the substrates where the sulfinyl moiety constitutes a part of the alcohol portion of the molecule. The first unsuccessful experiment in this area was reported by Burgess *et al.* [30] who found that the irreversible acylation of 2-phenylsulfinylethanol poceeds with poor enantioselectivity (ee>20%). In turn, we were able to achieve certain enantiodiscrimination in the enzyme-mediated hydrolysis of the acetoxymethyl sulfoxide **28** (Scheme 15).[38]

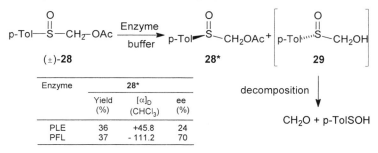

Enzyme	28*			decomposition
	Yield (%)	$[\alpha]_D$ (CHCl$_3$)	ee (%)	
PLE	36	+45.8	24	
PFL	37	- 111.2	70	

Scheme 15

In spite of the fact that half a substrate is irreversibly lost in this reaction due to the decomposition of the hydrolysis product, hydroxymethyl sulfoxide **29**, the sulfoxide **28***, obtained for the first time in an optically active form, is a very interesting compound since it may be considered as a chiral acyl anion equivalent (Scheme 16)[39].

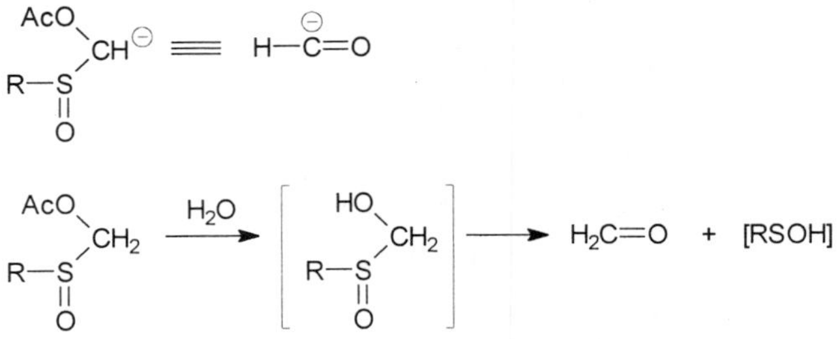

Scheme 16

We have also found that the prochiral substrate **30** undergoes enantioselective desymmetrization in the presence of PLE to give the monoacetate **31** with 67% ee (Scheme 17).[33] Interestingly, in this case from among a number of hydrolytic enzymes examined, only PLE proved to be effective, which may explain the lack of success of Burgess *et al.* [30] who screened only lipases.

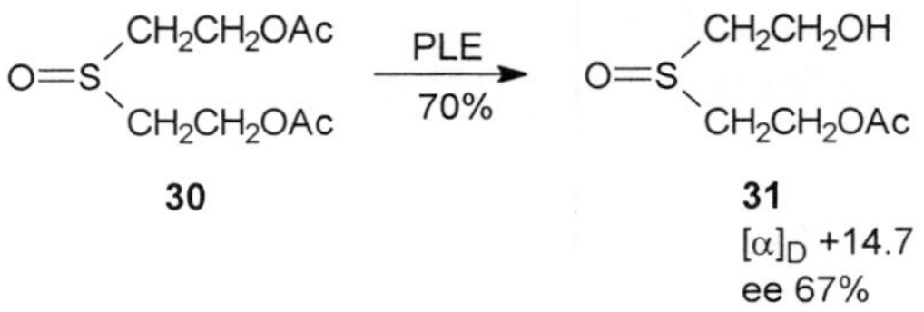

30

31
$[\alpha]_D$ +14.7
ee 67%

Scheme 17

In turn, Serreqi and Kazlauskas screened hydrolases for the kinetic resolution of the aryl alkyl sulfoxides with pendant acetoxy group and found cholesterol esterease (CE) to be the most enantioselective enzyme (Scheme 18) [40]. In all cases the *R* enantiomers were preferentially hydrolyzed, the effect of the change of substituents on the enantioselectivity being rather slight.

Recently, two attempts at the enzyme-promoted synthesis of an optically active platelet adhesion inhibitor (OPC-29030) **34** have been reported. In the first one the synthesis involved kinetic resolution of the precursor **32** by acetylation in the presence of various lipases (Scheme 19). The best, yet far from satisfactory, results were obtained with lipases AL and PL from *Alcaligenes* sp. (Meito) [41]. The unwanted sulfoxide (*R*)-**32*** was racemized using 3N HCl and re-used in the lipase-mediated transesterfication.

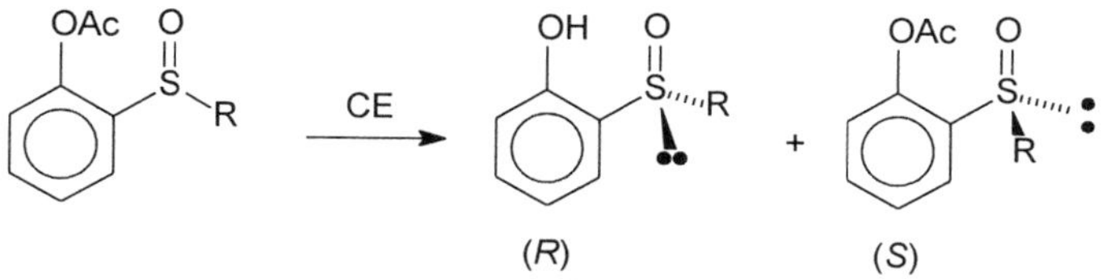

R=Me, CH₂Cl, n-Bu, Ph; ee (R) 62-80%, (S) 30-66%

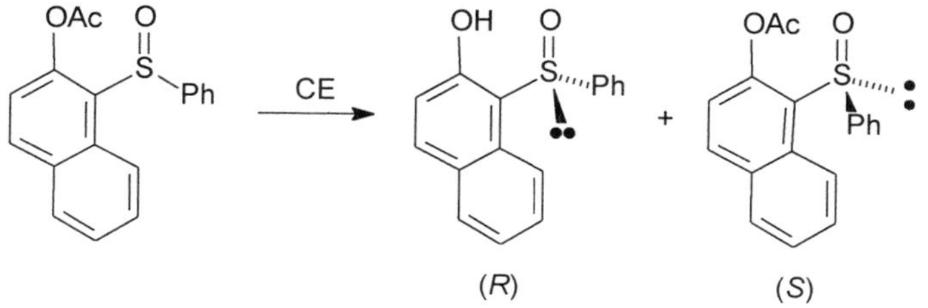

ee (R) 82%, (S) 62%

Scheme 18

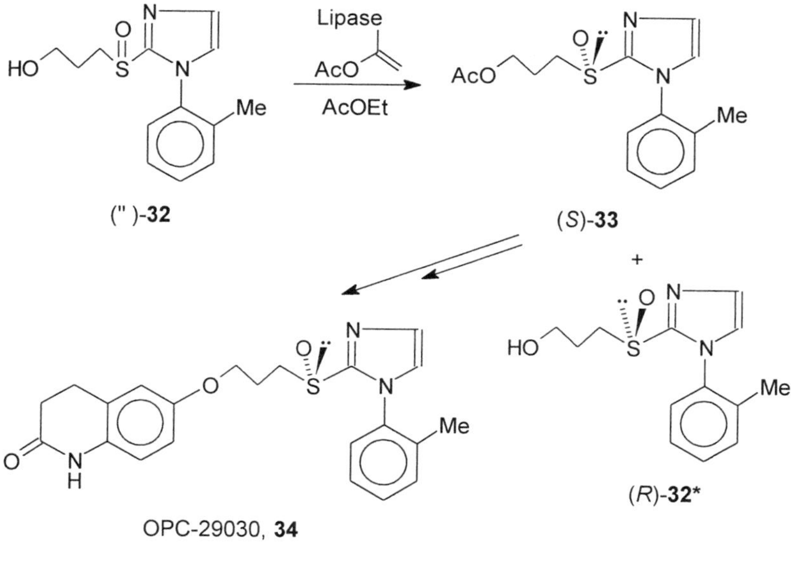

OPC-29030, **34**

Lipase	(S)-**33**		(R)-**32***	
	Yield (%)	ee (%)	Yield (%)	ee (%)
AL	41	63	59	43
PL	62	43	38	66

Scheme 19

Another approach, reported by the same group, utilized a different precursor, *viz.* **35**, whose enzyme-promoted acetylation proceeded with much higher enantioselectivity in spite in the fact that the reaction site (the benzylic hydroxyl group) was very remote (Scheme 20) [42]. Best results were obtained with Novozym 435 (an immobilized lipase from *Candida antarctica*) in chlorinated solvents. Repeated acetylations, performed on a preparative scale with enantiomerically enriched substrates gave the desired enantiomer with ee>95%.

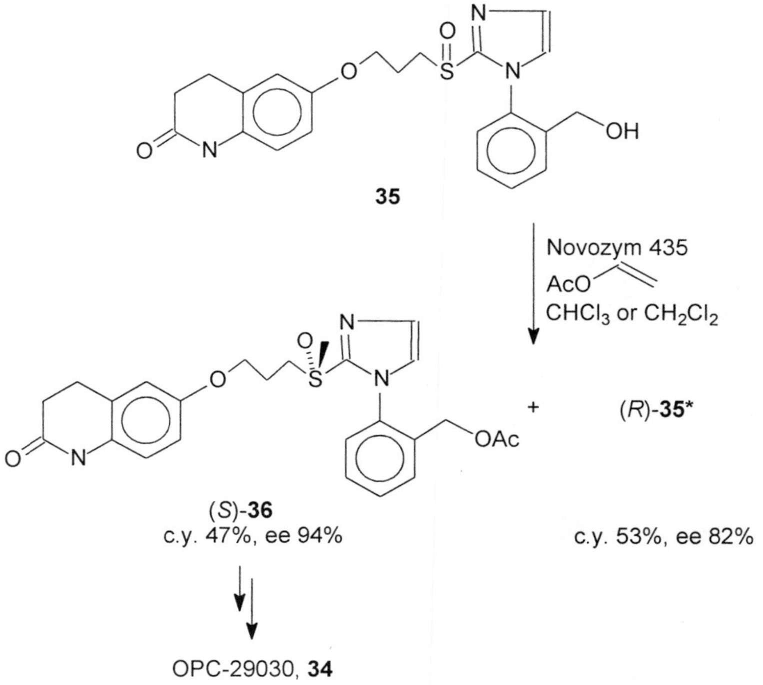

35

Novozym 435
AcO
CHCl$_3$ or CH$_2$Cl$_2$

+ (*R*)-**35***

(*S*)-**36**
c.y. 47%, ee 94%

c.y. 53%, ee 82%

OPC-29030, **34**

Scheme 20

Certain hydrolytic enzymes were found to exhibit a high diastereoselectivity towards epimeric β-hydroxysulfoxides **37** and **38**. Enzymatic acetylation enabled one to obtain the desired epimer with de >90% (Scheme 21) [43].

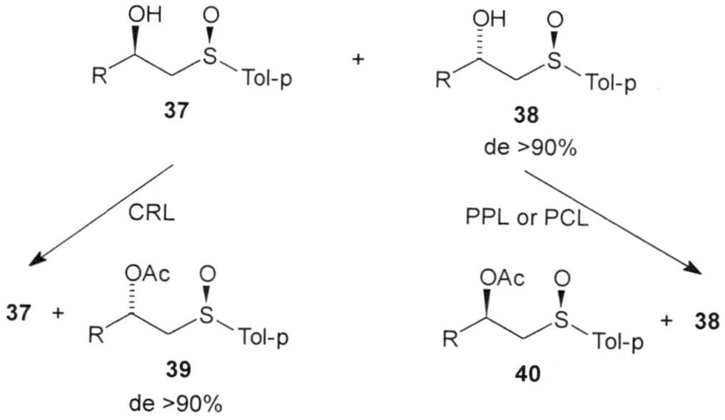

Scheme 21

3.1.2 Cyclic Sulfoxides

To check whether the enzymatic approach can be used for other types of sulfoxides, we turned our attention to 2-alkoxycarbonylthiopyran 1-oxides **41**, cyclic sulfoxides, easily obtainable as single *trans* diastereomers by the method of Zwanenburg *et al.* [44] A series of these sulfoxides were subjected to hydrolysis under kinetic resolution conditions in the presence of various hydrolytic enzymes to give both the unreacted esters and the corresponding acids in enantiomerically enriched forms (Scheme 22). However, the reaction proved to be very strongly dependent on the structural characteristics of the substrates. Several most important features of this reaction are listed below.

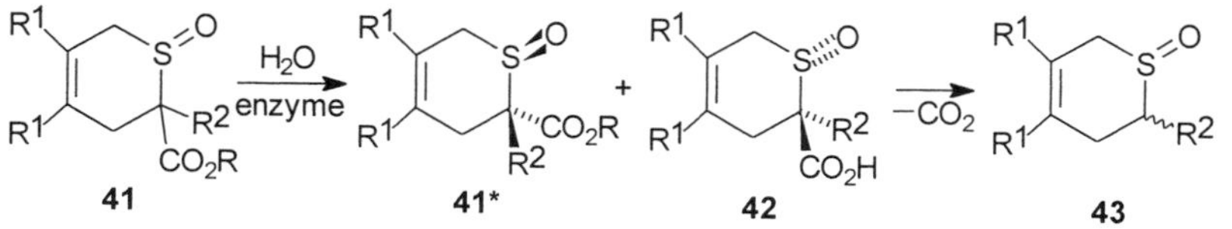

Enzyme: PLE, α-CT, PPL
R^1=H, Me R^2=CO_2R, CN, Ph, Me, H
ee: low to moderate; in a few cases >90%

Scheme 22

First of all, when the substituent at C-2 is an electron withdrawing group (R^2=CO_2R, CN), the acid formed undergoes spontaneous decarboxylation to give an epimeric mixture of sulfoxides **43**. On the other hand, the substrates with R^2=Ph are completely unreactive, irrespective of the enzyme and conditions used (most probably due to the steric hindrance exerted by the large phenyl group) [45,46,47]. In turn, in the case of the

2-methyl derivatives (R^2=Me), the presence or absence of the methyl groups at C-4 and C-5 (R^1=Me or H) becomes more pronounced: the former one is a better substrate for α-chymotrypsin (α-CT) than for PLE, while the latter one is totally unreactive in the presence of α-CT, but gives products with high ee in the presence of PLE. Finally, enzymatic hydrolysis of the substrates with an acidic hydrogen at C-2 (R^2=H) gives the products as a mixture of *cis* and *trans* diastereomers, each pair of epimers having almost the same optical purity. This observation allows the assumption that only one of the diastereomers undergoes hydrolysis, the other one being epimerized prior to the reaction [47]. X -ray crystallographic analyses revealed that the substrates **41** with R^2=CN [46] and R^2=H [47] adopt in the solid state a different conformation than those with R^2=Ph [46] and R^2=Me [47]. In the former case the ester group and the sulfinyl oxygen occupy equatorial positions whereas in the latter one both these groups are axial. However, these results do not allow to fully explain the differences in reactivity observed [47].

Interestingly, our preliminary studies of the use of-2-hydroxymethyl derivatives of the cyclic sulfoxides **44** as substrates for enzymatic acetylation have shown that, although the reaction is very slow (ca. 3 weeks), it gives the desired product **45** in reasonable yield and with an ee up to 50% (Scheme 23) [48]. It is noteworthy that **44** is an analog of **41** (R^2=Ph), which is entirely unreactive in the presence of any enzyme (*vide supra*).

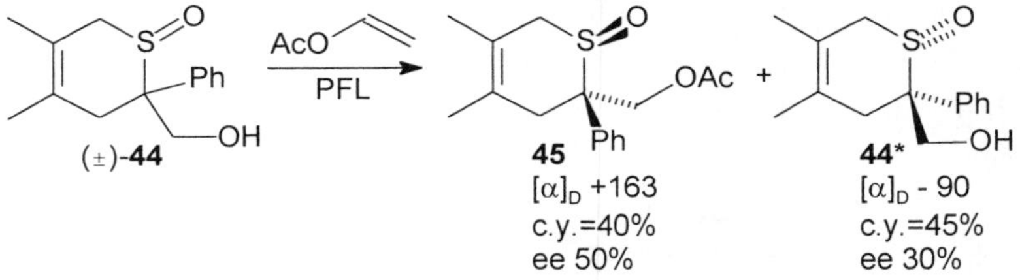

Scheme 23

3.2. OPTICALLY ACTIVE SULFOXIMINES

Chiral sulfoximines play an important role in asymmetric and stereoselective synthesis, since they can fulfill the functions of both chiral auxiliaries and chiral catalysts [26]. Although several methods of preparation of enantiomerically pure sulfoximines have been developed, we were the first to investigate the enzymatic kinetic resolution of racemic sulfoximines. We have subjected two types of sulfoximines, viz. **46** and **49** to hydrolysis in the presence of PLE. The hydrolysis was reasonably fast for **46** but very slow for **49** and gave both the unreacted esters and the products of hydrolysis and subsequent spontaneous decarboxylation with moderate enantioselectivity (Scheme 24) [49].

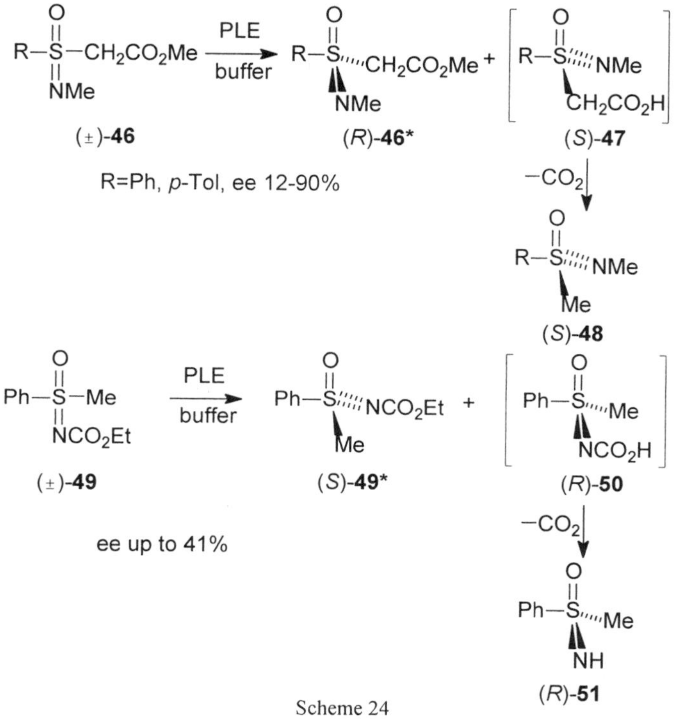

Scheme 24

In some cases the reaction was deliberately stopped at the conversion of less than 50% which enabled us to obtain desired products with an ee up to 90%. It is noteworthy that a regularity was observed in terms of the absolute configuration of the enantiomers preferentially hydrolyzed: in the case of C-alkoxycarbonyl derivatives **46** it is the S enantiomer, whereas in the case of the N-alkoxycarbonyl analog **49**, it is the R enantiomer. To explain this regularity, we have again applied the Jones' PLE active site model. Thus, for **46** the model predicts that the absolute configuration of the enantiomer preferentially hydrolyzed should be S (Figure 4), which is in agreement with the experimental result. On the other hand, the preferred binding of R-**49** is clearly visible in Figure 5: in this case the alkoxycarbonyl group, which must be located within the spherical locus of serine to undergo hydrolysis, does not allow the S=N group to be located in the front polar pocket (P_F), as it was for **46**. Hence, for **46** and **49** the enantiomers of opposite spatial arrangement are preferentially bound in the PLE active site.

186

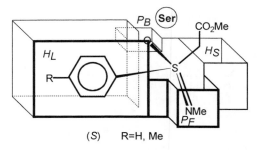

(S)   R=H, Me

Figure 4: Binding orientation of (S)-(+)-**46** in the active site of PLE.

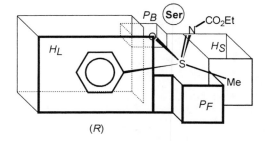

(R)

Figure 5: Binding orientation of (R)-(-)-**49** in the active site of PLE.

4.   Synthesis of Other Chiral Heteroorganic Compounds

4.1.   OPTICALLY ACTIVE SILANES

Fukui *et al.* screened over 30 different hydrolases for the enantioselective acylation of racemic hydroxymethylsilanes **52** and found that only a crude papain preparation exhibited high enantioselectivity (Scheme 25) [50]. It is noteworthy that vinyl acetate which is commonly used as an efficient acyl donor, was in this reaction inferior to 4-phenyl-pentanoic acid. As expected, the ee of the products decreased as the number of the methylene groups between the silicon atom and the hydroxyl group increased and dropped to zero when n=3. The absolute configuration of the products was not reported.

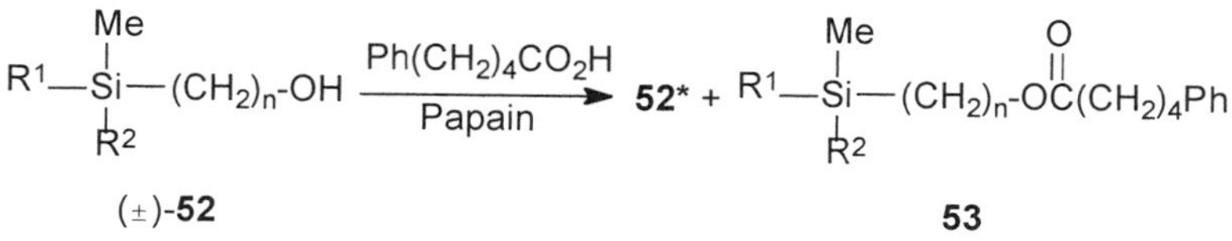

$$R^1-\underset{\underset{R^2}{|}}{\overset{\overset{Me}{|}}{Si}}-(CH_2)_n\text{-}OH \xrightarrow[\text{Papain}]{Ph(CH_2)_4CO_2H} \textbf{52*} + R^1-\underset{\underset{R^2}{|}}{\overset{\overset{Me}{|}}{Si}}-(CH_2)_n\text{-}O\overset{\overset{O}{\|}}{C}(CH_2)_4Ph$$

(±)-**52**   **53**

n=1, R^1=Ph, p-Tol, p-FC$_6$H$_4$, n-Hexyl; R^2=Et, n-Hexyl; ee 30-99%
n=2, R^1=Ph; R^2=Et, ee 43-59%

Scheme 25

In turn, Djerourou and Blanco described the asymmetric synthesis of isobutyryloxy-methyl(hydroxymethyl) silanes **55** *via* a lipase-promoted acylation of prochiral substrates **54** (Scheme 26) [51]. Among the several lipases screened two, *viz.* lipase from *Candida cylindracea* (CCL) and from *Chromobacterium viscosum* exhibited the highest, yet opposite stereoselectivity. As acylating agents, methyl isobutyrate or acetoxime isobutyrate were used. Interestingly, an unusually low temperature (down to - 20°C) had to be used for the CCL-catalyzed reactions, as the enantioselectivity significantly dropped with increasing temperature.

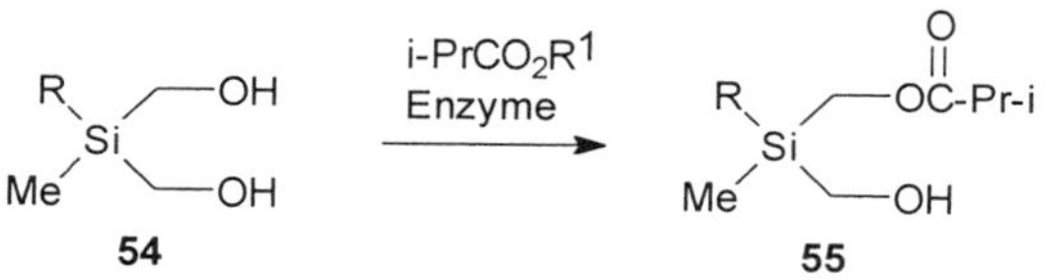

$$\underset{Me}{\overset{R}{\diagdown}}Si\overset{\diagup OH}{\diagdown OH} \xrightarrow[\text{Enzyme}]{i\text{-}PrCO_2R^1} \underset{Me}{\overset{R}{\diagdown}}Si\overset{\diagup O\overset{\overset{O}{\|}}{C}\text{-}Pr\text{-}i}{\diagdown OH}$$

54   **55**

R=Ph, n-Octyl, yield up to 80%, ee up to 75%

Scheme 26

4.2. OPTICALLY ACTIVE GERMANES

Only two papers have been published on the enzymatic synthesis of optically active, Ge-chiral derivatives, both concerning the use of prochiral substrates. In earlier work, immobilized PLE was applied by Tacke *et al.* to promote acetylation of bis(hydroxymethyl)methyl-phenylgermane **56** (R^1=Me) with vinyl acetate as a solvent and acyl donor [52]. Recently, it was reported by the same group that each enantiomer of hydridogermane monoacetates **57** (R^1=H) was obtained using either acetylation of the bis(hydroxymethyl) derivative **56** (R^1=H) or hydrolysis of the corresponding diacetate **58** (R^1=H). In both methods Porcine pancreatic lipase (PPL) was used as an enzyme and, obviously, each reaction led to a different enantiomer of **57** (Scheme 27) [53].

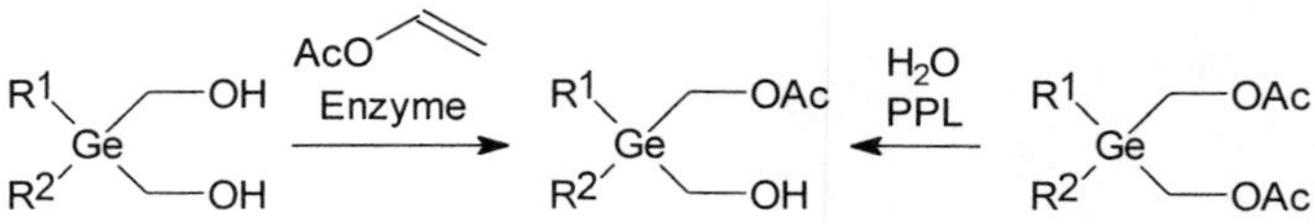

188

Enzyme: immobilized PLE, R^1=Me, R^2=Ph, ee 50%
Enzyme: PPL, R^1=H, R^2=Ph; p-FC$_6$H$_4$, ee 84÷94%

Scheme 27

5. References

1. Drauz, K. and Waldman, H. (eds.) (1995) *Enzyme Catalysis in Organic Chemistry - A Comprehensive Handbook*, VCH, Weinheim.
2. Faber, K. (1997) *Biotransformations in Organic Chemistry*, 3rd Edition, Springer Verlag, Berlin.
3. Koskinnen, A.M.P. and Klibanov, A.M. (eds.) (1996) *Enzymatic Reactions in Organic Media*, Blackie Academic and Professional, London.
4. Colonna, S., Gaggero, N., Richelmi, C. and Pasta, P., Enantioselective Oxidations Catalyzed by Peroxidases and Monooxygenases, This book, pp. 133 – 160 and references therein.
5. Abo, M., Okubo, A. and Yamazaki, S. (1997) Preparative Asymmetric Deoxygenation of Alkyl Sulfoxides by *Rhodobacter sphaeroides* f.sp. *denitrificans, Tetrahedron: Asymmetry* 8, 345-348.
6. Reid, T.W., Stein, T.P. and Fahrney, D. (1967) The Pepsin-Catalyzed Hydrolysis of Sulfite Esters. II. Resolution of Alkyl Phenyl Sulfites, *J. Am. Chem. Soc.* **89**, 7125-7126.
7. Dudman, N.P.B. and Zerner, B. (1973) Enzymatic Preparation of an Optically Active Phosphotriester, Asymmetric Only at Phosphrus, *J. Am. Chem. Soc.* **95**, 3019-3021.
8. Lewis, V.E., Donarski, W.J., Wild, J.R. and Raushel, F.M. (1988) Mechanism and Stereochemical Course at Phosphorus of the Reaction Catalyzed by a Bacterial Phosphotriesterase, *Biochemistry* **27**, 1591-1597.
9. Kiełbasiński, P. (1999) Enzyme-promoted Syntheses of Chiral, Non-racemic Heteroorganic Compounds with a Sole Stereogenic Center on the Heteroatom, *Rev. Heteroatom Chem.* **19**, 143-172.
10. Pietrusiewicz, K.M. and Zabłocka, M. (1994) Preparation of Scalemic P-Chiral Phosphines and Their Derivatives, *Chem. Rev.* **94**, 1375-1411.
11. Kiełbasiński, P., Żurawiński, R., Pietrusiewicz, K.M., Zabłocka, M. and Mikołajczyk, M. (1994) Enzymatic Resolution of Racemic Phosphinylacetates Having a Stereogenic Phosphorus Atom, *Tetrahedron Lett.* **35**, 7081-7084.
12. Kiełbasiński, P., Żurawiński, R., Pietrusiewicz, K.M., Zabłocka, M. and Mikołajczyk, M. (1998) Synthesis of P-Chiral, Non-racemic Phosphinylacetates via Enzymatic Resolution of Racemates, *Polish J. Chem.* **72**, 564-572.
13. For recent applications see; Omelańczuk, J. and Mikołajczyk, M. (1996) Chiral t-Butylphenyl-phosphinothioic Acid: A Useful Chiral Solvating Agent for Direct Determination of Enantiomeric Purity of Alcohols, Thiols, Amines, Diols, Aminoalcohols and Related Compounds, *Tetrahedron: Asymmetry* 7, 2687-2694 and references therein.

14. Korpiun, O., Lewis, R.A., Chickos, J. and Mislow, K. (1968) Synthesis and Absolute Configuration of Optically Active Phosphine Oxides and Phosphinates, *J. Am. Chem. Soc.* **90**, 4842-4846.

15. Farnham, W.B., Murray, R.K. and Mislow, K. (1970) Sterospecific Alkylation of Menthyl Phenylphosphinate, *J. Am. Chem. Soc.* **92**, 5809-5810.

16. Kiełbasiński, P., Góralczyk, P., Mikołajczyk, M., Wieczorek, M.W. and Majzner, W.R. (1998) Enzyme-promoted Kinetic Resolution of Racemic, P-Chiral Phosphonyl and Phosphorylacetates, *Tetrahedron: Asymmetry* **9**, 2641-2650.

17. Gałdecki, Z., Główka, M.L. Musierowicz, S. and Michalski, J. (1980) Stereochemistry of Phosphinylacetic Acids. Determination of Absolute Configuration of Optically Active O-Methylphenylphosphinylacetic Acid by the X-ray Method, *Polish J. Chem.* **54**, 539-542.

18. Toone, E.J., Werth, M.J. and Jones, J.B. (1990) Active-Site Model for Interpreting and Predicting the Specificity of Pig Liver Esterase, *J. Am. Chem. Soc.* **112**, 4946-4952.

19. Provencher, L. and Jones, J.B. (1994) A Concluding Specification of the Dimensions of the Active Site Model of Pig Liver Esterase, *J. Org. Chem.* **59**, 2729-2732.

20. Mikołajczyk, M., Drabowicz, J. and Kiełbasiński, P. (1995) Formation of C-P Bonds in Helmchen, G., Hoffmann, R.W., Mulzer, J. and Schaumann, E. (eds.) *Houben-Weyl: Methods of Organic Chemistry*, vol. E21e: *Stereoselective Synthesis*, G. Thieme Verlag, Stuttgart, pp. 5701-5731.

21. Kiełbasiński, P., Omelańczuk, J. and Mikołajczyk, M. (1998) Lipase-promoted Kinetic Resolution of Racemic, P-Chiral Hydroxymethylphosphonates and Phosphinates, *Tetrahedron: Asymmetry* **9**, 3283-3287.

22. Spangler, L.A., Mikołajczyk, M., Burdge, E.L., Kiełbasiński, P., Smith, H.C., Łyżwa, P., Fisher, J.D. and Omelańczuk, J. (1999) Synthesis and Biological Activity of Enantiomeric Pairs of Phosphosulfonate Herbicides, *J. Agr. Food Chem.* **47**, 318-321.

23. Serreqi, A.N. and Kazlauskas, R.J. (1994) Kinetic Resolution of Phosphines and Phosphine Oxides with Phosphorus Stereocenters by Hydrolases, *J. Org. Chem.* **59**, 7609-7615.

24. Baba, N., Ikeda, K. and Iwasa, J. (1992) Kinetic Partial Resolution of Racemic 4-Hydroperoxycyclophosphamide by Lipase in Vinyl Acetate, *Bull. Inst. Chem. Res. Kyoto Univ.* **70**, 326-328; *Chem. Abstr.* **118**, 191838z (1993).

25. Kiełbasiński, P. and Mikołajczyk, M. unpublished results.

26. Mikołajczyk, M., Drabowicz, J. and Kiełbasiński, P. (1997) *Chiral Sulfur Reagents: Applications in Asymmetric and Stereoselective Synthesis*, CRC Press, Boca Raton.

27. Drabowicz, J., Kiełbasiński, P. and Mikołajczyk, M. (1994) Synthesis of Sulphoxides and Appendix to Synthesis of Sulphoxides, in S. Patai and Z. Rappoport (eds.), *Syntheses of Sulphones, Sulphoxides and Cyclic Sulphides*, J.Wiley and Sons, Chichester, pp. 109-388.

28. Ohta, H., Kato, Y. and Tsuchihashi, G. (1986) Enzyme-mediated Synthesis of Optically Active ω-Arenesulfinylalkanoic Esters, *Chem. Lett.* 217-218.

29. Burgess, K. and Henderson, I. (1989) A Facile Route to Homochiral Sulfoxides, *Tetrahedron Lett.* **30**, 3633-3636.

30. Burgess, K., Henderson, I. and Ho, K.-K. (1992) Biocatalytic Resolutions of Sulfinylalkanoates: A Facile Route to Optically Active Sulfoxides, *J. Org. Chem.* **57**, 1290-1295.

31. Cardellicchio, C., Naso, F. and Scilimati, A. (1994) An Efficient Biocatalyzed Kinetic Resolution of Methyl (Z)-3-Arylsulphinylpropenoates, *Tetrahedron Lett.* **35**, 4635-4638.

32. Allenmark, S.G. and Andersson, A.C. (1993) Lipase-Catalyzed Kinetic Resolution of a Series of Esters Having a Sulfoxide Group as the Stereogenic Centre, *Tetrahedron: Asymmetry* **4**, 2371-2376.

33. Mikołajczyk, M., Kiełbasiński, P., Żurawiński, R., Wieczorek, M.W. and Błaszczyk, J. (1994) A Novel Enzymatic Approach to the Synthesis of Chiral Sulfoxides: Enzymatic Hydrolysis of Prochiral Sulfinyldicarobxylates, *Synlett* 127-129.

34. Wieczorek, M.W. and Błaszczyk, J., Mikołajczyk, M., Kiełbasiński, P. and Żurawiński, R. (1994) Crystal and Molecular Structure and Determination of Absolute Configuration of Two Enantiomers of Carbomethoxymethyl Carboxymethyl Sulfoxide ($C_5H_8O_5S$), *Heteroatom Chem.* **5**, 167-173.

35. Tamai, S., Miyauchi, S., Morizane, C., Miyagi, H., Shimizu, H., Kume, M., Sano, S., Shiro, M. and Nagao, Y. (1994) Enzymatic Hydrolyses of the σ-Symmetric Dicarboxylic Diesters Bearing a Sulfinyl Group as the Prochiral Center, *Chem. Lett.* 2381-2384.

36. Kiełbasiński, P. (2000) On the Applicability of the Jones Active Site Model of Pig Liver Esterase to Chiral and Prochiral Sulfinyl Substrates, *Yetrahedron:Asymmetry*, **11**, 911-915.

37. Drabowicz, J., Dudziński, B., Mikołajczyk, M., Wieczorek, M.W. and Majzner, W.R. (1998) Crystal and Molecular Structure of the Levorotatory (1R, 2S)-Ephedrinium (S)-t-Butylsulfinylacetate: A Definite Proof of the Absolute Configuration of Enantiomeric t-Butyl Methyl Sulfoxide, *Tetrahedron: Asymmetry* **9**, 1171-1178.

38. Kiełbasiński, P. and Mikołajczyk, M. unpublished results.

39. For other examples of chiral acyl equivalents see: Kiese, N., Urai, I. and Yoshida, I. (1998) Formation and Reaction of 2-Metallated N-Boc-4,4-Dimethyl-1,3-oxazolidines in the Presence of (-) Sparteine: New Chiral Formyl Anion Equivalents, *Tetrahedron: Asymmetry* **9**, 3125-3128 and references therein.

40. Serreqi, A.N. and Kazlauskas, R.J. (1995) Kinetic Resolution of Sulfoxides with Pendant Acetoxy Groups Using Cholesterol Esterase: Substrate Mapping and an Empirical Rule for Chiral Phenols, *Can. J. Chem.* **73**, 1357-1367.

41. Morite, S., Matsubara, J., Otsubo, K., Kitano, K., Ohtani, T., Kawano, Y. and Uchida, M. (1997) Synthesis of a Key Intermediate, (*S*)-2-[(3-Hydroxypropyl)-sulfinyl]1-(o-tolyl)imidazole, for the Platelet Aggregation Inhibitor, OPC-29030 via Lipase-catalyzed Enantioselective Transesterification, *Tetrahedron: Asymmetry* **8**, 3707-3710.

42. Kitano, K., Matsubara, J., Ohtani, T., Otsubo, K., Kawano, Y., Morita, S. and Uchida, M. (1999) An Efficient Synthesis of Optically Active Metabolites of Platelet Adhesion Inhibitor OPC-29030 by Lipase-Catalyzed Enantioselective Transesterification, *Tetrahedron Lett.* **40**, 5235-5238.

43. Medio-Simon, M., Gil, J., Aleman, P., Varea, T. and Asensio, G. (1999) Selective Lipase-Catalyzed Acylation of Epimeric α-Sulfinyl Alcohols: an Efficient Method of Separation, *Tetrahedron: Asymmetry* **10**, 561-566.

44. Lenz, B.G., Regeling, H., van Rozendaal, H.L.M. and Zwanenburg, B. (1985) Synthesis of α-Oxo Sulfines from Enol Silyl Ethers and Thionyl Chloride, *J. Org. Chem.* **50**, 2930-2934.

45. Kiełbasiński, P., Mikołajczyk, M., Zwanenburg, B. and de Laet, R.C. (1994) Kinetic Resolution of Cyclic Sulfoxides by Enzyme-Promoted Hydrolysis of 2-Alkoxy-carbonyl-3,6-Dihydro-2H-thiapyran 1-Oxides, *Phosphorus, Sulfur and Silicon* **95-96**, 495-496.

46. Wieczorek, M.W., Błaszczyk, J., Dziuba, B., Kiełbasiński, P., Mikołajczyk, M., Zwanenburg, B. and Rewinkel, J.B.M. (1995) Crystal and Molecular Structure of Cyclic Sulfoxides: 2-Cyano-2-ethoxycarbonyl-3,6-dihydro-4,5-dimethyl-2H-thiapyran 1-Oxide and 2-Phenyl-2-methoxycarbonyl-3,6-dihydro-2H-thiapyran 1-Oxide, *Heteroatom Chem.* **6**, 631-638.

47. Kiełbasiński, P., Zwanenburg, B., Damen, T.J.G., Wieczorek, M.W., Majzner, W.R. and Bujacz, G. (1999) Kinetic Resolution of Racemic Cyclic Sulfoxide Using Hydrolytic Enzymes, *Eur. J. Org. Chem.* 2573-2578.

48. Kiełbasiński, P. unpublished results.

49. Kiełbasiński, P. (1999) The First Enzymatic Preparation of S-Chiral Non-racemic Sulfoximines, *Polish J. Chem.* **73**, 735-738.

50. Fukui, T., Kawamoto, T. and Tanaka, A. (1994) Enzymatic Preparation of Optically Active Silylmethanol Derivatives Having a Stereogenic Silicon Atom by Hydrolase-catalyzed Enantioselective Esterification, *Tetrahedron: Asymmetry* **5**, 73-82.

51. Djerourou, A.H. and Blanco, L. (1991) Synthesis of Optically Active 2-Sila-1,3-propanediols Derivatives by Enzymatic Transesterification, *Tetrahedron Lett.* **32**, 6325-6326.

52. Tacke, R., Wagner, S.A. and Sperlich, J. (1994) Synthese von (-)-(Acetoxymethyl) (hydroxymethyl)-methyl(phenyl)german [(-)-MePhGe(CH$_2$OAC)(CH$_2$OH)] durch eine Esterase-katalysierte Umesterung: Die erste enzymatische Synthese eines optisch aktiven Germans, *Chem. Ber.* **127**, 639-642.

53. Tacke, R., Kosub, U., Wagner, S.A., Bertermann, R., Schwarz, S., Merget, S. and Gunther, K. (1998) Enzymatic Transformation and Liquid-Chromatographic Separation as Methods for the Preparation of the (*R*)- and (*S*)-Enantiomers of the Centrochiral Hydridogermanes p-XC$_6$H$_4$(H)Ge(CH$_2$OAC)CH$_2$OH (X=H, F), *Organometallics* **17**, 1687-1699.

BIODEGRADATION OF HYDROLYZED CHEMICAL WARFARE AGENTS BY BACTERIAL CONSORTIA

JOSEPH J. DEFRANK*, MARK GUELTA, STEVEN HARVEY
U.S. Army Edgewood Chemical Biological Center
Aberdeen Proving Ground, Maryland, USA 21010-5424

ILONA J. FRY
Geo-Centers, Inc.,Gunpowder, Maryland, USA 21010

JAMES P. EARLEY
SBR Technologies, Inc.,South Bend, Indiana, USA 46628

F. STEPHEN LUPTON
AlliedSignal Inc.,Des Plaines, Illinois, USA 60017-5016

1. Introduction

The U.S. Army has custody of chemical weapons (CW) containing nerve and blister (vesicant) agents located in eight sites in the continental United States and at Johnston Island, a small island in the Pacific Ocean. The eight sites in the continental U.S. are: Aberdeen Proving Ground, Maryland; Anniston Army Depot, Alabama; Blue Grass Army Depot, Kentucky; Newport Army Ammunition Plant, Indiana; Pine Bluff Arsenal, Arkansas; Pueblo Depot Activity, Colorado; Tooele Army Depot, Utah; and Umatilla Depot Activity, Oregon. The stockpile, which totals approximately 32,000 tons consists primarily of three agents, the nerve agents GB (sarin) and VX, and the blister agent HD (sulfur mustard). The structures of these agents are shown in Figure 1. The agents are stored in steel 1-ton bulk containers as well as in munitions such as bombs, rockets, artillery shells, and mines. All the munitions and agents are at least 25-50 years old.

The Army has had an ongoing need for disposal of surplus and obsolete weapons. Before 1969, CW munitions were disposed of by open-pit burning,

B. Zwanenburg et al. (eds.), Enzymes in Action, 193–209.
© 2000 *Kluwer Academic Publishers. Printed in the Netherlands.*

evaporative "atmospheric dilution," burial, and deep ocean dumping [18]. The ocean disposal of agents was prohibited by the Marine Protection, Research & Sanctuaries Act in 1972.

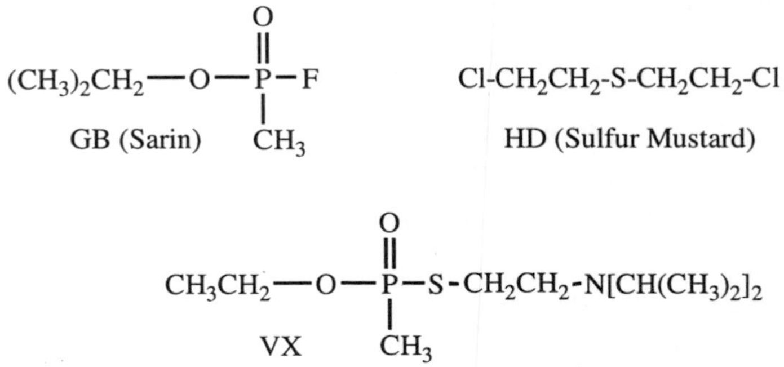

$$(CH_3)_2CH_2 - O - \overset{\overset{\displaystyle O}{\|}}{\underset{\underset{\displaystyle CH_3}{|}}{P}} - F \qquad Cl\text{-}CH_2CH_2\text{-}S\text{-}CH_2CH_2\text{-}Cl$$

GB (Sarin) \qquad\qquad HD (Sulfur Mustard)

$$CH_3CH_2 - O - \overset{\overset{\displaystyle O}{\|}}{\underset{\underset{\displaystyle CH_3}{|}}{P}} - S\text{-}CH_2CH_2\text{-}N[CH(CH_3)_2]_2$$

VX \qquad CH_3

Figure 1: Primary Agents in U.S. Chemical Weapons Stockpile.

In the 1970's, the principal methods employed for agent destruction were neutralization via alkaline hydrolysis for GB and incineration for HD. In 1982, the Army selected incineration as the preferred technology for destruction of obsolete components of the CW stockpile. The Chemical Stockpile Disposal Program was initiated in 1985, when Public Law 99-145 directed the Department of Defense (DOD) to destroy at least 90% of this stockpile by September 30, 1994. As the program proceeded, its pace was slower than anticipated, and its date for completion was revised several times. In 1988, Congress extended the completion date to 1997, and in 1990 to July, 1999.

International events soon had an impact on this program. In 1990, the United States and the former Soviet Union signed a memorandum of agreement to cease chemical weapons production, dispose of inventories, share disposal technology, and develop inspection procedures. In addition, on September 3, 1992, the Conference on Disarmament approved the Chemical Weapons Convention (CWC), signed on January 13-15, 1993, by the United States and several other nations, forbidding the development, production, stockpiling, or use of chemical weapons. Upon ratification, the deadline (which supersedes previous deadlines) for stockpile destruction will be December 31, 2004 or later. In the case of the United States, the ratification of the CWC on April 30, 1997, made the deadline April 30, 2007.

The Army plan to use incineration for the destruction of the CW stockpiles lead to considerable local and national opposition by citizens, environmental groups, and politicians. As a result, under Public Law 102-484, the Army was directed by Congress to report on Alternative Technologies for CW stockpiles. The Army requested the

National Research Council Committee on Review and Evaluation of the Army Chemical Stockpile Disposal Program to reevaluate the status of incineration. An additional NRC committee was also formed, the Committee on Alternative Chemical Demilitarization Technologies, to examine alternatives for CW disposal. Based on the two NRC studies [19, 20], the Army initially developed an Alternative Technology Plan to examine alternatives for two sites. The two sites, Aberdeen Proving Ground and Newport Army Ammunition Plant had only bulk agent (HD and VX respectively), thus the process would not be complicated by having to deal with explosives and propellants. Among the technologies to be examined was the use of chemical neutralization followed by biodegradation.

2.   Hydrolyzed HD Destruction

Sulfur mustard, often referred to as "mustard gas," is 2,2'-dichlorodiethylsulfide (Chemical Abstract Services number 505-60-2, military designation, HD). Mustard is an oily liquid with a freezing point of ~14°C and a boiling point of 217°C. Because mustard is nearly insoluble in water, it is very persistent in the environment and can contaminate soils and surfaces for long periods of time. Mustard affects the eyes and lungs and blisters the skin. It causes severe chemical burns and painful blisters, and is lethal at high dosages, especially if inhaled. Symptoms are generally delayed for several hours after exposure [3, 13, 21]. It was developed in Germany and used extensively during World War I where it caused 80% of the gas casualties. It was also used by Italy in Ethiopia in 1936, by Japan against China in World War II, by Egypt in Yemen in the early 1960's, and by Iraq against Iran during the 1980's.

The advantages of using hydrolysis as a neutralization reaction preceding biodegradation of HD include: an aqueous medium; complete dechlorination; and products (primarily alcohols) that are amenable to biodegradation. Also, the hydrolysis reaction does not add any additional carbon that would require subsequent biological removal. Alkaline hydrolysis has been previously utilized for the destruction of Canadian HD stockpiles [22]. Biodegradation has widespread application in municipal and industrial wastewater treatment plants, is conducted at ambient temperatures, and offers a favorable mass balance for the process.

The mechanism of HD hydrolysis has been the subject of considerable study beginning just after World War I [1, 2, 12, 16, 26-28]. It has been demonstrated that HD reacts through a series of sulfonium ion intermediates to produce thiodiglycol as shown in Scheme 1. However, additional ether or thioether products can also form depending on the actual conditions of the reaction. Because HD is insoluble in water, the reaction involves hydrolysis at the water/organic interface with the products then dissolving in the water [27]. Therefore, agitation during hydrolysis is a critical factor in carrying out this reaction. Both the rate of mass transfer and the rate of hydrolysis can be accelerated at elevated temperatures.

Thiodiglycol (TDG), the primary hydrolysis product of HD, is a water miscible, non-toxic liquid. However, it is listed as a Schedule Two compound (chemical weapon precursor) in the CWC and therefore must be destroyed as part of any demilitarization processes. Earlier studies with *Alcaligenes xylosoxydans* ssp. *xylosoxydans* demonstrated that TDG serve as sole carbon and energy source for microbial growth [15].

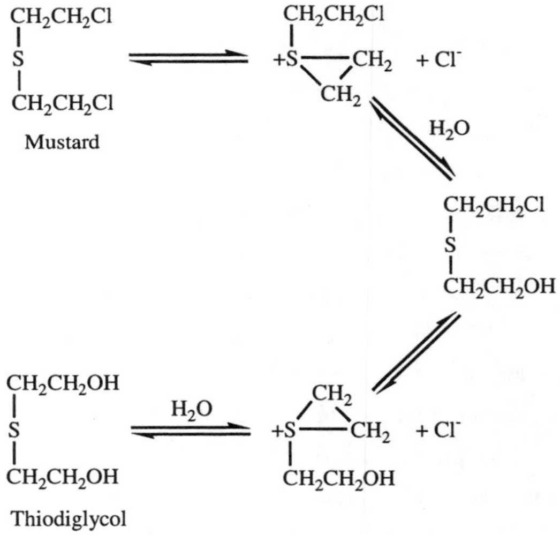

$$CH_2CH_2Cl\text{—}S\text{—}CH_2CH_2Cl \;(\text{Mustard}) \rightleftharpoons {}^{+}S(CH_2CH_2Cl)(CH_2\text{-}CH_2) + Cl^{-}$$

$$\xrightarrow{H_2O} CH_2CH_2Cl\text{—}S\text{—}CH_2CH_2OH$$

$$\rightleftharpoons {}^{+}S(CH_2CH_2OH)(CH_2\text{-}CH_2) + Cl^{-} \xrightarrow{H_2O} CH_2CH_2OH\text{—}S\text{—}CH_2CH_2OH \;(\text{Thiodiglycol})$$

Scheme 1: Hydrolysis Mechanism for HD.

The efficiency of conversion of HD to TDG has been shown to be dependent on temperature and whether the base used for neutralizing the acid is added before or after the reaction [11]. Elevating the temperature of alkaline hydrolysis reaction from 30 to 90°C increased the TDG concentration from 62 to 84%. Running the hydrolysis in hot water alone (with base neutralization after cooling) further increased the TDG level to >95%. The hydrolysis reaction in either caustic or water went to completion with HD not observed with a detection limit of 160 ppb.

For the biodegradation of hydrolyzed HD, the use of sequencing batch reactors (SBRs) was selected. SBRs offer several important operational advantages for the treatment of chemical waste [for a review see 14]. They are very efficient. The SBR models as a continuous flow stirred tank reactor (CFSTR) followed by a plug flow reactor, offering the ideal volumetric reactor configuration for unsteady state activated sludge systems. They operate in the batch mode, which not only offers a kinetic advantage over conventional CFSTRs (therefore smaller size), but, for hazardous waste treatment, they permit batch analysis and toxicity testing prior to discharge. They use a single tank for both treatment and settling, therefore no secondary clarifier is required. Finally, they are robust and flexible. SBRs are intentionally operated over a range of substrate concentrations, pH conditions and oxygen concentrations, thereby allowing selection of a very diverse and robust population of microorganisms. Operational strategies can also be varied to accomplish carbon, nitrogen and/or phosphorus removal.

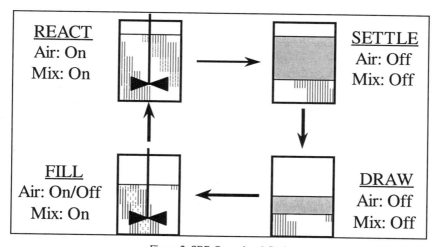

Figure 2: SBR Operational Cycle.

The reactor cycle (24 hour) used in these studies normally consisted of: 4 hour aerated FILL; 18 hour aerated REACT; 1.5 hour SETTLE; and 0.5 hour DRAW. The total organic carbon (TOC) loading in the feed was 3400-3800 mg/L. The hydraulic residence time (HRT) was 16 days and the original biomass for the SBRs was activated sludge from the Back River Waste Water Treatment Plant (Baltimore, MD). Trace metals were provided in a modified Wolin Salts Solution [24] in addition to NH_4Cl and KH_2PO_4 as nitrogen, potassium and phosphorus sources. The hydrolyzed HD was the sole source of carbon and sulfur. The SBR was operated at a pH of between 6.5 and 8.5. Buffering was primarily accomplished through the addition of $NaHCO_3$ to the feed, although a NaOH solution was occasionally used on demand when the pH dropped below 6.5.

In an SBR with caustic hydrolyzed HD, the TDG was typically removed to >99%, however, the effluent TOC showed a gradual increase until it reached ~1000 mg/L. At this point, the feed was changed to the water hydrolyzed HD. After 146 days of operation the reactor had significantly lower levels of effluent TOC and suspended solids. MICROTOX analysis of the effluent showed it to be non-toxic (100% effluent produced no changes in the luminescence levels of the test bacteria). In addition, toxicity characteristic leachate protocol (TCLP) analysis of the effluent found none of the restricted organics and metals at levels of concern. A mass balance calculated per ton of HD is shown in Table 1.

TABLE 1. Summary of Process Mass Balance [11]

Input (tons)		Output (tons)	
HD	1.00	CO_2	0.55
NaOH	1.11	H_2O*	0.60
O_2 (air)	0.93	Biomass**	0.30
NH_4Cl	0.16	Na_2SO_4	0.88
KH_2PO_4	0.04	NaCl	0.90
		KCl	0.01
Total Input:	3.24	Total Output:	3.24

* Represents only the water actually produced from the oxidation processes
** Dry weight of biomass

Based on results at the laboratory and bench scale, the neutralization/biodegradation method was selected by the Army for the destruction of the HD stockpile at APG. An $86 million demonstration unit was approved and is in the permitting process. Site clearing for the demonstration unit and full-size facility has been initiated. The official groundbreaking ceremony for the facility was held on June 26, 1999. A $306 million contract has been awarded for the construction, operation, and dismantling of the full-size facility. The goal of the program is to meet the April 30, 2007 deadline for destruction as mandated by the CWC.

The actual pathways by which TDG is metabolized have not been the subjects of extensive investigation. Some preliminary studies [29] with *Alcaligenes xylosoxydans* ssp. *xylosoxydans* indicated that this organism oxidized TDG by means of sequential NAD-dependent butanol dehydrogenase reactions to give thiodiglycolic acid (TDGA) as shown in Scheme 2. NMR analysis of the cell-free dehydrogenase reactions yielded two oxidation products from TDG: hydroxyethylthioacetic acid and TDGA. The identity of these products was confirmed by mass spectrometry analysis. Therefore, it appears that in this organism the biodegradation of TDG proceeds via an oxidative pathway that first yields a mono-acid product and then the di-acid product. The fate of the TDGA has not been determined.

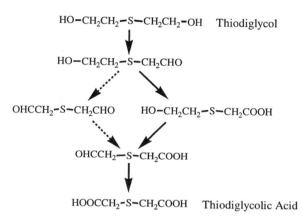

Scheme 2: Thiodiglycol Metabolism in *Alcaligenes* [29].

In the SBR studies with the microbial consortia and earlier work with *Pseudomonas picketii*, the transient appearance of thiodiglycol sulfoxide was occasionally observed [10]. Whether the sulfoxide is an actual metabolic intermediate in the degradation of TDG or the product of a side reaction is unknown. However, by whatever means it is produced, the sulfoxide is the subject of further degradation since it is not found in the final effluent from the SBR.

3.   Hydrolyzed VX Destruction

The nerve agent VX can be initially hydrolyzed by either the addition of an equimolar amount of water [25] or by caustic hydrolysis. The proposed mechanism for the VX/water reaction is shown in Scheme 3.

The addition of an equimolar amount of water (~10% by volume) to VX initiates this autocatalytic reaction. The reaction is irreversible and at room temperature requires 30-60 days to reach completion. The reaction mixture begins as a two-phase system but gradually is converted into a single yellowish organic phase. When conducted with a small volume of VX the reaction proceeds smoothly without the need for mixing. However, since the reaction is somewhat exothermic, with large volumes of VX (such as

200

in ton containers) some initial mixing is required to eliminate the formation of "hot spots" that could be hazardous. Although the principal products are ethyl methylphosphonic acid (EMPA) and the diisopropylaminoethyl thiol (VX-thiol), some additional reactions occur to give an array of products as shown in Scheme 4.

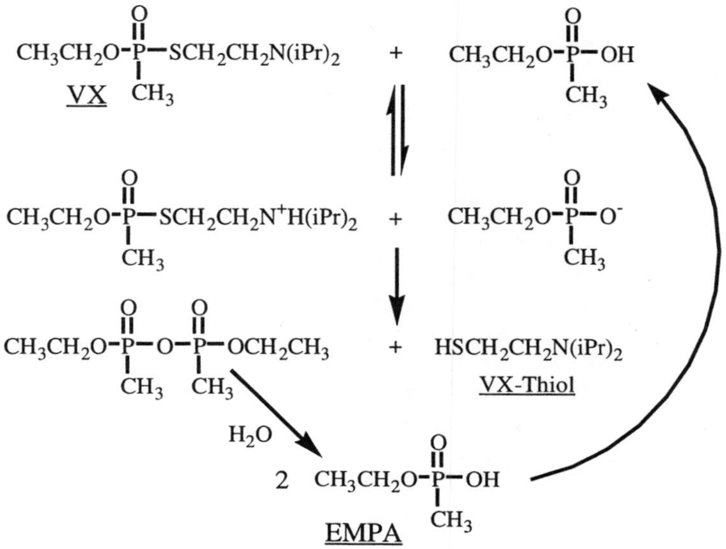

Scheme 3: VX/Water Reaction Mechanism (adapted from [25]).

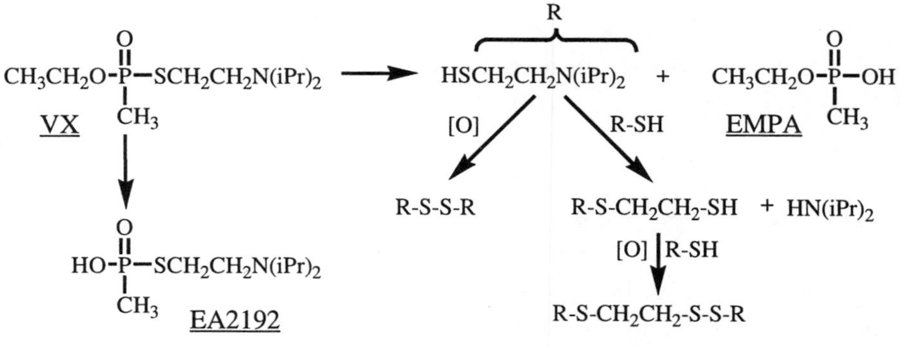

Scheme 4: Additional Products from VX/Water Reaction.

Of particular concern is the formation of EA2192, which is nearly as toxic as VX, is very stable, and is much more soluble in water than VX. The VX-thiol is rapidly oxidized in the presence of air to the disulfide (R-S-S-R) which is the primary organosulfur (OS) product.

The alternative method of VX hydrolysis is through the use of caustic. Typically, in this process, one part VX is added to two parts NaOH solution with a slight molar excess of NaOH. The reaction is run at 90°C for two hours. The production of EA2192 is greatly reduced under these conditions and any that does form is rapidly hydrolyzed. The VX-thiol also undergoes fewer side reactions so that the primary OS compound in the final product is the disulfide. The advantage of this system is that the absence of EA2192 makes the product much less toxic and safer to handle. The disadvantage is that it is a two-phase system with the small (3-5%) organic phase largely consisting of the disulfide. Since VX would preferentially dissolve in the organic disulfide, vigorous mixing is essential to ensure that the hydrolysis reaction goes to completion.

In regards to biodegradation, the major challenges to be faced are the breaking of the carbon-phosphorus bond of EMPA and dealing with the unusual tertiary amine of the VX-thiol and its related products. There have been no studies found in the literature that deal with the biodegradation of the type of tertiary amines such as those obtained from VX hydrolysis. The degradation of organophosphorus (OP) compounds is fairly well understood and the proposed pathway for EMPA degradation is shown in Scheme 5.

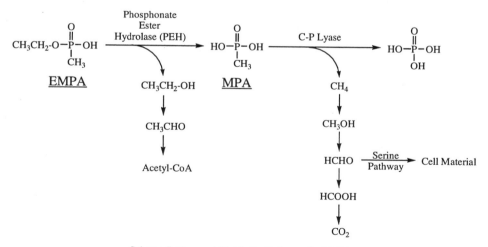

Scheme 5: Proposed Metabolic Pathways for EMPA.

A number of Phosphonate Ester Hydrolases (PEHs) have been identified and purified from bacterial isolates. Several of these will be discussed later in this chapter. In the case of EMPA the PEH reaction is very rapid, with C-P Lyase reaction being the rate-limiting step. C-P Lyase has been the subject of a number of studies [Metcalf,

Schowanek] and is known to be strongly inhibited by the presence of very low levels of inorganic phosphate. The methane that is released from methylphosphonic acid (MPA) could either be removed in the exhaust gas or be utilized by methanotrophic bacteria in the consortium.

As with the biodegradation of HD, the SBR was selected for use in treating the VX hydrolysis products. The cycle used for VX consisted of: 2 hour anoxic FILL; 3 hour aerated FILL; 17 hour aerated REACT; 1.5 hour SETTLE; and 0.5 hour DRAW. In order to have as diverse a population of microorganisms, the SBR was inoculated with activated sludge from: Back River Waste Water Treatment Plant (Baltimore, MD); Dupont Chambers Works (Penns Grove, NJ); Aberdeen Proving Ground Waste Water Treatment Facility (Aberdeen, MD); Zeneca Pesticide Plant Water Treatment Facility (Bucks County, AL); and Texas A&M University Experimental Water Treatment Facility (College Station, TX). In addition, microorganisms from various enrichment studies were routinely added to the reactors to potentially enhance their performance.

Initial studies were conducted with VX/Water as sole carbon and phosphorus source in the feed to the SBR [4]. The EMPA was rapidly hydrolyzed to MPA and ethanol, the latter being rapidly metabolized. However, very little MPA appeared to be degraded. It became clear that there was a great excess of phosphorus for the amount of available carbon. For that reason, a number of auxiliary carbon sources were added in subsequent experiments (with both VX/Water and VX/Caustic). These included acetate, dextrose, isopropanol, methanol, and glycerol. The VX hydrolysate was generally provided at 0.05-0.1% as sole phosphorus and sulfur source. With appropriately high levels of extra carbon, the OP compounds could be effectively mineralized. In general, a C:P ratio in excess of 100:1 would be necessary to get complete mineralization. Alternatively, if members of the consortium had a constitutive C-P Lyase or one that was not sensitive to P_i inhibition, the extra carbon could be eliminated. Obviously, the extra carbon significantly increases the cost and scale of the biodegradation process. If a sufficiently high strength (high carbon, low P) industrial waste stream was available, it could provide an inexpensive source of the extra carbon.

Efficient degradation of the OS compounds was not achieved in the SBRs. The best results obtained ranged from 40-60% removal [4]. The more numerous OS by-products in the VX/Water hydrolysate made for more difficult biodegradation. Inorganic sulfate was occasionally detected in the effluent, but it did not always correlate with the level of OS degradation. There did seem to be some correlation between the C:N ratio and the level of OS degradation, but the tertiary amine did not appear to be able to provide all the necessary nitrogen required for growth. In order to determine the pathway by which the OS compounds were being degraded samples of the SBR consortia were tested with two possible metabolic products of the VX-thiol. Neither diisopropylaminoethanol nor diisopropylaminoethylsulfonic acid were found to support growth. Presumably, the VX-thiol is not metabolized through these intermediates.

In regards to the destruction of the VX stockpile at Newport, IN, the current plans involve neutralization by caustic hydrolysis followed by supercritical water oxidation. While biodegradation could potentially be optimized, the time constraints placed on the demilitarization program by the CWC do not allow for the long-term research efforts that would be required.

4. Hydrolyzed GB Destruction

The third primary agent in the U.S. stockpile, GB (sarin), was subjected to caustic hydrolysis and biodegradation in an SBR system [6, 7]. As with VX, the main challenge with degrading the hydrolysis product from GB, isopropyl methylphosphonic acid (IMPA) was dealing with the C-P bond. Supplemental carbon sources utilized in the course of the studies included dextrose, methanol and hydrolyzed HD. In the latter studies, the GB/Caustic was the sole phosphorus source and the hydrolyzed HD sole sulfur and primary carbon source. The C:P ratio used in the studies was 64:1. The nitrogen source used was originally NH_4Cl, but switched to NH_4OH when the hydrolyzed HD (with its high chloride content) was provided.

Unlike the degradation of EMPA from VX, the hydrolysis of IMPA was the rate-limiting step. The MPA produced was immediately metabolized and did not accumulate in the bioreactor. Even though the TDG from the hydrolyzed HD was totally mineralized, the level of OP degradation was generally fairly poor (40-60%). As with VX, increasing the C:P ratio could improve the OP degradation, but would also increase the size and cost of the operation.

In spite of the somewhat poor performance, the GB-SBR cultures did provide some interesting enzymes. A strain of methylotrophic bacterium *Methylobacterium radiotolerans* GB21 was isolated from the SBR shortly after inoculation from several VX SBRs and an aerobic digester that was treating the effluent from a number of reactors [6, 7]. It was found to have at least two phosphonate ester hydrolases (PEH) with very different substrate preferences. One of the enzymes (I-PEH) will hydrolyze IMPA, diisopropyl phosphate, and to a much lesser degree, diisopropyl methylphosphonate (DIMP). The second enzyme, E-PEH, has a preference for EMPA, but also has activity against a chromogenic analog p-nitrophenyl methylphosphonic acid (NPMPA). Most importantly, the E-PEH was able to hydrolyze the P-S bond of EA2192, the highly toxic product from VX [5]. It is believed that this is the first enzyme described that has significant EA2192 activity. Both of these PEHs are being purified and further characterized. Although they may not play a role in the destruction of the chemical agent stockpiles, they may be very useful in the development of enzyme-based decontamination systems.

5. Immobilized Cell Bioreactors

All the studies described above involved the use of SBRs for the biodegradation of the hydrolyzed agents. An alternative approach is the use of biofilm systems in which the microorganisms grow attached to a solid support. AlliedSignal Environmental Systems (part of AlliedSignal, Inc.) owns and operates mobile bioreactor units for onsite neutralization of various contaminated waste streams. These reactors, commonly called Immobilized Cell Bioreactors (ICBs), use a stationary support of polyurethane foam blocks on which to grow the biomass. Cylindrical polypropylene spacers are also used to allow for more efficient aeration and liquid circulation. AlliedSignal provided laboratory-scale ICBs for evaluation with hydrolyzed HD and VX. These units had a total volume of approximately 1-liter and a working volume of from 600-800 ml. Studies were undertaken with the ICBs and both HD and VX hydrolysates [8].

5.1. HD-ICB OPERATIONS

An ICB was seeded with biomass from a HD SBR and allowed to acclimate in a batch feed mode. The feed for the HD-ICB consisted of (per liter): HD hydrolysate (3.8%), 333 ml; NH_4Cl, 1.2 g; KH_2PO_4, 0.35 g; Wolin Salts Solution [Wolin], 10 ml; and $NaHCO_3$, 5.0 g (for pH control). The feed was adjusted to a final pH of 7.5 with HCl prior to use. In the initial batch operations, a recirculating loop and the addition of NaOH provided additional pH control. Once sufficient biomass had been produced and continuous operations were begun, the pH was controlled entirely by adjusting the level of sodium bicarbonate in the feed. Pumping air through a frittered glass disk at the base of the ICB provided continuous aeration.

The initial working volume of the HD-ICB was 750 ml. As the biomass developed, the working volume decreased to the point that, when the studies were concluded, it was down to an estimated 287 ml. The feed rate to the ICB ranged from 100 to 300 ml/day giving a hydraulic residence time (HRT) of from 1.5 to 7.5 days. Very efficient mineralization of the TDG was achieved as long as the HRT was kept at greater than 3 days. In the laboratory unit, higher feed rates (and lower HRT) resulted in the biomass nearly plugging up the reactor. This resulted in anoxic or anaerobic zones where biodegradation of the TDG ceased. In full scale units (using larger foam blocks and spacers and higher velocity liquid flow and aeration) this plugging would probably not occur. Under all conditions in this study, the biomass stayed firmly attached to the support. The total suspended solids (TSS) in the reactor effluent were generally at or below standard detection limits. Overall, once established, the HD-ICB system was very efficient and required little daily maintenance. It would therefore lend itself to pump and treat operations or continuous-flow waste streams. Further studies will be necessary to determine whether biomass buildup would reach a point were it would hinder performance and if the same level of TDG mineralization could be maintained over long periods of time.

5.2.   VX-ICB OPERATIONS

The VX-ICB was also seeded from an existing SBR system. The feed for the ICB was the same as for the SBR with VX/Caustic as the sole phosphorus and sulfur source. The feed rate was maintained at 50 ml/day for the entire study. At that time the extra carbon source for the SBR was isopropanol. From the SBR studies, it was known that while isopropyl alcohol was readily metabolized, it did not generate much biomass. Therefore, the adaptation period for the VX-ICB was considerably longer than for the HD-ICB. At several points during this adaptation period the feed was augmented with the addition of glycerol, glycerol plus molasses, and molasses plus inorganic phosphate, in an effort to increase the biomass.

Although the level of biomass production in the VX-ICB was always much lower than in the HD-ICB, the TSS was always higher. It appears that the biomass that was generated did not adhere to the support as readily. The performance of the VX-ICB for OP degradation was comparable or somewhat worse that the SBR system, the highest levels being in the 30-40% range. As with the SBR systems, it is presumed that increasing the auxiliary carbon source (increasing the C:P ratio), and switching to one that would give a much higher cell yield could improve this. Alternatively, a secondary stirred tank reactor could be added (with the additional carbon) to complete the OP mineralization.

On the other hand, unlike the SBR, the level of OS mineralization was surprisingly good. In general, the OS compounds were nearly always 95-100% degraded. There are two mysteries in regards to this performance that still remain to be solved. The first is the actual metabolic pathways for the degradation of these materials. Possible metabolic intermediates were not observed in the ICB effluent. The second is the fate of the sulfur. While the initial level of inorganic sulfate in the ICB effluent corresponded to the amount of sulfur in the feed, this dropped considerably thereafter. This is in spite of the fact that the OS degradation remained at ~100%. Since the system was well aerated, it does not appear to be likely that it could have been converted to hydrogen sulfide or other gaseous form. Alternatively, it was thought that the activated carbon on the surface of the polyurethane foam might have preferentially adsorbed the VX-thiol (disulfide). However, if it was only adsorption without degradation, the carbon should have become saturated very early in the operation of the ICBs (which initially ran for 190 days).

More recent studies coupled the ICBs with a secondary stirred tank reactor (STR). The OS compounds continued to be totally removed in the ICB. By the addition of dextrose to the STR as a primary carbon source, the OP compounds are also fully degraded.

5.3. LARGE-SCALE ICB OPERATIONS

The successor to the Alternative Technologies program is the Assembled Chemical Weapons Assessment (ACWA) program. Unlike the former which only dealt with bulk agents (HD and VX), the ACWA program deals with weaponized agents and all the materials that may be involved in their destruction (explosives, propellants, etc.). As with the bulk agent program, the agent would first be chemically neutralized and the hydrolysate processed further. A number of different technologies are being evaluated, one of which is the use of ICBs followed by an oxidative polishing step. A 1000-gallon ICB unit was set up by AlliedSignal to process hydrolyzed HD (primary carbon source) plus hydrolyzed Tetrytol (a mixture of the explosives Tetryl and TNT, and primary nitrogen source). After an acclimation phase the ICB was run at 200 gallons per day for a 40-day operational trial. The final report from this trial has not been completed. However, while the biological system worked well for TDG degradation, the extent of Tetrytol degradation has not been determined. Whether or not this technology will be further evaluated for use in stockpile destruction is unknown at this time [8]. The main technical problem encountered dealt with pH control due to the high level of sulfate produced. It should be possible to deal with this by minor process and operational modifications.

6. Conclusion

Chemical neutralization combined with biodegradative processes can play a substantial role in destroying stockpiles of chemical warfare agents. This approach appears to be most readily adaptable to dealing with sulfur mustard (HD). Just such a method has been successfully demonstrated and is being put into operation at Aberdeen Proving Ground to destroy the 1600 tons of the agent located there.

For dealing with the nerve agents GB and VX, the potential use of biodegradation is dependent on finding a more efficient means of dealing with the C-P bond in these organophosphonates. If the PEHs and C-P Lyase (or an enzyme similar to it) could be used to treat the hydrolyzed agents in some type of flow-through reactor, the products could be fed to a bioreactor (SBR or ICB or combined systems). This would eliminate the need to add the extra carbon that greatly increases the cost and scale of the current systems. Although the use of biological systems for the destruction of the nerve agent stockpiles is not currently being considered, these studies have lead to the discovery of a number of very interesting enzymes. These may play an important role in agent decontamination on the battlefield, after terrorist attacks, or in dealing with spills or leaks that may occur at demilitarization facilities.

7. References

1. Bartlett, P.D. and Swain, C.G. (1949) Kinetics of hydrolysis and displacement reaction of β,β'-dichlorodiethyl sulfide (mustard gas) and β-chloro-β'-hydroxydiethyl sulfide (mustard chlorohydrin), *J. Am. Chem. Soc.* **71**, 1406-1415.
2. Bush, V. (1946) *Summary of Technical Report of Division 9, National Defense Research Committee*, Columbia University Press, Washington, DC.
3. Compton, J.A.F. (1987) Military Chemical and Biological Agents: Chemical and Toxicological Properties, Telford, Caldwell, NJ
4. DeFrank, J.J., Fry, I.J., Earley, J.P., and Irvine, R.L. (1997) Biodegradation of VX/water hydrolysate, *U.S. Army Edgewood Research, Development and Engineering Center Technical Report*, ERDEC-TR-429.

208

5. Fry, I.J. Personal communication.

6. Fry, I.J., DeFrank, J.J., Frost, C.M., and Earley, J.P. (1997) Sequencing batch reactor biodegradation of hydrolyzed sarin, in ERDEC-SP-048, *Proceedings of the 1996 ERDEC Scientific Conference on Chemical and Biological Defense Research*, pp. 361-367.

7. Fry, I.J., Earley, J.P., and DeFrank, J.J. (1998) Process biodegradation of hydrolyzed sarin by the GB2 microbial consortium, in ERDEC-SP-063, *Proceedings of the 1997 ERDEC Scientific Conference on Chemical and Biological Defense Research*, pp. 629-633.

8. Guelta, M.A. Personal communication.

9. Guelta, M.A. and DeFrank, J.J. (1998) Performance of immobilized cell bioreactors for treatment of HD and VX hydrolysates, *U.S. Army Edgewood Research, Development and Engineering Center Technical Report*, ERDEC-TR-497.

10. Harvey, S.P. Personal communication.

11. Harvey, S.P., Szafraniec, L.L., Beaudry, W.T., Earley, J.P., and Irvine, R.L. (1998) Neutralization and biodegradation of sulfur mustard, in Sikdar, S.K. and Irvine, R.L. (eds.), *Bioremediation: Principles and Practice, Vol. II: Biodegradation Technology Developments*, Technomics Publishing Inc., Lancaster, PA, pp. 615-636.

12. Helfrich, O.B. and Reid, E.E. (1920) Reactions and derivatives of β,β'-dichloroethyl sulfide, *J. Am. Chem. Soc.* **42**, 1208-1232.

13. International Agency for Research on Cancer (IARC) (1975) IARC monographs on the evaluation of carcinogenic risk of chemicals to man. Vol. 9, Some Aziridines, N-, S- and O-Mustards and Selenium, IARC, Lyon.

14. Irvine, R.L. and Ketchum, L.H., Jr. (1989) Sequencing batch reactors for biological wastewater treatment, in *CRC Critical Reviews in Environmental Control*, **18**, 255-294.

15. Lee, T.-s., Pham, M.Q., Weigand, W.A., Harvey, S.P., and Bentley, W.E. (1996) Bioreactor strategies for the treatment of growth-inhibitory waste: An analysis of thiodiglycol degradation, the main hydrolysis product of sulfur mustard, *Biotechnol. Prog.* **12**, 533-539.

16. McManus, S.P., Neamati-Mazrach, N., Hovanes, B.A., Paley, M.S., and Harris, J.M. (1985) Hydrolysis of mustard derivatives. Failure of the Raber-Harris probe in predicting nucleophilic assistance, *J. Am. Chem. Soc.* **107**, 3393-3395.

17. Metcalf, W.W. and Wanner, B.L. (1991) Involvement of the *Escherichia coli* phn (psiD) gene cluster in assimilation of phosphorus in the form of phosphonates, phosphites, P_i esters, and P_i, *J. Bacteriol.* **173**, 587-600.

18. National Research Council (1984) *Disposal of Chemical Munitions and Agents*, National Academy Press, Washington, DC.

19. National Research Council (1993) *Alternative Technologies for the Destruction of Chemical Agents and Munitions*, National Academy of Sciences, Washington, DC.

20. National Research Council (1994) Recommendations for the disposal of chemical agents and munitions.

21. Papirmeister, B. *et al.* (1991) Medical defense against mustard gas, in *Chemistry of Sulfur Mustard*, CRC Press, Boston, Chapter 5.

22. Reichert, C. (1975) Study of mustard destruction by hydrolysis, *Defense Research Establishment Suffield Ralston Technical Note*, DRES-TN-329.

23. Schowanek, D. and Verstraete, W. (1990) Phosphonate utilization by bacterial cultures and enrichments from environmental samples, *Appl. Environ. Microbiol.* **56**, 895-903.

24. Wolin, E.A., Wolin, M.J., and Wolf, H.S. (1963) Formation of methane by bacterial extracts, *J. Biol.*

Chem. **238**, 2882-2886.

25. Yang, Y.-C. (1995) Chemical reactions for neutralizing chemical warfare agents, *Chemistry and Industry* May 1 Issue, 334-337.

26. Yang, Y.-C., Szafraniec, L.L., Beaudry, W.T., and Ward, J.R. (1987) Direct NMR measurements of sulfonium chlorides produced from the hydrolysis of 2-chloroethyl sulfides, *J. Org. Chem.* **52**, 1637-1638.

27. Yang, Y.-C., Szafraniec, L.L., Beaudry, W.T., and Ward, J.R. (1988) Kinetics and mechanism of the hydrolysis of 2-chloroethyl sulfides, *J. Org. Chem.* **53**, 3293-3297.

28. Yang, Y.-C., Ward, J.R., Wilson, R.B., Burrows, W., and Winterle, J.S. (1987) On the activation energy for the hydrolysis of bis-(2-chloroethylethyl) sulfide, II, *Thermochim. Acta* **114**, 313-317.

29. Zulty, J.J. Personal communication.

BIODEGRADATION OF ORGANOPHOSPHORUS NERVE AGENTS BY SURFACE EXPRESSED ORGANOPHOSPHORUS HYDROLASE

W. CHEN*, R.D. RICHINS, P. MULCHANDANI, I. KANEVA, AND A. MULCHANDANI

Department of Chemical and Environmental Engineering,
University of California, Riverside, CA 92521, USA.

Abstract. Organophosphorus hydrolase (OPH) isolated from *Pseudomonas* and *Flavobacterium* has been shown to degrade organophosphates, and immobilized OPH has been applied as a means to detoxify organophosphate wastes. However, construction of an enzyme reactor is often very labor intensive and could suffer from diffusional limitations. In this chapter, we will describe a novel method for the biodegradation of organophosphate nerve agents using *Escherichia coli* with surface-exposed OPH as "live" biocatalysts. An *lpp-ompA-opd* fusion was used to direct OPH onto the surface. The resulting strains degraded parathion and paraoxon very effectively without the transport limitation observed in cells expressing OPH intracellularly. Cultures with surface-expressed OPH also exhibited a very long shelf-life, retaining 100% activity over a period of one month. Large-scale detoxification of organophosphates was investigated in both batch and sequence-batch immobilized cell bioreactors with the novel biocatalysts immobilized on non-woven polypropylene fabric. In batch operations, 100% of paraoxon was degraded in less than 2 h. The immobilized cells retained almost 100% activity during the initial six repeated cycles and close to 90% activity even after 12 repeated cycles. In addition to paraoxon, other commonly used organophosphate pesticides, such as diazinon, coumaphos and methyl parathion were hydrolyzed efficiently. The cell immobilization technology developed here paves the way for an efficient, simple and cost effective method for detoxification of organophosphate nerve agents.

B. Zwanenburg et al. (eds.), Enzymes in Action, 211–221.
© 2000 *Kluwer Academic Publishers. Printed in the Netherlands.*

212

1. Introduction

Synthetic organophosphorus compounds, which are among the most toxic substances known, are used extensively as agricultural and domestic pesticides including insecticides, fungicides, and herbicides [21]. In the United States annually over 40 million kg of organophosphate pesticides are consumed, while another 20 million kg are produced for export. The use of these pesticides, though very important to the success of the agricultural industry, affects the environment. One of the most serious problems is the approximately 400,000 litres of cattle dip wastes containing approximately 1,500 mg/l of the organophosphate insecticide coumaphos that are generated yearly along the Mexican border from a USDA program designed to control disease-carrying cattle ticks [17]. Because of the growing public concern regarding the residues of these pesticides in food products and contamination of water supplies, there is a great need to develop safe, convenient, and economically-feasible methods for the detoxification of organophosphorus pesticides.

Large stockpiles of extremely toxic organophosphate-based chemical warfare agents such as sarin, soman, and VX (LCt_{50} inhalation of 50-200 mg.min m^{-3} and LD_{50} skin exposure of 10-4000 mg/individual) are stored around the world, and in United States alone, the stockpile is estimated at 25 million kg [5]. Under a recently ratified international chemical weapons treaty, the U.S. must destroy all its chemical weapon (CW) agents within the next 10 years. Moreover, the terrorist attack on the Tokyo subway system and the chronic health problems experienced by Gulf War Veterans due to their suspected exposure to nerve gases, has focused the attention on the danger of these compounds.

Currently, the only EPA approved method for destroying these organo-phosphorus nerve agents is by incineration. This method, however, is very costly, as large amount of energy is required to reach the high temperatures necessary to destroy chemical weapons. Also, the potential for air pollution has made incineration extremely unpopular in communities where these incinerators are operating or will be operated.

Due to the problems associated with pollutant treatment by conventional methods, increasing consideration has been placed on the development of alternative, economical and reliable biological treatment processes, which might draw upon genetic engineering methods. Clearly the more important challenges in bioremediation are to better exploit existing biodegradative routes or to develop novel catabolic capabilities for pollutant elimination. Organophosphorus hydrolase (OPH) isolated form *Pseudomonas diminuta* MG and *Flavobacterium* sp has been shown to degrade organophosphate pesticides (P-O and P-S bond hydrolysis) and nerve gases (P-F and P-CN bond cleavage) [1, 15, 19]; enzymatic hydrolysis is at least 40-2450 times faster than chemical hydrolysis [20]. OPH has a broad substrate specificity and is able to catalyze the hydrolysis of organophosphate pesticides, such as methyl and ethyl parathion, paraoxon, dursban, coumaphos, cyanophos and diazinon and chemical warfare agents, such as sarin and soman [7, 8]. OPH is very high efficient in hydrolyzing pesticides such as paraoxon and coumaphos or neurotoxins such as diisopropyl fluorophosphate (DFP) and the nerve gas sarin. Several OPH mutants have also been generated by site-directed mutagenesis with increased activity against soman (257Leu and 257Val) and VX (254 Arg) [16]. While the hydrolyzed products of organophosphates such as parathion and coumaphos are 3-4 orders of magnitude less toxic, hydolyzed product of CW nerve agents such as soman and sarin are essentially nontoxic [26].

Since the economics of culturing the native soil bacteria in bioreactors is not very attractive due to slow specific growth rates, the *opd* gene has been cloned into *E. coli* [6], insect cell (fall armyworm) [22], and *Streptomyces* [24]. Both native and recombinant OPH, immobilized onto nylon (membrane, powder and tubing), porous glass and silica beads, have been applied as enzyme reactors for the detoxification of organophosphates in bulk liquid phase [2, 3, 20]. Moreover, the gas-phase degradation of paraoxon was recently demonstrated in an immobilized OPH reactor [25]. This example provides the possibility of utilizing OPH for the gas-phase detoxification of nerve agents in general. However, in most cases, the cost of this detoxification process depends heavily on the cost of purifying OPH. This problem can be eliminated if whole cells (either growing or non-growing), rather than enzymes, are immobilized onto the support (such as in an immobilized-cell bioreactor). The use of immobilized cells in a bioreactor does have disadvantages. One serious potential problem is the mass transport limitation of substrates and products across the cell membrane, since the outer membrane can act as a permeability barrier which may inhibit substrates from interacting with the enzymes resided within the cell. Recent reports have shown that the transport of organophsophate nerve agents is indeed restrictive [9, 13]. The resistance to mass transport can be reduced by treating cells with permeabilizing agents such as EDTA, DMSO, tributyl phosphate etc. However, not all enzymes are amenable to this treatment, and immobilized viable cells cannot be subject to permeabilization. One potential solution is to anchor and display the OPH on the cell surface, thereby obviating the need for enzyme purification and eliminating the transport limitation.

The ability to display heterologous proteins on the surface of microorganisms is generating intriguing opportunities such as recombinant bacterial vaccines, whole-cell adsorbents, and recombinant whole-cell biocatalysts. For example, expression of β-lactamase on the cell surface of *E. coli* eliminates diffusional limitations in the hydrolysis of β-lactams by whole cells, resulting in 50-fold higher rates of reaction compared with cells expressing the same level of enzyme in the periplasmic space [10]. We have recently demonstrated that active OPH can be anchored to the cell surfacea gene fusion system consisting of the signal sequence and the first nine amino acids of lipoprotein (Lpp) joined to a transmembrane domain from Outer Membrane Protein A (OmpA) has been described to anchor ОПH onto the cell surface [23]. Cells with surface-expressed OPH degraded parathion and paraoxon very effective without the diffusional limitation observed in cells expressing OPH intracellularly, and also exhibited a very long shelf life, retaining 100% activity over a period of one month. Immobilization of these live biocatalysts onto solid supports was demonstrated to provide an attractive and economical means for pesticide detoxification in place of immobilized enzymes or immobilized whole cell expressing OPH intracellularly, affording no diffusional barrier, smaller labor cost, and the potential for easy regeneration.

2. Surface Expression of OPH

Organophosphorus hydrolase was displayed and anchored onto the surface of *Escherichia coli* using an Lpp-OmpA fusion system [23], which has been used previously for displaying β-lactamase [10], Cex exoglucanase [11] and Fv antibody [12] fragments onto the cell surface. Two recombinant plasmids were created with one encoding an Lpp-OmpA-OPH fusion enzyme containing the native OPH signal sequence (pOP131) and one without the signal sequence (pOP231). Production of the fusion proteins in the membrane fractions was verified by immunoblotting with OmpA antisera. Proteins of approximately 55 and 53 kDa were detected from cells carrying constructs pOP131 and pOP231, respectively. Even though immunoblot analysis indicates that the quantity of Lpp-OmpA-OPH recombinant protein present in JM105(pOP131) and JM105(pOP231) cells is essentially the same, the activity of JM105(pOP231) was only about 5-10% that of JM105(pOP131) - indicative of the importance of the signal sequence in expressing active OPH on the surface.

Protease accessibility experiments were carried out to ascertain the presence of enzymatically-active OPH on the cell surface. More than 80% of the OPH activity was demonstrated to be located on the cell surface as OPHs were degraded by proteases only accessible to surface proteins. Cells expressing OPH on the cell surface degraded parathion and paraoxon very effectively without any diffusional limitation across the cell membrane. Whole cell with surface-expressed OPH retained 90% activity compared to cell lysates, while whole cells of JM105(pWM513) expressing OPH intracellularly only retained 10% activity of cell lysates. The low percentage of whole cells activity in JM105(pWM513) probably arises from a substantial diffusional barrier for parathion and paraoxon across the cell membrane, since the addition of a permeabilizing agent, EDTA, increased whole cell activity to a level similar to the lysates.

Although recombinant cells expressing OPH constitutively onto the cell surface degraded parathion very effectively, they were relatively unstable. A gradual decline in OPH activity was observed over a period of three months. It was reasoned that constitutive expression of a surface protein leads to high structural instability of the expression vector. To address this problem, a new plasmid (pOPK132) was constructed to express OPH onto the cell surface under control of a tightly regulated *tac* promoter system. The resulting plasmid was shown to be very stable and *E. coli* cells transformed with it degraded parathion and paraoxon very effectively. With similar levels of OPH (normalized by the OPH activity in the cell lysates), whole cells of JM105(pOPK132) had 7-fold higher activity than whole cells of JM105(pWM513) (Figure 1). Therefore, an improved "whole cell" technology for organophosphorus pesticide biodegradation has been realized.

3. Factors Influencing OPH Activity on the Cell Surface

Although recombinant cells with surface-expressed OPH were shown to degrade parathion and paraoxon very effectively without the diffusional limitation observed in cells expressing OPH intracellularly, the precise conditions for surface targeting or pesticide degradation were not fully understood. To gain a better understanding of this new system and to explore its potential for pesticide detoxification, we have investigated several factors influencing OPH expression and parathion degradation [14]. Production of active OPH onto the cell surface was found to be highly host-specific; high rate of parathion degradation was observed from strains JM105 and XL1-Blue, which regulated production of the OPH fusion very tightly. However, in the absence of ampicillin selection, plasmids were only favorably maintained in strain XL1-Blue. OPH activity was shown to be highly dependent on growth conditions. Optimal OPH activity was observed when cells were grown in LB-buffered medium at 37°C. OPH activity was further improved by supplementing the growth medium with cobalt chloride, which favors the formation of the metal active center. It was found that the timing of cobalt addition also influenced parathion degradation. Since the OPH domain is anchored on the cell surface, we argue that $CoCl_2$ is readily available to the histidine cluster of surface-exposed OPHs and uptake may not be necessary. Maximum OPH activity was demonstrated by adding cobalt to induced cultures during the late stationary phase (Figure 2). The resulting cultures grown under the optimized conditions had a 8-fold increase in parathion degradation.

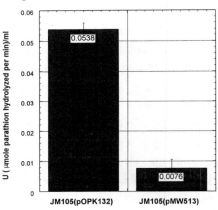

Figure 1: Whole cell activities with OPH expressed on the surface [JM105(pOPK132)] and intracellularly [JM105(pWM513)]. Whole cell activities were compared with equivalent levels of OPH activity (0.061 U/ml) from the cell lysates.

4. Detoxification of Organophosphorus Nerve Agents by Immobilized *E. coli* with Surface-Expressed OPH

Enzymatic degradation of organophosphorus nerve agents by organophosphorus hydrolase has been a subject of considerable attention during the past ten years. However, practical applications of large-scale enzymatic degradation has always been limited by the cost and stability of OPH. The availability of *E. coli* with surface-expressed OPH has opened new avenues for developing novel bioprocesses based on immobilized whole cell technology.

Immobilization of whole cells onto a solid support by adsorption is a mild, simple and non-specific process. Simultaneous immobilization and cultivation was found unsuitable in this study because (1) the cells are immobilized as a slimy film that adsorbed/bound very weakly, (2) the slimy layer was easily disturbed/dislodged from the surface by mere lifting of the support from the medium, and (3) a majority of the cell growth was in the medium as evident from the highly turbid medium. Since our prior research with non-growing (freely suspended in buffer in the absence of nutrients) *E. coli* with surface-expressed OPH were extremely stable and retained 100% of activity for more than a month [4], the possibility of immobilizing precultivated resting-cells on the non-woven fabric support was investigated. Preliminary results indicated that the non-growing cells adsorbed tightly to the non-woven fabric support and this method was adopted for cell immobilization in this research.

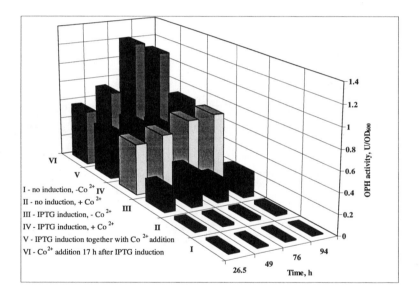

Figure 2. Effect of cobalt addition on OPH activity. Cultures of XL1-Blue:pOPK132 were grown in LB-buffered medium at 37°C, and samples were taken at various time points.

Since surface adsorption of cells is a function of a number of parameters depending on pH and ionic strength of the buffer, 11 different support materials were evaluated based on cell loading (g cells/cm^2 support) and the potential for regeneration. 909 PP white non-woven fabric was shown to provide the highest cell loading (~50 mg dry weight/12.5 cm^2) [18]. Cell immobilization was also function of surface area, with

218

increasing cell loading on larger support matrices (12 mg of cells on a 2.25 cm² matrix as compared to 50 mg on a 12.5 cm² matrix). The scanning electron micrograph of the cell support (Fig. 3) shows that the support matrix was covered completely and uniformly by the adsorbed cells. Cell immobilization was a strong function of pH, with the highest cell loading observed at pH 8.0.

The ability of the immobilized cells to degrade various organophosphates was evaluated in batch and sequence batch reactors. Immobilized cells with surface-expressed OPH effectively degrade diazinon, methyl parathion, paraoxon, and coumaphos with nearly 90-100% conversion. The rate of pesticides degradation was in line with the kinetic ability of OPH to degrade these substrates. In batch systems, almost 100% of paraoxon and diazinon was degraded in less than 3.5 h (Fig. 4a). We have also tested the ability of immobilized cells to carryout repeated cycles of degradation in sequence batch reactor. The immobilized cells remained stable and degraded 80% of paraoxon in 2 h even after 12 consecutive repeated sequence batch operation. (Fig. 4b).

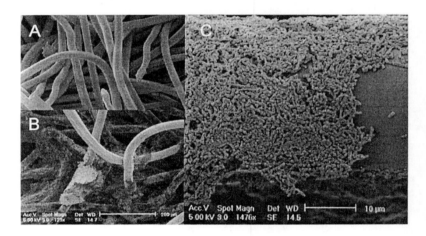

Figure 3. Scanning electron microscopy of 909 PP polyester in the absence (A) and presence (B&C) of *E. coli* cells with surface-expressed OPH after 48 hr incubation.

a)   b)

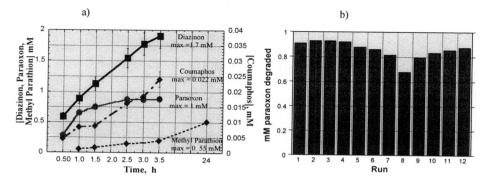

Figure 4. (a) Detoxification of organophosphates by immobilized cells with surface-exposed OPH. (b) Repeated paraoxon detoxification by immobilized cells.

5.   Conclusions

Enzymatic degradation of organophosphorus nerve agents by OPH has been a subject of considerable attention during the past decade. However, the practical applications of large-scale enzymatic degradation has always been limited by the cost and stability of OPH. We have successfully engineered *E. coli* cells with active OPH anchored on the cell surface. Cells displaying catalytically active OPH are in effect "live" immobilized-enzyme particels and could be employed for organophosphate hydrolysis in a similar fashion to conventional immobilized OPH without tedious purification. The availability of this organism has opened new avenues for developing bioprocesses based on immobilized whole cells.

A simple adsorption procedure was developed for the immobilization of non-growing cells with surface-expressed OPH. The method of cell immobilization developed here has advantages because there is no need for aseptic environment after culturing and scale-up versions of the present setup in either batch or continuous mode is feasible.

6.   Acknowledgement

We thank the U.S. EPA and NSF for their financial support.

7. References

1. Attaway, H, Nelson, J.O., Baya, A.M., Voll, M.J., White, W.E., Grimes, D.J. and Colwell, R.R. (1987) Bacterial detoxification of diisopropyl fluorophosphate, *Appl. Environ. Microbiol.* **53**, 1685-1689.

2. Caldwell, S.R. and Raushel, F.M. (1991) Detoxification of organophosphate pesticides using an immobilized phosphotriesterase from *Pseudomonas diminuta*, *Biotechnol. Bioeng.* **37**, 103-109.

3. Caldwell, S.R. and Raushel, F.M. (1991) Detoxification of organophosphate pesticides using a nylon based immobilized phosphotriesterase from *Pseudomonas diminuta*, *Appl. Biochem. Biotechnol.* **31**, 59-73.

4. Chen, W. and Mulchandani, A. (1998) The use of "live biocatalysts" for pesticide detoxification, *Trends Biotechnol.* **16**, 71-76.

5. Compton, J.A. (1988). *Military Chemical and Biological Agents*, Telford Press, NJ.

6. Dave, K.I, Miller, C.E. and Wild, J.R. (1993) Characterization of organophosphorus hydrolases and the genetic manipulation of the phosphotriesterase from *Pseudomonas diminuta*, *Chem.-Biol. Interactions.* **87**, 55-68.

7. Dumas, D.P., Durst, H.D., Landis, W.G., Raushel, F.M. and Wild, J.R. (1990) Inactivation of organophosphorus nerve agents by the phosphotriesterase from *Pseudomonas diminuta*, *Arch. Biochem. Biophys.* **227**, 155-159.

8. Dumas, D.P., Wild, J.R. and Raushel, F.M. (1989) Diisopropyl fluorophosphate hydrolysis by a phosphotriesterase from *Pseudomonas diminuta*, *Biotech. Appl. Biochem.* **11**, 235-243.

9. Elashvili, I., Defrank, J.J. and Culotta, V.C. (1998) *phnE* and *glpT* genes enhance utilization of organophosphates in Escherichia coli K-12, *Appl. Environ. Microbiol.* **64**, 2601-2608.

10. Francisco, J.A., Earhart, C.F. and Georgiou, G. (1992) Production and fluorescene-activated cell sorting of *Escherichia* coli expressing a functional antibody fragment on the external surface, *Proc. Natl. Acad. Sci.* **89**, 2713-2717.

11. Francisco, J.A., Stathopoulos, C., Warren, R.A.J., Kilburn, D.G. and Georgiou, G. (1993) Specific adhesion and hydrolysis of cellulose by intact *Escherichia coli* expressing surface anchored cellulase or cellulose binding domains, *Bio/Technol.* **11**, 491-495.

12. Francisco, J.A., Campbell, R., Iverson, B.L. and Georgiou, G. 1993. Production and fluorescene-activated cell sorting of *Escherichia coli* expressing a functional antibody fragment on the external surface, *Proc. Natl. Acad. Sci. USA* **90**:10444-10448.

13. Hung, S-C. and Liao, J C. (1996) Effects of ultraviolet light irradiation in biotreatment of organophosphates, *Appl. Biochem. Biotechnol.* **56**, 37-47.

14. Kaneva, I., Mulchandani, A. and Chen, W. (1998) Factors influencing parathion detoxification by recombinant *Escherichia coli* with surface-expressed organophosphorus hydrolase, *Biotechnol. Prog.* **14**, 275-278.

15. Karns, J.S., Muldoon, M.T., Mulbury, W.W., Derbyshire, M.K. and Kearney, P.C. (1987) Use of microorganisms and microbial systems in the degradation of pesticides, in H.M LeBaron, R.O Mumma, R.C. Honeycutt and J.H. Duesing (eds.), *Biotechnology in Agricultural Chemistry. ACS Symposium Series. 334*, American Chemical Society, Washington, D.C, pp. 156-170.

16. Lai, K., Grimsley, J.K., Kuhlmann, B.D., Scapozza, L., Harvey, S.P., DeFrank, J.J., Kolalowski, J.E. and Wild, J.R. (1996) Rational enzyme design: Computer modeling and site-directed mutagenesis for the modification of catalytic specificity in organophosphorus hydrolase, *Chimia* **50**, 430-431.

17. Mulbury, W.W., Del Valle, P.L. and Karns, J.S. (1996) Biodegradation of organophosphate insecticide coumaphos in highly contaminated soil and in liquid wastes, *Pestic. Sci.* **48**, 149-155.

18. Mulchandani, A., Kaneva, I. and Chen, W. (1999) Detoxification of organophosphate nerve agents by immobilized *Escherichia coli* with surface-expressed organophosphorus hydrolase, *Biotechnol. Bioeng.* **63**, 216-223.

19. Munnecke, D.M. and Hsieh, D.P.H. (1976) Pathways of microbial metabolism of parathion, *Appl. Environ. Microbiol.* **31**, 63-69.

20. Munnecke, D.M. (1979) Hydrolysis of organophosphate insecticides by an immobilized-enzyme system, *Biotechnol. Bioeng.* **21**, 2247-2261.

21. Munnecke, D.M. (1980) Enzymatic detoxification of waste organophosphate pesticides, *J. Agric. Food Chem.* **28**, 105-111.

22. Phillips, J.P., Xin, J.H., Kirby, K., Milne, C.P., Krell, P. and Wild, J.R. (1990) Transfer and expression of an organophosphate insecticide-degrading gene from *Pseudomonas* in *Drosophila melanogaster*, *Proc. Natl. Acad. Sci.* **87**, 8155-8159.

23. Richins, R., Kaneva, I., Mulchandani, A. and Chen, W. 1997. Biodegradation of organophosphorus pesticides by surface-expressed organophosphorus hydrolase, *Nature Biotechnology* **15**, 984-987.

24. Steiert, J.G., Pogell, B.M., Speedie, M.K. and Laredo, J. (1989) A gene coding for a membrane-bound hydrolase is expressed as a secreted soluble enzyme in *Streptomyces lividans, Bio/Technol.* **7**, 65-68.

25. Yang, F., Wild, J.R. and Russell, A.J. (1995) Nonaqueous biocatalytic degradation of a nerve gas mimic, *Biotechnol. Prog.* **11**, 471-474.

26. Yakovlevsky, K., Govarhan, C. Grimsley, J., Wild, J.R. and Margolin, A. (1997) Cross-linked crystals of organophosphorus hydrolase: An efficient catalyst for the breakdown of the O.P. compounds, 213th American Chemical Society National Meeting, April 13-17, San Francisco.

ACTIVE SITE MODIFICATIONS OF ORGANOPHOSPHORUS HYDROLASE FOR IMPROVED DETOXIFICATION OF ORGANOPHOSPHORUS NEUROTOXINS

JANET K. GRIMSLEY[1], BARBARA D. DISIOUDI[2], THOMAS R. HOLTON[1], JAMES C. SACCHETTINI[1], & JAMES R. WILD[1]
[1]Texas A&M University, Department of Biochemistry and Biophysics, College Station, TX 77843-2128, and [2]Scott Foresman, Educational Technology, 1900 East Lake Avenue, Glenview, IL 60025.

Abstract: Amino acid substitutions within the active site of the dimeric metalloenzyme Organophosphorus Hydrolase (OPH) result in a striking enhancement in the hydrolysis of certain chemical warfare agents and their analogues. These changes alter the metal content of the enzyme and we suggest that changes in metal requirements improve the catalytic characteristics by allowing greater structural flexibility and access of larger substrates to the active site. Crystallographic and three-dimensional modeling analyses have suggested that removing steric hindrances in the vicinity of the binding pocket could further enhance the effectiveness of OPH to hydrolyze VX. These studies also suggest that the hydrogen-bonding network supplying support and stability to the active site deserve a critical analysis for further catalytic improvements. The broad substrate specificity and hydrolytic efficiency of OPH and the ability to genetically engineer the enzyme for specific target organophosphate neurotoxins have provided realistic OPH-based technologies for detoxification of these compounds, including enzyme immobilization on various matrices, discriminative detection, and clinical therapy. A series of optimized enzymes for individual substrates can be envisioned that would maximize degradative activity under a particular environmental situation. The capacity for further improvement is remarkable, and the opportunity for a variety of biotechnological applications is quite pronounced.

B. Zwanenburg et al. (eds.), Enzymes in Action, 223–242.
© 2000 *Kluwer Academic Publishers. Printed in the Netherlands.*

1. Introduction

Organophosphorus compounds (OPs) are a broad class of neurotoxic chemicals that may be used as pesticides or chemical warfare agents (CWAs). Their toxicity is determined by the reactivity of the phosphoryl center, which can bind and inactivate acetylcholinesterase (AChE) and other enzymes in the target organism. AChE triggers the rapid breakdown of the signal chemical acetylcholine to choline. Neurotoxic OP compounds covalently bind to AChE through a phosphorylated intermediate. Inhibition is a progressive process and depends on the concentration of nerve agents, time of exposure, and nature of any clinical prophylaxis. Besides AChE related effects, OP-induced delayed neuropathy may result from events following initial cholinesterase inhibition (1-3). This neuropathy is not apparent until long after exposure and leads to the degeneration of axons in the peripheral nervous system and spinal cord followed by breakdown of myelinated fibers and muscle (4,5). The severe nature of these effects and the obligation to eliminate chemical warfare stockpiles have stimulated the search for strategies for effective management of exposure and toxicity.

In phosphotriester organophosphates, all four atoms bound to the phosphate are oxygen, but other subclasses substitute S, F, C, and/or N in one or more of the positions occupied by oxygen. The toxicity of OPs is dramatically reduced by either chemical or enzymatic hydrolysis. Several enzymes have been identified that are capable of this hydrolysis and they are compared in Table 1 (6-13). The focus of this work is organophosphorus hydrolase (OPH, phosphotriesterase, E.C. 8.1.3.1), which catalyzes the hydrolysis of many organophosphates and is the only OP-degrading enzyme which has been shown to hydrolyze the P—S bond of OP substrates (6-8).

TABLE 1. A comparison of some OP degrading enzymes

enzyme	bond cleaved by enzyme				
	P—C	P—O	P—F	P—S	P—CN
OPH	-	+	+	+	+
OPAA-2	-	+	+	-	+
human paraoxonase	+	+	+	-	+
squid DFPase	-	-	+	-	-

The gene encoding OPH, *opd* (*organophosphate degrading*) was originally isolated from native plasmids of *Pseudomonas diminuta* MG and *Flavobacterium* (14,15). The gene was subsequently cloned and the protein overexpressed in *Escherichia coli* (16-18). The *opd* gene was also introduced into a variety of other organisms to explore issues such as posttranslational modification (19); organophosphate resistance in selected beneficial insects (20), and the potential use of the fungus *Gliocladium virens* for bioremediation (21).

OPH from *Pseudomonas diminuta* MG detoxifies a variety of organophosphorus neurotoxins by hydrolyzing P-O, P-CN, P-F, and P-S bonds (6-9). The broad substrate range of OPH is shown in Figure 1 and includes several insecticides, chemical warfare agent analogs and neurotoxic chemical warfare agents (8,9,22,23). The catalytic specificities (k_{cat}/K_m) for this family of compounds range from diffusion limited (e.g. paraoxon, P-O bond) to rates which are eight orders of magnitude lower (e.g. acephate, P-S bond). Enzymatic specificities of OPH for representative OP compounds are shown in Table 2. The broad substrate specificity and high catalytic turnover rate for several OP compounds make OPH an ideal candidate for use in degradation of OP compounds and therapeutic treatment of OP poisoning.

Crystallographic investigations have provided detailed structural information on OPH (24-26). The enzyme is a homodimer with equal active sites at the C-terminus of each monomer. The molecular architecture of each monomer is a distorted α/β barrel with eight parallel β-strands forming the barrel and linked on the outer surface by 14 α helices (Figure 2). The enzyme requires divalent metal cations for activity, however, several metals including Co^{2+}, Cd^{2+}, Ni^{2+} and Mn^{2+} can be substituted for the native Zn^{2+} with varying effects on enzymatic activity (27,28) (Table 3).

Various histidine residues are involved in maintaining active site geometry, binding metal ions, and participating in the catalytic mechanism. The crystal structures of the Cd^{2+} and Zn^{2+}-substituted enzymes have established the identity of the metal binding ligands (25,26). The more buried zinc (M_1) interacts with histidine (His)55, His57, aspartate(Asp)301, and two bridging ligands, a water molecule and a carbamylated lysine(Lys169), in a distorted trigonal bipyramidal geometry (Figure 3). The more solvent exposed zinc (M_2) interacts with His201, His230, in addition to the two bridging ligands (Lys169 and a water molecule) in a distorted tetrahedral geometry. The highly positive character of the metal ions and the histidines is counteracted by the presence of five aspartate residues and the carbamoylated lysine. Particularly noteworthy is the participation of some of these residues in an intricate hydrogen-bonding network with metal binding residues.

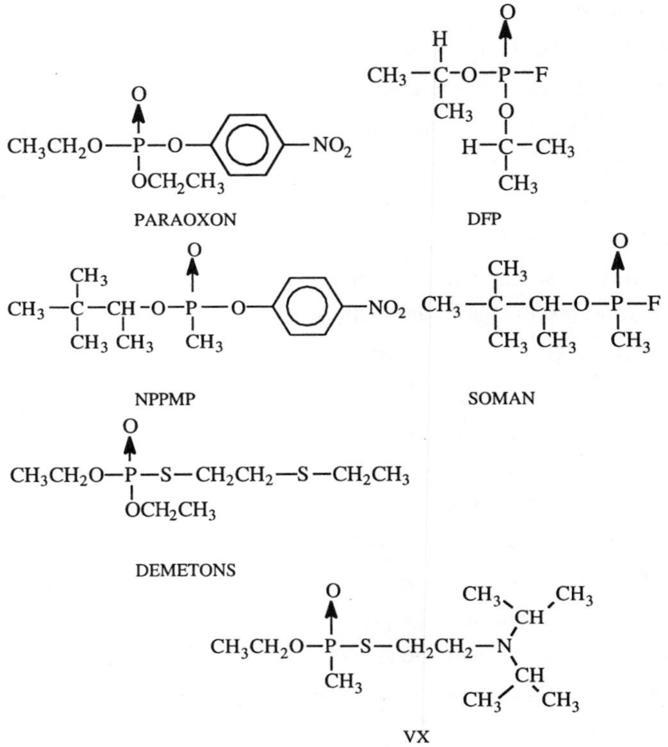

Figure 1: Representative Substrates of OPH

TABLE 2.

substrate	k_{cat}/K_M $(M^{-1}sec^{-1})$
paraoxon	1×10^8
diisopropyl fluorophosphonate(DFP)	1×10^7
sarin	8×10^4
p-nitrophenyl-o-pinacolyl-methylphosphonate (NPPMP)	2×10^4
soman	1×10^4
phosalone	2×10^2
demeton S	7×10^2
0-ethyl S-(2-diisopropylaminoethyl) -methylphosphonothioate (VX)	7×10^2
tetriso	4×10^2
acephate	2×10^1

The conformational stability of OPH has also been determined (29). Urea- or guanidinium HCl-induced unfolding of Zn^{2+} -OPH occurred as a three-state process which includes a partially folded, catalytically inactive dimeric intermediate. The overall conformational stability of OPH was estimated to be 40 kcal/mol, which is the highest stability reported for a dimeric protein, and the apparent T_m was 75 °C.

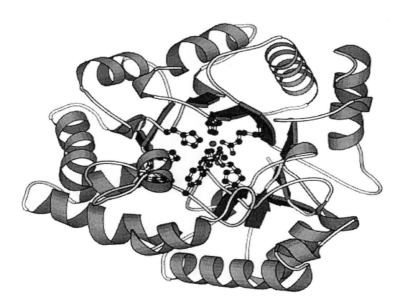

Figure 2: Ribbon drawing of OPH. The drawing was generated from the crystal structure (26) and depicts one active site with the ball and stick structures representing the active site residues. The small circles are the metal atoms.

TABLE 3. The effects of metals on OPH activity

Metal Ion	Substrate	Units/mg OPH
none	*paraoxon*	1760
Zn	*paraoxon*	2080
Co	*paraoxon*	8020
Ni	*paraoxon*	3320
Cd	*paraoxon*	4100
Mn	*paraoxon*	3020
Cu	*paraoxon*	1890
Fe	*paraoxon*	50
Co	DFP	625
Zn	DFP	215
Co	Sarin	323
Zn	Sarin	200

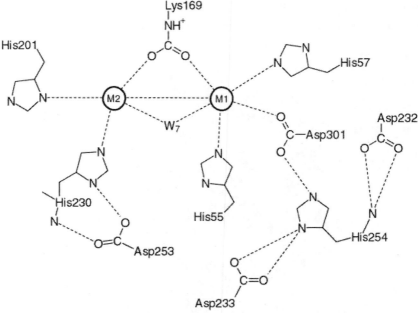

Figure 3. Schematic of the active site of OPH surrounding the zinc ions. M1 and M2 designate the buried and more solvent exposed metals, respectively. W_7 designates the catalytic water molecule.

2.   Enzyme Engineering

Considering the broad range of substrate accommodations, OPH is an ideal candidate for programmed alterations in specificity and catalytic activity. Investigations to identify amino acid residues responsible for binding divalent metal cations generated many enzymes with substitutions for the original histidine residues at positions 254 and 257 (30-33). It has been proposed that these residues influence catalysis by interacting with active site residues and the substrate in the binding pocket (26). The histidine at position 254 was replaced with arginine (H254R) and the one at position 257 was replaced with leucine (H257L) independently to form the single site-modified enzymes. The double modification was also constructed to incorporate both changes (H254R/H257L) (34). These three altered enzymes achieved a 2-to 30-fold increase in substrate specificity (k_{cat}/K_m) for demeton S (P-S bond), an analogue for the chemical warfare agent VX. In contrast, the substrate specificity for diisopropyl fluoro-phosphonate (DFP, P-F bond) was substantially decreased for each of these enzymes. However, H257L and H254R/H257L showed an 11- and 18-fold increase, respectively, in specificity for NPPMP, the analogue for thechemical warfare agent soman. The enzymatic characteristics of these enzymes are summarized in Table 4. These results demonstrate that it is possible to significantly enhance the ability of OPH to hydrolyze P-S bond substrates. Although neither His254 nor His257 is directly involved in metal ligation or take part in the reaction chemistry, amino acid substitutions at either (or both) position would be expected to disrupt the numerous hydrogen bonding and electrostatic interactions associated with the side chains of the native residues (Figure 3) (26). In addition, His254 stacks with both His257 and primary metal ligand His230 in a manner that suggests that these three residues are involved in stabilizing van der Waals stacking interactions (Figure 4). Even subtle changes to these intricate stabilizing interactions should impart a new flexibility to the active site that could be responsible for the enhanced catalysis of the altered enzymes for some larger substrates, such as demeton S and NPPMP, allowing them ready access to the binding pocket. These disruptions would also decrease the catalytic efficiency for smaller substrates, such as DFP, which is exactly what these data demonstrate. In summary, alterations to residues immediately surrounding the active site of OPH result in enzymes with dramatic alteration in metal binding and specificity for alternate substrates.

Although native OPH has two metals at each active site (four per dimer), all three of these altered enzymes possessed only one metal per active site (two per dimer) while retaining considerable enzymatic activity for the preferred phosphotriester (P-O bond) substrate, paraoxon (5-100% k_{cat}) (34). The reaction mechanism of the native enzyme is postulated to occur via an S_N2-type mechanism whereby an activated water molecule attacks the electrophilic phosphorus center of the substrate (35). The activation of the nucleophilic water molecule is thought to occur through interactions with the binuclear metal center and not via a traditional enzymatic base. As a result, there would seem to be an absolute requirement for two metal atoms in the active site to precisely position the activated hydroxide. The three enzymes discussed above (H254R, H257L and H254R/H257L) would have severely diminished enzymatic activity if this proposed mechanism is correct. Residual activity of these enzymes would presumably result from an asymmetry in the homodimer such that one of the individual monomers would have two metals with enhanced catalytic capabilities and the other would have no metals. Monomers with one metal would not be active, or at best retain only 50% of the native activity. This, of course, is not the case as indicated for the H257L enzyme, which retains 100% activity against paraoxon (Table 4). In order to reconcile the data with the mechanism, effects of metal substitution on catalysis were investigated (34). Metals were removed from the purified altered enzymes, as well as the wild type enzyme, and these apoenzymes were then systematically reconstituted with either Zn^{2+} or Co^{2+}. The assays were performed with both a rapidly hydrolyzed substrate (paraoxon) and a slowly reactive substrate (demeton S).

TABLE 4. The kinetic parameters of OPH enzymes

	Paraoxon	NPPMP	DFP	DemetonS
WT OPH				
k_{cat} (per second)	15000±300	23±2	75±6	4.2±0.1
K_M (mM)	0.12±0.01	1.3±0.3	0.96±0.1	4.8±0.2
k_{cat}/K_M ($M^{-1}s^{-1}$)	1.3e8	1.7e4	7.8e4	8.7e2
K_i (mM)	17±1	----	23±8	----
H254R				
k_{cat}	680±10	2.5±0.1	0.41±0.01	16±0.7
K_M	0.015±0.001	0.47±0.07	0.087±0.01	4.4±0.4
k_{cat}/K_M	4.4e7	5.3e3	4.7e3	3.6e3
K_i	1.8±0.1	----	----	----
H257L				
k_{cat}	16000±600	260±40	19±1	3.3±0.05
K_M	0.30±0.03	1.4±0.3	0.73±0.1	1.9±0.9
k_{cat}/K_M	5.3e7	1.9e5	2.6e4	1.8e3
K_i	15±3	4.4±2	21±7	----
H254R/H257L				
k_{cat}	1400±40	54±3	0.79±0.06	68±5
K_M	0.036±0.003	0.18±0.02	0.20±0.06	2.5±0.5
k_{cat}/K_M	4.0e7	3.0e5	4.0e3	2.7e4
K_i	4.4±0.5	1.7±0.2	----	----

Maximal catalysis of paraoxon by wild type OPH occurred when 2 equivalents or moles of metal, either Zn^{2+} or Co^{2+}, were provided per monomer, as described in earlier reports (27,30,34). Additional metal did not affect catalysis (Figure 5A). Data for the H254R/H257L enzyme, which is representative of the monometal variants, indicates that although up to 3 equivalents of metal were provided for reconstitution, the kinetic profile levels off at 1 equivalent of metal per mole of monomer for both paraoxon and demeton S (Figure 5). When the metal-reconstituted wild type OPH was used in comparable kinetic saturation experiments with the phosphorothiolate substrate demeton S, the profile levels off at 1 equivalent of metal per mole of monomer (Figure 5B). Although the wild type enzyme was shown by flame atomic absorption

231

spectroscopy to sequester up to 2 equivalents of metal per monomer (34), the association of a single metal per monomer produced maximal hydrolysis (Figure 5B), and higher concentrations of metal did not enhance catalysis of demeton S. The difference in metal requirements for catalysis of the two different reactions suggests that the catalytic mechanism for hydrolysis of paraoxon, and the alternative substrate, demeton S, may be different. It is probable that the loss of one metal from the active sites of the OPH variants causes subtle rearrangement and enlargement of the binding pocket allowing increased rates of hydrolysis for larger substrates and diminished activity for smaller substrates.

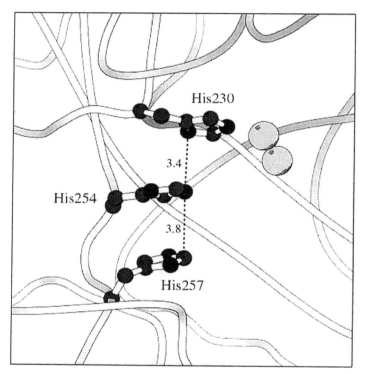

Figure 4: Interactions of secondary ligand His254 with His257 and metal ligand230. Interatomic distances are given in angstroms.

Molecular modeling of the enzymes resulting from changes at position 254 and 257 was performed using the Zn^{2+}-liganded OPH (26) as the structural template. Models of OP substrates were then docked into the active site of the enzyme in an effort to discern the differences in accommodation among the modified binding pockets. Figure 6 depicts the crystallographically determined conformation of the native active site of OPH with bound inhibitor, diethyl-4-Methylbenzylphosphonate. If this structure is compared with the structural model of the H254R/H257L double mutation (Figure 7) subtle yet striking differences are observed. For example, the substitution for leucine at position 257 has apparently made more room for the substrate. The methylbenzyl group of the inhibitor sits very close to His257 in the native enzyme yet appears to be nestled more comfortably next to Leu257 in the modified OPH binding pocket. The models indicate that the enzyme could more easily accommodate bulky OP substrates by expansion of the active site boundaries.

According to the modelling data, the k_{cat} increase is due to the positive charge of the arginine residue that is at a favorable distance from the ester S atom of demeton S or VX (Figure 7). Therefore, the electron on the S atom will be more localized and will not be shared with the P atom of the substrate. This causes the P atom to be more positively charged which favors a nucleophilic attack. In other words, Arg254 can favorably interact with the ligand effectively polarizing the P-S bond of the substrate. In addition, a careful analysis of the hydrogen bond network surrounding the active site indicates that some hydrogen bonds are missing in the modified OPH enzymes. Among these are bonds between His254 and Asp233 and a bond between His257 and the backbone of Asn300. Since both of these hydrogen bonds are missing in the H254R/H257L enzyme, the rigidity of the active site could very well be compromised allowing an easier induced fit for longer and bulkier molecules. Hence, the loss of hydrogen bonding potential within the active site may be very important for the structure and function of OPH and must be considered in protein engineering experiments.

Strikingly, preliminary crystallographic data have confirmed the modeling predictions that active site expansion is a characteristic of the H254 and H257 modified enzymes (unpublished observations). A three dimensional view into the active site of the H254R enzyme definitely displays a wider "mouth" or entrance into the binding pocket if compared directly to that of the native, Zn^{2+}-liganded wild type enzyme. The exact position(s) of the metal ions in these OPH variants are still somewhat speculative, but only one metal is positioned precisely within the active site. This is also encouraging since the possibility of these enzymes being "monometal" (one metal/active site) appears to be substantiated with structural information.

The only other example of an improved rate of catalysis by OPH was demonstrated by Watkins et al. (36). The engineering of OPH for enhanced catalysis of DFP was stimulated by the Zn^{2+}-structure with a substrate analog bound at the active site. The binding pocket surrounding the potential leaving group of the substrate is lined with hydrophobic residues (Trp131, Phe132, Leu271, Phe306, and Tyr309). Because

DFP has a fluoride leaving group, replacement of one side chain with a residue capable of hydrogen bond formation and proton donation was predicted to enhance catalysis. Mutations to active site residues Phe132 and Phe306, including some double mutations, succeeded in enhancing k_{cat} up to tenfold. However, there was no improvement in K_m, and in fact the higher K_m values caused the specificity of these enzymes (k_{cat}/K_m) for DFP to be improved only twofold over that of wild type OPH.

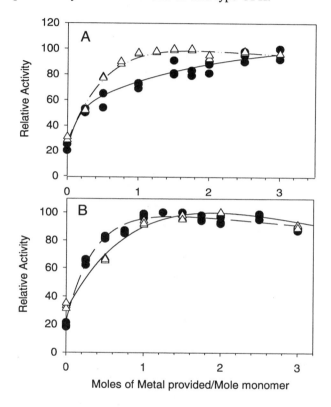

Figure 5: A) Relative activity of metal-reconstituted OPH samples with 3 mM paraoxon. Triangles denote H254R/H257L data and circles denote data for the wild type enzyme. B) Relative activity of metal-reconstituted OPH samples with 8 mM demetonS. Triangles denote H254R/H257L data and circles denote data for the wild type enzyme.

234

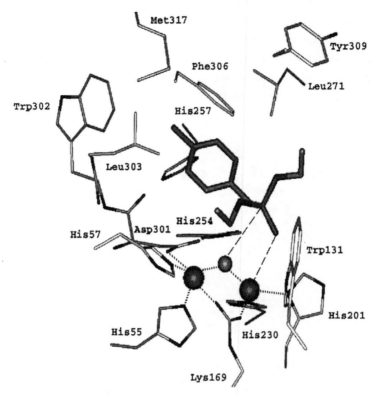

Figure 6. Model of the OPH active site found in the wild type enzyme. Large spheres represent metal atoms and small spheres represent water molecules. The ball and stick structure in the center of the figure represents the bound inhibitor, diethyl-4-Methylbenzylphosphonate. Lines with small dots indicate interaction between protein and the metals as well as metals and water. Lines with big dots show the interactions between the inhibitor and the catalytic water as well as the metal. Visualization of these molecules and substrates was done with the molecular modeling package INSIGHTII and the programs GRID and DOCK (Biosym Technologies, San Diego) were used for docking of substrates and orientation of specific atoms for enzyme/substrate interactions.

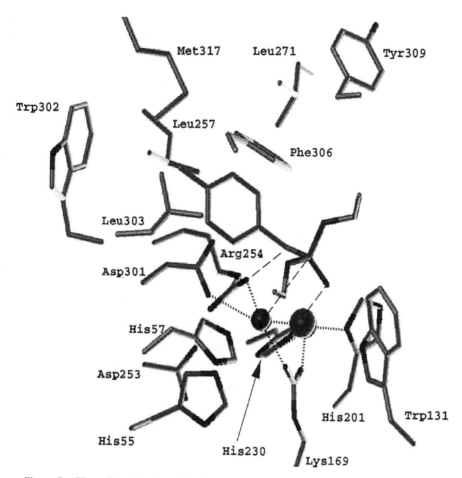

Figure 7. Three-dimensional model of the H254R/H257L active site with bound inhibitor diethyl-4-Methylbenzylphosphonate. Large sphere represents the single metal atom and small sphere represents the catalytic water molecule. The catalytic water lies at the same position as in the wild type enzyme. Lines with small dots indicate interaction between protein and the metal as well as metal and water. Hydrogen bond between the catalytic water and the protein is also displayed as small dot lines. Lines with big dots show the interactions between the ligand and the catalytic water as well as the metal. Putative interaction between Arg254 and the ligand is also given as a line with big dots.

3.   OPH-based technologies

The unique characteristics of OPH and the ability to genetically engineer improved catalytic characteristics make it a very realistic candidate for bioremediation and detection.   Technological applications range from immobilized cells and enzymes to prophylactic treatments against OP poisoning.   The potential and capacity of such OPH-based systems are quite remarkable as described below.

3.1.   ENZYME IMMOBILIZATION

OPH-based biosensors provide unique capabilities for recognition and discrimination of OP neurotoxins (37-40).   The basis for analysis derives from the enzymatic mechanism where the quantity of OP hydrolyzed corresponds with the quantity of protons release during the reaction.   Rainina *et al.* (37) created a biosensor for the detection of environmental OP contaminants using immobilized recombinant *Escherichia coli*, which expressed OPH.   A poly (vinyl) alcohol cryogel was used to entrap whole cells in a mechanically stable porous gel matrix.   Ten identical paraoxon samples read in succession gave values with 3.5% S.D. among them.   The paraoxon concentration coincided 97.5% with the concentration determined spectrophotometically.   The sensitivity of the cell-based biosensor was 0.25-250 ppm.

Simonian *et al.* (38) immobilized purified OPH on activated silica gel for use in a multienzyme biosensor.   This discriminative approach combines a highly sensitive AChE biosensor with immobilized OPH.   The value of this concept lies in the discrimination between carbamate and OP pesticides, and in the improvement of the accuracy of the traditional AChE-based biosensor.   Since OPH is capable of very specific hydrolysis of OP compounds, it is possible to remove them from mixed samples by enzymatic activity.   Preincubation of samples containing paraoxon with OPH totally eliminated the inhibition effect on AChE activity.   This approach provides a new perspective for discriminative determination using multicomponent or multienzyme arrays.   It has also been possible to use lyophilized OPH for OP-degradation applications.   Lyophilized enzyme in a continuous gas-phase reactor was remarkably active in the hydrolysis of paraoxon (41).   In addition, lyophilized OPH introduced into polyurethane foam retained 52% of the initial specific activity of the starting enzyme preparation (42).   These foams were produced using "First Defense" aqueous fire-fighting foam surfactant solution and the performance was highly reproducible and predictable in detoxifying OP nerve toxins deposited onto surfaces (43,44).   The application of these forms using multicomponent or multienzyme arrays with varying specificity will provide a safe, environmentally acceptable means of performing wide-area decontamination of nerve agents.

3.2. MEDICAL APPLICATIONS

The effectiveness of OP pesticides have contributed to both increased agricultural production (45) and decreased incidence of many insect-borne diseases such as malaria (46). However, unintentional overexposure has resulted in more than three million cases of pesticide poisoning annually worldwide (47). Moreover, some highly toxic OP compounds have been developed and manufactured explicitly for use as chemical weapons against humans. OP toxicity is greatly diminished upon cleavage of the phosphotriester bond (45). For this reason, OPH has been explored as an antidote or prophylactic measure for OP poisoning. Mice treated with OPH by intravenous injection survived multiple challenges with paraoxon or diethylfluorophosphate (DEFP) at concentrations of 7.3 and 2.9 x LD_{50}, respectively (48). The greater protection afforded against paraoxon intoxication was attributed to the faster OPH catalysis of the P-O substrate. Other investigators gauged the efficacy of OPH injection by measuring paraoxonase activity in the blood of mice subjected to doses of paraoxon (49). OPH injection enhanced paraoxonase activity of mouse serum 2 to 5-fold and prevented cholinesterase inhibition following DFP, sarin or soman exposure (50).

Pretreatment of animals with OPH protected animals from even higher doses of OP compounds. Mice with OPH administered intravenously ten minutes prior to intraperitoneal challenge with paraoxon, DFP, sarin, or soman tolerated paraoxon doses that were 33-fold higher than those tolerated by untreated mice (50,51). When OPH was used in conjunction with reversible acetylcholinesterase inhibitors, experimental mice tolerated four times the lethal dose of sarin (52). In some cases the improved efficacy of OPH in protecting animals from paraoxon or DFP intoxication when compared to soman or sarin intoxication was again attributed to the differences in enzyme kinetics among OP substrates. This discrepancy emphasizes the need to continue to improve catalytic efficiency and specificity of OPH for poorer substrates.

As an alternative to direct injection of OPH for therapeutic treatment of OP intoxication, the enzyme has also been delivered to mice in "carrier" red blood cells (CRBCs) (53). The enzyme, entrapped in the cells by hypotonic dialysis, protected mice from OP challenge. The LD_{50} of paraoxon in CRBC-treated mice was increased 126-fold; CRBCs used in conjunction with classical OP antidotes atropine and pralidoxime

(2-PAM) increased the LD_{50} of paraoxon more than 1000-fold (54). More recently, sterically stabilized liposomes encapsulating OPH were employed as a carrier model to antagonize the toxic effects of paraoxon providing a striking enhancement of protective effects (55). In addition, OPH has been immobilized onto a hollow-fiber reactor designed for extracorporeal blood circulation and is currently under experimentation for post-exposure detoxification (56).

4. Summary and Conclusions

OPH has broad substrate specificity and good catalytic rate for many neurotoxic organophosphate compounds and excellent inherent stability. Because of the good affinity (low K_m) of OPH for many of its substrates, systems to detect and/or remediate low concentrations of OP neurotoxins in the environment have been developed. Alterations to the metal content of the enzyme as well as mutations to specific active site residues have further enhanced the specificity of native OPH. These have productively enhanced the ability of the enzyme to degrade specific OP compounds. The inherent stability of OPH has been enhanced by immobilization of the enzyme in a number of matrices. Immobilization has been shown to enhance the shelf life and thermal stability of OPH, making the enzyme a better remediation agent for long-term applications and use in extreme environments. In addition, potential uses of this enzyme as an antidotal and/or prophylactic measure against OP poisoning have been investigated. Animals provided with OPH through injection or in carrier red blood cells were protected from intoxication with paraoxon, DFP, sarin, and soman. The native characteristics of OPH and engineered enhancement of the enzyme make it an attractive system for use in degradation of OP compounds and therapeutic treatment of OP poisoning.

5. References

1.   Barrett, D.S. and F.W. Oehme.   (1985) A review of organophosphorus ester-induced delayed neurotoxicity, *Vet.and Hum. Toxicol.* **27**, 22-37.

2.   Johnson, M.K. (1975) The delayed neuropathy caused by some organophosphorus esters: mechanism and challenge, *CRC Crit. Rev. Toxicol.* **3**, 289-316.

3.   Johnson, M.K. (1987) Receptor or enzyme: The puzzle of NTE and organophosphate-induced delayed polyneuropathy, *Trends Pharmacol* **8**, 174-179.

4.   Cavanagh, J.B. (1954) The toxic effects of tri-ortho-cresyl phosphate on the nervous system, and experimental study in hens, *J. Neurol. Neurosurg. Psychiat.* **17**, 163-172.

239

5. Sprague, G.L. and Bickford, A.A. (1981) Effect of multiple diisopropylfluorophosphate injections in hens: behavioral, biochemical and histological investigation, *J. Toxicol. Environ. Health* **8**, 973-988.

6. Chae, M.Y., Postula, J.F., and Raushel, F.M. (1994) Stereospecific enzymatic hydrolysis of phosphorus-sulfur bonds in chiral organophosphate triesters, *Bioorg. Med. Chem. Lett.* **4**, 1473-1478.

7. Lai, K., Stolowich, N.J., and Wild, J.R. (1995) Characterization of P-S bond hydrolysis in organo-phosphorothioate pesticides by organophosphorus hydrolase, *Arch. Biochem. Biophys* **318**, 59-64.

8. Kolakowski, J.E., DeFrank, J.J., Harvey, S.P., Szafraniec, L.L., Beaudry, W.T., Lai, K. and Wild, J.R. (1997) Enzymatic hydrolysis of the chemical warfare agent VX and its neurotoxic analogues by organophosphorus hydrolase, *Biocatal. Biotransform.* **15**, 297-312.

9. Dumas, D.P., Durst, H. D., Landis, W.G., Raushel, F.M., and Wild, J.R. (1990) Inactivation of organophosphorus nerve agents by the phosphotriesterase from *Pseudomonas diminuta*, *Arch. Biochem. Biophys.* **277**, 155-159.

10. DeFrank, J.J., Beaudry, W.T., Cheng, T.-C., Harvey, S.P., Stroup, A.N., and Szafraniec, L.L. (1993) Screening of halophilic bacteria and *Alteromonas* species for organophosphorus hydrolyzing enzyme activity, *Chem.-Biol. Interact.* **87**, 141-148.

11. Cheng, T.-C., Harvey, S.P., and Chen, G.L. (1996) Cloning and expression of a gene encoding a bacterial enzyme for decontamination of organophosphorus nerve agents and nucleotide sequence of the enzyme, *Appl. Environ. Microbiol.* **62**, 1636-1641.

12. Hassett, C., Richter, R.J., Humbert, R., Chapline, C., Crabb, J.W., Omiecinski, C.J., and Furlong, C.E. (1991) Characterization of cDNA clones encoding rabbit and human serum paraoxonase: the mature protein retains its signal sequence, *Biochemistry* **30**, 10141-10149.

13. Hoskin, F.C.G. and Roush, A.H. (1982) Hydrolysis of nerve gas by squid-type diisopropyl phosphorofluoridate hydrolyzing enzyme on agarose resin, *Science* **215**, 1255-1257.

14. Mulbry, W.W., Karns, J.S., Kearney, P.C., Nelson, J.O., McDaniel, C.S., and Wild, J. R. (1986) Identification of a plasmid-borne parathion hydrolase gene from *Flavobacterium* sp. by Southern hybridization with *opd* from *Pseudomonas diminuta*, *Appl. Environ. Microbiol.* **51**, 926-930.

15. Harper, L.L., McDaniel, C.S. Miller, C.E., and Wild, J.R. (1988) Dissimilar plasmids isolated from *Pseudomonas diminuta* MG and a *Flavobacterium* sp. (ATCC 27551) contain identical *opd* genes, *Appl. Environ. Microbiol.* **54**, 2586-2589.

16. McDaniel, C.S., Harper, L.L., and Wild, J. R. (1988) Cloning and sequencing of a plasmid-borne gene (*opd*) encoding a phosphotriesterase, *J. Bacteriol.* **170**, 2306-2311.

17. Mulbry, W.W. and Karns, J.S. (1989) Parathion hydrolase specified by the *Flavobacterium opd* gene: relationship between the gene and the protein, *J. Bacteriol.* **171**, 6740-6746.

18. Serdar, C.M., Murdock, D.C., and Rohde, M.F. (1989) Parathion hydrolase gene from *Pseudomonas diminuta* MG: subcloning, complete nucleotide sequence, and expression of the mature portion of the enzyme in *Escherichia coli*, *Bio/Technol.* **7**, 1151-1155.

19. Steiert, J.G., Pogell, B.M., Speedie, M.K., and Laredo, J. (1989) A gene coding for a membrane-bound hydrolase is expressed as a secreted, soluble enzyme in *Streptomyces lividans*, *Bio/Technol.* **7**, 65-68.

20. Phillips, J.P., Xin, J.H., Kirby, K., Milne, J. C. P., Krell, P., and Wild, J.R. (1990) Transfer and expression of an organophosphate insecticide-degrading gene from *Pseudomonas* in *Drosophila melanogaster*, *Proc. Natl. Acad. Sci. U. S. A.* **87**, 8155-8159.

21. Dave, K.I., Lauriano, C., Xu, B., Wild, J.R., and Kenerley, C.M. (1994) Expression of organophosphate hydrolase in the filamentous fungus *Gliocladium virens*, *Appl. Microbiol. Biotechnol.* **41** 352-358.

240

22. Dumas, D.P., Caldwell, S.R., Wild, J.R., and Raushel, F.M. (1989) Purification and properties of the phosphotriesterase from *Pseudomonas diminuta*, *J. Biol. Chem.* **264**, 19659-19665.

23. Hoskin, F.C.G., Walker, J.E., Dettbarn, W.-D., and Wild, J.R. (1995) Hydrolysis of tetriso by an enzyme derived from *Pseudomonas diminuta* as a model for the detoxication of *O*-ethyl *S*-(2-diisopropylaminoethyl) methylphosphonothiolate (VX), *Biochem. Pharmacol.* **49**, 711-715.

24. Benning, M.W., Kuo, J.M., Raushel, F.M., and Holden, H.M. (1994) Three-dimensional structure of Phosphotriesterase: An enzyme capable of detoxifying organophosphate nerve agents, *Biochemistry* **33**, 15001-15007.

25. Benning, M.W., Kuo, J.M., Raushel, F.M., and Holden, H.M. (1995) Three-dimensional structure of the binuclear metal center of Phosphotriesterase, *Biochemistry* **34**, 7973-7978.

26. Vanhooke, J.L., Benning, M.M., Raushel, F.M., and Holden, H.M. (1996) Three-dimensional structure of the zinc-containing Phosphotriesterase with the bound substrate analog diethyl 4-methylbenzylphosphonate, *Biochemistry* **35**, 6020-6025.

27. Omburo, G.A., Kuo, J.M., Mullins, L.S., and Raushel, F.M. (1992) Characterization of the zinc binding site of bacterial phosphotriesterase, *J. Biol. Chem.* **267**, 13278-13283.

28. Lai, K., Dave, K.I., and Wild, J.R. (1994) Enzymatic decontamination of organophosphorus chemical agents by genetic and biochemical manipulation of organophosphorus hydrolase, in D. A. Berg, J. D. Williams, Jr., and P. J. Reeves (eds.), *Proceedings of the 1993 ERDEC Scientific Conference on Chemical Defense Ressearch 16-19 November 1993*, U.S. Army, Aberdeen Proving Ground, MD, pp.887-894.

29. Grimsley, J.K., Scholtz, J.M., Pace, C.N., and Wild, J.R. (1997) Organophosphorus hydrolase is a remarkably stable enzyme that unfolds through a homodimeric intermediate, *Biochemistry* **36**, 14366-14374.

30. Lai, K., Dave, K.I., and Wild, J.R. (1994) Bimetallic binding motifs in organophosphorus hydrolase are important for catalysis and structural organization, *J. Biol. Chem.* **269**, 16579-16584.

31. Kuo, J.M., and Raushel, F.M. (1994) Identification of the histidine ligands to the binuclear metal center of phosphotriesterase by site-directed mutagenesis, *Biochemistry* **33**, 4265-4272.

32. Lai, K. Modification and characterization of the neurotoxic substrate specificity of organophosphorus hydrolase, Ph.D. Dissertation, Texas A&M University, 1994.

33. Kuo, J.M. Investigations of the active site and metal centers of phosphotriesterase, Ph.D. Dissertation, Texas A&M University, 1995.

34. diSioudi, B., Grimsley, J.K., Lai, K., and Wild, J.R. (1999) Modification of near active site residues in organophosphorus hydrolase reduces metal stoichiometry and alters substrate specificity, *Biochemistry* **38**, 2866-2872.

35. Lewis, V.E., Donarski, W.J., Wild, J.R., and Raushel, F.M. (1988) Mechanism and stereochemical course at phosphorus of the reaction catalyzed by a bacterial phosphotriesterase, *Biochemistry* **27**, 1591-1597.

36. Watkins, L.M., Mahoney, H.J., McCulloch, J.K., and Raushel, F.M. (1997) Augmented hydrolysis of diisopropyl fluorophosphate in engineered mutants of phosphotriesterase, *J. Biol. Chem.* **272**, 25596-25601.

37. Rainina, E.I., Efremenco, E.N., and Varfolomeyev, S.D. (1996) The development of a new biosensor based on recombinant *E. coli* for the direct detection of organophosphorus neurotoxins, *Biosens. Bioelectron.* **11**, 991-1000.

38. Simonian, A. L., Rainina, E.I., and Wild, J.R. (1997) A new approach for discriminative detection of organophosphate neurotoxins in the presence of other cholinesterase inhibitors, *Anal. Lett.* **30**, 2453-2468.

39. Mulchandani, A., Mulchandani, P., Kaneve, I., and Chen, W. (1998) Biosensor for direct determination of organophosphate nerve agents using recombinant Escherichia coli with surface-expressed organophosphorus hydrolase. 1. Potentiometric microbial electrode, *Anal. Chem.* **70**, 4140-4145.

40. Mulchandani, A., Kaneve, I., and Chen, W. (1998) Biosensor for direct determination of organophosphate nerve agents using recombinant Escherichia coli with surface-expressed organophosphorus hydrolase. 2. Fiber-optic microbial biosensor, *Anal. Chem.* **70**, 5042-5046.

41. Yang, F., Wild, J.R., and Russell, A.J. (1995) Nonaqueous biocatalytic degradation of a nerve gas mimic, *Biotechnol. Prog.* **11**, 471-474.

42. LeJeune, K.E. and Russell, A.J. (1996) Covalent binding of a nerve agent hydrolyzing enzyme within polyurethane foams, *Biotechnol. Bioeng.* **51**, 450-457.

43. LeJeune, K.E., Mesiano, A.J., Bower, S.B., Grimsley, J.K., Wild, J.R., and Russell, A.J. (1997) Dramatically stabilized phosphotriesterase-polymers for nerve agent degradation, *Biotechnol. Bioeng.* **54**, 105-114.

44. LeJeune, K.E., Wild, J.R., and Russell, A.J. (1998) Nerve agents degraded by enzymatic foams, *Nature* **395**, 27-28.

45. Ecobichon, D.J. (1991) Toxic effects of pesticides, in: M.O. Amdur, J. Doull, and C.D. Klaassen, (eds.), *Casarett and Doull's Toxicology: The Basic Science of Poisons*, McGraw-Hill, New York, pp. 565-622.

46. Matsumura, F. (1985) Toxicology of Insecticides, Plenum Press, New York.

47. World Health Organization (WHO) in collaboration with United Nations Environment Programme, Public Health Impact of Pesticides Used in Agriculture, (1990) World Health Organization, Geneva, p.128.

48. Ashani,Y., Rothschild, N., Segall, Y., Levanon, D., and Raveh, L. (1991) Prophylaxis against organophosphate poisoning by an enzyme hydrolysing organophosphorus compounds in mice, *Life Sci.* **49**, 367-374.

49. Kaliste-Korhonen, E., Ylitalo, P., Hänninen, O., and Raushel, F.M. (1993) Phosphotriesterase decreases paraoxon toxicity in mice, *Toxicol. Appl. Pharmacol.* **121**, 275-278.

50. Tuovinen, K., Kaliste-Korhonen, E., Raushel, F.M., and Hänninen, O. (1994) Phosphotriesterase - a promising candidate for use in detoxification of organophosphates, *Fundam. Appl. Toxicol.* **23**, 578-584.

51. Tuovinen, K., Kaliste-Korhonen, E., Raushel, F.M., and Hänninen, O. (1996) Protection of organophosphate-inactivated esterases with phosphotriesterase, *Fundam. Appl. Toxicol.* **31**, 210-217.

52. Tuovinen, K., Kaliste-Korhonen, E., Raushel, F.M.,and Hänninen, O. (1999) Success of pyridostigmine, physostigmine, eptastigmine and phosphotriesterase treatments in acute sarin intoxication, *Toxicology* **134**, 169-178.

53. Pei, L., Omburo, G., McGuinn, W.D., Petrikovics, I., Dave, K., Raushel, F.M., Wild, J.R., DeLoach, J.R., and Way, J.L. (1994) Encapsulation of phosphotriesterase within murine erythrocytes, *Toxicol. Appl. Pharmacol.* **124**, 296-301.

54. Pei, L. Petrikovics, I., and Way, J.L. (1995) Antagonism of the lethal effects of paraoxon by carrier erythrocytes containing phosphotriesterase, *Fundam. Appl. Toxicol.* **28**, 209-214.

55. Petrikovics, I., Hong, K., Omburo, G., Hu, QZ., Pei, L., McGuinn, W.D., Sylvester, D., Tamulinas, C., Papahadjopoulos, D., Jaszberenyi, J.C., and Way, J.L. (1999) Antagonism of paraoxon intoxication by

recombinant phosphotriesterase encapsulated within sterically stabilized liposomes, *Toxicol. Appl. Pharmacol.* **156**, 56-63.

56. Masson, P., Josse, D., Lockridge, O., Viguie, N., Taupin, C., and Buhler, C. (1998) Enzymes hydrolyzing organophosphates as potential catalytic scavengers against organophosphate poisoning, *J. Physiology (Paris)* **92**, 357-362.

HYDROLYSIS OF ORGANOPHOSPHORUS COMPOUNDS BY BACTERIAL PROLIDASES

TU-CHEN CHENG AND JOSEPH J. DEFRANK*
U.S. Army Edgewood Chemical Biological Center
Aberdeen Proving Ground, Maryland, USA 21010-5423

1. Introduction

Numerous organophosphorus (OP) compounds of importance in agriculture, medicine, military defense, and research have been shown to be potent inhibitors of cholinesterases and other enzymes with active serine residues in their active sites. Enzymes that catalyze the hydrolysis of OP compounds have been under investigation for over 50 years. These enzymes result in the detoxification of a variety of these highly toxic compounds, including chemical warfare (CW) nerve agents and pesticides. For military operations, enzyme-based decontamination systems offer considerable potential for replacing current materials that are toxic, highly corrosive, flammable, and a danger to the environment.

2. Organophosphorus Compounds

Although commonly referred to as nerve gases, chemical warfare (CW) nerve agents are actually liquids at ambient conditions. There are two general classes of nerve agents. The G-type agents (Figure 1) were first developed in Germany in the 1930s. In the 1950s, the V-type nerve agents (Figure 2) were developed in England. The V-type agents are both more toxic and more persistent than the G-type agents. A review of chemical warfare agents and their properties can be found in [27].

The primary G-type nerve agents are tabun or GA (N,N-dimethylethyl phosphoroamidocyanidate), sarin or GB (O-isopropyl methylphosphonofluoridate), soman or GD (3,3-dimethyl-2-butyl methylphosphonofluoridate) and GF (cyclohexyl

B. Zwanenburg et al. (eds.), Enzymes in Action, 243–261.
© 2000 *Kluwer Academic Publishers. Printed in the Netherlands.*

methylphosphonofluoridate. Also shown in Figure 1 is DFP (diisopropylfluoro-phosphate) which is a commonly used simulant or model compound for the G-type nerve agents.

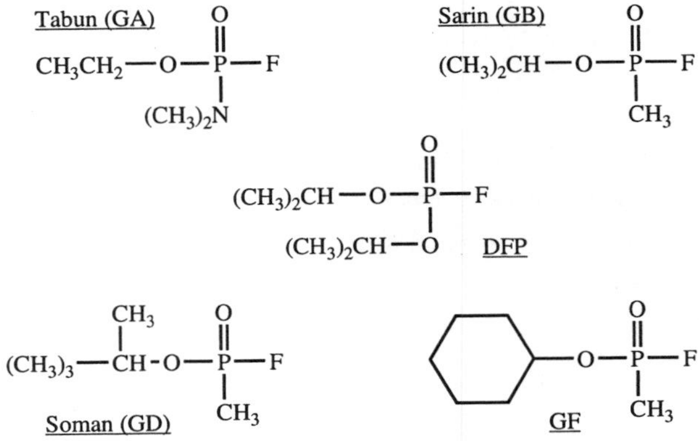

Tabun (GA)
$$CH_3CH_2-O-\overset{\overset{\displaystyle O}{\|}}{\underset{\underset{\displaystyle (CH_3)_2N}{|}}{P}}-F$$

Sarin (GB)
$$(CH_3)_2CH-O-\overset{\overset{\displaystyle O}{\|}}{\underset{\underset{\displaystyle CH_3}{|}}{P}}-F$$

$$(CH_3)_2CH-O-\overset{\overset{\displaystyle O}{\|}}{\underset{\underset{\displaystyle (CH_3)_2CH-O}{|}}{P}}-F \quad \underline{DFP}$$

$$(CH_3)_3-\overset{\overset{\displaystyle CH_3}{|}}{CH}-O-\overset{\overset{\displaystyle O}{\|}}{\underset{\underset{\displaystyle CH_3}{|}}{P}}-F$$
Soman (GD)

GF

$$\bigcirc-O-\overset{\overset{\displaystyle O}{\|}}{\underset{\underset{\displaystyle CH_3}{|}}{P}}-F$$

Figure 1: DFP and G-Type Nerve Agents.

There are two primary variants of the V-type nerve agents, the United States version of VX, (*S*-2(diisopropylamino)ethyl *O*-ethyl methylphosphonothioate) and the Russian version, R-VX, (*S*-2(diethylamino)ethyl *O*-isobutyl methylphosphonothioate). R-VX is an analogue and isomer of VX.

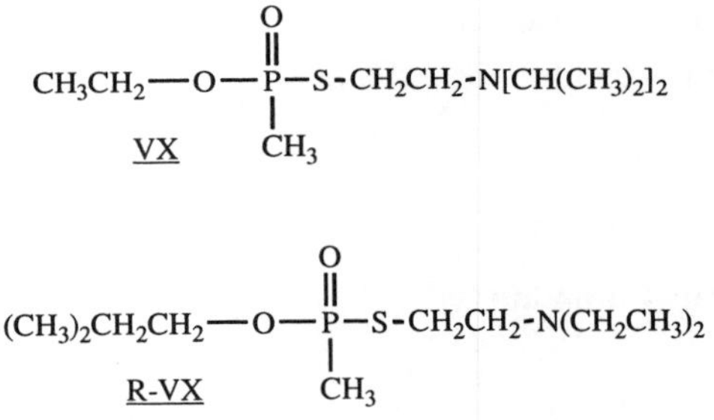

$$CH_3CH_2-O-\overset{\overset{\displaystyle O}{\|}}{\underset{\underset{\displaystyle CH_3}{|}}{P}}-S-CH_2CH_2-N[CH(CH_3)_2]_2$$
VX

$$(CH_3)_2CH_2CH_2-O-\overset{\overset{\displaystyle O}{\|}}{\underset{\underset{\displaystyle CH_3}{|}}{P}}-S-CH_2CH_2-N(CH_2CH_3)_2$$
R-VX

Figure 2: V-Type Nerve Agents.

Figure 3 shows some of the organophosphorus pesticides that have received the most attention in the literature. Although these pesticides are considered toxic, their degree of toxicity varies considerably as is shown in Table 1.

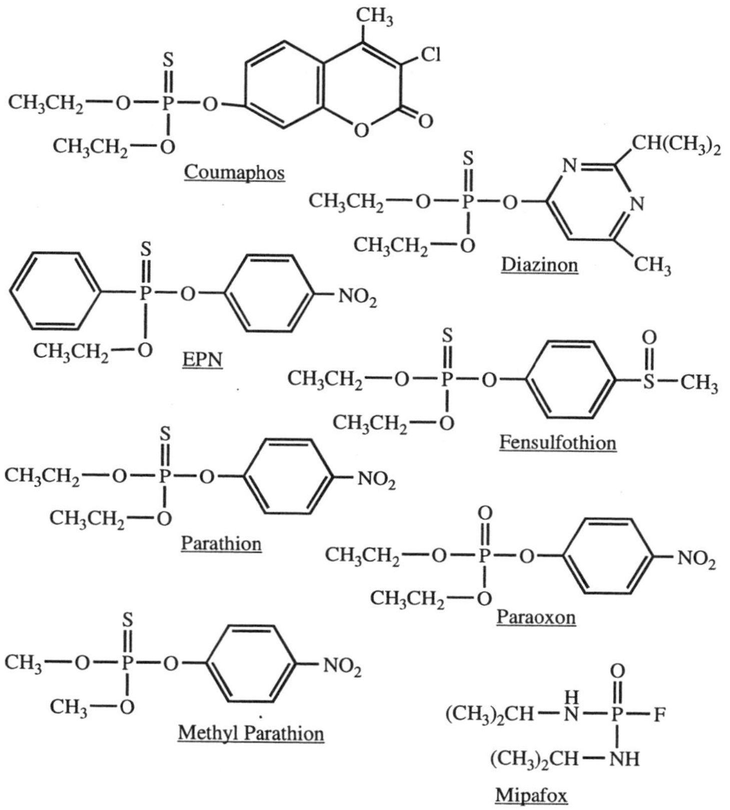

Figure 3: Organophosphorus Pesticides Substrates.

As a point of contrast, some of the more toxic, naturally occurring biomaterials have also been included in Table 1. While extremely dangerous, the toxicity of even the nerve agents pale in comparison to some natural toxins, the toxicity of botulinum toxin being three orders of magnitude greater than VX.

Although less toxic than VX, the G-type nerve agents are at least an order of magnitude more toxic than DFP and paraoxon, and several orders of magnitude more toxic than the pesticides.

TABLE 1. Toxicity's of Organophosphorus Compounds and Toxins [56,57]

Compound	Approx. LD_{50} (mg/kg, i.v.)
Diazinon	150-600
Coumaphos	90-110
EPN	35-45
Methyl Parathion	14
Parathion	13
Fensulfothion	5-10
Paraoxon	0.5
DFP	0.3
Sarin	0.01
Soman	0.01
Tabun	0.01
VX	0.001
Palytoxin	0.00015
Botulinum toxin	0.000001

3.   Organophosphorus Enzymes

The earliest work dealing with enzymes capable of hydrolyzing organophosphorus esters was reported by Mazur [54] in 1946. He described the hydrolysis of DFP and its detoxification by enzymes isolated from human and rabbit tissues. Using partially purified enzyme preparations from rabbit kidney, he was able to determine that the activity was not related to phosphatase, cholinesterase or esterase.

During the 1950's three investigators, Aldridge [4], Augustinsson [8-14], and Mounter [58, 60-62] carried out much of the work in this field. Aldridge reported on what he designated an A-serum esterase from mammalian sources that could hydrolyze paraoxon. This enzyme, more recently referred to as a phosphotriesterase or paraoxonase, was shown to differ from the phosphatases in that phosphatases only hydrolyze monoesters of orthophosphoric acid. In addition, he showed that his A-serum esterases could be stereospecific, hydrolyzing (+)-sarin, but not (-)-sarin.

Augustinsson confirmed the findings of Aldridge and extended the work to include enzymes that would hydrolyze organophosphorus compounds such as tabun. He determined that this phosphorylphosphatase or tabunase cleaved the P-CN bond of tabun to release hydrocyanic acid. Using a combination of electrophoretic separation, substrate specificities, and sensitivity to inhibition, he concluded that there were three types of esterases in plasma: arylesterase (aromatic esterase, A-esterase), aliesterase (carboxylesterase, B-esterase, "lipase") and cholinesterase. He noted that there were species variations with respect to the properties of individual enzymes. He also confirmed the observation made by Aldridge that the hog kidney phosphorylphosphatases showed stereoselectivity against tabun.

During that time period, Mounter was attempting to further purify and characterize the enzyme from rabbit kidney originally reported by Mazur. He referred to this enzyme as dialkylfluorophosphatase (DFPase, fluorophosphatase). After partial purification by ethanol fractionation, the enzyme was shown to be activated by Co^{2+} and Mn^{2+} ions. Reagents that reacted with metal ions, sulfhydryl or carbonyl groups were found to inhibit DFPase activity. Mounter was the first to report on the DFPases in microorganisms [59, 63]. Of the bacteria tested, the greatest activity was observed with the Gram-negative *Proteus vulgaris* and *Pseudomonas aeruginosa*, which were also Mn^{2+} stimulated. Based upon enzyme activity with different metal ions and inhibitors, it was also demonstrated that there were a number of different DFPases. Studies conducted with preparations from *Escherichia coli, Pseudomonas fluorescens, Streptococcus faecales*, and *Propionibacterium pentosaceum* showed that while differences were observed in the relative hydrolysis rates of a variety of organophosphorus compounds, they were comparable to those observed with the hog kidney enzyme.

During the late 1950's and early 1960's, a number of additional groups became involved in the investigation of the enzymatic hydrolysis of DFP, paraoxon, sarin and tabun [1-3, 26, 38, 43, 52, 53]. While considerable advances were made during this time in comprehending the diversity of enzymes, one of the most significant events was the beginning of the research efforts of Hoskin in this field. Of particular importance is his work, beginning in 1966, in the purification and characterization of the DFPase from squid [38, 39, 42]. The significance of the squid enzyme lies in the fact that its chemical and biological properties were different from all the other DFPases. To this day it appears to still be a fairly unique enzyme.

The interest in microbial enzymes for the degradation of organophosphorus compounds received a boost in the early 1970's by the isolation of bacteria capable of growing on a variety of pesticides. The initial report was by Sethunathan and Yoshida who isolated a diazinon-degrading *Flavobacterium* species from rice paddy soil [77].

Cell-free extracts of this organism could also hydrolyze the insecticides chlorpyrifos [diethyl (3,5,6-trichloropyridyl) phosphorothionate] and parathion, the aromatic or heterocyclic products of which were not further metabolized. In 1972, Rosenberg and Alexander [73] described two *Pseudomonas* isolates capable of hydrolyzing a variety of organophosphorus compounds and using the products as sole phosphorus source. In 1973, a pseudomonad capable of hydrolyzing parathion and utilizing the *p*-nitrophenol product as a source of carbon and nitrogen was isolated[81]. Somewhat later, Daughton and Hsieh isolated a *Pseudomonas stutzeri* capable of hydrolyzing parathion from a chemostat culture [28] in 1977. Strains of *Bacillus* and *Arthrobacter* that can hydrolyze parathion were also described [69] as well as a *Pseudomonas* capable of utilizing isophenfos as sole carbon and energy source [70]. With the exception of the *Flavobacterium*, little if any additional information has been reported about any of these organisms or their enzymes.

In addition to the *Flavobacterium* mentioned above, the other major parathion-degrading bacterium described in the literature is *Pseudomonas diminuta* MG, which was isolated in 1976 by Munnecke [66, 67]. By far, these two enzymes have been the most widely studied and are the subject of several papers at this conference. Therefore, they will not be described in any detail here. However, it is worthwhile to note that while the two organisms containing the enzymes were isolated on opposite sides of the world and the plasmids that contain their genes are highly dissimilar, the genes (and enzymes) themselves are virtually identical [36, 55, 65, 75-77]. These phosphotriesterases or parathion hydrolases are also the only well-characterized enzymes known to catalytically hydrolyze the P-S bond of V-agents [45, 71].

Beginning in the 1980's, the research efforts in this field have been divided into two major areas: the isolation and characterization of microorganisms (and their enzymes) capable of growth on a variety of organophosphorus pesticides; and the more random search for organisms that possess enzymes capable of hydrolyzing DFP and the related nerve agents. Examples of the former are the isolation of *Pseudomonas alcaligenes* C_1 which can hydrolyze and grow on fensulfothion [79]; the isolation of additional *Pseudomonas* sp. and other unidentified bacteria that hydrolyze and grow on parathion and/or methyl parathion [18, 64]; and the isolation of three distinct bacteria capable of metabolizing coumaphos [80].

In the search for nerve agent degrading enzymes, more recent investigations have gone in a number of directions. Landis and co-workers examined the ciliate protozoan *Tetrahymena thermophila* [46-48] and the clam *Rangia cuneata* [5]. Partial purification of extracts from *Tetrahymena* revealed that this organism had at least five enzymes with DFPase activity and molecular weights ranging from 67-96,000 Daltons. The rate ratios for soman and DFP hydrolysis as well as the effect of Mn^{2+} on activity varied considerably from one enzyme to another. Preliminary investigations on the clam also resulted in the detection of several DFPases with differing substrate specificity and metal stimulation. Of particular interest was the presence of an enzyme in the clam digestive gland that appeared to have significant activity on the DFP analog mipafox. Most enzymes described to date are either indifferent to mipafox or subject to fairly strong competitive inhibition.

Little *et al.*[51] reported on the characterization of an enzyme from rat liver that has a substrate preference of sarin > soman > tabun > DFP, but with no activity on paraoxon. The enzyme had a molecular weight of 40,000 and was stimulated by Mg^{2+} [51]. Unlike the *E. coli* enzyme, all the stereoisomers of soman were hydrolyzed at equal rates [15].

A screen of 18 Gram-negative bacterial isolates by Attaway *et al.* [7] resulted in the finding that, while all showed at least some activity on DFP, only parathion hydrolase producing cultures showed significant levels.

In the mid 1980's one of the authors (JJD) began an investigation on thermophilic bacteria as sources of DFPases. A Gram-positive, aerobic, spore-forming, rod-shaped, obligate thermophile was isolated from the soil of Aberdeen Proving Ground and found to possess activity against DFP and soman [29]. This isolate, designated as JD-100, was tentatively identified as a strain of *Bacillus stearothermophilus*. Its temperature range for growth was from 40 to 70°C, with an optimum in the vicinity of 55°C. While the crude cell extracts of JD-100 had low levels of DFP activity [29], the purified enzyme showed no detectable DFP activity but still retained soman hydrolyzing activity [25, 41]. The enzyme had a molecular weight of 82-84,000 and showed considerable stimulation by Mn^{2+} (~80-fold). It was unaffected by mipafox and degraded all the stereoisomers of soman.

4.   Nomenclature

As illustrated in the discussion above, the nomenclature of these enzymes has been unsystematic and confusing. In general, the names utilized have been representative of the particular substrate used by an individual investigator. Hence, the literature is filled with references to enzymes such as phosphorylphosphatase, fluorophosphatase, DFPase, paraoxonase, parathion hydrolase, phosphotriesterase, phosphofluorase, somanase, sarinase, and tabunase. In 1992, the Nomenclature Committee of the International Union of Biochemistry and Molecular Biology developed a new nomenclature that could be used until the natural substrates and functions of these enzymes could be determined. Under the general category of Phosphoric Triester Hydrolases (EC 3.1.8), the enzymes are divided into two subgroups. The first, EC 3.1.8.1, Organophosphate Hydrolase (also paraoxonase and phosphotriesterase) is for those enzymes with paraoxon and other P-esters (P-O bonds) as preferred substrates. The second group, EC 3.1.8.2, Diisopropyl-Fluorophosphatase (also Organophosphorus Acid Anhydrolase or OPAA) is for enzymes with a preference for OP compounds with P-F or P-CN bonds.

5.   Halophilic Bacterial OPAA

The work that provides the bulk of this chapter began in the late 1980's with the isolation of a variety of halophilic bacteria. These isolates were obtained from water and soil samples of salt springs in the state of Utah. In particular, one isolate, designated JD6.5, was obtained from Grantsville Warm Springs which has a relatively constant temperature of 24-32°C and 25,000 ppm dissolved solids (96% NaCl). The isolate was a Gram-negative, aerobic short rod, and an obligate, moderate halophile. It required at least 2% NaCl for growth, with an optimum between 5-10% NaCl. Fatty acid analysis of the isolate identified it as a strain of *Alteromonas*, a fairly common genus of marine bacteria. An intracellular OPAA enzyme from *Alteromonas* sp. JD6.5 was purified and characterized [32]. It is a single polypeptide with molecular weight of 58,500. It has a pH optimum of 8.5 and temperature optimum of 50°C. Maximum activity is seen with either Mn^{2+} or Co^{2+}. The enzyme was inhibited by iodoacetic acid, p-chloromercuri-benzoate, and N-ethylmaleimide, indicating that a sulfhydryl group is essential for activity. It is subject to competitive, reversible inhibition by mipafox and is significantly stimulated by NH_4^+ ions (3-5 fold). Its catalytic activity on nerve agents and related compounds is shown in Table 2.

The activity on soman corresponds to a 10^9-fold reaction acceleration in comparison to its spontaneous hydrolysis rate. It was also one to two orders of magnitude greater than any other enzyme known.

TABLE 2. OP Substrate Specificity for *Alteromonas* sp. JD6.5 OPAA [23]

Substrate*	k_{cat} (sec^{-1})
GD (soman)	3145
DFP	1820
GF	1654
GB (sarin)	611
Paraoxon	124
GA (tabun)	85
VX	0

* The activity on the substrates with fluoride leaving groups was measured with a fluoride ion-selective electrode method [42]. Activity on paraoxon was determined by measurement of the increase in absorbance at 405 nm representing the release of the *p*-nitrophenol group. For GA (tabun), cleavage of the P-CN bond was determined by P^{31}-NMR.

Having identified isolate JD6.5 as a strain of *Alteromonas*, a number of other *Alteromonas* strains were obtained from the laboratory of Dr. Rita Colwell, University of Maryland, and the American Type Culture Collection. These strains were evaluated for enzyme activity against DFP and the nerve agents [31]. Several showed high levels of activity and two of the enzymes were purified and characterized [22, 23]. Table 3 shows how these enzymes compare to the JD6.5 OPAA in physical and biochemical properties. As can be seen, there are significant similarities as well as differences with these enzymes. In order to produce the enzymes in larger quantities, the genes for *A.* sp. JD6.5 and *A. haloplanktis* were cloned into *E. coli* and expressed [21, 23]. In addition, the gene sequences were determined and translated into an amino acid sequence. The 10,000 Dalton molecular weight difference between the *A.* sp. JD6.5 and *A. haloplanktis* OPAA's was due to the presence of an extended C-terminal region in the *A.* sp. JD6.5 enzyme. The two enzymes were found to have a 77% amino acid homology. If the extended C-terminus of *A.* sp. JD6.5 was excluded, the homology increased to ~90%.

Previously, it had been assumed that the natural function of the OPAA's would have something to do with phosphorus metabolism (phospholipase, phosphodiesterase, etc.). Therefore, it came as a considerable surprise when the results of screening the amino acid sequence of *A.* sp. JD6.5 against the NCBI protein data base revealed a high degree of homology (48%) to the *E. coli* X-Pro dipeptidase. Two other matches were for *E. coli* aminopeptidase P (31% homology) and *Lactobacillus sake* dipeptidase (19% homology). There was no homology observed between the OPAA and the *Flavobacterium/Pseudomonas diminuta* phosphotriesterase (OPH).

252

TABLE 3. Comparison of *Alteromonas* OPAA's

OPAA Property	*A.* sp. JD6.5	*A. haloplanktis*	*A. undina*
Molecular Weight	60,000	50,000	53,000
pH Optimum	8.5	7.5	8.0
Temperature Opt. (°C)	50	40	55
Metal Requirement	Mn = Co	Mn	Mn
Substrate Specificity (k_{cat})			
DFP	1820	691	1403
GA	85	255	368
GB	611	308	426
GD	3145	1667	2826
GF	1654	323	1775

X-Pro dipeptidase, also known as Prolidase (EC 3.4.13.9), is a ubiquitous enzyme that hydrolyzes dipeptides with a prolyl residue in the carboxyl-terminal position. They are usually activated by Mn^{2+}, possibly thiol dependent, and usually do not act on tri- or tetrapeptides or dipeptides with proline at the N-terminus. They generally have a molecular weight of 40-50,000, a temperature optimum between 40 and 55°C, and a pH optimum between 6.5 and 8.0. All these properties are very similar to those of the *Alteromonas* OPAA's. The three OPAA's listed in Table 3 were also tested against a variety of di- and tripeptides by measuring the release of amino acids by a modified Cd-ninhydrin method [33]. The results of these assays are shown in Table 4. The results clearly indicate that the *Alteromonas* OPAA's are prolidases rather than aminopeptidases.

Another property of the prolidases is that they contain a conserved region with three histidine residues (Table 5a, *A.* sp. JD6.5 amino acid residues 331-348) as well as two smaller, highly conserved regions nearer the C-terminus, the larger of which is shown in Table 5b (*A.* sp. JD6.5 amino acid residues 371-388).

The residues that are identical in all the genes are boxed. Highly conserved or conservative substitutions are in uppercase bold text. Residues in lowercase are mismatched.

TABLE 4. *Alteromonas* OPAA Dipeptidase Profile [23]

Substrate	Specific Activity (μmoles/min/mg protein)		
	A. sp. JD6.5	*A. haloplanktis*	*A. undina*
Leu-Pro	636	988	810
Ala-Pro	510	725	658
Leu-Ala	82	63	220
Gly-Glu	<1	<1	1391
Met-Asn	<1	<1	410
Ala-Ala	<1	<1	105
Pro-Leu	<1	<1	<1
Pro-Gly	<1	<1	<1
Gly-Pro-Ala	<1	<1	<1
Ala-Pro-Phe	<1	<1	<1

TABLE 5a. Conserved Region #1 in Prolidases/Peptidases [16, 17, 21, 23, 24, 34, 68]

A. sp. JD6.5 OPAA (Prolidase)	P	H	G	L	G	H	H	I	G	L	Q	V	H	D	V	G	G	F
A. haloplanktis OPAA (Prolidase)	P	H	G	L	G	H	H	L	G	A	G	V	H	D	V	G	G	F
Halophile JD30.3 OPAA	P	H	G	L	G	H	f	I	G	L	G	V	H	D	V	G	G	F
E. coli PepQ (Prolidase)	P	H	G	I	G	H	p	I	G	L	G	V	H	D	V	A	G	F
E. coli Aminopeptidase P	m	H	G	L	s	H	w	L	G	L	d	V	H	D	V	G	V	
Methanococcus jannaschii (Prolidase)	i	H	s	L	G	H	g	V	G	L	e	V	H	E	e	p	r	l
Streptomyces lividans Aminopep. P-1	l	H	G	t	G	H	m	L	G	m	d	V	H	D	c	A	A	a
Human Prolidase	p	H	G	L	G	H	f	L	G	I	d	V	H	D	V	G	G	Y

TABLE 5b. Conserved Region #2 in Prolidases/Peptidases

A. sp. JD6.5 OPAA (Prolidase)	k	I	E	A	N	Q	V	F	T	I	E	P	G	L	Y	F	I	D
A. haloplanktis OPAA (Prolidase)	l	I	E	k	N	Q	V	F	T	I	E	P	G	L	Y	F	I	D
Halophile JD30.3 OPAA	d	V	E	A	g	Q	V	F	T	I	E	P	G	L	Y	v	V	D
E. coli PepQ (Prolidase)	i	L	q	p	g	m	V	l	T	I	E	P	G	I	Y	F	I	E
E. coli Aminopeptidase P	v	L	E	p	g	m	V	l	T	V	E	P	G	L	Y	I	A	p
Methanococcus jannaschii (Prolidase)	l	L	k	e	g	m	V	v	T	I	E	P	G	L	Y	l	k	D
Streptomyces lividans Aminopep. P-1	t	L	E	p	g	m	V	l	T	V	E	P	G	L	Y	F	q	a
Human Prolidase	h	L	q	p	g	m	V	l	T	V	E	P	G	I	Y	F	I	D

Hoskin and Walker [44] examined several nerve gas/DFP hydrolyzing enzymes to determine whether any of these might also be prolidases. Rather than measure dipeptide hydrolysis directly, they reasoned that if a DFP-hydrolyzing enzyme also hydrolyzes

Leu-Pro, then Leu-Pro should inhibit the hydrolysis of DFP. Their results are shown in Table 6.

TABLE 6. Effect of Leu-Pro on DFP Hydrolysis by Purified Nerve Gas/DFP Hydrolyzing Enzymes in Relation to Other Properties [44]			
Enzyme Source	% Inhibition by Leu-Pro	Soman/DFP Hydrolysis ratio	Mn^{2+} Stimulation
Loligo pealei (squid)	-3, -5*	0.2-0.25	0
Ps. diminuta OPH	0, 2, 3	0.125	0
Hog kidney	93, 93	~5	~5X
E. coli OPAA	72, 76	~50	~5X
Alteromonas	55, 68	~2	Yes

* Negative values = stimulation

This result demonstrated that the hog kidney and the *E. coli* OPAA's most likely are prolidases and that the squid-type OPAA and the *Pseudomonas diminuta* OPH are not. This does not eliminate the possibility that the squid-type enzyme may be a peptidase with different substrate specificity. However, in the case of OPH, the fact that it has no sequence homology suggests that it has a totally different natural function than the prolidases. Until the sequence of the squid-type enzyme has been published, a similar statement can not be made.

Since the *Alteromonas* prolidases have high level activity against the G-type nerve agents, the question arises whether other prolidases have OPAA activity as well. In the case of mammalian enzymes, partially purified human prolidase and porcine liver prolidase were obtained from Dr. Lin Liu, ChemGen Corporation and Sigma Chemical Co. respectively. Both had low levels of activity against DFP and G-agents, in the range of 1/200-1/500th of that seen with the OPAA's [23]. Their activity on X-Pro dipeptides was comparable to that observed with the OPAA's.

At the other extreme, a preparation of recombinant prolidase from the hyperthermophile *Pyrococcus furiosus* was obtained from Dr. Michael Adams, University of Georgia [35]. It was determined that it had measurable, but low levels of DFP activity at 80°C. The enzyme activity was measured well below the optimum temperature of this enzyme (100°C), but the maximum that the fluoride electrode could tolerate. For safety reasons, the tests were not repeated with the nerve agents. This enzyme is a homodimer with 39.4kDa subunits and a pH optimum of 7.0. In regards to metal requirements, Co^{2+} is preferred 4:3 over Mn^{2+}. The preferred substrates are Met-Pro and Leu-Pro. Comparison of the amino acid sequence of this enzyme with other prolidases found the greatest degree of similarity with that from another archeon, *Methanococcus jannaschii* (69%). Considerable similarity was also found with other prolidases: *Lactobacillus delbrueckii* (61%), *Haemophilus influenzae* (58%), *E. coli* (56%), human (53%), and *Alteromonas* sp. JD6.5 (51%). Based on the crystal structure of the *E. coli* methionine aminopeptidase, five amino acids (Asp97, Asp108, His171, Glu204, and Glu235) were shown to coordinate the binding of two Co^{2+} ions per active site [72]. All these prolidases conserved the same five residues, even though some of the enzymes are monomers instead of dimers. In the case of *Alteromonas* sp. JD6.5, the conserved residues are: Asp244, Asp255, His336, Glu381, and Glu420.

The question naturally comes up as to why prolidases are such efficient catalysts for the hydrolysis of organophosphorus compounds, in particular, the nerve agents. Molecular modeling studies comparing the structures of soman and Leu-Pro have been carried out [6]. It was determined that the three-dimensional structure and the electrostatic density maps of the materials look nearly identical. The organophosphorus compounds such as soman fit into the active site of the enzyme in an orientation that allows the hydrolysis of the target P-F, P-CN, or P-O bond.

6.   Production and Application

The *Alteromonas* sp. JD6.5 prolidase has proven to be quite amenable to produce through recombinant DNA technology. In initial studies the *opaA* gene encoding the *A. sp.* JD6.5 prolidase was cloned into pBluescript SK+ (pTC6513) and expressed in *E. coli*. The expressed enzyme constituted about 5% of the total cell protein. To further enhance production, the gene was cloned downstream of a strong *trc* promoter in a high-level, regulated expression vector, pSE420 (Invitrogen, San Diego, California). After induction with IPTG, the enzyme was produced at levels up of to 50-60% of total cell protein for a yield in shake-flask culture of 150-200 mg/L [24].

In order to be useful in the decontamination of nerve agents in military operations or after terrorist attacks, the enzyme-based formulation needs to be stable for long periods of time and easy to use. The *A. sp.* JD6.5 prolidase has been lyophilized in the presence of trehalose (α-D-glucopyranosyl-α-D-glucopyranoside) and stored for extended periods of time at room temperature with no apparent loss of activity [20]. In the absence of trehalose, lyophilized enzyme lost >90% of its activity.

The enzyme-based decontamination formulation is planned to be reconstituted in whatever water-based system the user has available. The systems being considered include fire-fighting foams and sprays, aqueous degreasers, laundry detergent, aircraft deicing solutions, etc. Table 7 shows the effect of a variety of these types of materials on the activity of the prolidase. The materials were evaluated at the shown concentrations. Both the prolidase and OPH have been shown to be active in generated foams [19, 74]. The use of foams offers several advantages. The foams are generally made up of surface-active agents that may help in the solubilization of the substrates, and the foam will stick to vertical surfaces for sufficient time to allow the enzyme action. As can be seen in Table 7, there is considerable variation in activity, but even in the systems where inhibition occurs, the residual activity may be sufficient to carry out the necessary decontamination. It should be noted that the enzyme was not optimized for use in any of these materials and that considerable enhancement may be possible.

TABLE 7. Effect of Aqueous Degreaser, Foams, and Other Materials on DFP Hydrolysis by *Alteromonas* sp. JD6.5 Prolidase [24]

Matrix (Source)	Property	Conc. Used (%)	Specific Activity (U/mg)*
Control (buffer only)	—	—	1950
AFC-380 (Sandia National Lab.; NM)	modified fire-fighting foam	6	1050
AFFF (3M; St. Paul, MN)	fire-fighting foam	6	460
BioSolve (Westford; Westford, MA)	fire-fighting wetting agent	6	1030
ColdFire (FireFreeze; Rockaway, NJ)	fire-suppressing agent	10	2340
Silv-EX (Ansul; Marinette, WI)	fire-fighting foam	6	320
Blue Base (Neutron; Torrance, CA)	degreaser	8	140
BV406LF (FireFreeze; Rockaway, NJ)	degreaser/cleaner	10	1430
Green Thunder (Jackson, MI)	degreaser/cleaner	10	250
SC-1000 (Gemtek; Phoenix, AZ)	wetting agent/ degreaser	5	650
Star Clean. Miracle (Hudsonville, MI)	wetting agent/oil removing	30	670
Supersolve (Gemtek; Phoenix, AZ)	wetting agent/ degreaser	10	440
Odor Seal (FireFreeze; Rockaway, NJ)	wetting agent/odor removing	10	1980

* The reaction medium contained 50 mM $(NH_4)_2CO_3$ (pH 8.7), 0.1 mM $MnCl_2$, 3 mM DFP, and 0.3-0.4 units of enzyme in a total volume of 2.5 ml. One unit (U) of enzyme activity is defined as the release of 1.0 μmoles of F^- min^{-1}.

In addition to these liquid matrices, the prolidase and OPH have been immobilized in polyurethane foams where they retained significant activity [19, 74]. The immobilized enzymes were considerably more stable than the free enzyme. This offers the potential use of the enzymes in sponges or wipes for the decontamination of personnel (including casualties) and small sensitive equipment.

7. Future Directions

In regard to the prolidases in particular, efforts will focus on searching for new or modified enzymes with activity against the V-type nerve agents and the organophosphorus pesticides. In addition to broader specificity, other enzyme properties that will be needed for decontamination applications include: activity at lower pH to extend the range of activity; activity at lower temperatures for use in cold weather operations; and enhanced stability to metals, solvents, detergents, and other inhibitors or denaturants.

Future applications could involve the use of enhanced human prolidase for prophylactic or therapeutic use. Alternatively, prolidases from bacteria could be rendered immunologically "invisible" to allow their use *in vivo*. In addition, immobilized enzymes could also find use in self-decontaminating materials and the treatment of contaminated air or water streams.

Because the nerve agents or pesticides do not inactivate the prolidases (and related enzymes), they offer the potential for use in detection systems. Such systems would not only be able to trigger an alarm in the presence of the agents, but also would indicate when the conditions are "all clear." The key to this application is the enzyme/sensor interface that provides for high sensitivity but few, if any, false alarms.

In conclusion, a class of enzymes that has no natural role in phosphorus metabolism appears to have considerable potential for dealing with some of the most toxic materials created by mankind.

8. References

1. Adie, P.A. (1956) Purification of sarinase from bovine plasma, *Can. J. Biochem. Physiol.* **34**, 1091-1094.
2. Adie, P.A., Hoskin, F.C.G., and Trick, G.S. (1956) Kinetics of the enzymatic hydrolysis of sarin, *Can J. Biochem. Physiol.* **34**, 80-82.
3. Adie, P.A. and Tuba, J. (1958) The intracellular localization of liver and kidney sarinase, *Can J. Biochem. Physiol.* **36**, 21-24.
4. Aldridge, W.N. (1953) Serum esterases 1. Two types of esterase (A and B) hydrolyzing p-nitrophenyl acetate, propionate and butyrate and a method for their determination, *Biochem. J.* **53**, 110-124.
5. Anderson, R.S., Durst, H.D., and Landis, W.G. (1988) Initial characterization of the organophosphate acid anhydrase activity in the clam *Rangia cuneata*, *Comp. Biochem. Phys.* **91C**, 575-579.
6. Ashman, W.P., Personal communication.
7. Attaway, H., Nelson, J.O., Baya, A.M., Voll, M.J., White, W.E., Grimes, D.J., and Colwell, R.R. (1987) Bacterial detoxification of diisopropyl fluorophosphate, *Appl. Environ. Microbiol.* **53**, 1685-1689.

8. Augustinsson, K.-B. (1954) The enzymatic hydrolysis of organophosphorus compounds, *Biochem. Biophys. Acta.* **13**, 303-304.

9. Augustinsson, K.-B. and Heimburger, G. (1954) Enzymatic hydrolysis of organophosphorus compounds I, Occurrence of enzymes hydrolyzing dimethylamido-ethoxy-phosphoryl cyanide (tabun), *Acta Chem. Scand.* **8**(5), 753-761.

10. Augustinsson, K.-B. and Heimburger, G. (1954) Enzymatic hydrolysis of organophosphorus compounds II, Analysis of reaction products in experiments with tabun and some properties of blood plasma tabunase, *Acta Chem. Scand.* **8**(5), 762-767.

11. Augustinsson, K.-B. and Heimburger, G. (1954) Enzymatic hydrolysis of organophosphorus compounds IV, Specificity studies, *Acta Chem. Scand.* **8**(9), 1533-1541.

12. Augustinsson, K.-B. and Heimburger, G. (1955) Enzymatic hydrolysis of organophosphorus compounds V, Effect of phosphorylphosphatase on inactivation of cholinesterases by organophosphorus compounds *in vitro*, *Acta Chem. Scand.* **9**(2), 310-318.

13. Augustinsson, K.-B. and Heimburger, G. (1955) Enzymatic hydrolysis of organophosphorus compounds VI, Effect of metallic ions on the phosphorylphosphatases of human and swine kidney, *Acta Chem. Scand.* **9**(3), 383-392.

14. Augustinsson, K.-B. and Heimburger, G. (1957) Enzymatic hydrolysis of organophosphorus compounds VII, The stereospecificity of phosphorylphosphatases, *Acta Chem. Scand.* **11**(8), 1371-1377.

15. Broomfield, C.A., Personal communication.

16. Bult, C.J., White, O., Olsen, G.J., Zhou, L.X., Fleischmen, R.D., Sutton, G.C., Blake, J.A., Fitzgerald, L.M., Clayton, R.A., Gocayne, J.D., Kerlavage, A.R., Dougherty, B.A., Tomb, J.F., Adams, M.D., Reich, C.I., Overbeek, R., Kirkness, E.F., Weinstock, K.G., Merrick, J.M., Glodek, A., Scott, J.L., Geoghagen, N.S.M., Weidman, J.F., Fuhrmann, J.L., Nguyen, D., Utterback, T.R., Kelley, J.M., Peterson, J.D., Sadow, P.W., Hana, M.C., Cotton, M.D., Roberts, K.M., Hurst, M.A., Kaine, B.P., Borodovsky, M., Klenk, H.P., Fraser, C.M., Smith, H.O., Woese, C.R., and Venter, J.C. (1996) Complete genome sequence of the methanogenic archaeon, *Methanococcus jannaschii, Science* **273**, 1058-1072.

17. Butler, M.J., Aphale, J.S., DiZonno, M.A., Krygsman, Walczyk, E., and Malek, L.T. (1994) Intracellular aminopeptidases in *Streptomyces lividans* 66, *J. Ind. Microbiol.* **13**, 24-29.

18. Chaudry, G.R., Ali, A.N., and Wheeler, W.B. (1988) Isolation of a methyl parathion-degrading *Pseudomonas* sp. that possesses DNA homologous to the *opd* gene from a *Flavobacterium* sp., *Appl. Environ. Microbiol.* **54**, 288-293.

19. Cheng, T.-c., Personal communication.

20. Cheng, T.-c. and Calomiris, J.J. (1996) A cloned bacterial enzyme for nerve agent decontamination, *Enz. Microb. Tech.* **18**, 597-601.

21. Cheng, T.-c., Harvey, S.P., and Chen, G.L. (1996) Cloning and expression of a gene encoding a bacterial enzyme for decontamination of organophosphorus nerve agents and nucleotide sequence of the enzyme, *Appl. Environ. Microbiol.* **62**, 1636-1641.

22. Cheng, T.-c., Harvey, S.P., and Stroup, A.N. (1993) Purification and properties of a highly active organophosphorus acid anhydrolase from *Alteromonas undina., Appl. Environ. Microbiol.* **59**, 3138-3140.

23. Cheng, T.-c., Liu, L., Wang, B., Wu, J., DeFrank, J.J., Anderson, D.M., Rastogi, V.K., and Hamilton, A.B. (1997) Nucleotide sequence of a gene encoding an organophosphorus nerve agent degrading

enzyme from *Alteromonas haloplanktis*, *J. Ind. Microbiol. Biotechnol.* **18**, 49-55.

24. Cheng, T.-c., Rastogi, V.K., DeFrank, J.J., and Sawiris, G.P. (1998) G-type nerve agent decontamination by *Alteromonas* prolidase, *Annals of the New York Academy of Sciences* **864**, 253-258.

25. Chettur, G., DeFrank, J.J., Gallo, B.J., Hoskin, F.C.G., Mainer, S., Robbins, F.M., Steinmann, K.E., and Walker, J.E. (1988) Soman-hydrolyzing and –detoxifying properties of an enzyme from a thermophilic bacterium, *Fund. Appl. Toxicol.* **11**, 373-380.

26. Cohen, J.A. and Warringa, M.G.P.J. (1957) Purification and properties of dialkylfluorophosphatase, *Biochem. Biophys. Acta* **26**, 29-39.

27. Compton, J.A.F. (1987) *Military Chemical and Biological Agents, Chemical and Toxicological Properties*, The Telford Press, Caldwell, NJ.

28. Daughton, C.G. and Hsieh, D.P.H. (1977) Parathion utilization by bacterial symbionts in a chemostat, *Appl. Environ. Microbiol.* **34**, 175-184.

29. DeFrank, J.J. (1986) Unpublished data.

30. DeFrank, J.J. (1991) Organophosphorus cholinesterase inhibitors: Detoxification by microbial enzymes, in J.W. Kelly and T.O. Baldwin (eds.), *Applications of Enzyme Biotechnology*, Plenum Press, New York, pp. 165-180.

31. DeFrank, J.J., Beaudry, W.T., Cheng, T.-c., Harvey, S.P., Stroup, A.N., and Szafraniec, L.L. (1993) Screening of halophilic bacteria and *Alteromonas* species for organophosphorus hydrolyzing enzyme activity, *Chem.-Biol. Interactions* **87**, 141-148.

32. DeFrank, J.J. and Cheng, T.-c. (1991) Purification and properties of an organophosphorus acid anhydrase from a halophilic bacterial isolate, *J. Bacteriol.* **173**, 1938-1943.

33. Doi, E., Shibata, D., and Matoba, T. (1981) Modified colorimetric ninhydrin methods for peptidase assay, *Anal. Biochem.* **118**, 173-184.

34. Endo, F., Hanoue, A., Nakai, H., Hata, A., Indo, Y., Titani, K., and Matsuda, I. (1989) Primary structure and gene location of human prolidase, *J. Biol. Chem.* **264**, 4476-4481.

35. Ghosh, M., Grunden, A.M., Dunn, D.M., Weiss, R., and Adams, M.W.W. (1998) Characterization of native and recombinant forms of an unusual cobalt-dependent proline dipeptidase (prolidase) from the hyperthermophilic archeon *Pyrococcus furiosus*, *J. Bacteriol.* **180**, 4781-4789.

36. Harper, L.L., McDaniel, C.S., Miller, C.E., and Wild, J.R. (1988) Dissimilar plasmids isolated from *Pseudomonas diminuta* MG and *Flavobacterium* sp. (ATCC 27551) contain identical *opd* genes, *Appl. Environ. Microbiol.* **54**, 2586-2589.

37. Hoskin, F.C.G. (1956) The enzymatic hydrolysis products of sarin, *Can. J. Biochem. Physiol.* **34**, 75-79.

38. Hoskin, F.C.G. (1969) Possible significance of "DFPase" in squid nerve, *Biol. Bull.* **137**, 389-390.

39. Hoskin, F.C.G. (1971) Diisopropylphosphorofluoridate and tabun: Enzymatic hydrolysis and nerve function, *Science* **172**, 1243-1245.

40. Hoskin, F.C.G. (1976) Distribution of diisopropylphosphorofluoridate-hydrolyzing enzyme between sheath and axoplasm of squid giant axon, *J. Neurochem.* **26**, 1043-1045.

41. Hoskin, F.C.G., Chettur, G., Mainer, S., Steinmann, K.E., DeFrank, J.J., Gallo, B.J., Robbins, F.M., and Walker, J.E. (1989) Soman hydrolysis and detoxication by a thermophilic bacterial enzyme, in E. Reiner, W.N. Aldridge, and F.C.G. Hoskin (eds.), *Enzymes Hydrolysing Organophosphorus Compounds*, Ellis Horwood, Chichester, England, pp. 53-64.

42. Hoskin, F.C.G. and Roush, A.H. (1982) Hydrolysis of nerve gas by squid type diisopropylphosphoro-fluoridate hydrolyzing enzyme on agarose resin, *Science* **215**, 1255-1257.

260

43. Hoskin, F.C.G. and Trick, G.S. (1955) Stereospecificity in the enzymatic hydrolysis of tabun and acetyl-β-methylcholine chloride, *Can. J. Biochem. Physiol.* **33**, 963-969.

44. Hoskin, F.C.G. and Walker, J.E. (1998) A closer look at the natural substrate for a nerve-gas hydrolyzing enzyme in squid nerve, *Biol. Bull.* **195**, 197-198.

45. Kolakowski, J.E., DeFrank, J.J., Harvey, S.P., Szafraniec, L.L., Beaudry, W.T., Lai, K., and Wild, J.R. (1997) Enzymatic hydrolysis of the chemical warfare agent VX and its neurotoxic analogues by organophosphorus hydrolase, *Biocat. and Biotrans.* **15**, 297-312.

46. Landis, W.G, Haley, D.M., Haley, M.V., Johnson, D.W., Durst, H.D., and Savage, R.E. (1987) Discovery of multiple organofluorophosphate hydrolyzing activities in the protozoan *Tetrahymena thermophila, J. Appl. Toxicol.* **7**(1), 35-41.

47. Landis, W.G, Haley, M.V., and Johnson, D.W. (1986) Kinetics of the DFPase activity in *Tetrahymena thermophila, J. Protozool.* **33**, 216-218.

48. Landis, W.G., Savage, R.E., and Hoskin, F.C.G. (1985) An organofluorophosphate-hydrolyzing activity in *Tetrahymena thermophila, J. Protozool.* **32**(3), 517-519.

49. LeJeune, K.E., Mesiano, A.J., Bower, S.B., Grimsley, J.K., Wild, J.R., and Russell, A.J. (1997) Dramatically stabilized phosphotriesterase-polymers for nerve agent degradation, *Biotechnol. Bioeng.* **52**, 105-114.

50. LeJeune, K.E. and Russell, A.J. (1999) Biocatalytic nerve agent detoxification in fire fighting foams, *Biotechnol. Bioeng.* **62**, 659-665.

51. Little, J.S., Broomfield, C.A., Boucher, L.J., and Fox-Talbot, M.K. (1986) Partial characterization of a rat liver enzyme that hydrolyzes sarin, soman, tabun and DFP, *Fed. Proc.* **45**(4), 791.

52. Main, A.R. (1960) The purification of the enzyme hydrolyzing diethyl *p*-nitrophenyl phosphate (paraoxon) in sheep serum, *Biochem. J.* **74**, 10-20.

53. Main, A.R. (1960) The differentiation of the A-type esterases in sheep serum, *Biochem. J.* **75**, 188-195.

54. Mazur, A. (1946) An enzyme in animal tissue capable of hydrolyzing the phosphorus-fluorine bond of alkyl fluorophosphates, *J. Biol. Chem.* **164**, 271-289.

55. McDaniel, C.S., Harper, L.L., and Wild, J.R. (1988) Cloning and sequencing of a plasmid-borne gene (*opd*) encoding a phosphotriesterase, *J. Bacteriol.* **170**, 2307-2311.

56. McEwen, F.L. and Stephenson, G.R. (1979) *The Use and Significance of Pesticides in the Environment*, Wiley-Interscience, New York.

57. Middlebrook, J.L. and Dorland, R.B. (1984) Bacterial toxins: Cellular mechanisms of action, *Microbiol. Rev.* **48**, 199-221.

58. Mounter, L.A. (1955) The complex nature of dialkylfluorophosphatases of hog and rat liver and kidney, *J. Biol. Chem.* **215**, 705-709.

59. Mounter, L.A., Baxter, R.F., and Chanutin, A. (1955) Dialkylfluorophosphatases of microorganisms, *J. Biol. Chem.* **215**, 699-704.

60. Mounter, L.A., and Dien, L.T.H. (1956) Dialkylfluorophosphatase of kidney V. The hydrolysis of organophosphorus compounds, *J. Biol. Chem.* **219**, 685-690.

61. Mounter, L.A., Dien, L.T.H., and Chanutin, A. (1955) The distribution of dialkylfluorophosphatases in the tissues of various species, *J. Biol. Chem.* **215**, 691-697.

62. Mounter, L.A., Floyd, C.S., and Chanutin, A. (1953) Dialkylfluorophosphatase of kidney I. Purification and properties, *J. Biol. Chem.* **204**, 221-232.

63. Mounter, L.A., and Tuck, K.D. (1956) Dialkylfluorophosphatases of microorganisms II. Substrate specificity studies, *J. Biol. Chem.* **221**, 537-541.

64. Mulbry, W.W. and Karns, J.S. (1989) Purification and characterization of three parathion hydrolases from Gram-negative bacterial strains, *Appl. Environ. Microbiol.* **55**, 289-293.

65. Mulbry, W.W., Karns, J.S., Kearney, P.C., Nelson, J.D., McDaniel, C.S., and Wild, J.R. (1986) Identification of a plasmid-borne parathion hydrolase gene from *Flavobacterium* sp. by Southern hybridization with *opd* from *Pseudomonas diminuta*, *Appl. Environ. Microbiol.* **51**, 926-930.

66. Munnecke, D.M., (1976) Enzymic hydrolysis of organophosphate insecticides, a possible pesticide disposal method, *Appl. Environ. Microbiol.* **32**, 7-13.

67. Munnecke, D.M. (1980) Enzymatic detoxification of waste organophosphate pesticides, *Agric. Food Chem.* **28**, 105-111.

68. Nakahigashi, K., and Inokuchi, B. (1990) Nucleotide sequence between the *fad*B gene and the *rrn*A operon from *Escherichia coli*, *Nucleic Acids Res.* **18**, 6439.

69. Nelson, L.M. (1982) Biologically-induced hydrolysis of parathion in soil: isolation of hydrolyzing bacteria, *Soil Biol. Biochem.* **14**, 219-222.

70. Racke, K.D. and Coats, J.R. (1987) Enhanced degradation of isofenphos by soil microorganisms, *J. Agric. Food Chem.* **35**, 94-99.

71. Rastogi, V.K., DeFrank, J.J., Cheng, T.-c., and Wild, J.R. (1997) Enzymatic hydrolysis of Russian-VX by organophosphorus hydrolase, *Biochem. Biophys. Res. Comm.* **241**, 294-296.

72. Roderick, S.L. and Matthews, B.W. (1993) Structure of the cobalt-dependent methionine aminopeptidase from *Escherichia coli* – a new type of proteolytic enzyme, *Biochemistry* **32**, 3907-3912.

73. Rosenberg, A. and Alexander, M. (1979) Microbial cleavage of organophosphorus insecticides, *Appl. Environ. Microbiol.* **37**, 886-891.

74. Russell, A.J., Personal communication.

75. Serdar, C.M. and Gibson, D.T. (1985) Enzymatic hydrolysis of organophosphates: cloning and expression of a parathion hydrolase gene from *Pseudomonas diminuta*, *Bio/Technology*, **3**, 567-571.

76. Serdar, C.M., Gibson, D.T., Munnecke, D.M., and Lancaster, J.H. (1982) Plasmid involvement in parathion hydrolysis by *Pseudomonas diminuta*, *Appl. Environ Microbiol.* **44**, 246-249.

77. Serdar, C.M. Murdock, D.C., and Rohde, M.F. (1989) Parathion hydrolase gene from *Pseudomonas diminuta* MG: Subcloning, complete nucleotide sequence, and expression of the mature portion of the enzyme in *Escherichia coli*, *Bio/Technology* **7**, 1151-1155.

78. Sethanathan, N. and Yoshida (1973) A *Flavobacterium* that degrades diazinon and parathion, *Can. J. Microbiol.* **19**, 873-875.

79. Sheela, S. and Pai, S.B. (1983) Metabolism of fensulfothion by a soil bacterium, *Pseudomonas alcaligenes* C$_1$, *Appl. Environ. Microbiol.* **46**, 475-479.

80. Shelton, D.R. and Somich, C.J. (1988) Isolation and characterization of coumaphos metabolizing bacteria from cattle dip, *Appl. Environ. Microbiol.* **54**, 2566-2571.

81. Siddaramappa, R., Rajaram, K.P., and Sethunathan, N. (1973) Degradation of parathion by bacteria isolated from flooded soil, *Appl. Microbiol.* **26**, 846-849.

ANCILLARY FUNCTION OF HOUSEKEEPING ENZYMES: FORTUITOUS DEGRADATION OF ENVIRONMENTAL CONTAMINANTS

R. SHANE GOLD, MELINDA E. WALES, JANET K. GRIMSLEY & JAMES R. WILD*

Texas A&M University, Department of Biochemistry and Biophysics, College Station, TX 77843-2128

Abstract: Most of the environmental contamination of greatest concern around the world today involves xenobiotic materials which have only been produced or used in industrial/agricultural production over the past several decades. These include various organic chemical solvents, heavy metals, neurotoxic pesticides, halogenated aromatic compounds, explosives and carcinogenic industrial chemicals which may or may not have natural counterparts. These have been synthesized with increasing diversity and volume over the last fifty to seventy years. Although their introduction into the environment is lamentable, and the effects they can have on the environment devastating, many different types of microbial systems have already acquired the ability to chemically modify many of these compounds. The likelihood that totally new enzymes with specific detoxification activities could emerge in such a brief period of time is problematic. It is much more likely that enzymes already possessed by microbial communities, possibly used for maintenance functions, can bee recruited to address the new challenge. For example, FMN oxidoreductases may serve to reduce trinitrotoluene and a human lipid-phosphoesterase or a bacterial peptidase can degrade organo-phosphate pesticides. Five categories of reactions wherein housekeeping enzymes may be able to catalyze ancillary reactions involving previously unknown chemicals include 1) general oxidoreductases, 2) denitrases, 3) phosphoesterases, 4) ring cleavage enzymes and 5) metal-sequestering proteins. Once a potential metabolic route is established, directed evolution based on random mutation or substitutional recombination may serve to enhance activity if a competitive metabolic edge could result.

B. Zwanenburg et al. (eds.), Enzymes in Action, 263–286.
© 2000 *Kluwer Academic Publishers. Printed in the Netherlands.*

1. Introduction

One of the most impressive characteristics of enzymes is the specificity of their action, which for some enzymes is almost absolute. The specificity they exhibit is two-fold, they are typically specific both to the type of chemical reaction as well as to the physical nature of the substrate subjected to that reaction. In 1956, an International Commission on Enzymes devised a system for classifying enzymes based on the reaction catalyzed. All enzymes are in one of six broad classes, each of which has subclasses and sub-subclasses (Table 1).

TABLE 1. Classification of enzymes and their corresponding action

Enzyme Class	Action
Oxidoreductases	Catalyze oxidation-reduction reactions
Transferases	Catalyze the transfer of functional groups between donors and acceptors
Hydrolases	Catalyze the transfer of functional groups to H_2O
Lyases	Catalyze the addition or removal of the elements of H_2O, CO_2 or NH_4^+
Isomerases	Catalyze a change in the geometric or spatial configuration of a molecule
Ligases	Catalyze the joining together of two molecules with the accompanying hydrolysis of a high-energy bond

The concept of an enzyme as a rigid template was formalized in the 1890's by Emil Fisher when experimentation indicated that glycosidases are highly stereospecific with respect to their substrates. He proposed that a similar specificity existed in the structure of the enzymes, adopting the "lock and key" analogy to describe the precise fit which was thought to have to occur between substrate and active site for the reaction to take place. An extension to the "lock and key" model came with the realization that enzymes are not rigid templates but rather highly flexible and conformationally dynamic chemical catalyst. This led to the "induced fit" hypothesis of Koshland in which substrate binding alters the conformation of the protein, so that the protein and the substrate fit each other more precisely. In either case, the spatial arrangement of the active site determines not only which compounds can fit stereospecifically into the site but also the nature of the catalysis. Hypothetically, any compound that can fit into the active site and interact with the vital catalytic residues in a thermodynamically and kinetically favorable manner should be able to serve as a substrate for the reaction. Many enzymes, such as fumarase, have absolute specificity for a single substrate and product. Other enzymes have a less limited specificity, e.g. trypsin, which hydrolyzes peptide, amide or ester bonds of lysine or arginine. Still others have a broad relative substrate specificity, such as leucine aminopeptidase which hydrolyzes many "-L-amino acid amides and dipeptides at different rates. The rate of reaction for these potential substrates would be expected to vary depending on the number of specificity determinants they accommodate. When an enzyme is said to be specific for a particular substrate, that often simply means that the reaction rate is higher for the "good" substrate than it is for a poor one.

2.   Metabolism Overview

Many enzymes exhibit considerable substrate ambiguity and, for those enzymes involved in the bioremediation of environmental contaminants, the ability to accept alternative substrates may be of central importance. Many of the compounds presently contaminating the environment have only been extant for a few decades. It is not likely that novel enzymes and enzyme systems have evolved to accommodate these new chemicals in such a short period of time. It is much more likely that native enzymes are able to accept the newly introduced compound into their active site and to utilize that compound as an alternative substrate for the reaction. The source for these new catalytic reactions must lie within the enzymes of primary and secondary metabolism. Within the 2000-3000 reactions of primary (or intermediate) metabolism lies the ability to convert glucose and inorganic ions into a living cell. In contrast to primary metabolism, secondary metabolism is not essential for growth and reproduction, but is most often responsible for interactions between the organism and its environment (Table 2). While most of the enzymes of primary metabolism are present in all living organisms, the enzymes of secondary metabolism are restricted to a small subset of all living organisms. In general, organisms in which secondary metabolism occurs inhabit resource-rich environments. Consistent with this, Archea, existing in harsh conditions which demand metabolic economy, have not been reported to commonly produce secondary metabolites. In the domain of Eubacteria, these eclectic metabolic activities are unevenly distributed and appear to be absent from anaerobes and chemolithotrophs. The most proficient group within the Eubacteria appears to be soil microorganisms, which often live in an extremely competitive environment (1).

   Like primary metabolites, most secondary metabolites are formed by enzymatic pathways. The genes encoding the enzymes are usually chromosomal, but a few have been shown to be plasmid borne. Regardless of their genetic location, they are usually clustered and are strongly regulated by nutrients, inducers, products, metals and/or growth rate.

TABLE 2. Reactions of primary metabolism (2) and some suggested biological functions for secondary metabolites (3).

Primary Metabolism	
Assembly reaction	the chemical modification, transport and assembly of macromolecules
Polymerization reactions	sequential linking of molecules into polymeric chains, e.g. protein, DNA, RNA, lipopolysaccharides, etc.
Biosynthetic reactions	cellular biosynthetic reactions which begin with one of just 12 precursor molecules and produce the building blocks for the polymerization reactions
Fueling reactions	produce the 12 precursor metabolites, the energy (ATP) and reducing power or biosynthesis
Secondary Metabolism	
Defense	antibiotics, toxins
Symbiosis	growth promoters, nodulation factors
Metal-transport	siderophores
Chemical Signals	luminescence, sporulation
Communication	pheromones and other chemical signals which coordinate sexual reproduction and differentiation

A third metabolic component of many organisms today is the degradation of xenobiotics. The reactions and pathways of primary and secondary metabolism provide hundreds of enzymes as a reserve catalytic pool for a cell to utilize to detoxify xenobiotics. The ability to utilize xenobiotics as substrates may be an ancillary function of native enzymes, and may be expected to fall into one of four categories: 1) those accepting substrates possessing primary structural similarities, 2) those accepting substrates possessing unrelated but similarly reactive structural groups, 3) those forcing reverse reactions, and 4) those with atypical reaction conditions, such as catalysis in an organic environment.

Many enzymes have been identified and are well characterized relative to their activity on xenobiotic compounds. The University of Minnesota maintains a database of microbial biocatalytic reactions and biodegradative pathways primarily for xenobiotic, chemical compounds (http://www.labmed.umn.edu/umbbd/index.html) (4). Table 3 gives some examples of xenobiotic compounds, their associated enzymatic activities and the proposed housekeeping function.

TABLE 3. Examples of xenobiotics, their corresponding enzyme and its classification.

Compound[a]	Ancillary Function	Housekeeping Function	Ref.
TeCH	TeCH reductive dehalogenase	glutathione S-transferase	(5)
PETN	PETN reductase	FMN oxidoreductase	(6)
PCP	PCP 4-monooxygenase	tryptophan-2-monooxygenase	(7)

[a] TeCH, 2,3,5,6-tetrachlorohydroquinone; PETN, Pentaerythritol tetranitrate; PCP, pentachlorophenol;

3. The Phosphotriesterase (PTE) Family

Many organophosphate compounds (OPs) are potent cholinesterase inhibitors, accounting for their widespread use as insecticides and nerve agents. In addition to the nearly 3 million cases of pesticide poisonings world-wide each year attributed to OP pesticides (8), there are risks associated with the major international effort to destroy the approximately 25,000 tons of chemical agents (9). Organophosphorus hydrolase (OPH) is a bacterial enzyme that hydrolyzes a broad variety of organophosphate (OP) neurotoxins. The substrate range of OPH includes several insecticides, the neurotoxic chemical warfare agents and their analogs (10,11,12,13,14,15). Catalytic specificities for this family of compounds have been shown to range from rates that are diffusion limited (e.g. paraoxon, P–O bond, k_{cat}=10,000 sec^{-1}) to rates which are eight orders of magnitude lower (e.g. acephate, P–S bond). All of the phosphate triesters found to be substrates of OPH are synthetic compounds, and the identity of any naturally occurring substrate for the enzyme is unknown. The enzyme is a homodimer with equivalent active sites at the C-terminus of each monomer. The molecular architecture is a distorted α:β barrel with eight parallel β-strands forming the barrel and linked on the outer surface by 14 α-helices (16,17). Since the initial characterization of this enzyme, other enzymes have been identified which accomplish this hydrolysis for specific classes of OP compounds. Although such enzymes were classified as organophosphorous acid anhydrolases (OPAA) by the International Union of Biochemistry in 1992, no consensus nomenclature has emerged in the literature. The best characterized of these enzymes are compared in Table 4, and their original designation is retained in this text.

Of the four enzymes listed in Table 4, OPH, OPAA, and a variety of the serum paraoxonases have been cloned and sequenced. The gene encoding OPH, *opd*, was isolated on dissimilar plasmids from *Pseudomonas diminuta* MG and *Flavobacterium* sp. Strain ATCC 27551 (18, 19). The open reading frame contains 975 bases which

268

TABLE 4. Comparison of some OP degrading enzymes

Enzyme	Bond cleaved by enzyme				
	P–C	P–O	P–F	P–S	P–CN
OPH	-	+	+	+	+
OPAA	-	+	+	-	+
Human paraoxonase	+	+	+	-	+
Squid OPAA	-	-	+	-	-

encode a polypeptide of 325 amino acid residues with a molecular mass of 35 kDa. Subsequently, OPA anhydrolase (OPAA) enzymes have been identified in various halophilic and *Altermonas* species (20), with the enzymes purified and characterized from two *Altermonas* strains, *Altermonas* sp. strain JD6.5 and *A. haloplanktis* (21, 22, 23). While these enzymes utilize common substrates, OPAA has its best activity against soman (GD; *O*-pinacolylmethylphosphonofluoridate) and little activity against phosphotriesters, while OPH has diffusion limited catalytic activity against paraoxon. These enzymes exhibit no sequence similarity to each other (Figure 1; Table 5). When the OPH and OPAA amino acid sequences were used to search the NCBI database of orthologs, or genes in different species which evolved by speciation (http://www.ncbi.nlm.nih.gov/COGS), only OPAA was identified as a member of any functionally characterized group (amino acid transport and metabolism) based on strong sequence homology (50% identity) with the *E. coli pepQ* (proline dipeptidase, or prolidase) and related genes. Recent biochemical analysis has established OPAA to be a prolidase (EC 3.4.13.9), a dipeptidase which cleaves the C-N bond of dipeptides with a prolyl residue at the carboxy terminus (23). It has been suggested that the activity of OPAA on organophosphate compounds may be due to a fortuitous similarity between these compounds and X-Pro dipeptides (24) (Figure 2).

TABLE 5. Sequence pair distances, in percent similarity, using the Clustal method (26)with the PAM250 residue weight table.

		1	2	3	4
1	OPH[a]		--	--	--
2	OPAA			80.5	42.9
3	Ah-OPAA				49.2
4	Ec-prolidase				

[a] indicates a percent similarity of ≤10%, values which were also achieved by completely random, pairwise alignments.

OPAA   MNKLAVLYAEHIATLQKRTREIIERRNLDGVVFHSGQAKRQFLDDMYPFKVNPQFKAWLPVIDNPHCWIVANGTDKPKLIFYRPVDFWHKVPDE
Ah-OPAA   -E------------Q--T-C-Q-G-E-L-I----------------H-------------------V--S--------
Ec-prolidase -ES--S--KN------E--DALA-FK--ALLI--ELFNV---HP----------V--TQV-N--LLVD-VN---W--L---Y--N-EPL
OPH   -SIGTGDRINTVRGPITISEAGFTLTHEHICGSSA-FLRAWPEFFGSRKALAEKAVRGLRRARAAGVRTIVDVSTFDIGRDVSLLAEVSRAADVH

OPAA   PNEYWADYFDIELLVKPDQVEKLLPYDKARFAYIGEYLEVAQALGFELMNPEP.VMNFYHYHRAYKTQYELACMREANKIAVQGHKAARDAFFQ
Ah-OPAA   -RDF-E-----LQ------K---------SI----.-SI------L-YI------E-L-N--R---D------N
Ec-prolidase -TSF-TEDVEVIA-P-A-GIGS--AARGNIG---PVP-R-LQ--I-AS-INPKG-IDYL--Y-SF-E-------Q-M--N--R--EE--RS
OPH   IVAATGLW--PP-SMRLRS--E-TQFFLREIQ-GI-DTGIRAGIIKVATTGKATPFQELVLKAA-RASLATGVPVTHTAASQRDGEQQAAI-ES

OPAA   GKSEFEIQ.QAYLLATQHSENDNAYGNIVALNENCAILHYTHFDRVAPATHRSFLIDA...GANFNGYAADITRTYDFTGEGEFAELVATMKQHQ
Ah-OPAA   -G---D-----M-RQ--EMP------------------EPK--Q--N---.....---------KKQ----D--NA-TA--
Ec-prolidase -M--D-N.I--T-G-RDTDVP-S----HAAV---KI-HQ--EEM---L---EY----L--WSAKSDNDY-Q--KDVNDE-
OPH   EGLSPSRVCIGHSDD-DDLSYIT-LAARGY-IGLDH-P-SAIGLEDNASASALLG-RSWQTR-LLIKALI-QGYMKQIIVSNDWLFGFSSYVTNI

OPAA   IALCNQLAPGKLYGELHLDCHQRVAQTLSDFNIV.DLSADEIVAKGITSTFFPHGLGHHIGLQVHDVGGFMADEQGAHQEPPEGHPFLRCTRKIE
Ah-OPAA   -E-GKS-K--L---D---N-I--L---D--.K-P-A--ERQ----------L-A------R--T--A------L--
Ec-prolidase L--IATMKA-VS-VDY-IQF--I-KL-RKHQ-IT-M-EEAM-ENDL-GP-M--I--PL------A---Q-DS-T-LAA-AKY-Y----ILQ
OPH   MDVMDRVN-DGMAFIPLRVIPFLREKGVPQETLAGITVTNPARFLSPTLRAS

OPAA   ANQVFTIEPGLYFIDSLLGDLAATDNNQHINWDKVAELKPFGGIRIEDNIIVHEDSLENMTRELRARLJTTHSLRGLSAPQFSINDPAVMSEYSYP
Ah-OPAA   K--------------Q--K-F---E--EAF-----------------------NL.LL
Ec-prolidase PGM-L-----I---E--APWREGQFSK-F--Q-IEA---------------VVI--NNV------TGVMESWLIPAAPVTVVEEKKSRFITMLAHT

OPAA   SEPLSYEEEIKKSTFIVHVRTRRILVRRRTLSPILIAVTPMPAITAGLM
Ec-prolidaase DGVEAAKAFVESVRABHPDARHCVAWVAGAPDDSQQLGFSDDGEGKPLAQLMGSGVGEITAVVVRYYCGILLGTGGLVKAYGGGVNQA

Ec-prolidase VRQLTTQRKTPLTEYTLQCEYHQLTGIEALLGQCDGKIINSDYQAFVLLRVALPAAKVAEFSAKLADFSRGSLQLLAIEE

Figure 1. Sequence alignment of the *Altermonas* sp JD6.5 (OPAA) and the *A. haloplanktis* (Ah-OPAA) phosphotriesterase related proteins with the *E. coli* prolidase sequence (25). The *E. coli* (OPH) sequence is shown for comparison only since the absence of any significant homology precludes a meaningful alignment. -, residues identical to the OPAA sequence; . , gaps inserted to optimize alignments.

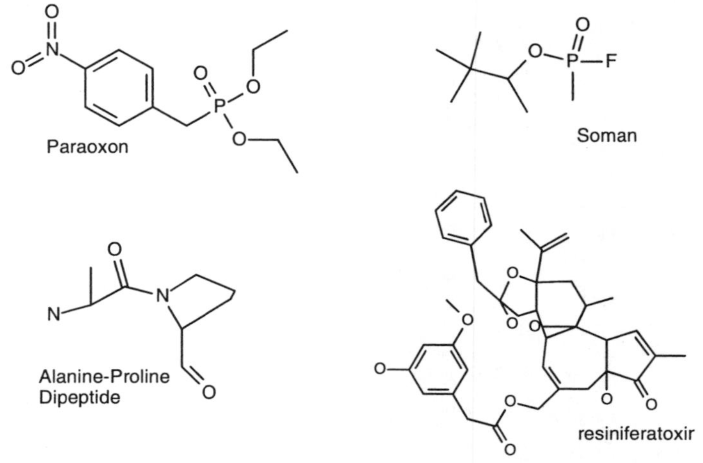

Figure 2. Substrates of OPH and related proteins

3.1.   THE SERUM PARAOXONASES (PON)

Human serum paraoxonase/arylesterase (PON1) (E.C. 3.1.8.1) is a calcium-dependent, HDL (High Density Lipoprotein) -associated protein that appears to have multiple roles *in vivo*. In one guise, PON1 hydrolyses organophosphate insecticides and nerve gases and is responsible for determining the selective toxicity of these compounds in mammals. It is a 354 amino acid enzyme which shows no homology with any of the bacterial phosphotriesterase-related proteins (27). It has three cysteine residues in positions 41, 283 and 352 and for years has been thought to be a cysteine esterase with mechanistic similarities to the serine esterases (cholinesterases and carboxyesterases)(28). However, while Cys 41 and Cys 352 form a disulfide bond essential for the paraoxonse/arylesterase activity of the enzyme, Cys-283, which has a free sulfhydryl, is not required for either of the activities (29). Additionally, PON1 requires two Ca^{2+} for stability and catalytic activity. Zn^{2+} and Cd^{2+} can replace Ca^{2+} and prevent irreversible denaturation, but neither metal confers catalytic activity (30,31). A combination of chemical modification and site-directed mutagenesis has identified twelve additional amino acids among the His and Asp/Glu residues which are essential for activity/Ca^{2+} binding (32)(Figure 3).

In addition to playing a significant role in OP detoxification, several lines of evidence have indicated that PON1 acts as an important guardian against cellular damage from oxidized lipids in the plasma low density lipoproteins (LDL) (33,34), and against bacterial endotoxins (35). PON1 is associated with HDL in serum and HDL levels are inversely related to the risk of developing atherosclerosis. That PON1 is the component of HDL responsible for the protective effect has been verified by the creation of PON1-knockout mice. In a co-cultured cell model of the artery wall, HDLs isolated from PON1-deficient mice were unable to prevent LDL oxidation, and both HDLs and LDLs isolated from PON1-knockout mice were more susceptible to oxidation than the lipoproteins from wild-type littermates. When fed on a high-fat, high-cholesterol diet, PON1-null mice were more susceptible to atherosclerosis than their wild-type littermates (33). PON's ability to protect LDL against oxidation is accompanied by inactivation of the enzyme, resulting from an interaction between the enzyme's free sulfhydryl group (Cys 283) and oxidized lipids (36). In a similar fashion, the HDL complex is protective against the development of endotoxemia during bacterial infections and it has been shown recently that the protein component responsible for the protective effect is PON1 (35).

Evaluation of the protective effects of PON1 in human populations is complex because not only does the level of enzyme vary between individuals, but PON1 is polymorphic as well. The two allozymes, A and B, differ in their substrate specificities and kinetic properties. Sequencing of the human *PON1* gene identified two genetic polymorphisms distinguished by amino-acid substitutions at positions 55 and 192 (37, 38). Pedigree analysis established the 192 polymorphism as concordant with the serum paraoxonase phenotypes (37), with the arginine 192 isoform (type B) hydrolyzing paraoxon more rapidly than the glutamine 192 isoform (type A)(39, 40). This effect of the PON1 polymorphism is reversed for the hydrolysis of diazoxon, soman and especially sarin, with the serum of type A homozygotes having almost 10-fold greater activity against sarin than the B homozygotes (41). Additionally, the B-isozyme is less effective than the A type at protecting LDL from oxidation (42). The arylesterase activity does not exhibit any functional polymorphism.

```
                          *                                                                                *  *
pon1   MAKLLALTLIVGMGLALFKNHRSSYQTRLNAFREVTPVELPNCNCNLVKGIETGSEDLEILPNGLAFISTGLKYPGIKSFDPSKPGKILLMDLNEKDPAVLEL*TTGS
pon2   -GR*VG*G-A-DRA--LGERLLALRN--K-S---ES-D---H-I----A----ID------F-V---F--LH--A-D---G--M---K-EK-RA*--R-SRG
pon3   LSLVEEM.........FLAFRE-V--SQ--E---PE--H-IEEL-S---ID--S--S---SE----MPN-A-DE---F-----QN-RAQA-E-S-G
OPH    MSIGTGDRINTVRGPITISEAGFTLTHEHICGSSAGFLRAWPEFFGSRKALAEKAVRGLRRARAAGVRTIVDVSTFDIGRDVSLLAEVSRAADVHIVAATGLWFD

              *                          *                                   *                        *
pon1   KFD*SSFNPHGISTFTDEDNTVYLLVVNHPDSKSTVELFKFQEEKSLLHLKTITHKLLPSINDIVAVGPEHFYATNDHYFADPYLRSWEMYLGLAWSNVVYYSP
pon2   .--LA--------I-*-D---F----EF-N---I---E-A-N-----K-E---V---I----A------S--F-KYL-T--N-H-A----
pon3   .--KEL------I-I-K----Y----H-K----I--E-QQR--V*---K-E-K-V---VL---Q---R----TNSL-SFF--I-D-R-TY-LF----
OPH    PPLSMRLRSVEELTQFFLREIQYGIEDTGIRAGIIKVATTGKATFPQELVKAAARASLATGVPVTTHTAASQRDGEQQAAIFESEGLSPSRVCIGHSDDTDDLS

                                                        *                        *     **
pon1   DKVRVVAEGFDFANGINISPDGKYVYIAELLAHKIHVYEKHANWTLTPLKSLDFNTLVDNISVDPVTGDLWVGCHPNGMRIFFYDSENPPASEVLRIQNILSEDP
pon2   *E-K-------S-------K--I-V-DI---E---L---P-MN--Q--V-ELD----L-I--SS-------QKL-V--PN---S--------C-K-
pon3   RE-K---K--CS----TV-A-Q----V-DVA--KN--IM---D-WD---Q--VIQ-G----LTVA-A--IILA-----P-KLLN-NP-D--G-------V--K-
OPH    YLTALAARGYLIGLDHIPHSAIGLEDNASASALLGIRSWQTRALLIKALIDQGYMKQILVSNDWLFGFSSVYTNIMDVMDRVNPDGMAFIPLRVIPFLREKGVPQ

            *
pon1   KVTVVYAENGTVLQGSTVAAVYKGKLLIGTVFHKALYCEL
pon2   T--T---N--S----S--S--D------LY-R------
pon3   R-ST---N--S---TS--S--H--I----X-T-----
OPH    ETLAGITVTNPARFLSPTLRAS
```

Figure 3. Alignment of the PON amino acid sequences. Due to space limitations, the 10 individual PON sequences are not shown, rather the individual sequences were used to generate a consensus within each group, and the alignment of the consensus sequences is shown here. The E. coli (OPH) sequence is show for comparison only since the absence of any significant homology precludes a meaningful alignment. -, residues identical to the PON1 sequence; ., gaps inserted to optimize alignments, * within a sequence, denotes residues where there was no consensus within that group; *, above the sequence indicates those residues thought to be important for activity or Ca^{2+} binding (32)

Although *PON1* shows no homology to any of the proteins with which it shares functional similarity [the cholinesterases, the bacterial PTEs (OPH and OPAA) or the mammalian DFPases], it is a member of a multigene family composed of three distinct, closely linked genes (*PON1, PON2* and *PON3*; Figure 3, Table 6). The *PON* genes are located on chromosome 7 in human and on chromosome 6 in the mouse. Turkey and chicken, like most birds, lack paraoxonase activity and are very susceptible to organophosphates. However, they have a *PON2*-like gene with approximately 70% identity with the human *PON* genes. The similarity within the PON genes strongly suggest that these related genes arose by gene duplication (27). Although the protein products of the *PON2* and *PON3* genes have not yet been isolated, tissue distribution of human PON2 transcripts were observed to be highest in liver, brain and heart tissue (43). Similar to PON1, PON2 contains a common polymorphism at codon 311 and some population studies indicate that the PON1 B allele and the PON2 S allele synergistically contribute to an increased risk for coronary heart disease (44).

3.2. THE PHOSPHOTRIESTERASE HOMOLOGY PROTEINS (PHP)

The phosphotriesterase homology proteins (PHP) provide an interesting subset of the phosphotriesterase related proteins and genes. Unlike the OPAAs and PONs, their inclusion in this broad family of related proteins stems from similarities in sequence rather than similarities in function (Table 7, Figure 4). The first PHP gene, *mpr56-1* for mouse *p*hosphotriesterase-*r*elated, was identified by its abnormal expression in injured and polycystic kidneys (25). It appears that the *mpr56-1* gene is developmentally regulated in the kidney, and that its expression is highly sensitive to injury induced dedifferentiation. A related gene was subsequently identified in the rat kidney (45). The rat phosphotriester-related gene, *rpr-1*, has 92% identity at the amino acid level with the mouse *mpr56-1*. The *rpr-1* gene was isolated by screening an expression library with resiniferatoxin, a plant derived vanilloid that is of pharmacological interest, suggesting that the PTE homologue can bind to the trialkoxycarbon structure found in resiniferatoxin (Figure 2)(45). Additionally, the *rpr-1* gene has been used as a probe to localize a human version of the gene to the short arm of chromosome 10 (25), distinct from the *pon* gene family on chromosome 7. Four other genes with homologous sequences have recently been identified in the genomic sequence databases for *E. coli, Mycobacterium tuberculosis, Deinococcus radiodurans* and *Mycoplasma pneumonia* (46,47,48,49), bringing the total PHP family to six. However, for all of these PHP proteins, the homology to the bacterial phosphotriesterase is the only clue to their function. There is not yet any experimental evidence that these proteins are themselves phosphotriesterases. In fact, studies with the *E. coli* PHP suggest otherwise, as no enzymatic activity was detected when this enzyme was tested for general esterase, aminopeptidase, sulfatase, phosphatase, carbonic anhydrase, phosphodiesterase and phosphotriesterase activities. (46)

TABLE 6. Sequence pair distances, in percent similarity, using the Clustal method (26) with the PAM250 residue weight table.

	2	3	4	5	6	7	8	9	10	11	12
1 OPH[a]	--	--	--	--	--	--	--	--	--	--	--
2 human pon1		85	82	80	63	64	62	67	66	59	56
3 rabbit pon1			81	79	62	62	62	65	65	58	57
4 mouse pon1				94	62	63	62	65	64	58	57
5 rat pon1					63	63	61	65	64	57	55
6 human pon2						91	85	72	72	63	60
7 canis pon2							85	71	72	62	60
8 mouse pon2								68	68	60	59
9 chick pon2									97	65	61
10 turkey pon2										64	61
11 human pon 3											80
12 mouse pon 3											

[a] -- indicates a percent similariy of <10%, values which were also achieved by completely random, pairwise alignments.

Buchbinder et al. (1998) have suggested that the PHP proteins may belong to the family of proteins from which phosphotriesterase evolved. Similar to OPH, the *E. coli* PHP (PHP-ec) structure consists of a long, elliptical α/β barrel with a binuclear zinc center in the cleft at the carboxy end of the barrel near the presumptive active site. Although the aspartate and all four histidine residues that coordinate Zn^{2+} in OPH are conserved, significant difference between the two structures are found in the region corresponding to the active site. The active site of OPH has been described with the substrate analog diethyl 4-methylbenzylphosphonate (50). The analog occupies a site near the binuclear metal center in a fairly hydrophobic pocket defined by Trp 131, Tyr 309, His 257, Leu 271, Phe 306 and Met 317. Of these residues, Met 317 is the only relatively conservative substitution (Leu 250) in the *E. coli* PHP. Phe 306, Tyr 309 and Leu 271 have no corresponding residues in PHP-ec(46).

TABLE 7. Sequence Pair distances, in percent similarity, using the Clustal method (6) with the PAM250 residue weight table.

	1	2	3	4	5	6	7
1 OPH		22.3	25.6	15.4	30.4	18.3	17.8
2 PHP-ec			24.3	18.2	25.7	23.3	22.3
3 PHP-dr				17.0	28.7	16.7	19.4
4 PHP-mp					16.3	15.4	17.0
5 PHP-mt						19.6	19.6
6 PHP-mm							92.3
7 PHP-r							

3.2.1 Evolution of the α/β Barrel Structure

In 1990, Farber and Petsko gave a general description of the 17 α/β barrel enzymes for which there was a known structure (51). These enzymes were characterized as having a single domain with eight parallel β-strands surrounded by seven or eight α-helices, giving a general secondary structure of $(\beta\alpha)_8$. The location of extra structural features in combination with the shape of the "barrel" was used to classify the α/β barrel structures into four families. By 1995, 11 new α/β barrel structures had been published and 5 of the structures defined two new families–including the phosphotriesterase and adenosine deaminase family (52). By this time, it had become clear that not all α/β barrel proteins were enzymes, and those that were catalyzed many different types of reactions. There are now currently over 84 different α/β barrel structures in the SCOP database (53) (*Structural Classification of Proteins*; http://scop.mrc-lmb.cam.ac.uk/scop). Urease has joined phosphotriesterase and adenosine deaminase as a member of the metal-dependent hydrolase family (54).

```
                              *  *
OPH     MSIGTGDRINTVRGPITISEAGFLTHEHICGSSAGFLRAWPEFFGSR.................KALAEKAVRGLRRARAAGVRTIVDVSTFDIG
PHP-ec  --FDPT-.....-Y--A--LHIDLS-KNN....VDC-.....................LDQYAFICQEMNDLMTR---NVIEMTNRYM-
PHP-dr  MTA.......QT-T-AVAAAQL-A--P--VIFGYP-YAGDVTLGPFDH-..................A-ALASCTETA-ALL-R-IQ-V-ATPN-C-
PHP-mp  -....KRFVR--L-D-DPKDL-ICDC-D-LIKNWGPEAKEH-D-VMLSNEAAI.................KECL.FVHHGGRSI-TMDPPN
PHP-mt  -P.....EL--A----DTADL-V--M---VFIMTEIAQNY-EAWGDE.........DKRVAGAIARLGELK-RGVDTIVDL-VIGL
PHP-mm  MS.SLSGKVQTVLGLVEPSQL-R--T---LTMTFDS-YCPP-PCHEVTSKEPIMMKNLFWIQKNPYSHRENLQLNQEVGAIREELLYFK-KGGGALVEN-TTGL
PHP-r   MS.SLSGKVQTVLGPVEPSQL-R--T---LTMAFDS-YCPP-PCQEAASREPIMMKNLFWIQKNPYSHQENLQLNQEVAVREELLYFK-KGGGAVVEN-TTGL

                                                                              *
OPH     RDVSLLAEVSRA.A.DVHIVAATGLWFDPPLSM........RLRSVEELTQFFLREIQYGIEDTGIRAGII.........KVATTGKA.TPFQELVLKAA
PHP-ec  -NAQFMLD-M-E..T.GINV--C--YYQ-A.......FFPEHVAT--Q--A-EMVD--EQ--DG-ELK----A.........EIG-SEGKI--LE-K-FI--
PHP-dr  -NPAF-R--E-..TGLQILC-TGFYYEGEGATTYFKF.RASLGDAES-IYEMMRT-VTE--AG-----V--.............L--SSRD-I--YEQ-FFR--
PHP-mp  VGRDVKRM-AI-EQLKGKLNIIMATGFHKA...AFYDKGSSWLAQVPVNEIVPM-VAEIEGM-LYNYS-PVVKRGKAKAGII-AG-GYA-IDRLELKA-E-V
PHP-mt  GRYIPRIARVA-A.TELN--V---YTYNDVPFYFHYLGPGAQLDGPEIMTDM-V-D-EH--A---K---L.........C--DEPGL--GV-R--R-V
PHP-mm  SRDVHTLKWLAEQ..TG---I-GA-FYV-ATHS.........AAT-AM---Q--DVLIN--LH-ADG-S-KC-V-G.........EIGCS.WPL-DSERKI-E-T
PHP-r   SRDVRTLKWLAEQ..TG---I-GA-FYV-ATHF.........AAT-AM---Q--DVLIS--LH-ADG-S-KC-V-G.........EIGCS.WPL-DSERK--Q-T

                            *
OPH     ARASLATGVPVTTHTAASQRDGEQQAAIFESEGLSPSRVCIGHSDDTD.DLSYLTALAARGYLIGLDHIPHSAIGLEDNASASALLGIRSWQTRALLIKALIDQ
PHP-ec  -L-HNQ--R-IS---SF-TM.-LE-L-LLQAH-VDL--TV--C-LK-.NLDNILKMIDL-AYVQF-T-GKNSYYPDEK.............-IAMLH--R-R
PHP-dr  --VQRE----II--QEGQ.Q-P---ELLT-LD-A-IM--M-GNT.-PA-HRETLRH-VS-AF-RIGLQ.................GMV-TPTDAB-LSVLTT-LGE
PHP-mp  -IT-IT--A--LV--QLGTM.AYEAAQHLIDF-VN-RK.ILS-LNKNPDEYY-AKIIREL-VTLCF-GPDRVKYYPDCL...............L-KH--Y-V-L
PHP-mt  -Q-HKR--A-IS---HAGL-R-LD-QR-FAE--VDL---V---CG-ST.-VG--EB-I-A-SYL-M-RFGVDV-S.............PF-D-VNIVARMCER
PHP-mm  -H-QAQL-C--II-PGRNPGAPF-IIR-LQEA-ADI-KTVMS-L-R-IF-KKE-LEF-QL-CYLEY-LFGTELLNYQLSPD..........IDMPDDNK-IRRVHF-V-E
PHP-r   -H-QAQL-C--II-PGRNPGAPF-IIRVLQEA-ADI-KTVMS-L-RSIF-KKE-LEF-QL-CYLEY-LFGTELLNYQLSPD..........IDLPDDNKGLGGVRF-VNE

OPH     GYMKQILVSNDWLFGFSSYVTNIMDVMDRVNPDGMAFIPLR..VIPFLREKGVPQETLAGITVTNPARFLSPTLRAS.
PHP-ec  -LLNRVML-M-..............-TRRSHLKANGGYGYDY-LTTF--Q--QS-FS-ADVDVMLRE--SQ-F.......Q
PHP-dr  --ADRL-L-H-SIWHLGRPPA-PEAALP.AVKDWHPLHISDDIL-D--RR-ITE-QVGQM--Q---LFG
PHP-mp  -FV-H-TLAL-.......AGR-LYQNT-.V.KKGRSALGLHICLNALFRSLK-WEFPMRPL-
PHP-mt  --HADKMVL-H-....ACC-FDALPBELVPV-MPNWHYLHIHNDVI-A-LKQHGVTDBQLHTMLVDNPRRIFERQ.GGYQ
PHP-mm  --EDR--MAH-...........-HTKHRLMKYGGHGYSHILTNIV-KMLLR-LTERV-DKILIE--KQW-T-....FK
PHP-r   --EDR--MAH-...........-HTKHRLMKYGVHGYSHILTNVV-KMLLR-LTERV-DKILRE--KQW-T-....FK
```

Figure 4. Sequence alignment of the phosphotriesterase-homology proteins (PHP): OPH, organophosphate hydrolase; PHP-ec, E. coli; PHP-dr, D. radiodurans; PHP-mp, M. pneumonia; PHP-mt, M. tuberculosis; PHP-mm, mouse; PHP-r, rat. -, residues identical to the OPH sequence; .., gaps inserted to optimize alignments. The conserved histidines/aspartic acid are marked with a *.

These three enzymes have a common structural core consisting of an ellipsoidal $(\beta\,\alpha)_8$ barrel with a conserved metal binding site at the C-terminal end of strands $\beta1$, $\beta5$, $\beta6$ amd $\beta8$ (Figure 5). They share a common reaction mechanism in which the metal ion(s) deprotonate a water molecule for nucleophilic attack on the substrate. The metal binding ligands, 4 histidines and 1 aspartic acid, are strictly conserved, although the type of metal bound is not. Phosphotriesterase binds two zinc ions with a carbamoylated lysine acting as a bridging ligand. This binding motif is shared by urease, with nickel substituting for zinc. Adenosine deaminase binds a single zinc.

Using the signature sequence of the conserved histidines/aspartic acid in combination with structure-based pattern identification, Holm and Sander (1997) have linked 10 additional protein families. (54) Members of this superfamily catalyze the hydrolysis of amide or amine bonds in a variety of different substrates. Although the histidines/aspartic acid are conserved, both in residues and context, the carbamoylated lysine appears unique to urease and phosphotriesterase, as the other members of the family have no invariant lysines in the same region. Searching the NCBI (55) for family members among the 21 completed genomes revealed the phylogenetic pattern given in Table 8 (not included in this table is a family of neuronal developmental proteins previously identified)(54). It is interesting to note that the two most widely distributed members are enzymatic components of primary metabolism involved in either *de novo* pyrimidine biosynthesis (dihydroorotase) or pyrimidine salvage/reutilization (cytosine deaminase). Dihydroorotase (DHOase) may provide a "snapshot" of evolution, as noncatalytic DHOase homologues have been identified in *Pseudomonas putida* and *Saccharomyces cervesiae* (56, 57). The proposed function for the cryptic DHOase of *P. putida* is to maintain the associated aspartate transcarbamylase activity by conserving its dodecameric structure (57). In yeast, the cryptic DHOase may perform a similar structural role as it is part of a multienzymatic structure with carbamoylphosphate synthetase and aspartate transcarbamoylase activity. The implication is that the $(\beta\alpha)_8$ structure, in addition to having tremendous evolutionary potential as a catalytic template, also has the potential for recruitment in a structural role.

4. The Evolution of Catalytic Function

It is generally noted that ancient organisms possessed small genomes with few gene products in comparison with contemporary organisms. Jensen (58) suggested this was possible because primitive enzymes had broad substrate specificity, allowing them to

278

react with a wide range of related substrates and produce a family of related products. Gene duplication could then have provided the genetic reservoir necessary for the narrowing of substrate specificity, leading to the collection of specialized enzymes seen in most organisms today. The increasing number of deduced amino acid sequences made available by the many genomic sequencing projects has revealed many large gene families. As can be seen within the family of phosphotriesterase-related proteins, predictions of homology relationships based on similarity in function are not always straightforward.

Protein homologs are termed paralogs if they are the product of gene duplication. The relationship is orthologous when homologs were produced by speciation. Thus, a family tree produced by multiple alignment of amino acid sequences should parallel the phylogenetic tree if a family of proteins contains only orthologs (55). In the case of the phosphotriesterase-related proteins, the *PON* family tree (Figure 6) roughly corresponds to the phylogenetic relationships of the organisms, suggesting that this is an orthologous family of paralogs in which the duplication event occurred before phylogenetic divergence. The PHP and the OPAA group appears to be orthologous, while the absence of any significant homology **between** the three groups (PON, OPAA and PHP) suggest that the shared function of OP hydrolysis is the result of convergent evolution.

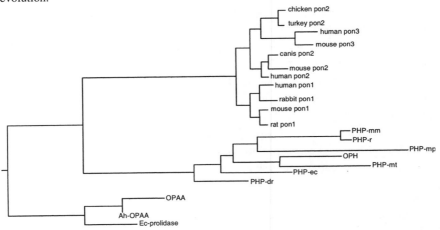

Figure 6. Dendrogram showing amino acid homology relationships of phosphotriesterase related proteins. Abbreviations are as previously described.

Since functions can be fulfilled in a number of ways, a variety of different proteins can acquired similar functions. However, with the synthesis of phosphotriester insecticides first reported in 1950 and the first gene cloned in 1989, this places a limit of 40-50 years on the evolution of OP hydrolyzing activity. While this may challenge traditional views on the pace of evolution, experimental studies with an evolved β-galactosidase operon in *E. coli* indicate it is certainly possible to acquire a novel function within a relatively short period of time given the proper environmental conditions. (59, 60). OPH provides an additional challenge in that not only did it evolve the OP hydrolyzing function in such a relatively short time, but it also reached what has been described as an evolutionary endpoint, catalysis at the limit of diffusion control. (61) This framework for the evolution of catalysis predicts that OPH should have absolute specificity for paraoxon. This is not the case as OPH has appreciable reaction rates with a variety of OP compounds.

In evaluating the evolutionary optimization of catalysis, it may be important to consider whether the substrate is a physiological substrate. Demetrius (62) suggest that there is a broad class of "evolutionarily unstable" enzymes which are characterized by variable rates and broad substrate specificity. Structurally, these enzymes are postulated to be rigid (undergoing little conformational change upon binding substrate) while their substrates are flexible. This characteristic allows the prediction that there can be non-physiological substrates which will be catalyzed at or near the limit of diffusion. While there is currently no experimental evidence suggesting this is the case for OPH, X-ray crystallography and magnetic resonance spectra of enzyme-substrate complexes should provide a means for characterizing enzyme-substrate interactions in these terms.

The distinction of physiological vs. non-physiological substrates is also important to consider when evaluating the evolutionary dynamics of catalysis. For most enzymes, the significant component for evolution of catalysis is predicted to be competition between physiological substrates and related compounds for the active site of the enzyme. However, with xenobiotics, significant evolutionary pressure may derive from competition between the various metabolic targets of the compound. For example, OP compounds bind and inhibit acetylcholinesterase in humans, and the evolutionary dynamics of the PON proteins may be determined by competition between acetylcholinesterase and the PON proteins. An obvious analogy is missing in bacterial cells, as they don't possess cholinesterases. However, the presence of the phosphotriesterase homology proteins provides an opportunity for directed evolution studies which may begin to answer some of these questions.

280

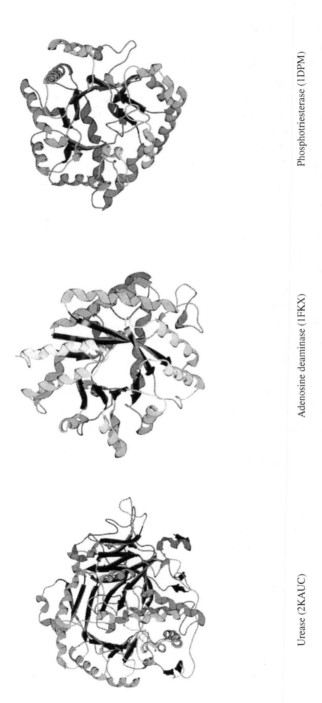

Urease (2KAUC)   Adenosine deaminase (1FKX)   Phosphotriesterase (1DPM)

Figure 5. Graphic representation of the thee dimensional structures of urease, adenosine deaminase and phosphotriesterase. The names of the coordinate files are given in parenthesis. Urease contains a "small domain" not contained in the other two structures.

TABLE 8. Members of the (β α)₈ metal-dependent hydrolase superfamily and their phylogenetic patterns[1].

	Archaea				Bacteria: Free-living								Bacteria: Parasitic						Euk
	Af	Mj	Mt	Ph	Aa	Tm	S	Ec	Bs	Mt	Hi	Hp	Mp	Bb	Tp	Ct	Cp	Rp	Sc
adenosine deaminase								x	x	x					x				x
AMP deaminase																			*
adenine deaminase	x	x	x					x	x					x					
cytosine deaminase	x	x	x	x	x	x	x	x	x	x	x	x				x	x	x	x
urease	x	x	x	x			x	x	x	x	x	x							
hydantoinase																			
dihydroorotase	x	x	x	x	x	x	x	x	x	x		x							x
aminoacylase				x		x	x	x	x		x								
imidazolonepropionase	x		x	x		x		x	x	x									
phosphotriesterase								x		x			x						
chlorohydrolase		*						*											*
formylmethanofuran dehydrogenase	x	x	x	x															

[1]The phylogenetic pattern was identified using the NCBI-COGs database (55) or as previously cited (54). *, The abbreviations are: Euk, Eukaryotic Af, Archaeoglobus fulgidus; Mj, Methanococcus jannaschii; Mt, Methanobacterium thermoautotrophicum; Ph, Pyrococcus horikoshii; Aa, Aquifex aeolicus; Tm, Thermotoga maritima; S, Synechocystis; Ec, Escherichia coli; Bs, Bacillus subtilis; Mt, Mycobacterium tuberculosis; Hi, Haemophilus influenzae; Hp, Helicobacter pylori; Mp, Mycoplasma pneumoniae; Bb, Borrelia burgdorferi; Tp, Treponema pallidum; Ct, Chlamydia trachomatic; Cp, Chlamydia pneumoniae; Rp, Rickettsia prowazekii; Sc, Saccharomyces cerevisiae

282

4.1.   SIGNIFICANCE FOR BIOCATALYSIS

Enzyme-based transformation of families of "fine chemical" has provided a wide spectrum of industrial and pharmacologically active products (see other chapters in this text). As with any area of science, the selection of enzymes has often been conservative and limited to standard and readily available systems. With the explosive revelation of new, yet related, enzymes throughout nature, it is now possible to expand the opportunities for biocatalysis and take advantage of new subtleties in the production of chemical agents. Furthermore, it is increasingly possible to genetically modify enzymes already in use in order to make them more effective with previously unusable substrates. In the future, it can be expected that it will be easy to build specific enzyme catalysts for the transformation of given "fine chemicals". The limitation on enzyme-based catalysis will not be the technology, but the innovation of the applicant.

5.   References

1.   Vining, L.C. (1992) Roles of secondary metabolites from microbes. In D.J. Chadwick, J. Whelan (eds.) *Secondary metabolites: Their function and evolution*, John Wiley & Sons, Chichester, pp. 184-194.
2.   Ingraham, J. L., Maaloe, O., and Neidhardt, F. C. (1983) Growth of the Bacterial Cell. Sinauer Associates, Inc, Sunderland, MA, p. 193.
3.   Davies, J. (1992) Introduction. In D.J. Chadwick, J. Whelan (eds.) *Secondary metabolites: their function and evolution*. John Wiley & Sons, Chichester, p. 171.
4.   Ellis, L.B., Hershberger, C.D., and Wackett, L.P. (1999) The University of Minnesota Biocatalysis/Biodegradation Database: specialized metabolism for functional genomics. *Nucleic Acids Res.*, **27**, 373-376.
5.   Orser, C.S., Dutton, J., Lange, C., Jablonski, P., Xun, L., and Hargis, M. (1993) Characterization of a Flavobacterium glutathione S-transferase gene involved reductive dechlorination. *J.Bacteriol.*, **175**, 2640-2644.
6.   Binks, P.R., French, C.E., Nicklin, S., and Bruce, N.C. (1996) Degradatin of Pentaerythritol Tetranitrate by *Enterobacter cloacae* PB2. *Appl. Environ. Microbiol.*, **62**, 1214-1219.
7.   Orser, C.S., Lange, C.C., , X.L., , Z.T.C., and , S.B.J. (1993) Cloning, sequence analysis, and expression of the Flavobacterium pentachlorophenol-4-monooxygenase gene in *Escherichia coli*. *J.Bacteriol.*, **175**, 411-416.
8.   World Health Organization. (WHO) in collaboration with United Nations Environment Programme, Public impact of pesticides used in agriculture. (1990) World Health Organization, Geneva, p. 128
9.   U.S. Congress, Office of Technology Assessment, Disposal of Chemical Weapons: Alternative Technologies-Background Paper. (1992) U.S. Government Printing Office, Washington, DC.

10. Dumas, D.P., Caldwell, S.R., Wild, J.R., and Raushel, F.M. (1989) Purification and properties of the phosphotriesterase from *Pseudomonas diminuta*. *J Biol. Chem*, **264**, 19659-19665.

11. Dumas, D.P., Wild, J.R., and Raushel, F.M. (1989) Diisoproplyfluorophosphate hydrolysis by a phosphotriesterase from *Pseudomonas diminuta*. *Biotechnol. Appl. Biochem.*, **11**, 235-243.

12. Dumas, D.P., Durst, H.D., Landis, W.G., Raushel, F.M., and Wild, J.R. (1990) Inactivation of organophosphorus nerve agents by the phosphotriesterase from *Pseudomonas diminuta*. *Arch. Biochem. Biophys.*, **277**, 155-159.

13. Kolakowski, J.E., DeFrank, J.J., Harvey, S.P., Szafraniec, L.L., Beaudry, W.T., Lai, K.H., and Wild, J.R. (1997) Enzymatic hydrolysis of the chemical warfare agent VX and its neurotoxic analogues by organophosphorus hydrolase. *Biocataysis &. Biotransformation* **15**, 297-312.

14. Lai, K., Stolowich, N.J., and Wild, J.R. (1995) Characterization of P-S bond hydrolysis in organophosphorothioate pesticides by organophosphorus hydrolase. *Arch. Biochem. Biophys.*, **318**, 59-64.

15. Rastogi, V.K., DeFrank, J.J., Cheng, T.C., and Wild, J.R. (1997) Enzymatic hydrolysis of Russian-VX by organophosphorus hydrolase. *Biochem.Biophys.Res.Comm.*, **241**, 294-296.

16. Benning, M.M., Kuo, J.M., Raushel, F.M., and Holden, H.M. (1994) Three-dimensional structure of phosphotriesterase: an enzyme capable of detoxifying organophosphate nerve agents. *Biochemistry* , **33**, 15001-15007.

17. Benning, M.M., Kuo, J.M., Raushel, F.M., and Holden, H.M. (1995) Three-dimensional structure of the binuclear metal center of phosphotriesterase. *Biochemistry*, **34**, 7973-7978.

18. Harper, L.L., McDaniel, C.S., Miller, C.E., and Wild, J.R. (1988) Dissimilar plasmids isolated from *Pseudomonas diminuta* MG and a *Flavobacterium* sp. (ATCC 27551) contain identical opd genes. *Appl. Environ. Microbiol.*, **54**, 2586-2589.

19. McDaniel, C.S., Harper, L.L., and Wild, J.R. (1988) Cloning and sequencing of a plasmid-borne gene (opd) encoding a phosphotriesterase. *J.Bacteriol.*, **170**, 2306-2311.

20. DeFrank, J.J., Beaudry, W.T., Cheng, T.C., Harvey, S.P., Stroup, A.N., and Szafraniec, L.L. (1993) Screening of halophilic bacteria and *Alteromonas* species for organophosphorus hydrolyzing enzyme activity. *Chem Biol Interact*, **87**, 141-148.

21. Cheng, T.C., Liu, L., Wang, B., Wu, J., DeFrank, J.J., Anderson, D.M., Rastogi, V.K., and Hamilton, A.B. (1997) Nucleotide sequence of a gene encoding an organophosphorus nerve agent degrading enzyme from *Alteromonas haloplanktis*. *J. Ind. Microbiol. Biotech.*, **18**, 49-55.

22. DeFrank, J.J. and Cheng, T.C. (1991) Purification and properties of an organophosphorus acid anhydrase from a halophilic bacterial isolate. *J.Bacteriol.*, **173**, 1938-1943.

23. Cheng, T.C., Harvey, S.P., and Chen, G.L. (1996) Cloning and expression of a gene encoding a bacterial enzyme for decontamination of organophosphorus nerve agents and nucleotide sequence of the enzyme. *Appl. Environ. Microbiol.*, **62**, 1636-1641.

24. Cheng, T.C., Rastogi, V.K., DeFrank, J.J., and Sawiris, G.P. (1998) G-type nerve agent decontamination by *Alteromonas* prolidase. *Ann. N. Y. Acad. Sci.*, **864**, 253-258.

25. Nakahigashi, K. and Inokuchi, H. (1990) Nucleotide sequence between the *fadB* gene and the *rrnA* operon from *Escherichia coli*. *Nucleic Acids Res.*, **18**, 6439

26. Higgins, D.G. and Sharp, P.M. (1988) CLUSTAL: a package for performing multiple sequence alignment on a microcomputer. *Gene*, **73**, 237-244.

27. Primo-Parmo, S.L., Sorenson, R.C., Teiber, J., and La Du, B.N. (1996) The human serum paraoxonase/arylesterase gene (PON1) is one member of a multigene family. *Genomics*, **33**, 498-507.

284

28. Augustinsson, K.B. (1968) The evolution of esterases. In N. van Thoai, J. Roche (eds) *Homologous Enzymes and Biochemical Evolution*, Gordon & Breach, New York, pp. 299-311.

29. Sorenson, R.C., Primo-Parmo, S.L., Kuo, C.L., Adkins, S., Lockridge, O., and La Du, B.N. (1995) Reconsideration of the catalytic center and mechanism of mammalian paraoxonase/arylesterase. *Proc. Natl. Acad. Sci. USA*, **92**, 7187-7191.

30. Kuo, C.L. and La Du, B.N. (1998) Calcium binding by human and rabbit serum paraoxonases. Structural stability and enzymatic activity. *Drug Metab.Dispos.*, **26**, 653-660.

31. Kuo, C.L. and La Du, B.N. (1995) Comparison of purified human and rabbit serum paraoxonases. *Drug Metab.Dispos.*, **23**, 935-944.

32. Josse, D., Xie, W., Renault, F., Rochu, D., Schopfer, L.M., Masson, P., and Lockridge, O. (1999) Identification of residues essential for human paraoxonase (PON1) arylesterase/organophosphatase activities. *Biochemistry*, **38**, 2816-2825.

33. Shih, D.M., Gu, L., Xia, Y.R., Navab, M., Li, W.F., Hama, S., Castellani, L.W., Furlong, C.E., Costa, L.G., Fogelman, A.M., and Lusis, A.J. (1998) Mice lacking serum paraoxonase are susceptible to organophosphate toxicity and atherosclerosis. *Nature*, **394**, 284-287.

34. Mackness, M.I., Mackness, B., Durrington, P.N., Fogelman, A.M., Berliner, J., Lusis, A.J., Navab, M., Shih, D., and Fonarow, G.C. (1998) Paraoxonase and coronary heart disease. *Curr. Opin.. Lipidol.*, **9**, 319-324.

35. La Du, B.N., Aviram, M., Billecke, S., Navab, M., Primo-Parmo, S., Sorenson, R.C., and Standiford, T.J. (1999) On the physiological role(s) of the paraoxonases. *Chem Biol Interact*, **119-120**, 379-388.

36. Aviram, M., Rosenblat, M., Billecke, S., Erogul, J., Sorenson, R., Bisgaier, C.L., Newton, R.S., and La Du, B. (1999) Human serum paraoxonase (PON 1) is inactivated by oxidized low density lipoprotein and preserved by antioxidants. *Free Radic. Biol. Med.*, **26**, 892-904.

37. Adkins, S., Gan, K.N., Mody, M., and La Du, B.N. (1993) Molecular basis for the polymorphic forms of human serum paraoxonase/arylesterase: glutamine or arginine at position 191, for the respective A or B allozymes. *Am. J. Hum. Genet.*, **52**, 598-608.

38. Furlong, C.E., Costa, L.G., Hassett, C., Richter, R.J., Sundstrom, J.A., Adler, D.A., Disteche, C.M., Omiecinski, C.J., Chapline, C., and Crabb, J.W. (1993) Human and rabbit paraoxonases: purification, cloning, sequencing, mapping and role of polymorphism in organophosphate detoxification. *Chem Biol Interact*, **87**, 35-48.

39. Humbert, R., Adler, D.A., Disteche, C.M., Hassett, C., Omiecinski, C.J., and Furlong, C.E. (1993) The molecular basis of the human serum paraoxonase activity polymorphism. *Nat. Genet.*, **3**, 73-76.

40. Smolen, A., Eckerson, H.W., Gan, K.N., Hailat, N., and La Du, B.N. (1991) Characteristics of the genetically determined allozymic forms of human serum paraoxonase/arylesterase. *Drug Metab. Dispos.*, **19**, 107-112.

41. Davies, H.G., Richter, R.J., Keifer, M., Broomfield, C.A., Sowalla, J., and Furlong, C.E. (1996) The effect of the human serum paraoxonase polymorphism is reversed with diazoxon, soman and sarin. *Nat. Genet.*, **14**, 334-336.

42. Aviram, M., Billecke, S., Sorenson, R., Bisgaier, C., Newton, R., Rosenblat, M., Erogul, J., Hsu, C., Dunlop, C., and La Du, B. (1998) Paraoxonase active site required for protection against LDL oxidation involves its free sulfhydryl group and is different from that required for its arylesterase/para-oxonase activities: selective action of human paraoxonase allozymes Q and R. *Arterioscler. Thromb. Vasc. Biol.*, **18**, 1617-1624.

43. Mochizuki, H., Scherer, S.W., Xi, T., Nickle, D.C., Majer, M., Huizenga, J.J., Tsui, L.C., and Prochazka, M. (1998) Human PON2 gene at 7q21.3: cloning, multiple mRNA forms, and missense polymorphisms in the coding sequence. *Gene*, **213**, 149-157.

44. Sanghera, D.K., Aston, C.E., Saha, N., and Kamboh, M.I. (1998) DNA polymorphisms in two paraoxonase genes (PON1 and PON2) are associated with the risk of coronary heart disease. *Am. J. Hum. Genet.*, **62**, 36-44.

45. Davies, J.A., Buchman,V.L., Krylova,O., and Ninkina, N.N.. (1997) Molecular cloning and expression pattern of rpr-1, a resiniferatoxin-binding , phosphotriesterase-related protein, expressed in rat kidney tubules. *FEBS Lett.*, **410**, 378-382.

46. Buchbinder, J.L., Stephenson, R.C., Dresser, M.J., Pitera, J.W., Scanlan, T.S., and Fletterick, R.J. (1998) Biochemical characterization and crystallographic structure of an *Escherichia coli* protein from the phosphotriesterase gene family. *Biochemistry*, **37**, 5096-5106.

47. Himmelreich, R., Hilbert, H., Plagens, H., Li, B.-C., and Herrmann, R. (1996) Complete sequence analysis of the genome of the bacterium *Mycoplasma pneumoniae*. *Nucleic Acids Res.*, **24**, 4420-4449.

48. White, O., Eisen, J.A., Heidelberg, J.F., Hickey, E.K., Peterson, J.D., Dodson, R.J., Haft, D.H., Gwinn, M.L., Nelson, W.C., Richardson, D.L., Moffat, K.S., Qin, H., Jiang, L., Pamphile, W., Crosby, M., Shen, M., Vamathevan, J.J., McDonald, L., Utterback, T., Zalewski, C., Makarova, K.S., Aravind, L., Daly, M.J., Minton, K.W., Fleischmann, R.D., Ketchum, K.A., Nelson, K.E., Salzberg, S., Smith, H.O., Craig, J., Venter, J.C., and Fraser, C.M. (1999) Genome Sequence of the Radioresistant Bacterium *Deinococcus radiodurans* R1. *Science*, **286**, 1571-1577.

49. Cole, S.T., Brosch, R., Parkhill, J., Garnier, T., Churcher, C., Harris, D., Gordon, S.V., Eiglmeier, K., Gas, S., Barry III, C.E., Tekaia, F., Badcock, K., Basham, D., Brown, D., Connor, R., Davies, R., Devlin, K., Feltwell, T., Gentles, S., Hamlin, N., Holroyd, S., Hornsby, T., Jagels, K., Krogh, A., McLean, J., Moule, S., Murphy, L., Oliver, S., Osborne, J., Quail, M.A., Rajandream, M.A., Rogers, J., Rutter, S., Seeger, K., Skelton, S., Squares, S., Sulston, J.E., Taylor, K., Whitehead, S., and Barrell, B.G. (1998) Deciphering the biology of *Mycobacterium tuberculosis* from the complete genome sequence. *Nature*, **393**, 537-544.

50. Vanhooke, J.L., Benning, M.M., Raushel, F.M., and Holden, H.M. (1996) Three-dimensional structure of the zinc-containing phosphotriesterase with the bound substrate analog diethyl 4-methylbenzylphosphonate. *Biochemistry*, **35**, 6020-6025.

51. Farber, G.K. and Petsko, G.A. (1990) The evolution of alpha/beta barrel enzymes. *TIBS*, **15**, 228-234.

52. Reardon, D. and Farber, G.K. (1995) The structure and evolution of alpha/beta barrel proteins. *FASEB J.*, **9**, 497-503.

53. Murzin, A.G., Brenner, S.E., Hubbard, T., and Chothia, C. (1995) Scop: a structural classification of proteins database for the investigation of sequences and structures. *J. Mol.Biol.*, **247**, 536-540.

54. Holm, L. and Sander, C. (1997) An evolutionary treasure: unification of a broad set of amido-hydrolases related to urease. *Proteins*, **28**, 72-82.

55. Tatusov,R.L., Koonin,E.V. and Lipman,D.J. (1997) A genomic perspective on protein families. *Science* **278**, 631-637.

56. Schurr, M.J., Vickrey, J.F., Kumar, A.P., Campbell, A.L., Cunin, R., Benjamin, R.C., Shanley, M.S., and O'Donovan, G.A. (1995) Aspartate transcarbamoylase genes of *Pseudomonas putida*: requirement for an inactive dihydroorotase for assembly into the dodecameric holoenzyme. *J.Bacteriol.*, **177**, 1751-1759.

57. Denis-Duphil, M. (1989) Pyrimidine biosynthesis in *Saccharomyces cerevisiae*: the ura2 cluster gene, its multifunctional enzyme product, and other structural or regulatory genes involved in *de novo* UMP synthesis. *Biochem.Cell Biol.*, **67**, 612-631.

58. Jensen, R.A. (1976) Enzyme recruitment in evolution of new function. *Annu. Rev. Microbiol..*, **30,** 409 - 25, 409-425.

59. Campbell, J.H., Lengyel, J.A., and Langridge, J. (1973) Evolution of a second gene for β-galactosidase in *Escherichia coli*. *Proc.Natl.Acad.Sci.USA* **70**, 1841-1845.

60. Hall, B.G. (1999) Experimental evolution of Ebg enzyme provides clues about the evolution of catalysis and to evolutionary potential. *FEMS Microbiol. Lett.*, **174**, 1-8.

61. Albery, W.J. and Knowles, J.R. (1976) Evolution of enzyme function and the development of catalytic efficiency. *Biochemistry*, **15**, 5631-5640.

62. Demetrius, L. (1998) Role of enzyme-substrate flexibility in catalytic activity: an evolutionary perspective. *J.Theor.Biol.*, **194**, 175-194.

FIELD-DEPLOYABLE AMPEROMETRIC ENZYME ELECTRODES FOR DIRECT MONITORING OF ORGANOPHOSPHATE NERVE AGENTS

ASHOK MULCHANDANI*, PRITI MULCHANDANI AND WILFRED CHEN
Department of Chemical and Environmental Engineering
University of California, Riverside, CA 92521, USA

JOSEPH WANG AND LIANG CHEN
Department of Chemistry and Biochemistry
New Mexico State University, Las Cruces, NM 880003, USA

Abstract: Amperometric enzyme electrodes for field, on-site and in-situ/on-line remote, monitoring of organophosphate nerve agents is reported. The sensors rely upon the coupling of the rapid biocatalytic hydrolysis of the organophosphates by the immobilized organophosphorus hydrolase and the sensitive and rapid anodic detection of p-nitrophenol product. The biosensors are fast responding, highly selective and show sub-micromolar lower detection limit for paraoxon and methyl parathion. The low detection limits coupled with high selectivity and faster and simplified operation are the benefits of the developed biosensors over the analogous enzyme inhibition biosensors make them ideal for field monitoring of organophosphate compounds.

1. Introduction.

Organophosphates (OPs), amongst the most toxic substances, are widely used as pesticides and insecticides in agriculture and as chemical warfare agents [1-3]. Because of the widespread use of these highly neurotoxic compounds in modern agricultural practices they are significant major environmental and food pollutants. Since OPs are found to be present in sampled soils, streams, ground and waste waters, rapid

B. Zwanenburg et al. (eds.), Enzymes in Action, 287–296.
© 2000 *Kluwer Academic Publishers. Printed in the Netherlands.*

determination of the source and magnitude of the pollutant is necessary in order to take preventive action [4]. Current analytical techniques such as gas and liquid chromatography, although very sensitive and reliable, have disadvantages. These techniques are time consuming, expensive, have to be performed by highly trained technician and are not suitable for rapid analyses under field conditions [4].

Biosensors technology is ideally-suited to provide solutions to these problems. Until recently, biosensors for OP monitoring have relied upon the use of inhibition enzyme electrodes, based on the modulated activity of cholinesterases (acetylcholinesterase or butyrylcholinesterase) [5]. These biosensors, although very sensitive, have major drawbacks. They have poor selectivity, are irreversible and because of a multistep analysis protocol are slow. Recently, biosensors for direct monitoring of OPs based on the use of organophosphorus hydrolase (OPH) was introduced by us [6-11]. OPH hydrolyzes a number of OP pesticides and insecticides (e.g., paraoxon, parathion, coumaphos, diazinon) and chemical warfare agents (e.g., sarin) [12]. The use of OPH is extremely attractive for the biosensing of OPs that act as substrates for the enzyme, rather than exerting an inhibitory action. Several types of OPH-based biosensors have been introduced recently, including potentiometric and optical [6-11].

In this chapter we report the development of two OPH-based amperometric enzyme electrodes that have potential for field monitoring of OP nerve agents.

2.   Experimental

2.1.   REAGENTS

Organophosphorus hydrolase (OPH) (7250 IU/mg protein, 15 mg protein/mL) was produced and purified according to the methods described by Mulchandani et al. [4]. Paraoxon and methyl parathion were obtained from Sigma Chemical Co. (St. Louis, MO) and Supelco Inc. (Belefonte, PA), respectively. Carbon ink (Electrodag 440B) and laser scribed alumina ceramic substrate were purcahsed from Acheson, Ontario, CA and Coors Ceramics, Golden, CO, respectively. The graphite powder (grade #38) was obtained from Fisher Scientific (Pittsburgh, PA), while Aldrich (Milwaukee, WI) provided the mineral oil and the Nafion (5%wt) solution. The Rio Grande river water samples were collected in Las Cruces, NM.

2.2.   ELECTRODE FABRICATION

2.2.1 Thick-Film Strip Electrode

A semi-automatic screen-printer (Model TF-100, MPM Inc., Franklin, MA) was used for fabricating the strip transducers. The carbon ink (Electrodag 440B, Acheson, Ontario, CA) was printed on a 1.33"x 4.00" laser scribed alumina ceramic substrate (Coors Ceramics, Golden, CO), through a patterned stencil to yield 10 strips (3.3x1.0 cm each) with a 30x2 mm printed carbon area. The ink drying was carried out subsequently at 120°C for 30 min. An insulating layer was then printed to cover most of the printed carbon area, exposing a 5x2 mm working-electrode area and a similar area (on the opposite side) that served as an electrical contact.

OPH, was immobilized by casting a 20 µL droplet, containing OPH (10 µL of 108 IU/µL) in Nafion (1%, in ethyl alcohol) onto the printed carbon surface, and allowing the solvent to evaporate. The enzyme-modified strip was subsequently dried completely at room temperature and kept in a refrigerator (at 4°C) until use.

2.2.2 Carbon Paste Electrode

The carbon paste electrode was prepared by thoroughly hand mixing 120 mg mineral oil and 180 mg graphite powder. The resulting paste was packed firmly into the electrode cavity (3 mm diameter, 1mm depth) of a 4-cm long Teflon sleeve. Electrical contact (to the inner part of the paste) was established via a stainless steel screw, contacting the brass screw within the female connector on the PVC housing. The paste surface was smoothed on a weighing paper. OPH was immobilized by casting a 10µl droplet, containing 5µl (of 108 IU/µl) OPH and 5µl Nafion (in 1% ethanol) onto the carbon surface, and allowing the solvent to evaporate. The OPH-modified carbon paste electrode was stored at 4°C until use.

The design of the remote sensing probe was similar to that previously reported remote phenol biosensor [13]. The electrode assembly, housed in a PVC housing tube, connected to a 15-m long shielded cable via three-pin environmentally-sealed rubber connectors. The assembly included the OPH-modified carbon-paste working electrode, a Ag/AgCl reference electrode (BAS, Model RE-4) and a platinum wire counter electrode. Two female coupling connectors, fixed with epoxy in the PVC tube, served for mounting the working and reference electrodes; brass screws, located within these connectors, provided the electrical contact.

2.3. APPARATUS

Experiments were performed with the BAS CV-27 voltammetric analyzer (Bioanalytical Systems, W. Lafayette, IN), in conjunction to the BAS X-Y-t recorder. For studies with the thick-film strip electrode, screen-printed working enzyme electrode, the Ag/AgCl (3M NaCl) reference electrode (BAS RE-1), and platinum wire counter electrode were introduced to the 10mL cell (Model CV-2, BAS) through holes in its Teflon cover. A magnetic stirrer and stirring bar provided the desired convective transport. In studies with the remote sensing electrode, experiments were performed in a 50 ml glass beaker, into which the probe (sensing) head was submersed.

2.4. PROCEDURE

Experiments with screen printed strip electrodes were performed in amperometric mode in solution stirred at 300 rpm. Unless otherwise stated, these experiments were performed by applying a potential of +0.85V and allowing the transient background current to decay to a steady-state value prior to additions of the OP substrate. Experiments with remote sensing electrode were performed in chronoamperometric mode in quiescent solution conducted by applying a potential step (from open circuit to +0.85V) and recording the resulting transient current. The quantitative information was obtained by sampling the current after 60 sec. All experiments were performed at room temperature.

3. Results and Discussion

3.1. OPH-MODIFIED CARBON-STRIP ELCTRODE FOR ON-SITE MONITORING

Figure 1 shows the hydrodynamic voltammogram (HDV) for paraoxon at the OPH-modified screen printed electrode. The oxidation of the enzymatically-liberated p-nitro-phenol starts at +0.75V, with the response rising sharply up to +0.95V, and leveling off thereafter. This HDV profile mirrors the HDV profile for *p*-nitrophenol (data not shown). All subsequent work was performed at +0.85V due to a small signal observed for the direct oxidation of paraoxon at the "enzyme free" strip using potential higher than +0.9V.

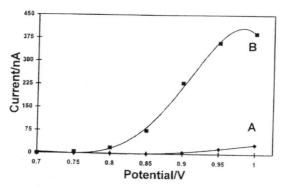

Figure 1: Hydrodynamic voltammogram for 2x10-5 M paraoxon at the Nafion-coated carbon strip (A) and OPH-modified carbon strip (B). OPH surface loading 1080 IU. Electrolyte, phosphate buffer (0.05 M, pH 7.4); stirring rate 600 rpm..

The current-time amperometric recordings obtained with the thick-film OPH-modified strip for paraoxon and methyl parathion are shown in Fig. 2. As the figure show (inset Fig. 2), the calibration plots were linear up to 20 μM paraoxon (slope, 1.67 nA/μM; correlation coefficient, 0.999) and 5 μM methyl parathion (slope of initial linear portion, 2.83 nA/μM) which then curved. The enzyme-modified strip electrodes had very good lower detection limits for both paraoxon ($9x10^{-8}$ M) and methyl parathion ($7x10^{-8}$ M). These lower detection values are significantly lower than the $2\text{-}5x10^{-6}$ M values for OPH-based potentiometric and fiber-optic biosensors [4-9] developed in our laboratory. The nanomolar detection limits of cholinesterase-inhibition biosensors are attained following prolonged incubation period of 10-30 min [5]. Further improvements in the sensitivity and lower detection limit of the amperometric OPH-modified enzyme electrode may be achieved through an electrocatalytic (accelerated) detection of the *p*-nitrophenol product. Alternatively, the device may be combined with common sample preparation/extraction (enrichment) protocols. OPH-modified carbon-strip electrode had a very fast response time, with ca 10 sec for attaining steady-state currents (Fig. 2). Such a response is significantly faster than that of current cholinesterase-inhibition amperometric OP biosensors which require addition of the substrate and an incubation period [4, 5].

292

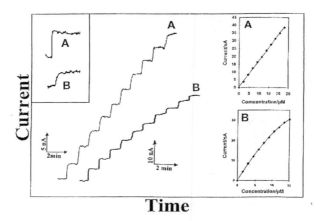

Figure 2: Current-time amperometric response to successive 2.3×10^{-6} M paraoxon (A) and 1.5×10^{-6} M of parathion increments. Also shown (inset) are calibration plots and the response to 4×10^{-7} M paraoxon (A) and 1.5×10^{-7} M methyl parathion. Operating potential, $+0.85$ V. Other conditions as in Fig. 1.

3.2. OPH-MODIFIED CARBON-PASTE ELCTRODE FOR REMOTE SENSING

The carbon strip electrode, although ideal for on-site measurements are not suitable for applications of remote or on-line/in-situ sensing that may be necessary for certain situations. Wang's group has developed a submersible amperometric electrode and demonstrated its application for in-situ biosensing of toxic phenolic [13] or peroxide [14] substances. Such remote biosensing technology offers a fast return of the chemical information in a timely, safe, and cost-effective manner [15]. We have combined the OPH enzyme with the submersible amperometric electrode connected to a 15 m long shielded cable for remote on-line/in-situ monitoring of OPs in untreated river water.

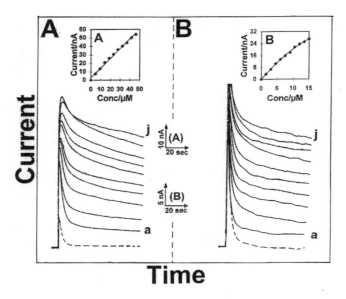

Figure 3: Chronoamperometric response to an untreated river water sample containing increasing levels of paraoxon (A) and methyl parathion (B) in 4.6x10⁻⁶M and 1.5x10⁻⁶M steps, respectively (a-j).Dotted lines represent the response of the unspiked samples. Potential step to +0.85V.

Figure 3 displays chronoamperograms for an unaltered river water sample containing increasing levels (a-j) of paraoxon and methyl parathion. As shown, the submersible bioprobe yielded well-defined amperometric signals that permits convenient quantitation of micromolar concentrations. Despite the 15 m long cable, and the full submersion of the sensor head, the noise level was very low. (Note the sensitive current scale.) The response of the unspiked water sample (dotted line) was also very small, indicating minimal interferences of coexisting oxidizable compounds. The detection limits of the bioprobe for methyl parathion and paraoxon, respectively, was 4×10^{-7}M and 9×10^{-7}M (S/N=3). The response was linear from 4.6-46x10⁻⁶M for paraoxon and up to 5×10^{-6}M for methyl parathion (see Fig. 3 insets). The slopes of the linear portions of these resulting plots corresponded to sensitivities of 2.14 (methyl-parathion) and 1.45 (paraoxon) nA/μM. Under the chronoamperometric conditions of Figure 3, 60 repetitive measurements were possible every hour, which will provide a rapid warning in case of sudden OP discharge. Such a frequent response cannot be attained with inhibition (acetylcholine esterase) biosensors due to their slow multi-step protocol, involving addition of the substrate, incubation period and enzyme regeneration. Submersible inhibition biosensors also require an internal reservoir for the substrate solution [16].

The submersible biosensor responded rapidly to dynamic changes in the concentration of both paraoxon and methyl-parathion with no apparent carry-over when going from low to high and back to low concentration. This demonstrated the ability of the bioprobe to respond rapidly to change in concentration of OP, a necessary attribute for rapid warning in the event of sudden discharge. After a short initial stabilization period (approximately 10 min.) the response of the bioprobe was unchanged over the duration (4 hours for paraoxon and 2.5 h for methyl parathion) of the test. Such a stability is highly desirable for in-situ monitoring. The relative standard deviations over these series are 3.9% for paraoxon and 5.8% for methyl parathion. This data also. These results also indicate no apparent surface fouling upon prolonged submersion in the river water sample and reflect the reusable character of OPH bioelectrodes (as opposed to inhibition biosensors) indicating a great promise for continuous monitoring of nerve agents.

Unlike laboratory-based sensing applications, where the solution conditions, pH and ionic strength, can be adjusted for optimal performance, submersible probes rely on the use of the natural conditions. For in-situ application, it is important to evaluate the influence of pH and ionic strength on the biosensor response. The results of these studies on OPH-modified electrode showed that while the electrode response was a weak function of the ionic strength, it was strongly influenced by the pH (Fig. 4). Field calibrations (using the target sample) or programmable pH compensation (common to water profilers) would be useful for addressing the observed pH dependence.

4. Conclusions

In conclusion, we have combined organophosphorus hydrolase with screen printed thick film electrode and submersible electrode technologies to develop amperometric biosensors suitable for direct, rapid, selective, sensitive and low cost field monitoring of organophosphate nerve agents. These biosensors can be potentially used for discreet and continuous in-situ field monitoring.

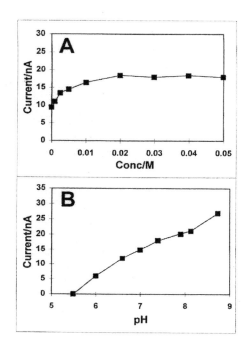

Figure 4: Effect of buffer concentration (A) and pH (B) upon the response to 2.3x10⁻⁵ M paraoxon. Potential step to +0.85V; medium: phosphate buffer of pH 7.4 (A) and 0.05M concentration (B).

5. Acknowledgements

This work was supported by grants from the US EPA and the National Science Foundation (to AM) and U.S. DOE WERC (to JW).

6. References

1. Food and Agricultural Organization of the United Nations, Rome. (1989) *FAO Prod. Yearb.*, **43**, 320.

2. United States Department of Agriculture. (1992). *In:* Agrcultural Statistics. United States Government Printing Office, Washington, D.C., pp. 395.

3. Compton, J.A. (1988) *In:* Military Chemical and Biological Agents. Telford Press, NJ, pp. 135.

4. Diehl-Faxon, J., Ghindilis, A.L., Atanasov, P. and Wilkins, E. (1996) Direct electron transfer based tri-enzyme electrode for monitoring of organophosphorus pesticides. *Sens. Actuators B* **35-36**, 448-457.

5. Paddle, B.M. (1996) Biosensors for chemical and biological agents of defence interest. *Biosens. Bioelectron.*, **11**, 1079-1113.

6. Mulchandani, P., Mulchandani, A., Kaneva, I. and Chen, W. (1998) Biosensor for direct determination of organophosphate nerve agents. 1. Potentiometric enzyme electrode. *Biosens. Bioelectron.***14**, 77-85.

7. Mulchandani, A., Mulchandani, P. Kaneva, I. and W. Chen. (1998) Biosensor for direct determination of organophosphate nerve agents using recombinant *Escherichia coli* with surface-expressed organophosphorus hydrolase. 1. Potentiometric microbial electrode. *Anal. Chem.* **70**, 4140-4145.

8. Mulchandani, A., Mulchandani, P., Chauhan, S., Kaneva, I. and Chen, W. (1998) A potentiometric microbial biosensor for direct determination of organophosphate nerve agents. *Electroanalysis* **10**, 733-737.

9. Mulchandani, A., Kaneva, I. and Chen, W. (1998) Microbial biosensor for direct determination of organophosphate nerve agents using recombinant *Escherichia coli* with surface-expressed organophosphorus hydrolase. 2. Fiber-optic microbial biosensor. *Anal. Chem.* **70**, 5042-5046.

10. Rogers, K., Wang, Y., Mulchandani, A., Mulchandani, P. and Chen, W. (1999) Development of a versatile organophosphorus hydrolase-based assay for organophosphate pesticides. Biotechnol. Prog. **15**, 517-522.

11. Mulchandani, A., Pan, S. and Chen, W. (1999) Fiber-optic biosensor for direct determination of organophosphate nerve agents. Biotechnol. Prog. **15**, 130-134.

12. Dave, K.I, Miller, C.E. and Wild, J.R. (1993) Characterization of organophosphorus hydrolases and the genetic manipulation of the phosphotriesterase from *Pseudomonas diminuta. Chem.-Biol. Interactions.* **87**, 55-68.

13. Wang, J. and Chen, Q. (1995) Remote electrochemical biosensor for field monitoring of phenolic compounds. *Anal. Chimica Acta* **312**, 39-44

14. Wang, J., Cerpia, G. and Chen, Q. (1996) Submersible bioprobe for continuous monitoring of peroxide species. *Electroanalysis* **8**, 124-127

15. Wang, J. (1997) Remote electrochemical sensors for monitoring inorganic and organic pollutants. *Trends Anal. Chem.* **16**, 84-88.

16. Wang, J., Tian, B. Lu, J., MacDonald, D., Wang, J. and Luo, D.(1998) Renewable-reagent enzyme inhibition sensor for remote monitoring of cyanide. *Electroanalysis* **10**, 1034-1037.

ENVIRONMENTAL BIOCATALYSIS

SERGEI VARFOLOMEYEV AND SERGEI KALYUZHNYI

Chemical Enzymology Department, Faculty of Chemistry,

The Lomonosov Moscow State University, 119899, Moscow, Russia

Environmental contamination is one of the major concerns of the second half of the 20th century. The growth of the amount of waste, including the highly toxic agents, is of anthropogenic origin. The dramatic boom of environmental contamination in the last half of the 20th century is definitely correlated to the increase in human population. The kinetic curve of the growth of human population is shown in Figure 1. In the last millennium, this population growth is not simply exponential but constantly accelerated, and adequately described by expression shown in equation (1),

$$N(t) = N_0 \exp\left[\frac{k_{app}}{k_0} \exp(k_0 t)\right] \qquad (1)$$

wherein N_0 is the population at the start of our era, a and b are the constants, t is the time expressed in years [1].

We have termed this dynamic law as hyperexponential (the exponential in the exponential).

In principle, the same dynamic law can be used to describe the growth of human waste activity. This process of growing waste production can be stopped whereby an eco equilibrium is reached, provided adequate efforts are made. The intensity of such efforts in a certain amount of time must fully take into account the hyperexponential law of development. Part of the solution of this enormous problem requires the following complementary measures: (i) Development of new technological processes which produce a minimal level of pollutants (xenobiotics) and (ii) Elaboration of efficient processes for the complete destruction of contaminants to environmentally acceptable products. In this chapter the focus is on the conversion of pollutants by biotechnological methods using enzymes present in microorganisms.

B. Zwanenburg et al. (eds.), Enzymes in Action, 297–310.
© 2000 *Kluwer Academic Publishers. Printed in the Netherlands.*

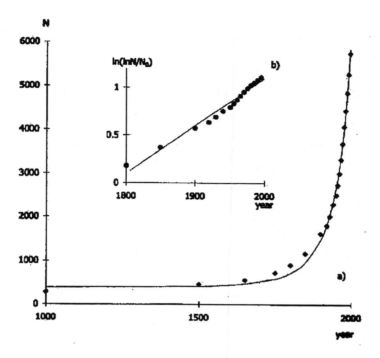

Figure 1: a). Dynamics of human population growth in macrohistorical scale. b). Linearization of Figure 1a data in the coordinates ensued from Eq. (1)

It is common knowledge that any waste, from a supertoxic chemical to metal, getting into the environment, ultimately undergoes destruction up to a complete mineralization. Highly unfortunate, such processes are often far to slow and in fact the underlying chemistry cannot be unravelled due to the very long induction periods. More recently, a notable progress has been made. As many of these processes occur by the action of enzymes, they can be referred to as "Environmental Biocatalysts". Studies in this area should be aimed at understanding the functions of microorganisms and enzymes in chemical and dynamic terms. Such knowledge provides the necessary background for achieving a notable intensification of waste destruction.

A special feature of biocatalytic destruction processes in Nature is that they ultimately lead to complete destruction to give basic components such as CO_2, NH_3, H_3PO_4, H_2S etc. These processes are usually called biomineralization, which are of great interest from an environmental point of view for the removal of xenobiotics. Such biomineralization can take place in the environment, for instance, for the degradation of petroleum and its products pollutants, but also in special set-ups as for example the anaerobic destruction of agricultural and food waste).

1. Kinetic behavior types

Typical patterns of time dynamics of pollutant concentration can be deduced from available experimental data [2, 3]. Such an analysis reveals a relation with the evolution of a biocatalytic system with time. This insight in evolution is of help to formulate schemes for the identification of waste destruction. Two types of kinetic behavior can be distinguished: (i) Destruction of pollutant is effected by enzymes present in microorganisms, whereby the decomposition of material is not an essential factor for the growth of microorganism population, i.e. the biocatalyst concentration remains constant in time, and (ii) The pollutant decomposition provides the microorganisms with the necessary ingredients such as carbon, nitrogen or phosphorus, thus the destruction is a necessary condition for the development of the microorganisms.

These two extreme cases can be described in kinetic terms as is outlined below. Destruction of pollutant by nongrowing microbial culture. This represents a simple kinetic case of transformation of substrate (S) into product (P), as shown in Scheme (2) and which is adequately described by first-order kinetics (equation 3).

$$S \xrightarrow{k_{obs}} P, \tag{2}$$

$$-\frac{ds}{dt} = \frac{dp}{dt} = k_{obs}S$$

$$S(t) = S_0 e^{-k_{obs}t} \; ; P(t) = S_0(1 - e^{-k_{obs}t}) \tag{3}$$

The observed rate constants k_{obs} depends on the biocatalyst concentration. Several examples of such kinetic behavior are reported [3]. In Figure 2 the kinetic curve of the decomposition of the chemical warfare agent Soman (an organophosphorus compound) by soil microorganisms is shown. The k_{obs} is 0.2 h^{-1}, and the half-life of Soman 3.45 h.

300

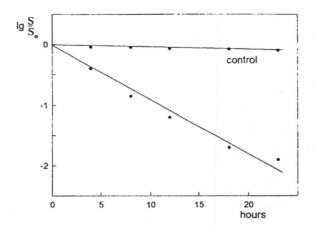

Figure 2: Kinetics of Soman degradation by soil microorganisms (experimental results in semilogarithmic coordinates) [3].

Similar kinetics are observed for a slow growth rate of microorganisms with a relatively high rate of substrate consumption. This is exemplified for the destruction of petroleum hydrocarbons by "Roder", a consortium of 2 cultures of *Rhodococcus* in soil (Figure 3). First-order kinetics are fairly well obeyed.

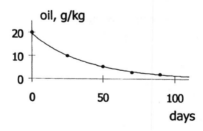

Figure 3: Kinetics of petroleum biodegradation by *Rhodococcus* culture (experimental results of V.P. Murygina).

<u>Destruction of pollutants by growing microbial cultures.</u> The kinetics of pollutant decomposition often depends on the kinetics of cell growth. This can adequately be described by the Monod equation (4), wherein M is the cell concentration (biomass).

$$\frac{dM}{dt} = \mu(S)M \ , \ \mu(S) = \frac{\mu_m S(t)}{K_s + S(t)} \tag{4}$$

Assuming that the amount of converted substrate and the accretion of biomass shows a linear relation (5),

$$dM = -Y_s d_s,$$ (5)

where Ys is the economical coefficient defining the amount of substrate accumulated as biomass. Integration of the equations (4), thereby assuming that the amount of biomass is much larger than the introduced amount of inoculate, gives equation (6), where M_0 is the initial amount of introduced, and M_m is the final amount of the accrued biomass.

$$\ln X - \frac{K_s}{S_0 + K_s} \ln(1 - ax) = \frac{\mu_m S_0}{K_s + S_0} t$$ (6)

$$X = \frac{M}{M_0}, \qquad a = \frac{M_0}{Y_s S_0} = \frac{M_0}{M_m}$$

Transformation of these equations to substrate dynamics need the insertion of linear correlation (7):

$$M = Y_s(S_0 - S) + M_0 \approx Y_s(S_0 - S)$$ (7)

Experimental results can now be analysed using the equations (8), whereby straight lines are expected.

$$\frac{\ln M / M_0}{t} = \mu + \varphi \frac{\ln(1 - M / M_m)}{t}$$

$$\frac{\ln M / M_0}{\ln(1 - M / M_m)} = \varphi + \mu \frac{t}{\ln(1 - M / M_m)}$$ (8)

$$\frac{\ln(1 - M / M_m)}{\ln(M / M_0)} = \frac{1}{\varphi} - \frac{\mu}{\varphi} \frac{t}{\ln(M / M_0)}$$

where $\mu = \dfrac{M_m S_0}{K_s + S_0}$; $\varphi = \dfrac{K_s}{S_0 + K_s}$

Alternatively, the experimental data can also be analyzed using the difference form of these equations, shown in (9), where at $t=t_i$ and $t=t_j$ the substrate and product concentrations equal to S_i and S_j, P_i and P_j, respectively, and P_m is the extreme amount of the formed product.

302

$$\frac{\ln\dfrac{S_0 - S_i}{S_0 - S_j}}{t_j - t_i} = \mu + \varphi \frac{\ln(S_j / S_i)}{t_j - t_i}$$

(9)

$$\frac{\ln(P_i / P_j)}{t_j - t_i} = \mu + \varphi \frac{\ln\dfrac{P_m - P_j}{P_m - P_j}}{t_j - t_i}$$

The kinetics of the degradation of 2,4-dichlorphenolacetic acid are shown in Figure 4, as an example of equation (9).

The most typical case upon destruction of highly toxic agents by growing microbial populations is the effect of inhibition by excess substrate (pollutant).

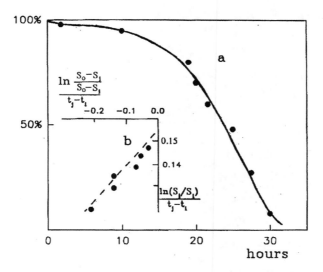

Figure 4: Kinetics of 2,4-dichlorphenoxyacetic acid degradation [3].

Strategies of pollutant destruction. The kinetically different mechanisms discussed above lead to two different approaches in the process intensification. In Figure 5 the kinetics for these two extreme cases of destruction are drawn schematically. Thus, when the process proceeds by the type involving growing biomass, process intensification is related to the acceleration of growth of the microorganism, which can be realized by addition of growth factors such as sources of phosphorus, nitrogen, microelements, etc. When on the other hand pollutant destruction follows first-order kinetics, then process intensification can be achieved by the introduction of an additional amount of destructing microorganism (Figure 5b). The destruction of petroleum pollutants shown in Figure 3 is a typical example, whereby addition of biocatalyst for several times is beneficial for the process. Table 1 shows the relevant data for the purification of river and lake surfaces in Western Siberia (results of V.P. Murygina). From these data it can be deduced that 2 to 3 inoculations fo biocatalyst reduce the amount of pollutant 10000 fold in a month.

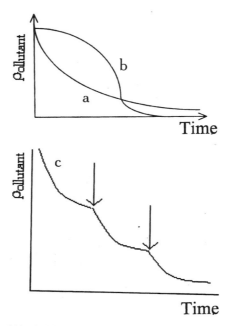

Figure 5: Two types of kinetic behavior - two strategies of pollutant destruction: a). biocatalytic exponential destruction of pollutant; b). biodestruction coupled with microorganism growth; c). acceleration of biodestruction via introduction of an additional portion of biocatalyst.

TABLE 1. *Destruction of petroleum pollutants by* Rhodococcus *culture.*

river		lake	
Days*	content of hydrocarbons mg/l	Days*	content of hydrocarcons, mg/l
0*	440	0*	151
14	52	10	78
28	0.04	24	6.56
		50	0.005

* The time of introduction of *Rhodococcus* culture

2. Adaptation of destructor microorganism.

The destruction of xenobiotics by microorganisms is often accompanied by adaptation of microorganisms to this pollutant. Such a phenomenon becomes manifest when during the first passage the kinetic curve of growth of the microorganisms shows a notable lag phase, sometimes a very long one. The next passage proceeds very rapidly with no lag phase, implying that adapted cells are now in action. The theoretical aspects in kinetic terms has been reported in two monographs [1, 2]. During the lag phase the synthesis of the necessary enzymes in the microorganism is taking place along with an adequate adaptation in the cell, whereby destruction of xenobiotics can be handled. The kinetic scheme derived from the consecutive reactions

$$ N \xrightarrow{\ k_a\ } N_a \xrightarrow{\ \mu\ } 2N_a $$

where k_a is the rate constant of the adaptation process and μ is the specific rate of the adapted cell growth. The time dependence of the number of cells in the system follows the equation (10)

$$ \ln \frac{N_a}{N_0} = \mu + \ln \left\{ \frac{k_a}{k_a + \mu} \left[1 + \frac{\mu}{k_a} \exp(-k_a + \mu)t \right] \right\} \qquad (10) $$

where N_0 is the starting concentration of cells. At long times of process performance, the equation is simplified into (11):

$$ \ln \frac{N_a}{N_0} = \mu t + \ln \frac{k_a}{k_a + \mu} \qquad (11) $$

The slope of the corresponding curve represents the specific rate of the cell growth and the intercept of the time axis the lag phase τ according to equation (12).

$$\tau = \frac{1}{\mu} \ln\left(1 + \frac{\mu}{k_a}\right) \qquad (12)$$

Thus, a success cell adaptation will notably reduce the τ value and this can be achieved by increasing the specific rate of microorganism growth. The procedure for cell adaptation involves the growth of cells in the presence of a xenobiotic using media with an effective co-substrate.

This approach is exemplified by the biomineralization of products arising from the chemical neutralization of mustard gas. This destruction of this chemical warfare agents proceeds in two steps, viz. a chemical neutralization of the active ingredient by treatment with ethanol amine and a subsequent biodegradation of the neutralization products as well as the excess of neutralization [4], according to Scheme 13.

$$(13)$$

Each of the compounds shown in the scheme underwent biomineralization upon the action of a methane generating anaerobic consortium of microorganisms. The relevant data are collected in Table 2. The data reveal that thiodiglycol is most resistant to biodegradation as only 42% was converted in half-year incubation, with a lag phase of almost two months. With the co-substrates acetic & propionic acid (last line in Table 2) there is no lag phase while the biodegradation was complete in about a month time.

TABLE 2.

pollutant products	lag-phase, days	incubation time, days	Biodegradability, %
ethanolamine	8	22	100
mercaptoethanol	0	55	80
ethyleneglycol	2	6	100
thiodiglycol	52	185	42
thiodiglycol+acetic and propionic acids	0	32	100

3. Microbial consortiums. Potentiation of "impossible" reactions.

Natural mineralization of organic compounds usually proceeds by the action of several microorganisms in a symbiotic consortium. Theoretical aspects of such consortiums are reported in monographs [1, 2]. A detailed study involves symbiotrophic consortiums where the product of one microorganism serves as substrate for the other. One of the most striking examples is the decomposition of organic compounds to methane and carbonic acid by an anaerobic microbial consortium. The microorganisms combined in a consortium show chemical conversions that would be impossible in one-step processes. Anaerobic methane producing bacteria is an impressive example of such formally not feasible processes.

Water as oxidant of carbon-carbon bonds. If one would ask the question whether water could serve as an oxidant of a carbon-carbon bond at normal pressure and temperature, the reply will surely be negative. For example, would the reaction of water with butyric acid to give acetic acid and hydride be possible? Under normal conditions this would thermodynamically not feasible.

$$(14)$$

When however, a microbial consortium is used an efficient biocatalytic conversion can be accomplished. Figure 6 shows the degradation of butyric acid by a microbiological consortium from a methane tank, whereby indeed two molecules of acetic acid are produced per molecule of substrate. This conversion proceeds strictly anaerobic and the only source of oxygen is water. However, a consortium of microbiological organisms is required for such an operation. By taking into account that the overall conversion consists of at least two, but probably more, chemical reactions, the thermodynamics can be understood as well.

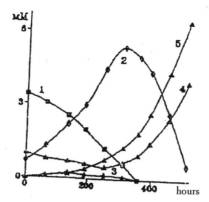

Figure 6: Kinetics of anaerobic conversion of butyric acid by microbial consortium of methane-producing bacteria: 1. butyric acid, 2. acetic acid, 3. hydrogen, 4. methane, 5. CO_2

Kinetic model of methanogenesis. The dynamics of methane formation by anaerobic methane-producing microbiological consortiums has been elaborated [5-8], by invoking several intermediates. Experimental results were appropriately fitted with the theoretically predicted curves, as shown in Figures 7 and 8 for two substrates, viz. lysine and butyric acid/ethanol. Similarly carbohydrates, proteins and a series of aliphatic acids and alcohols were investigated in this anaerobic microbiological breakdown to methane and carbon dioxide. The dynamics for all low molecular weight components formed in those processes were established. It was assumed that for the overall chemical conversion a minimum number of microorganisms is needed. For glucose these four chemical operations are shown in Scheme 15, whereby the microorganisms $X_1 - X_4$ are involved. The first organism splits glucose into propionic acid, ethanol, carbon dioxide and hydrogen, the second step (X_2) is an oxidation of ethanol and carboxylic acid. The third and fourth steps are producing methane. By comparing experimental and theoretical analyses a complete model was derived for the methanogenesis, which was quite helpful in predicting the dynamic behavior of various substrates.

308

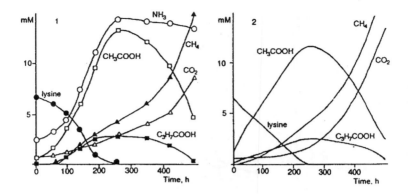

Figure 7: Kinetics of lysine conversion by methanogenic consortiums:
1). experimental results; 2). theoretical calculation.

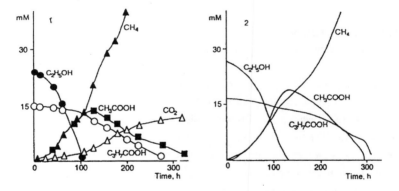

Figure 8: Kinetics of conversion of butyric acid + ethanol mixture by methanogenic consortiums:
1). experimental data; 2). theoretical calculations.

$$C_6H_2O_6 + 0.002H_2O \xrightarrow{x_1} 0.3HC_2H_5OH + 0.39HC_3H_7COOH$$
$$+ 1.14CO_2 + 0.82H_2$$

$$C_2H_5OH + H_2O \xrightarrow{x_2} CH_3COOH + 2H_2$$

$$C_3H_7COOH + 2H_2O \xrightarrow{x_2} 2CH_3COOH + 2H_2 \tag{15}$$

$$H_3COOH \xrightarrow{x_3} CH_4 + CO_2$$

$$H_2 + 0.25CO_2 \xrightarrow{x_4} 0.25CH_4 + 0.5H_2O$$

4. **Modeling bioreactors. The competition between sulfate-reducing and methane-producing bacteria in anaerobic reactors.**

Many ecobiocatalytic reactions are conducted in appropriately designed reactors. In designing the reactors and optimizing the processes, it is necessary to take into account the full array of complicated processes occurring in natural biocatalytic systems. Such undertaking will be exemplified by an anaerobic purification of waste water containing organic pollutants, which usually also contains sulfate ions. Under anaerobic conditions the sulfate reducing (SRB) and methane producing methanogenic bacteria (MB) are at work in a complex interaction scenery. Substrate competition is possible at two levels. First, the competition between SRB and acetogenic bacteria (AB) for fatty acids and ethanol. Second, the competition between SRB and methanogenic bacteria (MB) for acetate and hydrogen. The role of sulfate reduction is negliglible for sugars and amino acids as the substrates. Thermodynamic and Monod-kinetic data of SRB, AB and MB for growth on volatile fatty acids and hydrogen indicate that SRB should be able to outperform AB and MB. This prediction has been confirmed experimentally for hydrogen and fatty acids. The situation is different when acetate is used in anaerobic reactors. It was shown by several researchers that during the breakdown of sulfate-containing waste water, SRB can successfully compete with MB for acetate [9], whereas we showed that acetate is mostly degraded to methane [10, 11]. These differences can be explained by other factors influencing the outcome of the competition between SRB and MB, e.g. the $COD:SO_4^{2-}$ ratio (COD = chemical oxygen demand), the type of seed sludge, the sludge retention, H_2S inhibition, pH and the nutrient limitation [13].

It should be appreciated that a comprehensive mathematical modelling of anaerobic reactors is a rather complicated task because of the complexity of the system. Substantial progress could be made in this area using the pioneering work of Andrews [12]. The most advanced model for the competing sulfate reduction and methanogenesis has been reported [13-15], whereby most parameters are adequately accommodated. This model allows a prediction of the influence of various factors on the outcome of e.g. sulfate-fed granular sludge reactors, during the start-up phase as well as the further continuation of the anaerobic processes.

Our research clearly illustrates the pertinent role of Environmentally Biocatalysis in getting insight into effective environmental onctrol by technological processing.

310

5. Literature

1. Varfolomeyev, S.D. & Gurevich, K.G. (1999). *Biokinetics* (monograph) (Russ). Moscow, Fair Press, 737 pp
2. Varfolomeyev, S.D. & Kalyuzhnyi (1990). *Kinetic grounds of microbiological processes* (Russ). Moscow, Moscow University Publishing House, 311 pp
3. Varfolomeyev, S.D., Spivak, S.I. & Zavialova, N.V. (1997). *Perspectives in bioremediation. Technologies for environmental improvement.* Eds: J.R. Wild, S.D. Varfolomeyev & Scozzafava, A. (1997). Kluwer Academic Publishers, NATO ASI Series, 3. High Technology, 19, 39-56
4. Sklyar, V.I., Mosolova, T.P., Kucherenko, I.A., Degtyarova, N.N., Kalyuzhnyi S.V. & Varfolomeyev, S.D. (1999) *Appl.Biochem. Biotechnol.* **80**, 3
5. Kalyuzhnyi & Varfolomeyev, S.D. (1986). *Biotekhnologia* (Russ), **3**, 70-81
6. Kalyuzhnyi, S.V., Spivak, S.I. & Varfolomeyev, S.D. (1986). *Biotekhnologia* (Russ), **5**, 94-102
7. Kalyuzhnyi, S.V., S.I. Varfolomeyev, S.D. & Spivak, S.I. (1989). *Appl.Biochem.Biotechnol.*, **22**, 351-360
8. Kalyuzhnyi, S.V., Gachok, V.P., Skliar, V.I. & Varfolomeyev, S.D. (1001), *Appl.Biochem.Biotechnol.*, **28/29**, 183
9. Alphenaar, P.A., Visser, A. & Lettinga, G. (1993). *Biores.Technol.*, **43**, 249-258
10. Hocks FWJMM, Ten Hoopen HJG, Roels, J.A. & Kuenen, J.G. (1984). *Progr.Ind.Microbiol.*, **20**, 113-119
11. Mulder, A. (1984). *Progr.Ind.Microbiol.*, **20**, 133-143
12. Andrews, J.F. (1969). *J.Sanit.Engin.Div ASCE*, **95-Sal**, 95-110
13. Kalyuzhnyi S. & Fedorovich, V. (1997). *Wat.Sci.Technol.* **36**, 201-208
14. Kalyuzhnyi S. & Fedorovich, V. (1998). *Biores.Technol.* **65**, 227-242
15. Kalyuzhnyi S., Fedorovich, V., Lens, P. Hulshooff, Pol L. & Lettinga, G. (1998). *Biodegradation.* **9**, 187-199

BIOCATALYTIC SYNTHESIS OF ALKALOIDS AND CARBOHYDRATES: AN UPDATE

TOMAS HUDLICKY

Chemistry Department, University of Florida, Gainesville,
FL 32611-7200, USA

1. Introduction

Biocatalysis and its applications to synthesis has risen to prominence in recent years in both academic and industrial domains. Several recent reviews highlight the applications of enzymes to the production of enantiomerically pure synthons for manufacturing of fine chemicals.[1-5] Various lipases are now commonly used in organic synthesis for desymmetrization of meso compounds. On the other hand, oxidoreductases have not been used so extensively in organic synthesis because there are difficulties in handling such enzymes as isolated entities and due to the necessity of recycling cofactors. For this reason, whole-cell fermentation is the method of choice in the application of oxidoreductases to organic synthesis. Both bacterial cells (*E. coli* recombinant organisms) and yeast are now used for such reactions as epoxidation, diol formation, and Baeyer-Villiger oxidation.

This chapter highlights the potential and utility of prokaryotic dioxygenases that catalyze the formation of arene *cis*-dihydrodiols through the disruption of aromaticity, a process sometimes referred to as the "Birch Oxidation,"[6] Scheme 1. The use of arene *cis*-dihydrodiols is now well-established in asymmetric synthesis and the applications in natural product synthesis by our group, and those of Banwell, Johnson, and others are- and continue to be so numerous that further resistance to chemoenzymatic synthesis from "traditionalists" is inconsequential to its further development. (For a discussion of attitudes toward chemoenzymatic chemistry see reference 7). This update focuses on the parallel between the field of alkaloid synthesis and a new area of unnatural oligosaccharide synthesis. These two disciplines are intimately connected through the now well-established dihydroxylation of aromatics.

311

B. Zwanenburg et al. (eds.), Enzymes in Action, 311–321.
© 2000 *Kluwer Academic Publishers. Printed in the Netherlands.*

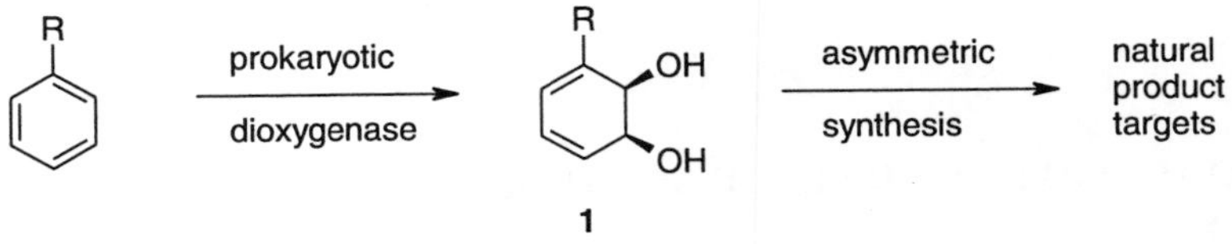

Scheme 1: Biocatalytic Asymmetric Synthesis.

2. Discussion

Thirteen years have passed since we entered into the area of chemoenzymatic synthesis. Our first project was the three-step conversion of toluene, an achiral molecule, to enone **4** via diol **3** whose protection, followed by ozonolysis, gave a ketoaldehyde which was cyclized to the enone (Scheme 2).[8] Such oxidative cleavage and recyclization was later exploited in design for carbohydrate synthesis.[7]

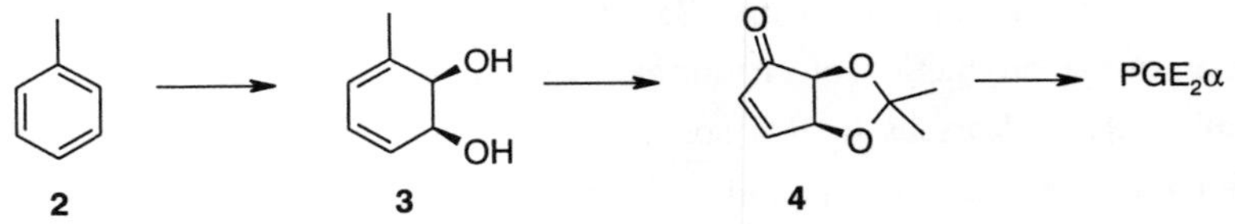

Scheme 2: From Toluene to Prostaglandins.

The enone has been converted to PGE$_2\alpha$ by Johnson.[9] At that time there was no shorter sequence available for the preparation of PGE series of prostaglandins-five operations in all. The complexity of projects grew as more arene cis-diols became available (there are now more than 250 known examples).[5] Over the years many complex products have been synthesized. A few of these are shown in Figure 1, along with the name of principal coworker, the number of steps, and the year of completion. The diversity of structures, most attained from bromobenzene, is formidable. This chapter offers an opportunity for an update in two diverse areas: synthesis of alkaloids by a multigenerational design and investigation of unnatural inositol oligomers.

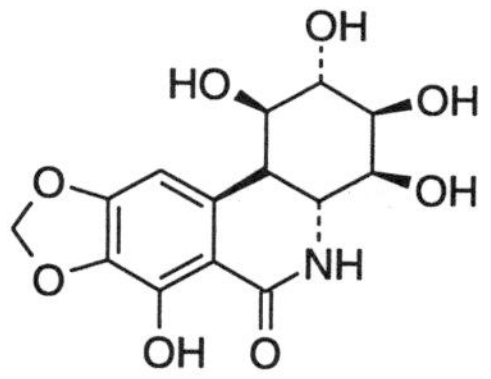

Pancratistatin
Xinrong and Konigsberger, 1995
13 steps

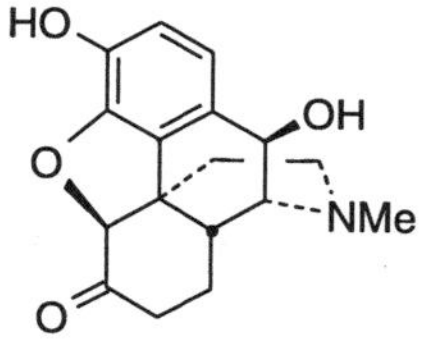

ent-morphinan
Butora, 1996
14steps

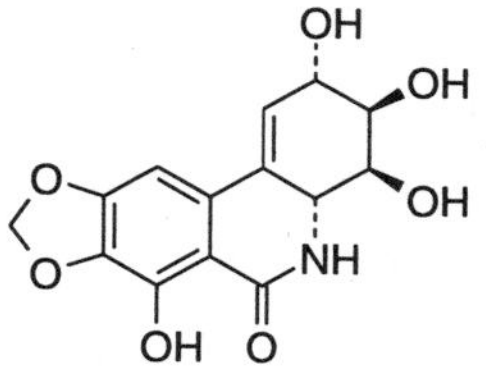

Narciclasine
Gonzalez and Martinot, 1999
8 steps

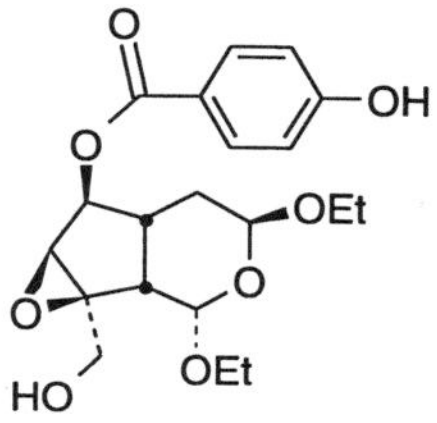

Specionin
Natchus, 1992
13 steps

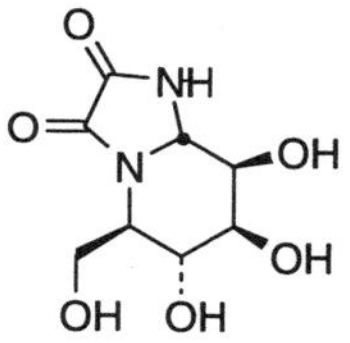

Kifunensine
Rouden, 1993
11 steps

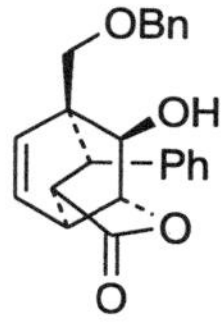

Zeylena
Seoane, 1989
10 steps

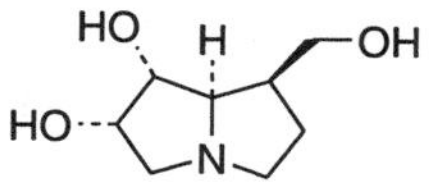

Trihydroxyheliotridane
Seoane, 1990
8 steps

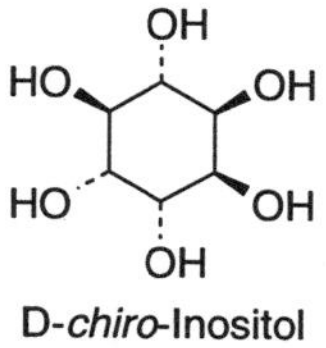

D-chiro-Inositol
Mandel, 1993
3 steps

Figure 1: Examples of Recently Synthesized Natural Products.

2.1.   ALKALOIDS SYNTHESIS.

We have focused on Amaryllidaceae and morphine alkaloids for a simple reason: it is easy to identify those chiral centers that would be directly installed during the enzymatic oxygenation in the target compounds. For the alkaloids of the pancratistatin type, the disconnection leads to an amino inositol that has to be coupled with an aromatic fragment, as indicated in Scheme 3.

These disconnections have been reduced to practice, and with each generation the overall efficiency of the synthesis has improved. For example, in the first generation preparation of pancratistatin many problems were encountered with manipulations of the robust tosylamide and benzamide moieties, at the expense of at least six synthetic operations.[10] The final transamidation did not work well as originally proposed. In the second-generation attempt this issue was addressed by using the aziridine protecting/activating group as the electrophilic center for the future amide.[11] This particular improvement resulted in an eleven-step synthesis of 7-deoxypancratistatin. Further ameliorations and current research led to the design that employed intramolecular opening of aziridine, as shown in Scheme 4.

2.2.   MORPHINE ALKALOIDS.

The diol unit in morphine was envisioned as being generated by biooxidation of a suitable aromatic precursor, whereas the catechol unit in ring A was seen as originating from an overoxidation of a diol by catechol dehydrogenase, the second enzyme in the pathway employed by the wild strain organism.[12] The strategies for morphine synthesis are shown in Scheme 5. In the tandem radical cyclization, low yields and low stereospecificity were observed.[12] This problem was solved (partially) by performing the radical cyclization in two steps. The isoquinoline derivatives were also obtained in low (2:1) selectivity with respect to the C-9 of morphine.[13] The selectivity issue was ultimately solved by cyclization of benzoates in either *cis* or *trans* series and iminium ion closures [14] in which anchimeric assistance of the benzoate groups led to control of the C-9 center, Scheme 6.

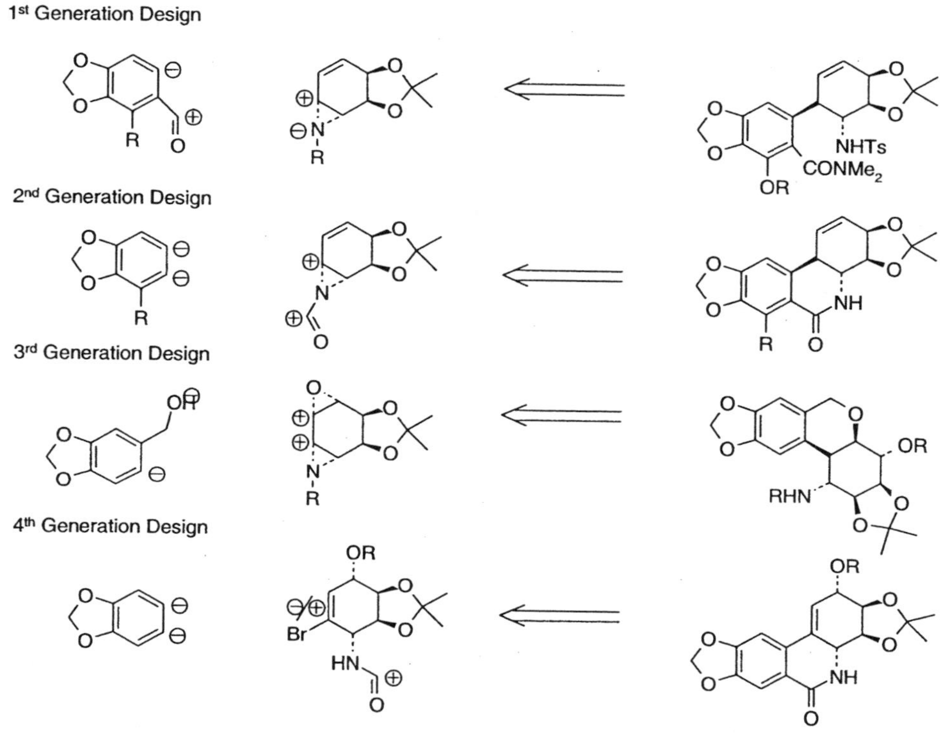
1st Generation Design
R
O
R
NHTs
CONMe2
OR
2nd Generation Design
R
O
NH
R
O
3rd Generation Design
OH
N
R
OR
RHN
4th Generation Design
OR
Br
HN
O
OR
NH
O

Scheme 3: Analysis of Design for Amaryllidaceae Alkaloids.

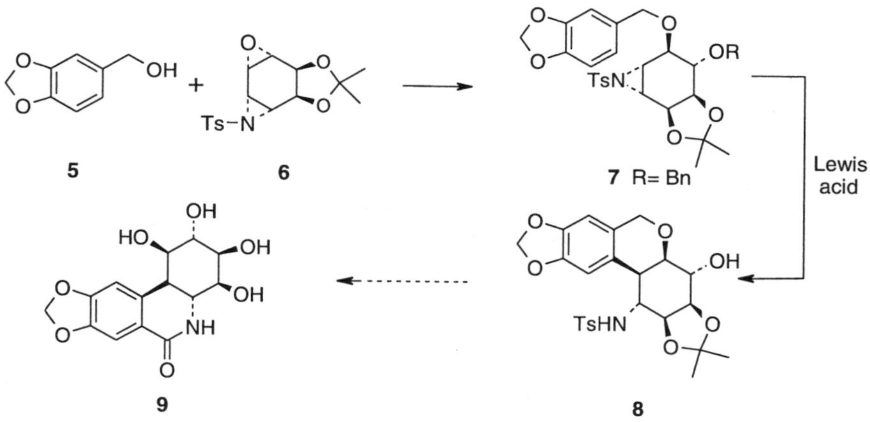
OH
Ts−N
5
6
OR
TsN
7 R= Bn
Lewis
acid
OH
HO
OH
OH
NH
O
9
OH
TsHN
8

Scheme 4: Intramolecular Opening of Aziridine-Approach to 7-deoxypancratistatin.

316

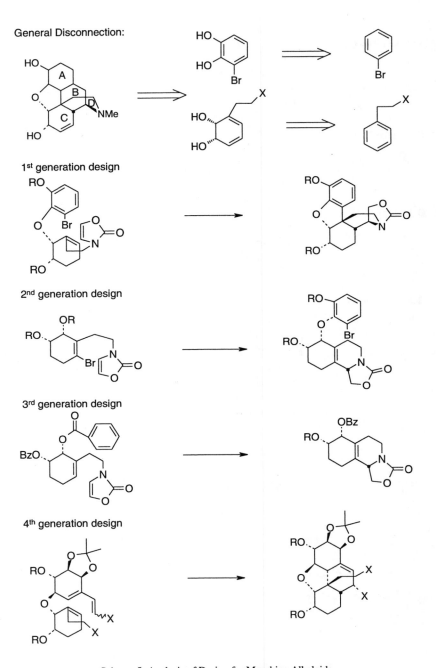

Scheme 5: Analysis of Design for Morphine Alkaloids.

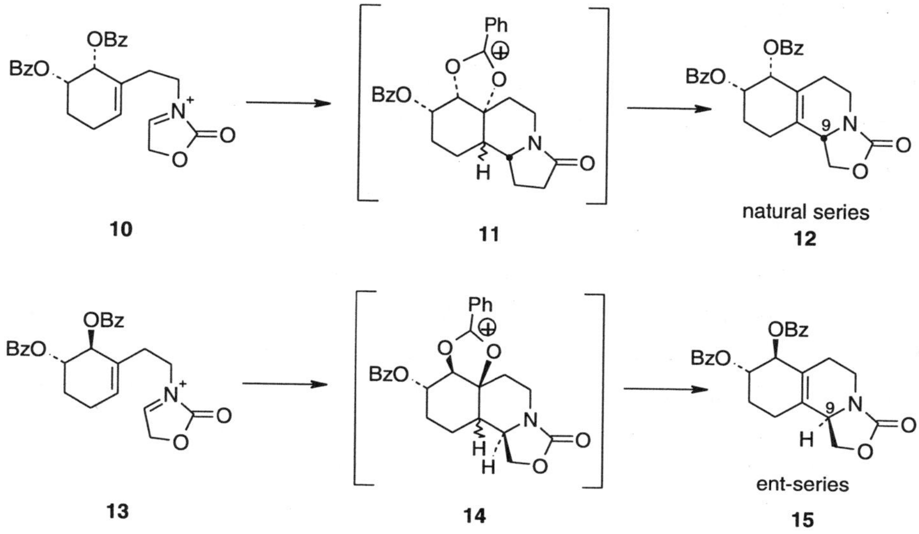

Scheme 6: Control of C-9 Stereochemistry in Morphine Intermediates.

Finally, an intramolecular Diels-Alder [15] approach was considered. Model studies led to complete control of all five chiral centers of morphine, Scheme 7. An unrelated approach employing the chelate-enolate Claisen rearrangement [16] also led to complete control of absolute stereochemistry at C9/C14 (morphine numbering), Scheme 8.

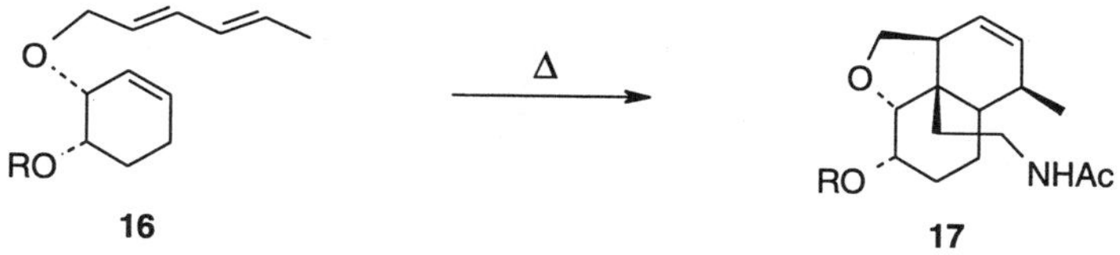

Scheme 7: Intramolecular Diels-Alder Approach to Morphine.

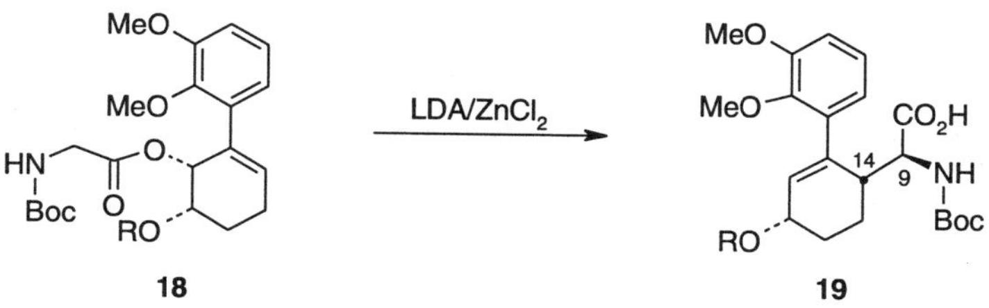

Scheme 8: Chelated-enolate Claisen Rearrangement Approach to Morphine.

318

2.3.   OLIGOSACCHARIDE SYNTHESIS.

The design for asymmetric synthesis of simple carbohydrates has been reduced to practice in our laboratory.[17] All types of monosaccharides have been synthesized from bromobenzene including aminosugars, aza sugars, inositols, amino inositols, deutero sugars, etc, as shown in Scheme 9. The synthetic rationale has been reviewed on several occasions.[5,7,17] Recently, a new class of compounds, oligomers of inositols, has been investigated.[18]   The rationale behind this investigation rests on the predictions of helical structures (β–turn) that are possible with tetrameric and higher 1,2–linked oligomers. To date, up to octameric units have been prepared in the homogeneous series from epoxides **20** and **21**, Scheme 10. The interesting molecular properties of these compounds, such as their assembly into secondary and tertiary structures, chelation of metal ions, and inhibition of gylcosidic enzymes, prompts interesting questions as to the future direction of research in this area.   Quite recently, oligomers derived from naphthalene diol have also been synthesized.[19] This project offers a perfect match of the projected targets with the technology based on dihydroxylation of arenes (it would be impossible to imagine another way to prepare these polyhydoxylated entities). The idea of heterogeneous oligomers, that is those containing alternating monomeric units of any of the nine inositols, would lead to synthesis of new materials containing an alphabet-letter-word system similar to that in peptides but mathematically infinitely more diverse. Prediction of combinatorial possibilities, studies of molecular shapes, and properties of oligomers such as these direct the future endeavours in this area.

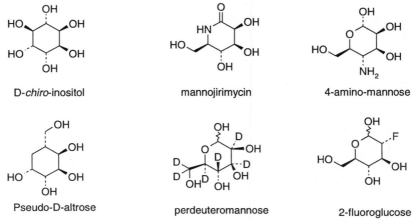

Scheme 9: Monosaccharides Synthesized from Halobenzenes.

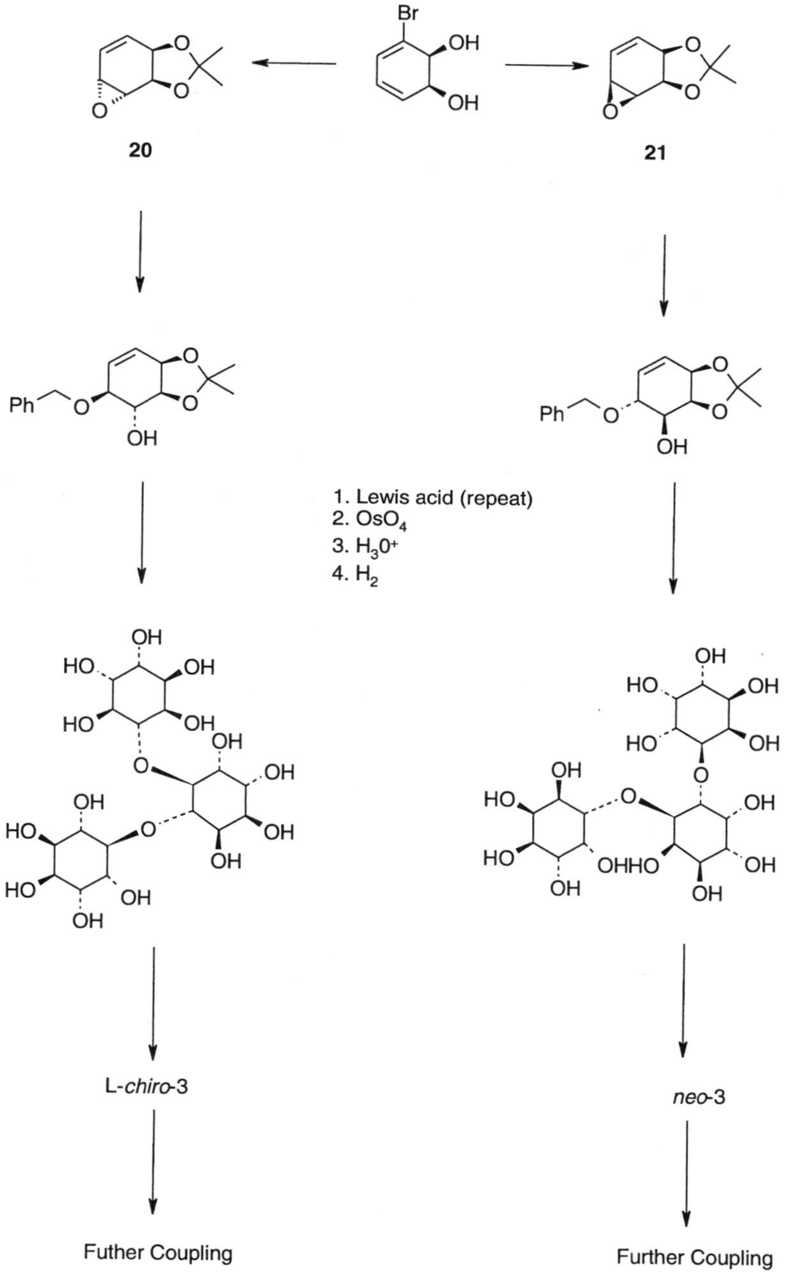
Br
OH
OH
20
21
Ph
O
OH
1. Lewis acid (repeat)
2. OsO4
3. H30+
4. H2
L-chiro-3
neo-3
Futher Coupling
Further Coupling

Scheme 10: Oligoinositols in the L-*chiro*-and *neo*-series

320

3. Acknowledgement.

The author is thankful to the coworkers whose names appear in the citations and without whose effort none of this work would have been possible. Special thanks are extended to David Gibson (U. of Iowa) for the provision of bacterial strains. Funding for the research was generously provided by National Science Foundation, Environmental Protection Agency, Mallinckrodt Specialty Chemicals Inc., TDC Research Inc., and the University of Florida.

4. References.

1. Faber, K. (1997) Biotransformations of non-Natural Compounds: State of the Art and Future Development, *Pure Appl. Chem.* **69**, 1613-1632

2. Roberts, S. M. J. (1998) Preparative Biotransformations: The Employment of Enzymes and Whole-Cells in Synthetic Organic Chemistry, *J. Chem. Soc. Perkin Trans.* **1**, 157-169.

3. Stewart, J. D. (1998) Baker's Yeast Reduction in Asymmetric Synthesis, *Current Opinion in Drug Discovery and Development* **1**, 278-289.

4. Schoffers, E., Golebiowski, A., and Johnson, C. R. (1996) Enantioselective Synthesis Through Enzymatic Asymmetrization, *Tetrahedron* **52**, 3769-3826.

5. Hudlicky, T.; Gonzalez, D.; and Gibson, D. T. (1999) Enzymatic Dihydroxylation of Aromatics in Enantioselective Synthesis: Expanding Asymmetric Methodology. *Aldrichimica Acta* **32**, 35-62.

6. This term was coined by Martin Banwell (Australian National University, Canberra) at the 8th Symposium on the Latest Trends in Organic Synthesis, Gainesville, October, 1998.

7. Hudlicky, T. (1996) Design Constraints in Practical Syntheses of Complex Molecules: Current Status, Case Studies with Carbohydrates and Alkaloids, and Future Perspective. *Chem. Rev.* **96**, 3-30.

8. Hudlicky, T.; Luna, H.; Barbieri, G.; and Kwart, L. D. (1988) Enantioselective Synthesis Through Microbial Oxidation of Arenes. 1. Efficient Preparation of Terpene and Prostanoid Synthons, *J. Am. Chem. Soc.* **110**, 4735-4741.

9. Johnson, C. R.; and Penning, T. D. (1988) Triply Convergent Synthesis of (-)-Prostaglandin-E_2 Methyl Ester, *J. Am. Chem. Soc.* **110**, 4726-4735.

10. Tian, X. R.; Hudlicky, T.; and Konigsberger, K. (1995) First Enantioselective Total Synthesis of (+)-Pancratistatin: An Unusual Set of Problems, *J. Am. Chem. Soc.* **117**, 3643-3644.

11. Hudlicky, T.; Tian, X. R.; Konigsberger, K.; Maurya, R.; Rouden, J.; and Fan, B. (1996) Toluene Dioxygenase-Mediated cis-Dihydroxylation of Aromatics in Enantioselective Synthesis. Asymmetric Total Syntheses of Pancratistatin and 7-Deoxypancratistatin, Promising Antitumor Agents, *J. Am. Chem. Soc.* **118**, 10752-10765.

12. Butora, G.; Hudlicky, T.; Fearnley, S. P.; Gum, A. G.; Stabile, M. R.; and Abboud, K. (1996) Chemoenzymic Synthesis of the Morphine Skeleton via Radical Cyclization and a C10-C11 closure, *Tetrahedron Lett.* **37**, 8155-8158.

13 Butora, G.; Hudlicky, T.; Fearnley, S. P.; Stabile, M. R.; Gum, A. G.; and Gonzalez, D. (1998)

Toward a Practical Synthesis of Morphine. The First Several Generations of a Radical Cyclization Approach, *Synthesis*, 665-681.

14. Bottari P.; Endoma M. A. A.; Hudlicky T.; Ghiviriga I.; and Abboud K. A. (1999) Intramolecular *N*-Acyliminium Ion-olefin Cyclization in the Synthesis of Optically Pure Isoquinoline Derivatives: Control of Stereochemistry and Application to Synthesis of Morphine Alkaloids. *Collect. Czech. Chem. Commun.* **64**, 203-216.

15. Butora G.; Gum A. G.; Hudlicky T.; and Abboud K. A. (1998) Advanced Intramolecular Diels-Alder Study Toward the Synthesis of (-)-Morphine: Structure Correction of a Previously Reported Diels-Alder Product, *Synthesis*, 275-278.

16. Gonzalez, D.; Schapiro, V.; Seoane, G.; Hudlicky, T.; and Abboud, K. (1997) Chemoenzymic Synthesis of Unnatural Amino Acids via Modified Claisen Rearrangement of Glycine Enolates. Approach to Morphine Synthesis, *J. Org. Chem.* **62**, 1194-1195.

17. Hudlicky, T.; Entwistle, D. A.; Pitzer, K. K.; and Thorpe, A. J. (1996) Modern Methods of Monosaccharide Synthesis from non-Carbohydrate Sources, *Chem. Rev.* **96**, 1195-1220.

18. Hudlicky, T.; Abboud, K. A.; Entwistle, D. A.; Fan, R.; Maurya, R.; Thorpe, A. J.; Bolonick, J.; and Myers, B. (1996) Toluene-Dioxygenase-mediated cis-Dihydroxylation of Aromatics in Enantioselective Synthesis. Iterative Glycoconjugate Coupling Strategy and Combinatorial Design for the Synthesis of Oligomers of nor-Saccharides, Inositols, and Pseudosugars with Interesting Molecular Properties, *Synthesis*, 897-911.

19. Desjardins, M.; Lallemand, M.; Freeman, S.; Hudlicky, T.; and Abboud, K. A. (1999) Synthesis and Biological Evaluation of Conduritol and Conduramine Analogs, *J. Chem. Soc., Perkin Trans. 1*, 621-628.

BIOCONVERSION OF PLANT MATERIALS: LIPIDS, PROTEINS AND CARBOHYDRATES

Surface Active Compounds from Renewable Resources

B. AHA,   M. BERGER,   B. JAKOB,   B. HAASE,   J. HERMANN,
G. MACHMÜLLER,   S. MÜLLER,   C. WALDINGER   AND
M.P. SCHNEIDER*

FB 9 - Bergische Universität - GH - Wuppertal
D - 42097 Wuppertal, Germany

1.   Introduction

Agricultural crops such as oil seeds and numerous cereals represent a considerable reservoir of useful and low cost raw materials like lipids (fats and oils, phospholipids etc.), proteins and carbohydrates.

By selective combination of their molecular constituents (i.e. fatty acids, glycerol, oligopeptides, amino acids and saccharides), using both chemical and biocatalytic methods, a variety of surface active biodegradable materials can be prepared (*Figure* 1). [1]

Lipases are well established biocatalysts for the enantio- and regioselective formation and hydrolysis of ester bonds in a wide variety of natural and unnatural substrates. They therefore seemed ideally suited also for the bioconversion of the above mentioned plant materials and the formation of combination products with surface active properties such as partial (mono-)glycerides, *N*-acylated amino acids and protein hydrolysates as well as sugar esters (

Figure 2).

B. Zwanenburg et al. (eds.), Enzymes in Action, 323–345.
© 2000 *Kluwer Academic Publishers. Printed in the Netherlands.*

324

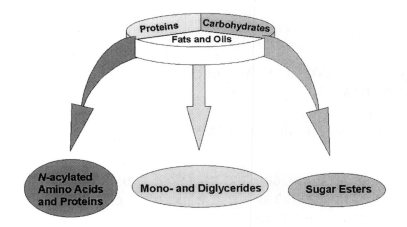

Figure 1. Surface active compounds from plant materials

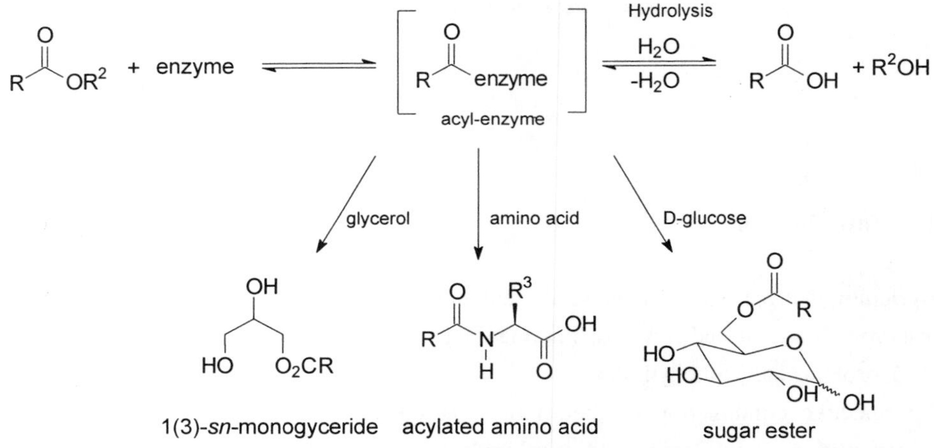

Figure 2. Surface active compounds *via* enzymatic acyl transfer

2. Partial glycerides

2.1. INTRODUCTION

Mixtures of long-chain monoglycerides and other partial glycerides of varying composition and constitution are widely employed as non-ionic surfactants (emulsifiers)

in the processing of foods and related applications [2], [3]. They also serve as starting materials for the preparation of numerous conjugates, e.g. with citric-, lactic-, tartaric- and acetic acid [4].

Isomeric mixtures of such "monoglycerides" are classically produced by alcoholysis of the corresponding triacylglycerols with two equivalents of glycerol in the presence of metal catalysts at temperatures of 210 - 240 °C [5], [6] (*Figure 3*).

The resulting mixtures contain only ca. 50 - 60 % of the desired monoglycerides as isomeric mixtures together with isomeric diglycerides, triglycerides and free fatty acids.

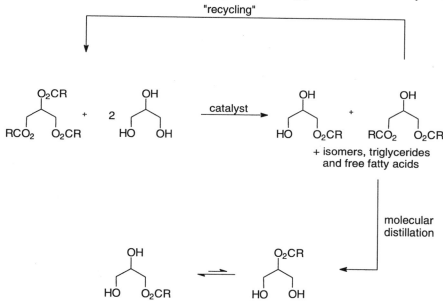

Figure 3. Technical synthesis of "monoglycerides"

Due to the high temperatures employed, these materials are usually coloured and not free of odour. "Monoglycerides" of higher chemical (not isomeric!) purity (> 90%) can only be obtained by cost- and energy intensive molecular distillation of these crude mixtures.

2.2. BIOTECHNOLOGICAL ROUTES TO MONO- AND DIGLYCERIDES

Enzymes catalyse a wide variety of organic reactions under mild conditions and frequently in a highly chemo- , regio- and stereoselective manner. This is particularly true for numerous lipases which are available from a wide variety of commercial sources. Consequently, the use of biocatalysts for the preparation of partial glycerides could provide numerous advantages as compared to the above described, conventional methods such as increased selectivity, higher product purity and quality, energy conservation and the avoidance of toxic catalysts (*Figure* 4).

- **starting materials: natural fats and oils**
- **natural biocatalysts: immobilized and recycled**
- **non-toxic solvents: recycled**
- **reaction conditions: mild, neutral pH**
- **energy consumption: low, room temperature or slightly above**
- **selectivities: high**
- **products: natural**
- **properties: pure isomers, colourless, odourless**

Figure 4. Biotechnological routes to regioisomerically pure mono- and diglycerides – Advantages

It is not surprising therefore, that a number of research groups became recently engaged in the synthesis of monoglycerides by lipase catalyzed reactions [7], [8], [9], [10]. Two general routes to the desired molecules are available (
Figure 5), namely, in principle hydrolysis/alcoholysis of triglycerides and esterification of glycerol.

For stoichiometric reasons it is obvious that only *alcoholysis of triglycerides with glycerol* or the *direct esterification of glycerol* can produce the desired molecules in high yields.

Thus under carefully controlled conditions and at very low temperatures the glycerolysis of triglycerides was reported to lead to high yields of these molecules – again as mixtures of regioisomers [11], [12]. Rapid acyl migrations, prevailing under aqueous or protic conditions, have prevented the enzymatic preparation of isomerically pure 1(3)-*sn*-monoacylglycerols.

Recently, we found that 1(3)-*sn*-monoacylglycerols are quite stable towards acyl group migrations in aprotic solvents with low water content (< 2 %) [13], [14]. Based on this observation that the synthesis of isomerically pure 1(3)-*sn*-monoacylglycerols is conceivable by direct enzyme catalyzed esterifications in such solvents.

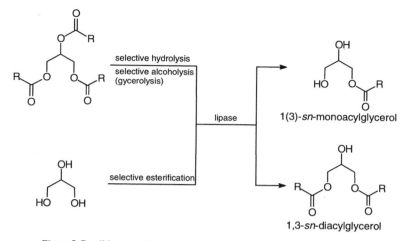

Figure 5. Possible enzymatic routes to isomerically pure mono- and diglycerides

Unfortunately, however, glycerol is immiscible with non polar organic solvents and all earlier attempts for its direct enzymatic esterification in these media have been unsuccessful (*Figure* 6).

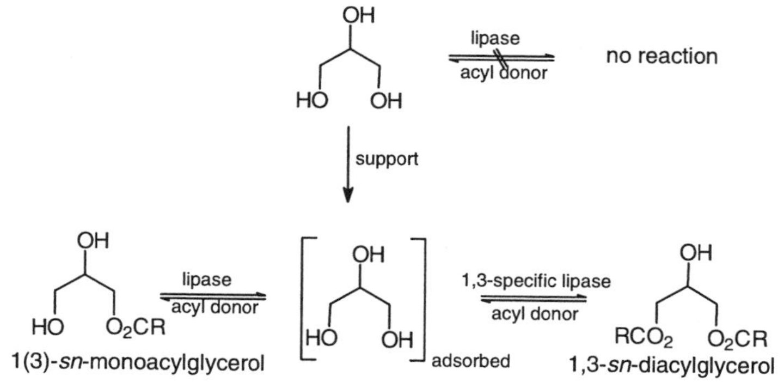

Figure 6. Enzymatic esterification of glycerol in organic media

This basic problem can easily be overcome by prior adsorption of glycerol onto a solid support. Presumably this process proceeds at an artificial liquid-liquid interphase on the support surface. Such conditions which generally somewhat similar to those of lipase catalysed transformations of glycerides, e.g. natural fats and oils.

Typically, glycerol and the corresponding inert solid support (silica gel, active carbon, etc) are mechanically mixed until free flowing "dry" powders are obtained which contain up to 50% of glycerol (w/w). This composite behaves completely different from glycerol itself. Suspensions of these powders in nonpolar organic solvents (e.g. *n*-hexane or *t*-

328

BuOMe) react as if the glycerol was homogeneously dissolved in these media. They are treated with the corresponding acyl donor (free fatty acid, fatty acid methyl ester, fatty acid vinyl ester, triglycerides or natural fats and oils) and a variety of lipases [15], [16]. The acyl transfer reactions proceed smoothly under these conditions and their progress can be monitored conveniently by TLC.

After completion of the esterification, both the enzyme and the solid support can be removed by simple filtration or centrifugation. After evaporation of the solvent the resulting glycerides were isolated in high yields and purified further, preferably by recrystallisation.

2.3. SYNTHESIS OF REGIOISOMERICALLY PURE 1,3-SN-DIGLYCERIDES

Using a molar ratio of glycerol and acyl donor of 1 : 2 a series isomerically pure 1,3-sn-diacylglycerols was prepared in very high yields and purities. Examples of such preparations are given in *Figure* 7 (using fatty acids), *Figure* 8 (using fatty acid methyl esters) and *Figure* 9 (using fatty acid vinyl esters).

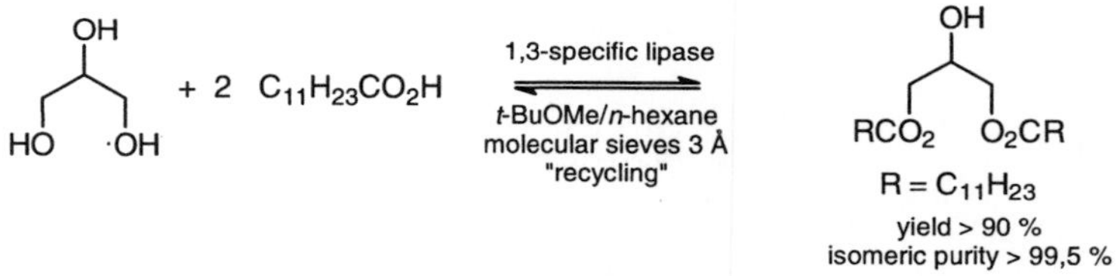

Figure 7. 1,3-Diglycerides from fatty acids

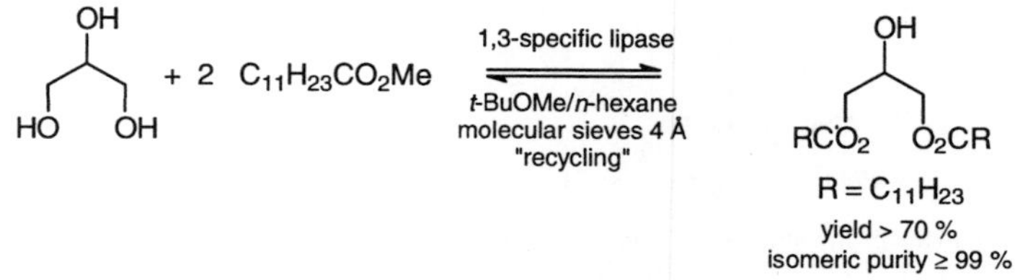

Figure 8. 1,3-Diglycerides from fatty acid methyl esters

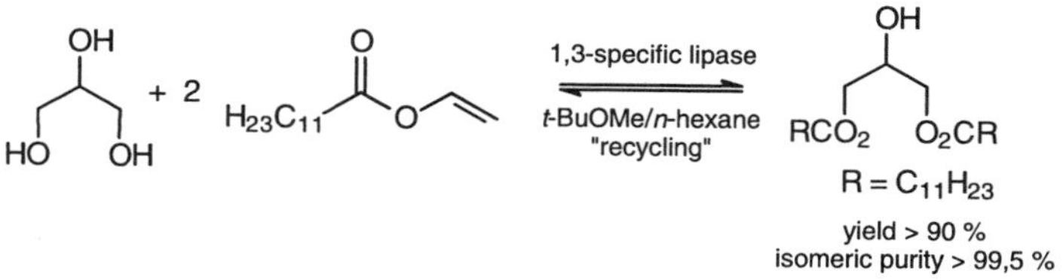

Figure 9. 1,3-Diglycerides from fatty acid vinyl esters

The data shown in Table 1 reveal that the method is well suited for the preparation of isomerically pure 1,3-*sn*-diacylglycerols with fatty acid residues having variable chain length and degree of unsaturation.

The purification process by recrystallisation is best suited for the preparation of saturated 1,3-*sn*-diacylglycerols of moderate (C_8) to long (C_{18}) chain length.

TABLE 1. Regioisomerically pure 1,3-*sn*-diglycerides

product	enzyme	yield [%]	purity [%]
1,3-*sn*-diacetin	A	75	94
1,3-*sn*-divalerin	A	75	93
1,3-*sn*-dicaproin	A	77	93
1,3-*sn*-dicaprylin	B	86	98
1,3-*sn*-dicaprin	B	84	99
1,3-*sn*-didec-9-enoin	B	84	99
1,3-*sn*-didecanoin	B	84	98
1,3-*sn*-didec-10-enoin	B	85	99
1,3-*sn*-dilaurin	B	85	>99
1,3-*sn*-dilaurin/myristin	B	72	99
1,3-*sn*-ditridecanoin	B	83	99
1,3-*sn*-dimyristin	B	82	98
1,3-*sn*-dipentadecanoin	B	84	>99
1,3-*sn*-dipalmitin	B	80	99
1,3-*sn*-distearin	B	81	99
1,3-*sn*-diolein	C	70	98
1,3-*sn*-dilinolein	E	70	97
1,3-*sn*-dierucain	D	71	96
1,3-*sn*-di-(*S*)-corioloin	E	65	98
1,3-*sn*-di-(*R*)-ricinolin	E	61	97
1,3-*sn*-di-(12-(*R*)-hydroxy)stearin	E	52	96

A = lipase from *Rhizopus delemar*
B = lipase from *Mucor miehei*
C = lipase from *Chromobacterium viscosum*
D = lipase from *Penicillium roquefortii*
E = lipase from *Rhizopus niveus*

2.4. ENZYME CATALYZED SYNTHESIS OF ISOMERICALLY PURE 1(3)-MONOGLYCERIDES

In principle, it should be possible to extend the above described method also to the production of regioisomerically pure 1(3)-*sn*-monoacylglycerols. Unfortunately, however, under these conditions these products were only produced in low to moderate yields. In view of the high regioselectivities displayed by many lipases for the primary positions of glycerides it should be possible to improve their yield by simply employing an excess of glycerol. As documented in *Figure* 10 an increase to yields of 60 - 70 % was indeed achieved in this way.

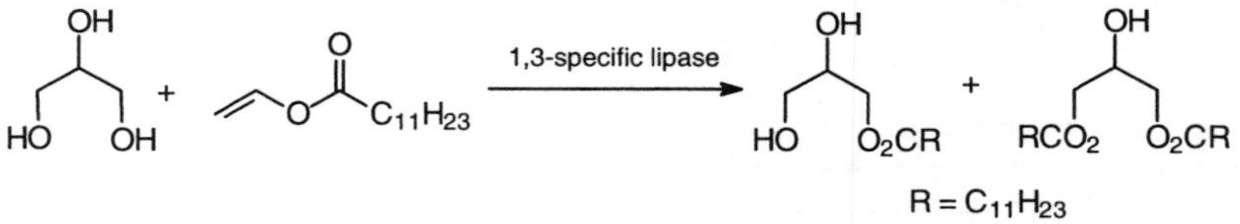

ratio			conversion [%]	(yield[%])	conversion [%]
1	:	1	30	(12)	70
2	:	1	45	(24)	55
5	:	1	70	(50)	30
10	:	1	65	(45)	35

$R = C_{11}H_{23}$

Figure 10. Approaches to 1(3)-*sn*-monoglycerides

However, even with a ten-fold excess of glycerol the content of 1(3)-*sn*-monoacylglycerols in the reaction mixture did not exceed 70 %. Apparently, the primarily produced monoacylglycerols are excellent substrates for the lipase catalysts and are rapidly converted into the corresponding 1,3-*sn*-diacylglycerols.

Another disadvantage is that no triacylglycerols, i.e. natural fats and oils can be employed as acyl donors because then mixtures of regioisomeric monoglycerides will be produced.

Clearly, a practical and useful method for the preparation of regioisomerically pure 1(3)-*sn*-monoacylglycerols had to fulfil the requirements outlined in *Figure* 11.

- **stoichiometric ratios of reactants**
- **fatty acid as acyldonors**
- **triglycerides as acyldonors**
- **natural fats and oils as acyldonors**
- **no or non-toxic organic solvents**
- **preparation in useful quantities**
- **high yields**
- **no purification steps**
- **potential for continuous process**

Figure 11. Isomerically pure 1(3)-*sn*-monoacylglycerols – process requirements

In order to achieve quantitative conversions of all employed reactants without the need of excess glycerol two major problems had to be solved, namely, a simple method for the continuous separation of the desired products from the reaction mixtures, and simultaneously an effective recycling method for all undesirable materials, including 2-monoacyl-glycerols, diacylglycerols and triacylglycerols.

Both goals were achieved simply as outlined in *Figure* 12 by a compartmental separation of the two steps of the process – synthesis and isolation. The enzymatic esterification is carried out in the reactor vessel with stoichiometric amounts of glycerol and the corresponding acyl donor under the desired reaction conditions. The thus obtained reaction mixture is circulated into a second vessel – termed separator – in which the desired monoacylglycerols are separated at lower temperatures.

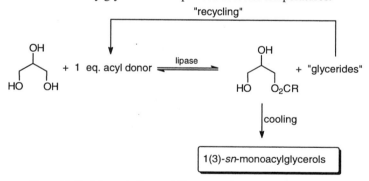

Figure 12. Regioisomerically pure 1(3)-*sn*-monoglycerides: synthetic process

A schematic representation of the employed equipment is shown in *Figure* 13.

Crucial for the success of this method is the choice of the solvent or solvent mixture in which the desired, more polar 1(3)-*sn*-monoacylglycerols are less soluble than all of the other products, at least at the chosen low temperatures. While the desired products are

332

thus precipitating out in the cooled separator, all other products as well as unreacted acyl donors, remain in solution and are fed back to the reactor, which contains both the enzyme and the glycerol on the solid support. High yields of monoglycerides are obtained this way using a variety of acyl donors.

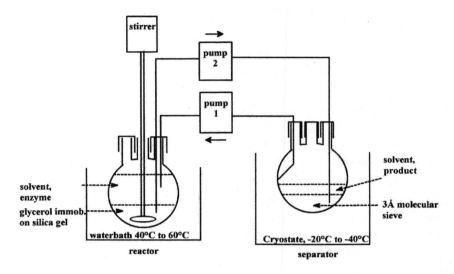

Figure 13. Synthesis of monoglycerides – the equipment

The success of the method was demonstrated with a series of acyl donors, such as free fatty acids, fatty acid methyl esters and vinyl esters (*Figure* 14 - *Figure* 16). Also synthetic triglycerides and natural fats and oils can be utilized (see below).

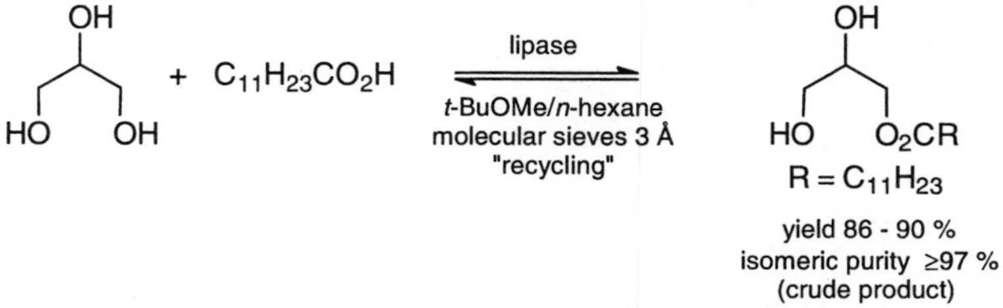

Figure 14. Regioisomerically pure 1(3)-*sn*-monoglycerides by lipase catalysed esterification of glycerol

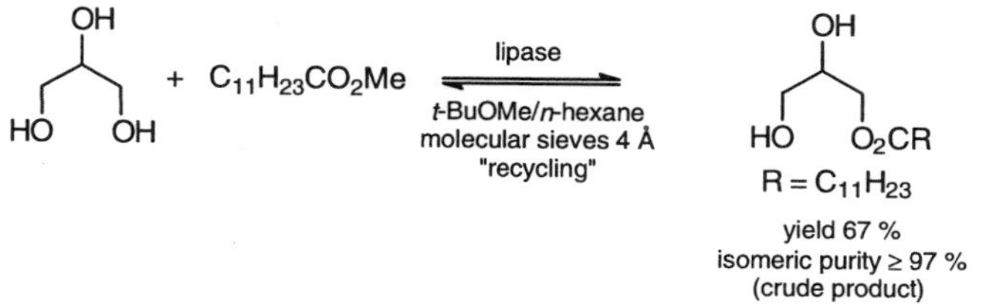

333

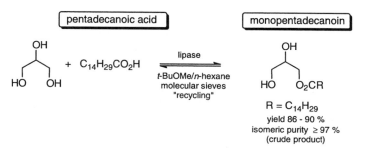

Figure 15. Regioisomerically pure 1(3)-*sn*-monoglycerides by lipase catalysed transesterification of glycerol

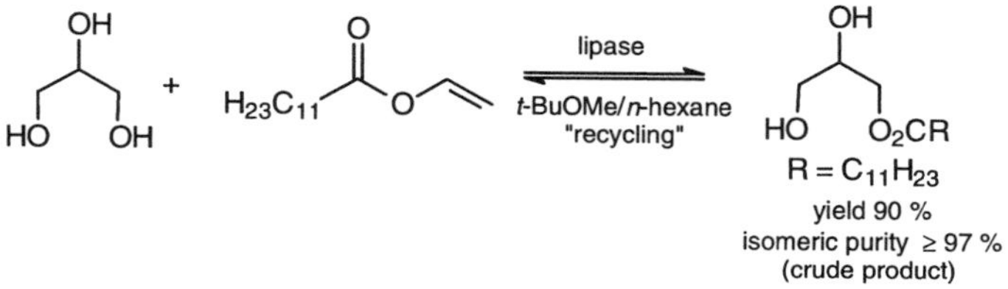

Figure 16. Regioisomerically pure 1(3)-*sn*-monoglycerides by lipase catalysed irreversible transesterification of glycerol

The method is not limited to natural fatty acids, also unnatural fatty acids can be employed. In *Figure* 17 the synthesis of monopentadecanoin, a product for cosmetic applications is shown.

Figure 17. Monopentadecanoin

Important from a practical point of view is that unpurified technical fatty acid mixtures can be used directly (*Figure* 18), leading to the corresponding monoglycerides in high yields and excellent chemical purities.

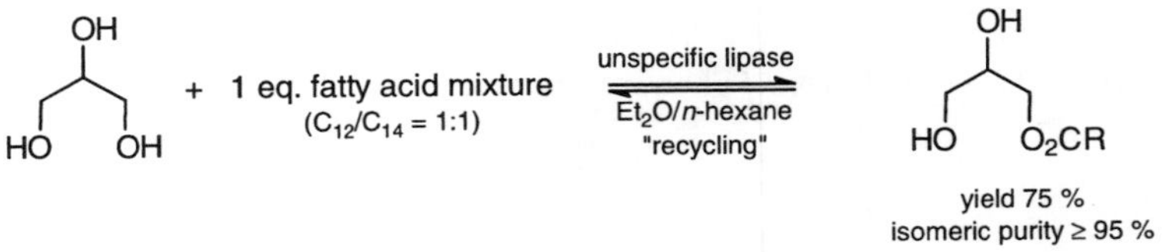

Figure 18. Monoglycerides from technical fatty acid mixtures

It is interesting to note that for the conversion of triglycerides (*Figure* 19) and natural fats and oils into regioisomerically pure 1(3)-monoglycerides, unspecific lipases are best suited. Obviously all acyl groups should be used by the employed biocatalysts, also those in the *sn*-2-positions of triglycerides.

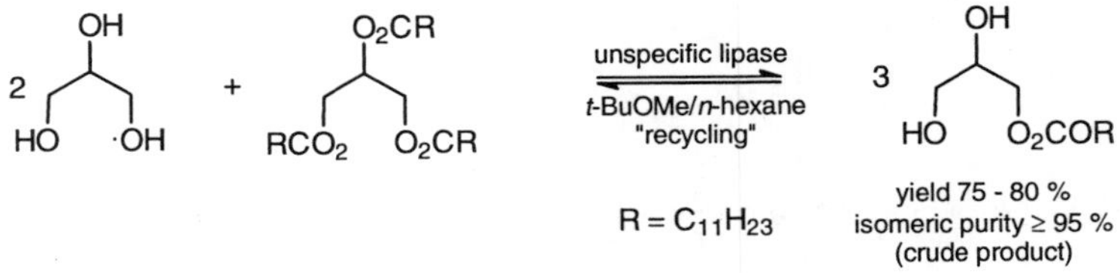

Figure 19. Monoglycerides by glycerolysis of triglycerides

The conversion of natural coconut oil or palm kernel oil leads to mixtures of the corresponding monoglycerides in high yields and purities without the need for further purification steps (*Figure* 20). The natural distribution of fatty acids, shown as triglycerides in the GC-diagram of the starting oil (*Figure* 21 a) is converted exclusively into the corresponding mixture of isomerically pure 1(3)-*sn*-monoacylglycerols (*Figure* 21 b) with only traces of the corresponding isomers or diglycerides.

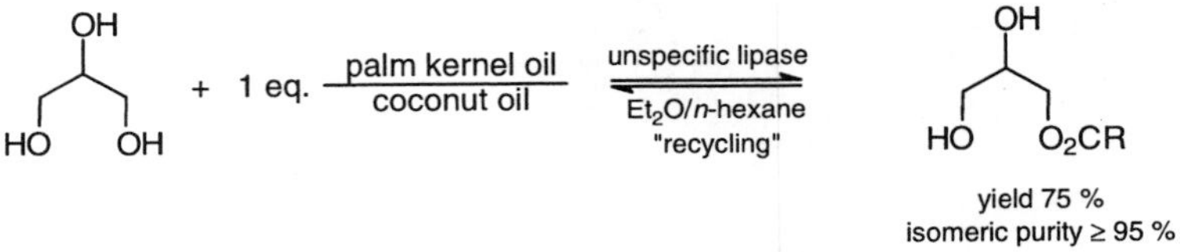

Figure 20. Monoglycerides by glycerolysis of native oils

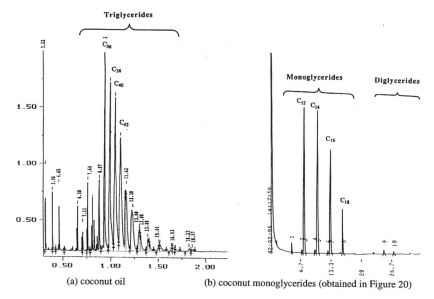

Figure 21. GC-diagrams of coconut oil and coconut monoglycerides.

A selection of the thus prepared 1(3)-*sn*-monoacylglycerols are collected in Table 2.

336

TABLE 2. Regioisomerically pure 1(3)-*sn*-monoglycerides

product	enzyme	yield [%]	purity (%)
1(3)-*sn*-monovalerin	A	57	93
1(3)-*sn*-(5′-brom)-monovalerin	A	52	95
1(3)-*sn*-(6′-brom)-monocaproin	A	54	95
1(3)-*sn*-monoaprylin	B	60	91
1(3)-*sn*-monocaprin	C	90	97
1(3)-*sn*-monodec-9-enoin	C	90	96
1(3)-*sn*-monoundecanoin	C	89	96
1(3)-*sn*-monoundec-10-enoin	C	88	96
1(3)-*sn*-mono-(10′-oxo)-undecanoin	B	60	91
1(3)-*sn*-monolaurin	C	90	98
1(3)-*sn*-mono-(12′-hydroxy)-dodec-9-enoin	B	42	95
1(3)-*sn*-monotridecanoin	C	91	97
1(3)-*sn*-monomyristin	C	91	96
1(3)-*sn*-mono-(13′-oxo)-tetradecanoin	B	52	92
1(3)-*sn*-monopentadecanoin	C	91	96
1(3)-*sn*-monopalmitin	C	90	95
1(3)-*sn*-monostearin	C	88	96
1(3)-*sn*-monoolein	D	65	99
1(3)-*sn*-monolinolein	F	79	96
1(3)-*sn*-monoerucain	F	71	98
1(3)-*sn*-mono-(*S*)-corioloin	E	70	95
1(3)-*sn*-mono-(*R*)-ricinolin	E	66	96
1(3)-*sn*-mono-(12-(*R*)-hydroxy)-stearin	E	56	98

A = lipase from *Rhizopus delemar*
B = lipase from *Mucor miehei* – Lipozym
C = lipase from *Pseudomonas sp.*
D = lipase from *Chromobacterium viscosum*
E = lipase from *Penicillium roquefortii*
F = lipase from *Rhizopus niveus*

2.5. MONOGLYCERIDE SULFATES

The thus prepared, isomerically pure monoglycerides are excellent precursors for monoglyceride sulfates, another class of skin friendly surfactants.

Using SO_3/pyridine – the usual laboratory method – a complete series of these products were prepared in high chemical purities (

Figure 22).

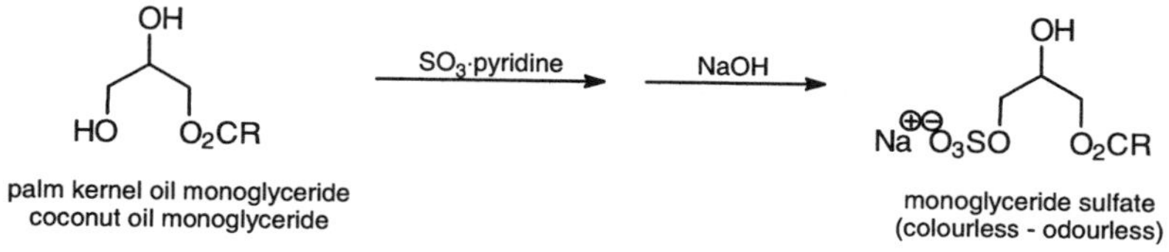

Figure 22. Monoglyceride sulfates from monoglycerides

2.6. CONCLUSION

The above described procedures for the synthesis of regioisomerically pure mono- and diglycerides provides ready access to these materials in good yields. Using stoichiometric quantitities of starting materials products of high chemical and isomeric purity are obtained without the need of additional purification steps. All products are completely colourless and odourless. They are therefore ideally suited for applications in food technology, cosmetic formulations and for pharmaceutical applications. From a regulatory point of view, it is important to mention, that the mono- and diglycerides obtained from natural fatty acids or fats and oils *via* the action of natural biocatalysts in non-toxic solvents could be termed "natural" in every respect. Approval of government bodies for the use of these materials should therefore pose less problems.

3. *N*-Acylated Amino Acids and Oligopeptides

3.1. INTRODUCTION

Numerous plant proteins and their enzymatic (protease) catalyzed hydrolyses provide access to mixtures of oligopeptides and amino acids which can be converted by chemical or enzymatic acylation into anionic or amphoteric surfactants which are mild and skin friendly materials for applications in the cosmetic field [17] (*Figure* 23).

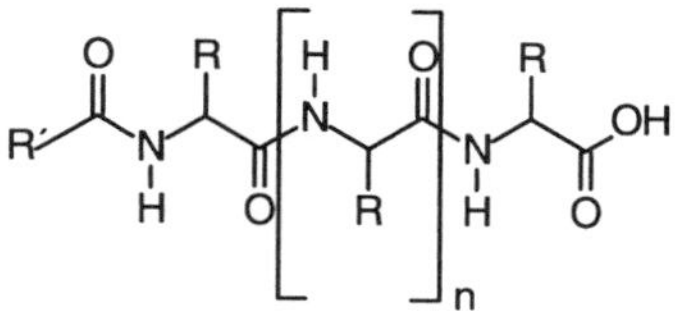

Figure 23. Acylated protein hydrolysate

338

In order to develop a reliable structure-function relationship regarding the surfactant properties of these materials we decided to synthesize a complete series of acylated aminoacids derived from all twenty native amino acids found in plant materials and using lauric acid (representative for the so-called lauric range oils) and oleic acid (main constituent of native oils such as rape seed and sunflower) as fatty acid components.

Although we are actively engaged in developing enzymatic routes towards N-acylated amino acids (*Figure* 24) at present no general method for this purpose is available. Therefore, we had to resort for the moment to purely chemical routes for their synthesis which are summarized in *Figure* 25.

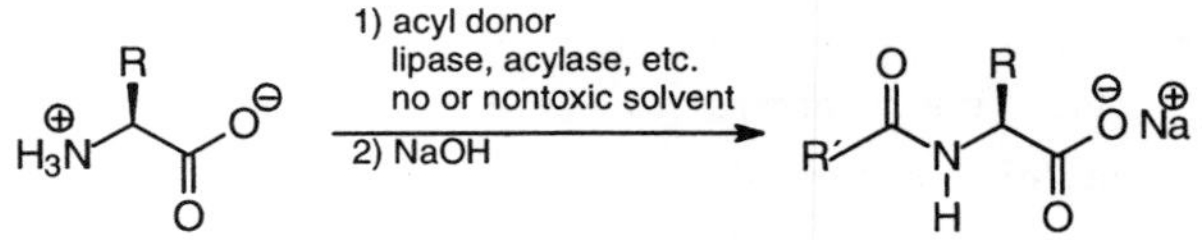

Figure 24. N-acylation of amino acids using biocatalysts - ideal reaction conditions

Using three methods, namely Schotten-Baumann and Einhorn reactions as well as the silylation method according to Kricheldorf, a complete range of these materials was prepared and characterised. Although the classical Schotten-Baumann-reaction is clearly the simplest method it did not provide the required chemically pure reference compounds which we needed for the determination of the surfactant parameters. For this reason the two alternative methods were mainly employed leading to the desired compounds in sufficient chemical purities.

The results revealed that oleic acid indeed can be used as hydrophobic building block for these surfactants, the CMCs are clearly much lower as compared to the corresponding lauric acid derivatives. While the foaming properties are good, the wetting properties are clearly less expressed (Table 3). The results suggest that the N-acyl amino acids can be made available for application in personal care products such as shampoos but also for hand washing liquids etc.

TABLE 3. N-Lauroyl and N-oleoyl amino acid salts – surface active properties.

N-lauroyl amino acids	N-oleoyl amino acids
high CMC	low CMC
good foaming properties	good foaming properties
good wetting properties	poor wetting properties

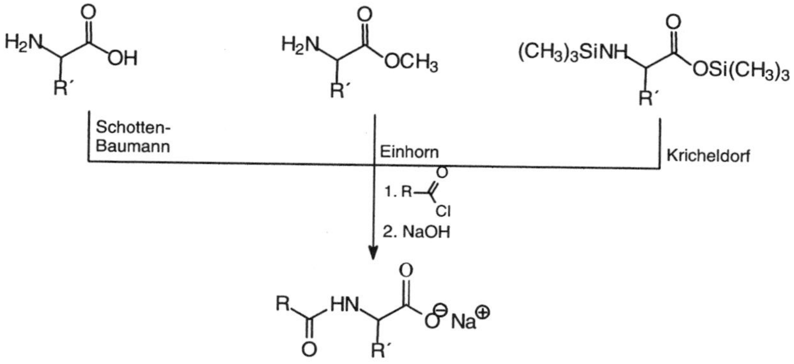

Figure 25. Chemical acylation of amino acids

It has been demonstrated that *N*-acylated amino acids are highly suited for the bioremediation of oil contaminated soils and probably also for the treatment of large scale oil spills [18].

Currently, we are extending the above experiments towards applications with enzymatically prepared protein hydrolysates from plant sources.

4. Sugar Esters

4.1. INTRODUCTION

Sugar esters of long chain fatty acids constitute an important class of biodegradable, non-ionic surfactants with applications as food emulsifiers and additives to detergent formulations. They are also of potential technical interest e.g. as detergents for metal surface cleaning processes. Selectively esterified monosaccharides are of interest to organic chemists as regioisomerically pure building blocks for synthetic applications and for biochemists as detergents for the solubilisation of membrane proteins [19].

The target molecules shown in *Figure* 26 were selected for the first series of experiments.

340

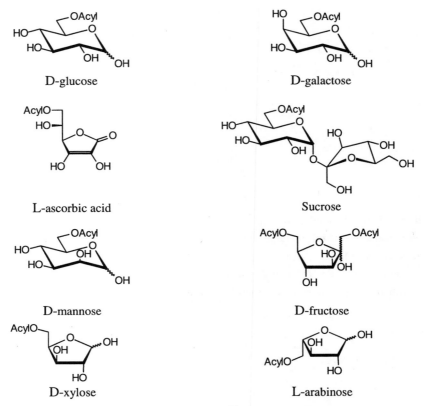

Figure 26. Sugar esters by lipase catalysed esterification of native saccharides

4.2. SYNTHESIS OF SUGAR ESTERS – CHEMICAL AND ENZYMATIC ROUTES

Due to the multifunctionality of native, i.e. unprotected sugars chemical esterifications are leading to ill characterised mixtures of regioisomers with variable degrees of esterification (*Figure* 27).

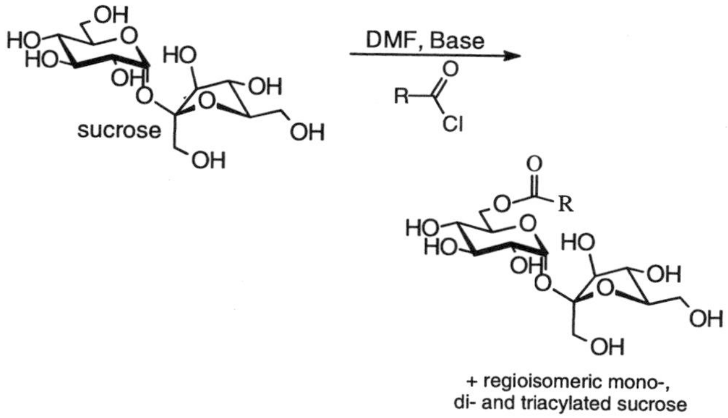

Figure 27. Chemical synthesis of sugar esters – sucrose

In contrast, the corresponding enzyme (protease, lipase) catalyzed transformations are often highly regioselective. Unfortunately, common to both chemical and enzymatic approaches, native sugars are usually only soluble in highly polar solvents like pyridine, DMSO, DMF or DMA [20] which are difficult to remove and – due to their toxicities – of low acceptance, especially in food related applications.

4.3.   THE STATE OF THE ART

Both problems - regioselectivity and insolubility - can be compromised by enzymatic esterifications using highly regioselective lipases (e.g. *Mucor miehei, Pseudomonas sp.* or *Candida antarctica*) [21] or temporary or permanent protection groups such as acetals [22], simple glycosides (*Figure* 28) [19] or solubilizing agents (e.g. phenylboronic acid, *Figure* 29) [23].

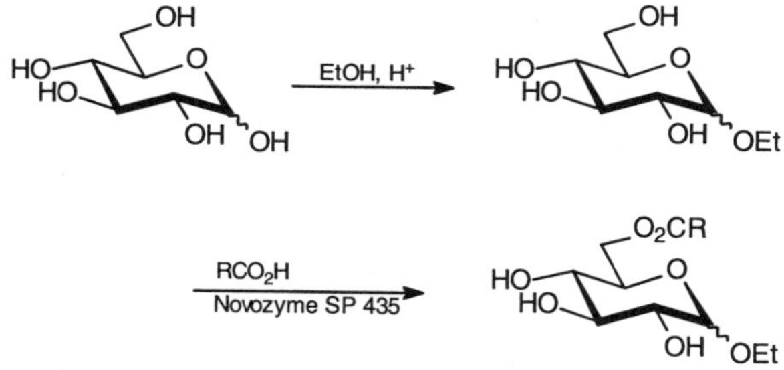

Figure 28. Acylation of D-glucose *via* formation of ethyl glucosides

In all of the above cases chemical derivatisations or the use of other chemical auxiliaries are required. Ideally, lipase catalysed esterifications of sugars should be carried out in non-toxic, aprotic organic solvents of low boiling points and the substrate sugars should be employed in their native form, i.e. without derivatisations or the addition of "solubilising agents".

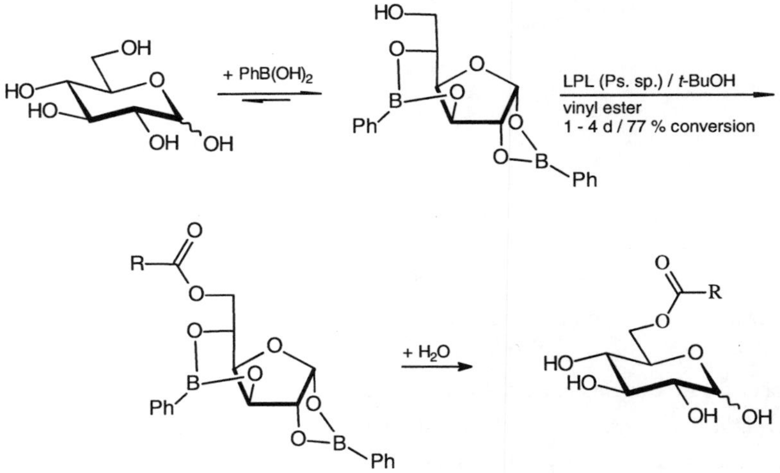

Figure 29. Acylation of insoluble sugars *via* solubilization using phenylboronic acid

4.4.   ENZYMATIC ESTERIFICATION OF NATIVE MONOSACCHARIDES

According to a literature report the native pentoses 2-deoxi-D-ribose and D-ribose - both soluble in THF - were esterified regioselectively in presence of the lipase from *Candida antarctica* B using propionic anhydride as acyl donor leading exclusively to the corresponding 5-*O*-acyl derivatives in their furanoid form with high yields (> 95 %) [24] (*Figure* 30).

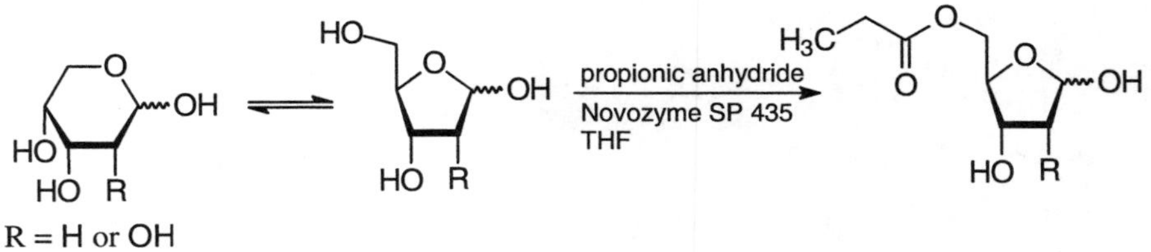

R = H or OH

Figure 30. Regioselective acylation of 2-deoxi-D-ribose and D-ribose.

Although native sugars such as D-glucose, D-galactose and D-mannose are practically insoluble in THF (glucose: ca. 20 mg/L, mannose: ca. 53 mg/L, galactose: ca. 3 mg/L) we were surprised to find that these sugars can indeed be acylated with high regioselectivity under similar conditions. The apparently insoluble sugars are converted into their esters in suspension. Under these conditions all three sugars were converted rapidly, regioselectively and with high yields (>85 %) into the corresponding 6-O-acyl derivatives [25], [26], [27] (*Figure* 31).

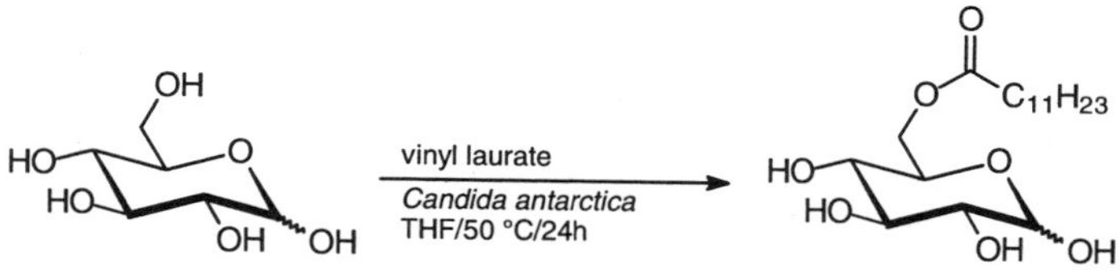

Figure 31. Regioselective acylation of D-Glucose

The method is not limited to specific acyl donors. In fact the whole spectrum of naturally occuring fatty acids was employed both as vinyl and methyl esters.

The rate of esterification proved to be strongly dependent on the employed solvents, whereby the highest (apparent) activities were observed in dioxane, monoglyme, diglyme, THF, dioxolane, sulfolane and *t*-BuOH (in decreasing order). Using the above described method, the sugar esters in *Figure* 26 were prepared.

Recently we discovered that this method can be employed successfully for a variety of structurally different monosaccharides (including amino sugars) and also anhydrosugars and disaccharides [27]. Further experiments along these lines are in progress. The interest in this topic is also apparent from activities in several other research groups [21], [28].

5.   Concluding Remark

The "combination" products, shown in *Figure* 1 and *Figure* 2, resulting from the bioconversion of plant materials display a wide range of useful surface active properties [29], [30].

344

6. References

1. We thank the *German Department of Agriculture* (BML) for financial support of this work.
2. Sonntag, N.O.V. (1982) *in Bailey's Industrial Oils and Fat Products*, Vol. 2, 4th edn., edited by Swern D., John Wiley and Sons, New York, 134.
3. Porter, M. R. (1988) *Handbook of Surfactants*, Blackie and Sons, Glasgow.
4. Aebi H., Baumgartner E., Fiedler H.P., Ohloff G. (1978) *Kosmetika, Riechstoffe und Lebensmittelzusatzstoffe*, G. Thieme Verlag, Stuttgart.
5. Sonntag, N.O.V. (1982) *J. Am. Oil Chem. Soc.* **59**, 795A.
6. Lauridsen, J. B. (1976) *J. Am. Oil Chem. Soc.* **53**, 400.
7. Hayes, D. G. and Gulari, E. (1991) *Biotechnol. Bioeng.* **38**, 507.
8. Weis A. (1990) *Fat. Sci. Technol.* **92**, 392.
9. Holmberg, K and Osterberg (1988) *J. Am Oil Chem. Soc.* **65**, 1544.
10. European Patent EP 0.191.217.
11. McNeil, G.P., Shimizu, S., Yamane, T. (1991) *J. Am. Oil Chem. Soc.* **68**, 1.
12. McNeil, G.P., Yamane, T. (1991) *J. Am. Oil Chem. Soc.* **68**, 6.
13. Berger, M., Schneider, M. P. (1991) *Biotechnol. Lett.* **13**, 333.
14. Robbins, S. J., Nicholson, S. H. (1987) *J. Am. Oil Chem. Soc.* **64**, 193.
15. Berger, M., Schneider, M.P. (1992) *J. Am. Oil Chem. Soc.* **69**, 961-965.
16. Berger, M., Laumen, K., Schneider, M.P. (1992) *J. Am. Oil Chem. Soc.* **69**, 955-960.
17. Spivak, J. D. (1976) *in Surfactant Science Series*, Vol VII, Part 1, Marcel Dekker, New York.
18. http://www.dechema.de/biotech/bodenr.htm, english version
19. Björkling, F., Godtfredsen, S. E., Kirk, O., Patkar, S. A., Andresen, O. (1992) *Microbial reagents in organic synthesis*, Servi, S. (ed.), Kluwer Academic publishers, Netherlands, 249-253.; Björkling, F., Godtfredsen, S. E., Kirk, O. (1989) *J. Chem. Soc. Chem. Commun.,* 934.; Adelhorst, K., Björkling, F., Godtfredsen, S.E., Kirk, O. (1990) *Synthesis,* 112.; European Patent 0394 280 B1 (18.08.1988) Novo Nordisk A/S; Andresen, O., Kirk, O. (1995) in *Proceedings Symposium Carbohydrate Bioengineering in Progress in Biotechnology* (Hrsg.: Petersen, S.B., Svensson, B., Pedersen, S.), Elsevier, Amsterdam, Lausanne, New York, Oxford, Shannon, Tokyo, 343-349.
20. Therisod, M., Klibanov, A. M. (1986) *J. Am. Chem. Soc.* **108**, 5638.; Riva, S., Chopineau, J., Kieboom, A.P.G., Klibanov, A.M. (1988) *J. Am. Chem. Soc.* **110**, 584.
21. Khaled, N., Montet, D., Pina, M., Graille, J. (1991) *Biotechnol. Letters* **13**, 167-172.; Scheckermann, C., Schlotterbeck, A., Schmidt, M., Wray, V., Lang, S. (1995) *Enzyme Microb. Technol.,* **17**, 57-162. ; Ducret, A., Giroux, A., Trani, M., Lortie, R. (1995) *Biotechnol. Bioeng.* **48**, 214-221.; Ducret, A., Giroux, A., Trani, M., Lortie, R. (1996) *J. Am. Oil Chem. Soc.* **73**, 109-113.
22. Seemayer, R. (1992) Ph. D. thesis, Wuppertal; Panza, L., Luisetti, M., Criati, E., Riva, S. (1993) *J. Carbohydr. Chem.* **12**, 125.
23. Ikeda, I., Klibanov, A.M. (1993) *Biotechnol. Bioeng.* **42**, 788.; Schlotterbeck, A., Lang, S., Wray, V., Wagner, F. (1993) *Biotechnol. Lett.* **15**, 61.
24. Prasad, A.K., Sørensen, M.D., Parmar, V.S., Wengel, J. (1995) *Tetrahedron Lett.* **36**, 6163-7.
25. Schneider, M., Haase, B. and Machmüller, G., German Patent application 1962 6943.1 from 4 July 1996;
26. Haase, B. (1997) Ph. D. thesis, Wuppertal

27. Machmüller, G., unpublished, manuscript in preparation.

28. Cao, L., Fischer, A., Bornscheuer, U.T., Schmid, R.D. (1997) *Biocatalysis and Biotransformation* **14**, 269; Cao, L., Bornscheuer, U.T., Schmid, R.D. (1996) *Fett/Lipid* **98**, 332.

29. Aha, B. (1999) Ph. D. thesis, Wuppertal

30. Aha, B., Haase, B., Herrmann, J., Machmüller, G., Waldinger, C. (1998) *Chemische Nutzung heimischer Pflanzenöle – Schriftenreihe „Nachwachsende Rohstoffe"* **12**, Landwirtschaftsverlag GmbH, Münster, 13-49.; Berger, M., Machmüller, G., Waldinger, C., Schneider, M.P. (1998) *Biokonversion nachwachsender Rohstoffe – Schriftenreihe „Nachwachsende Rohstoffe"* **10**, Landwirtschaftsverlag GmbH, Münster, 139-153.

ENZYMATIC REGIOSELECTIVE TRANSFORMATIONS IN NATURAL PRODUCTS

MIGUEL FERRERO AND VICENTE GOTOR

Departamento de Química Orgánica e Inorgánica, Facultad de Química, Universidad de Oviedo, Avenida Julián de Clavería, 8; 33006-Oviedo, Spain

Abstract: The synthetic potential of lipases in organic solvents in the area of natural products is described. Enzymatic regioselective acylations and alkoxycarbonylations of polyhydroxylated compounds such as carbohydrates, nucleosides, and steroids are described. In many cases, chemoenzymatic routes are used to prepare analogs of natural products or intermediates. Oxime esters or carbonates are used as acylating or alkoxy-carbonylating reagents in enzymatic processes of acylation or alkoxycarbonylation, respectively.

1. Introduction

Selective synthetic transformations of polyhydroxylated compounds such as carbo-hydrates, nucleosides, and steroids by means of enzymes are becoming increasingly popular. Monoesters of some monosaccharides and disaccharides have been obtained using lipases or proteases in polar organic solvents. Various reagents have been used in the enzymatic acylation of sugars, such as acids, activated esters, enol esters or oxime esters. A relevant transformation in the carbohydrate field is the regioselective alkoxycarbonylation, and in particular the enzymatic alkoxycarbonylation. These conversions have received relatively little attention so far.

In the case of nucleosides, selective reactions of their functional groups are of interest. Successful developments in this area may lead to new methods for the synthesis of nucleoside analogues. Indeed, nucleoside derivatives are compounds of great significance as they may show antineoplastic and antiviral activity. Even just a single

B. Zwanenburg et al. (eds.), Enzymes in Action, 347–363.
© 2000 *Kluwer Academic Publishers. Printed in the Netherlands.*

acylation can lead to compounds with antitumour activity. Similarly, the regioselective acylation of nucleosides represents a way of introducing protecting base-labile groups, an issue of relevance in oligonucleotide synthesis.

Selective protection and deprotection of steroids containing various functional groups may constitute a problem in this area. Development of selective reactions may lead to new synthetic methods for steroid analogues. In this chapter special attention is given to vitamin D derivatives, which are significant in the treatment of several human diseases.

The synthetic potential of lipases in organic solvents in the field of natural product synthesis has been a central issue in our research group. In this chapter the enzymatic acylation and alkoxycarbonylation of several carbohydrates, nucleosides and steroids, using lipases in organic solvents is reviewed.

2. Carbohydrates

Lipase-catalyzed acylation in organic solvents has been successfully applied to the regioselective acylation of carbohydrates. In general, these processes require the use of activated esters or special reaction conditions. Our interest in the development of an enzymatic method for the regioselective acylation of carbohydrates under mild reaction conditions led us to examine the reaction of α-D-glucopyranose and methyl α-D-glucopyranoside with oxime esters [1], which were used as irreversible acyl transfer agents whereby acetonoxime is the leaving group (Scheme 1).

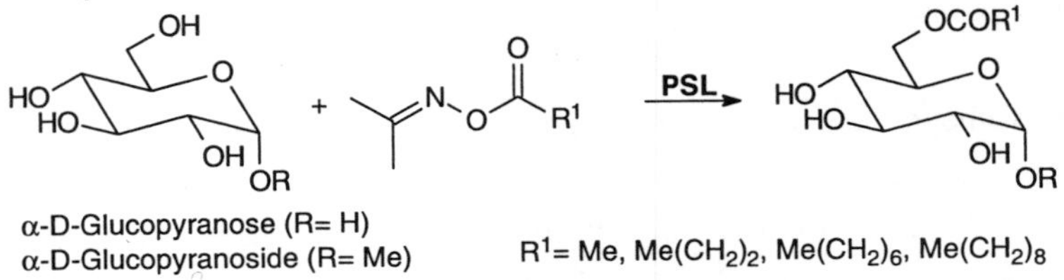

α-D-Glucopyranose (R= H)
α-D-Glucopyranoside (R= Me)

R^1= Me, Me(CH$_2$)$_2$, Me(CH$_2$)$_6$, Me(CH$_2$)$_8$

Scheme 1

After preliminary screening, we selected *Pseudomonas cepacia* lipase (PSL) as biocatalyst and pyridine or 3-methylpentan-3-ol as solvents. These sugars do not react with oxime esters in the absence of enzyme. The acylated sugars are obtained in yields higher than 90%. This strategy allowed the selective acylation of the primary hydroxyl group of the carbohydrates giving rise to 6-*O*-acyl esters. The length of the acyl moiety in the oxime ester used did not influence the selectivity, but long chain esters gave better the yields.

To study the scope and limitations of the enzymatic acylation of carbohydrates by oxime esters we applied the method to other hexoses, both aldoses and ketoses, as well as to some pentoses [2]. D-Galactose and D-mannose reacted with oxime esters in the presence of PSL in dry pyridine at room temperature and in an inert atmosphere to afford derivatives with an acylated primary hydroxyl group (Scheme 2). These results show that changes in the configuration at C-2 or C-4 in the hexose do not affect the preference of the enzyme for the primary hydroxyl group. The primary hydroxyl group of the ketohexose α-L-sorbose was similarly acylated to give 1-O-acyl derivatives. The best selective acylations of the primary hydroxyl functions of methylfuranosides were obtained with L-arabinose and D-ribose using PSL as catalyst (Scheme 2). A careful control of the temperature allowed us to isolate the 5-O-acyl furanoses as a mixture of α and β anomers. With moderate or long chain oxime esters (C_4, C_{10}) the process must be carried out at 0 °C, since at higher temperatures polyacylated compounds are formed. A different lipase, *Candida antarctica* lipase (CAL), similarly gave monoacylated products at room temperature. Two different monoacylated products were obtained when D-ribose was used with PSL as catalyst. In contrast, with CAL, monoacylation products of ribofuranose at O-5 were formed exclusively. This lipase promoted the formation of acylation products of D-lyxose at the anomeric hydroxyl group while monoacylation at O-5 was not observed. Analogously to the hexoses, higher reaction rates are observed for the more lipophilic esters (C_4, C_{10}), albeit with a somewhat decreased selectivity (Scheme 2).

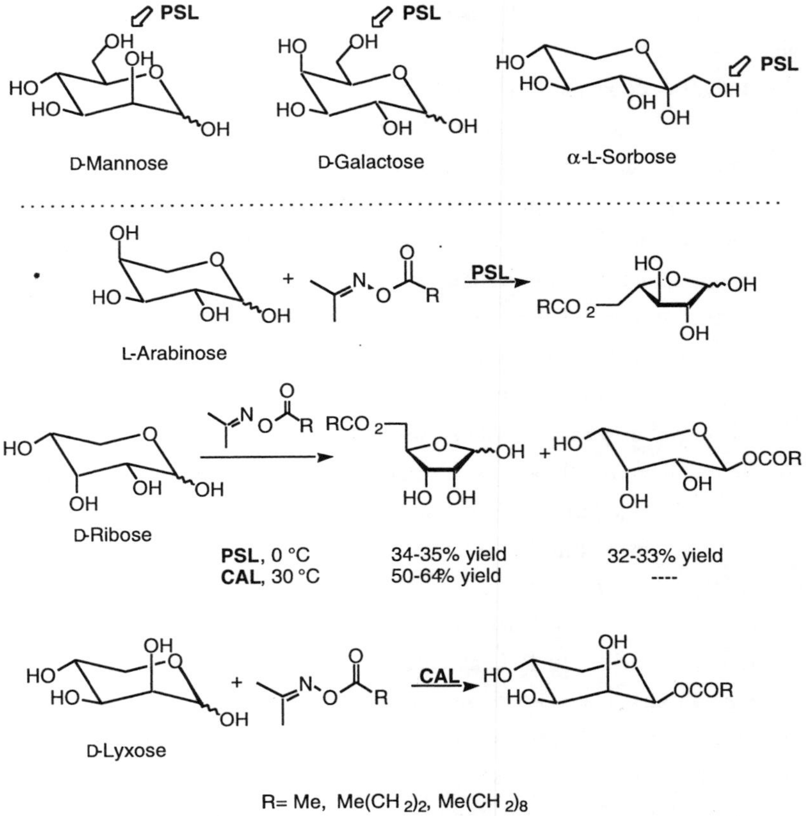

D-Mannose   D-Galactose   α-L-Sorbose

L-Arabinose

D-Ribose

PSL, 0 °C   34-35% yield   32-33% yield
CAL, 30 °C   50-64% yield   ----

D-Lyxose

R= Me, Me(CH$_2$)$_2$, Me(CH$_2$)$_8$

Scheme 2

Another relevant transformation in the carbohydrate field is the regioselective alkoxycarbonylation. Carbonates of primary and secondary alcohols are useful intermediates in carbohydrate chemistry, since their selective cleavage in the presence of other functional groups can readily be accomplished. Enzymatic alkoxycarbonylation has received little attention and has not been applied at all in the carbohydrate field. The satisfactory results obtained with oxime esters in the acylation processes described above, have encouraged us to elaborate alkoxycarbonylating reagents [3].

D-Galactopyranose, D-glucopyranose, D-mannopyranose, and ketohexose L-sorbopyranose reacted with acetone oxime carbonate esters in 1,4-dioxane at 60 °C in the presence of CAL to afford derivatives which were selectively carbonylated at the primary hydroxyl group. When acetone O-(allyloxycarbonyl)oxime and acetone O-(benzyloxycarbonyl)oxime were used, the allyl and benzyl groups of the reagents acted as leaving groups, as well as the oxime, to yield a mixture of both O-6-carbonylated compounds. The alkoxycarbonylation of pentoses (D-ribose, D- and L-arabinose) was achieved by CAL and a wide variety of carbonate moieties could be introduced (aliphatic, aromatic, olefinic) using this methodology, which is complementary to the chemical selective alkoxycarbonylation in solvents of low polarity (Scheme 3).

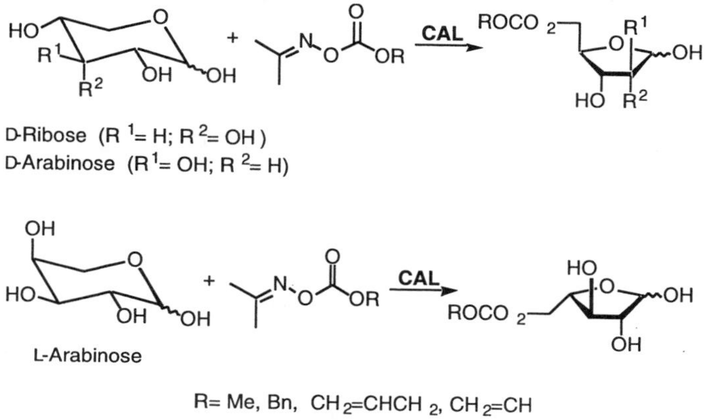

D-Ribose (R^1= H; R^2= OH)
D-Arabinose (R^1= OH; R^2= H)

L-Arabinose

R= Me, Bn, CH$_2$=CHCH$_2$, CH$_2$=CH

Scheme 3

As an extension on our program on the enzyme-catalyzed regioselective transformations of carbohydrates with oxime esters, we have studied the acylation and alkoxycarbonylation of 2-deoxy-D-*arabino*-hexopyranose, 2-deoxy-D-*lyxo*-hexopyranose, and the conformationally rigid 1,6-anhydro-β-D-glucopyranose (Scheme 4) [4]. The reason for this study is to check whether the absence of a hydroxyl group at C-2, and the rigid conformation, can affect the selectivity of the lipases. CAL was used for the regioselective acylation and alkoxycarbonylation of the primary hydroxyl group of 2-deoxy-D-*arabino*-hexopyranose and 2-deoxy-D-*lyxo*-hexopyranose, using oxime esters or carbonates, respectively. PSL also catalyzed the regioselective acylation of the primary hydroxyl group of both just-mentioned hexopyranoses. Moreover, this lipase was a suitable biocatalyst for the acylation of the secondary 3-OH of 6-*O*-(benzyloxy)carbonyl derivatives of 2-deoxy-D-hexoses (Scheme 4). Subsequent deprotection of the primary hydroxyl group by catalytic hydrogenation readily gave the 3-*O*-acyl derivatives of the former.

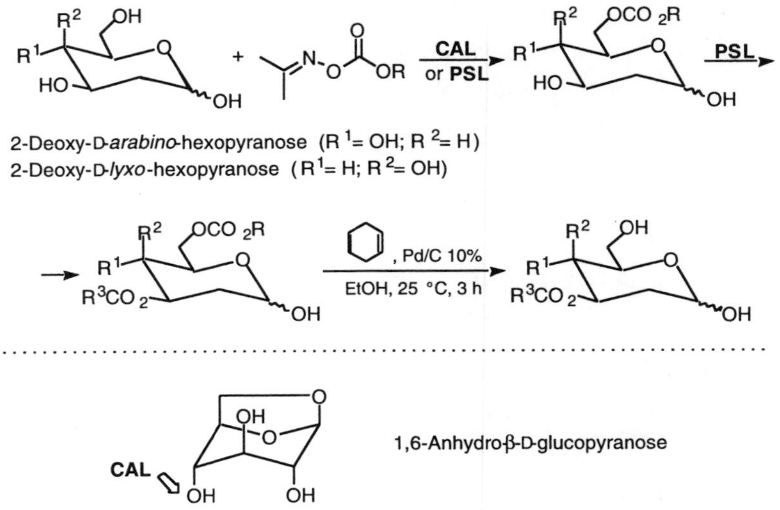

2-Deoxy-D-*arabino*-hexopyranose (R^1= OH; R^2= H)
2-Deoxy-D-*lyxo*-hexopyranose (R^1= H; R^2= OH)

Scheme 4

Regioselective acylation of the conformationally rigid carbohydrate 1,6-anhydro-β-D-glucopyranose was achieved using CAL in dioxane at room temperature. Derivatives resulting from regioselective monoacylation at O-4 were obtained. At higher temperature, the selectivity decreased and complex mixtures of all possible monoacylated derivatives were obtained. Experiments carried out with PSL also led to mixtures of monoacylated products.

3.   Nucleosides

Selective modification of nucleosides, which contain several functional groups with similar chemical reactivities represents an interesting challenge for organic chemists. For the purpose of designing new regioselective enzymatic transformations of nucleosides, we studied the lipase-mediated acylation of acyclonucleosides (Scheme 5) [5]. Thus, acyclonucleosides bearing different substituents in the 2-hydroxyethoxymethyl chain were prepared in almost quantitative yields using PSL and several ester derivatives using a transesterification reaction. This was the first example of the use of enzymes in acyclonucleoside chemistry that allowed a facile preparation of prodrugs of 4-quinolone acyclonucleosides, which were tested as antiviral agents [6].

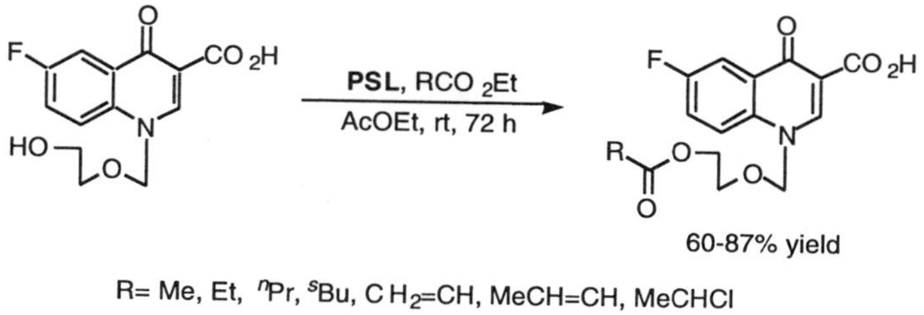

Scheme 5

In the preceding section we have used oxime esters as irreversible acyl transfer agents in regioselective enzymatic transesterification reactions fo the primary alcohol function in carbohydrates. In contrast, it was found that the regioselective acylation of 2'-deoxynucleosides occurs on the secondary hydroxyl group, instead of the more chemically reactive primary alcohol, when using PSL in pyridine [7]. This procedure turned out to be very versatile because acylation with various oxime esters containinring saturated or unsaturated chains, gave exclusively the 3'-O-acyl derivatives (Scheme 6). Also, N-acylation of 2'-deoxyadenosine was not observed. Similarly, CAL showed high regioselectivity toward the primary hydroxyl group of both deoxy- and ribonucleosides [8]. 2'-Deoxynucleosides, such as thymidine and 2'-deoxyadenosine, were acylated with oxime esters carrying saturated, unsaturated, aromatic, and functionalized chains, to give 5'-O-acylated derivatives along with small quantities of the 3'-O-acylated regioisomer. Uridine, adenosine, and inosine, as representative ribonucleosides, were acylated exclusively at the 5'-OH by using the same methodology.

Aminoacyl nucleosides and their oligonucleotide derivatives, are highly significant in biological chemistry, since the 3'-terminus of all t-RNAs contain the common CCA sequence and the 3'-terminal adenosine cis-diol system is a site of attachment of the amino acid. So aminoacylated derivatives of oligoribonucleotides are important tools in the study of the specific incorporation of amino acids into proteins. We have applied the method described above (Scheme 6) to prepare aminoacylated nucleosides in a regioselective manner by using the lipases PSL and CAL, with N-protected acetoxime aminoacyl esters as acylating agents [9]. It is noteworthy that, in contrast to other methods, the exocyclic amino function in adenine nucleosides remains unaffected. Amino acids protected with benzyloxycarbonyl groups are better substrates than those derivatized by the $tert$-butyloxycarbonyl group. α-Branched amino acids do not seem to meet the steric requirements of the enzyme active sites.

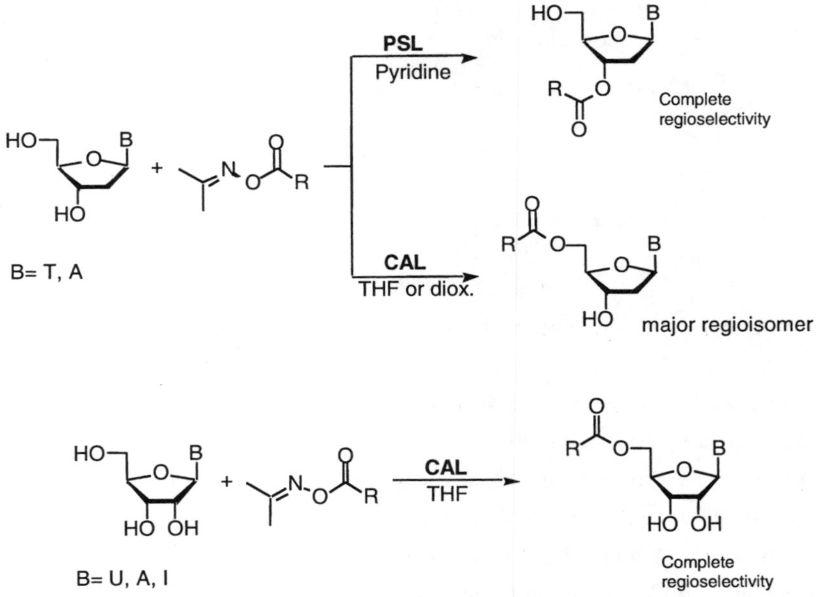

Acylation R= Me, ^{n}Pr, Me(CH_2)$_6$, Me(CH_2)$_8$, CH_2=CH, MeCH=CH, Ph, ClCH_2
Aminoacylation R= (CH_2)$_n$NHP [Gly (n=1), β-Ala (n=2); P= Cbz, Boc]
Alkoxycarbonylation R= MeO, BnO, CH_2=CH-O, CH_2=CH-CH_2O

Scheme 6

An important transformation in nucleosides is the selective alkoxycarbonylation of the sugar moiety in order to obtain nucleoside carbonates, which play a role in the synthesis of oligonucleotides and analogues. Scheme 6 summarizes a general, new and simple procedure for the synthesis of pyrimidine and purine 3'-carbonates from 2'-deoxynucleosides using PSL and *O*-[(alkoxy)carbonyl]oxime [10]. In this method, no protection of the primary hydroxyl group is necessary in contrast to preparation of these compounds. 5'-*O*-Carbonates of ribonucleosides and 2'-deoxyribonucleosides could be obtained by enzymatic alkoxycarbonylation with CAL and oxime carbonates [11], which in turn are easily prepared from chloroformates. Ribonucleosides yield two types of 5'-*O*-carbonates, depending on whether the alkoxy group or the acetone oxime moiety acted as the leaving group. For 2'-deoxynucleosides, the leaving group was always the acetonoxime moiety, giving rise to the regioselective formation of the corresponding 5'-*O*-alkyl carbonates, together with small amounts of 3'-*O*-regioisomer and diacylated compounds (Scheme 7).

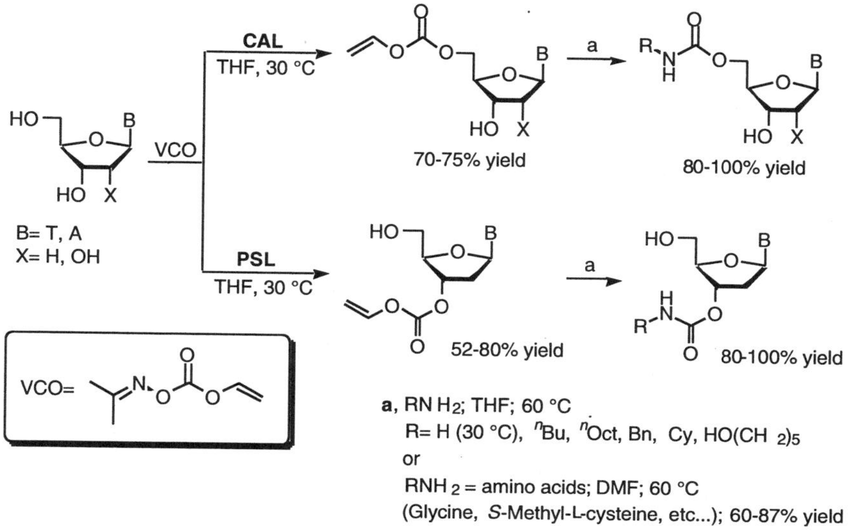

Scheme 7

The carbamate group occurs in several classes of anti-tumor agents and is also used to improve the permeation through biological membranes. By also taking into account that the carbamate group in nucleosides improves chemical and enzymatic stability, it was considered desirable to derive a simple and general synthesis of such nucleoside carbamates. Various lipases were tested to promote the reaction between vinyl carbamates and nucleosides, however, without positive result. One conceivable reason for this failure could be that some carbamates are good inhibitors of many serine hydrolases. As an alternative we considered to use a two-step procedure (Scheme 7). [12] The key step is the enzymatic synthesis of 5'-O- or 3'-O-vinyloxycarbonyl-nucleoside in a regioselective way, using CAL or PSL catalysis, respectively. In a second aminolysis step these carbonates give the corresponding urethanes. This methodology allows us to couple ammonia, amines, amino alcohols, and L-amino acids to 2'-deoxynucleosides (to 3'- or 5'-positions) or ribonucleosides (5'-position) [13].

Another issue of interest is the enzymatic differentiation between hydroxyl group with different spatial orientation [14]. Thus, 5'-O-acyl and 5'-O-alkoxycarbonyl derivatives of α-, anhydro-, *xylo*-, and *arabino*-nucleosides are obtained through a lipase-mediated reaction with CAL by using acetoxime butyrate or butyric anhydride, and benzyloxycarbonyl-O-acetoxime as acylating or alkoxycarbonylating agents (Scheme 8).

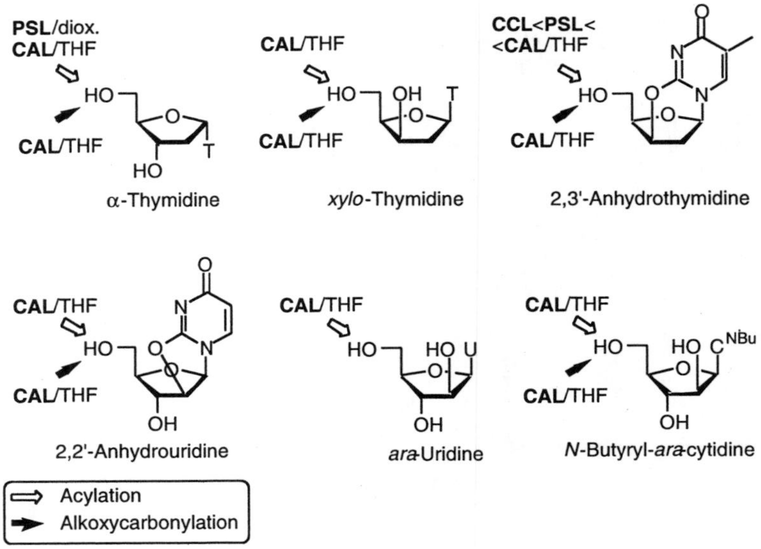

Scheme 8

For the synthesis of the highly hindered 3'-O-acetyl α- and *xylo*-thymidine [15], we first prepared the 5'-Cbz-derivatives, and then chemically performed an acylation reaction of the free 3'-OH hydroxyl group. Subsequent removal of the benzyloxy-carbonyl group gave the desired products (Scheme 9).

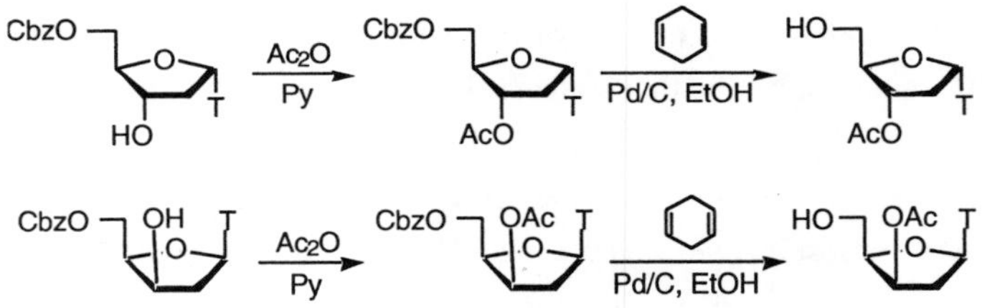

Scheme 9

Aminosugar nucleosides are known to possess strong antibacterial, anticancer, and inhibitory properties. Therefore, considerable effort has been devoted to the preparation of such compounds. Scheme 10 shows a novel and general chemoenzymatic procedure, developed by us, to obtain the 3'-amino-*xylo*-nucleosides [16]. This scheme is based on the 5' directed intramolecular nucleophilic substitution at the 3'-activated position of the nucleoside. The approach of the incoming group to this position takes place regio- and stereoselectively from the most hindered face of the nucleoside. This methodology is applicable to ribonucleosides and 2'-deoxyribonucleosides, irrespective of their base substituent.

A chemoenzymatic procedure for the first synthesis of 3'- and 5'-carbazoyl nucleoside derivatives is shown in Scheme 11 [17]. This process involves the regioselective enzymatic alkoxycarbonylation of nucleosides and the subsequent transformation with hydrazine into novel carbazoyl nucleoside derivatives. It should be noted that 3'-alkylidencarbazoyl 2'-deoxynucleosides, 5'-alkylidencarbazoyl 2'-deoxynucleosides and 5'-alkylidencarbazoyl ribonucleosides are interesting targets, since they combine structural features found in both therapeutic nucleoside derivatives and fungicide / herbicide nucleoside analogues.

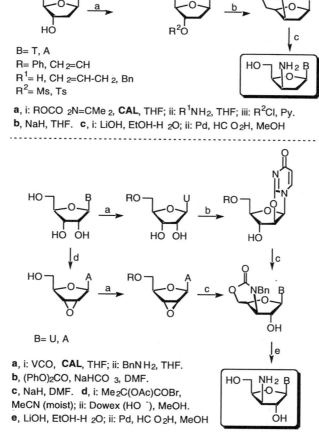

B= T, A
R= Ph, CH₂=CH
R¹= H, CH₂=CH-CH₂, Bn
R²= Ms, Ts

a, i: ROCO₂N=CMe₂, **CAL**, THF; ii: R¹NH₂, THF; iii: R²Cl, Py.
b, NaH, THF. c, i: LiOH, EtOH-H₂O; ii: Pd, HCO₂H, MeOH

B= U, A

a, i: VCO, **CAL**, THF; ii: BnNH₂, THF.
b, (PhO)₂CO, NaHCO₃, DMF.
c, NaH, DMF. d, i: Me₂C(OAc)COBr, MeCN (moist); ii: Dowex (HO⁻), MeOH.
e, LiOH, EtOH-H₂O; ii: Pd, HCO₂H, MeOH

Scheme 10

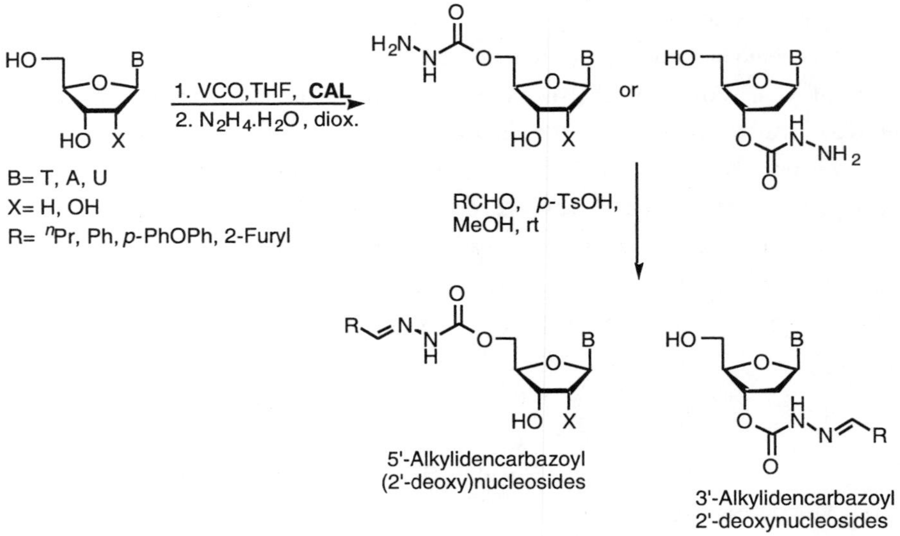

358

Scheme 11

Modulation of gene expression by antisense technologies requires the design of modified oligonucleotides showing enhanced cellular uptake, resistance toward degradation by nucleases, and appropriate hybridization to target natural oligonucleotides. Such modified oligonucleotides are now being actively investigated as a new generation of pharmaceuticals. Much effort has been directed toward the synthesis of analogues with an altered phosphodiester linkage, and one of the most important modifications is the complete substitution of the phosphate internucleoside bridge. Thus, several properties of the natural oligonucleotides have been improved for the potential therapeutic application of the antisense strategy. It is of interest to consider backbone linkages that contain carbazoyl groups, because this type of linkage may serve as an attractive surrogate for the phosphodiester linkage, being non-ionic, hydrolytically stable, and non-chiral.

The synthesis of backbone modified dinucleotide analogues is shown in Scheme 12 [18], in which the natural phosphodiester linkage is replaced by a 3'-5' carbazoyl linkage. The bridge was formed via a coupling reaction between an appropriate 3'-carbazoyl nucleoside analogue and an aldehyde nucleoside derivative.

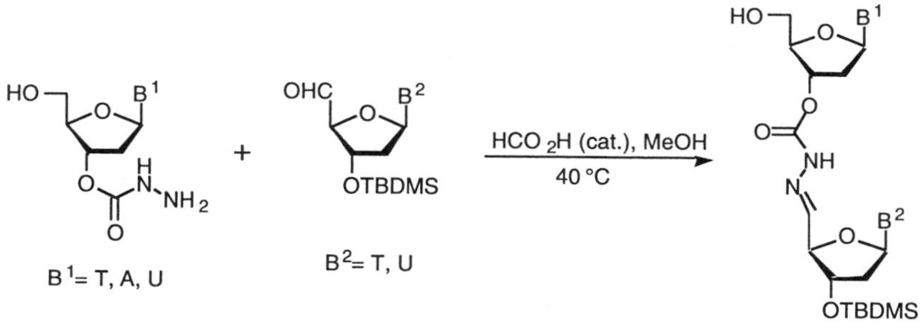

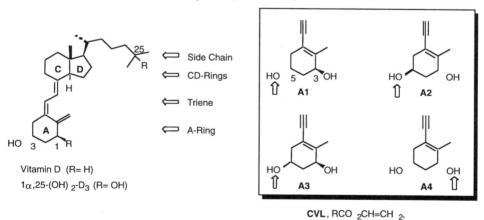

Scheme 12

4.   Steroids (Vitamin D)

Steroid research has expanded enormously in recent years with the discovery that several of their metabolites and analogues exhibit a much broader spectrum of biological activity than originally thought. Parallel to this, new enzymatic methodologies (especially in organic solvents) for the preparation of several precursors, have been developed [19]. We have centered our attention on the vitamin D field. Vitamin D_3, through its hormonally active form $1\alpha,25$-dihydroxyvitamin D_3 [$1\alpha,25$-(OH)$_2$-D_3], plays an important role in the endocrine system (Scheme 13).

Vitamin D (R= H)
$1\alpha,25$-(OH)$_2$-D_3 (R= OH)

Side Chain
CD-Rings
Triene
A-Ring

CVL, RCO$_2$CH=CH$_2$,
R= Me, Et, ^{n}Pr, ClCH$_2$, Ph,
not solvent or THF

Scheme 13

The biological profile of $1\alpha,25-(OH)_2-D_3$ and the medical promise of certain vitamin D analogues for treatment of human diseases such as cancer, AIDS, and Alzheimer's disease, are clear stimulants for the synthesis of potential chemotherapeutic vitamin D drugs. One area entails studies of analogues that bind to various receptors and other proteins. A-Ring modification of $1\alpha,25-(OH)_2-D_3$ is the second issue of interest together with side chain modifications. For the preparation of A-ring synthons, we wanted to selectively modify one of the secondary alcohols at the C-3 or C-5 position. These two hydroxyl groups have similar reactivities. Differentiation using enzymes was therefore considered. Evaluation of the enzymatic acylation [20] of the $1\alpha,25-(OH)_2-D_3$ A-ring enynes **A1–A4** (Scheme 13) revealed that *Chromobacterium viscosum* lipase (CVL) was the best enzyme for effecting practical levels of regioselectivity. It was found that only the *cis* diol **A4** was acylated at the allylic C-3 hydroxyl, while the other stereosiomers were acylated at C-5.

Given the difference in reaction ratios observed for the enantiomers **A1** and **A2** as well as for **A3** and **A4**, CVL could be useful in the kinetic resolution of racemic vitamin D A-ring fragment shown in Scheme 14. If successful, this would have a great advantage for the preparation of large amounts of that synthon. Indeed CVL, the resolution of (±)-vitamin D A-ring synthon precursor gave the natural enynol (−)-isomer nicely in only one step (Scheme 14). In contrast, Okamura [21] resolved these compounds via the tedious formation of its diastereoisomeric *(1S,1'S)* and *(1R, 1'S)* carbamates derived from *(S)*-naphthyl isocyanate, followed by separation by HPLC and then deprotection of the *(1S,1'S)* carbamate. Using

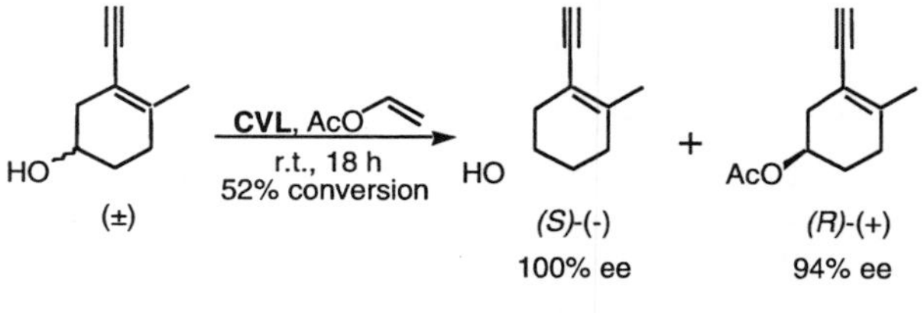

Scheme 14

The resolution using CVL was carried out at room temperature with vinyl acetate acting as both acylating agent and solvent. After 18 hours, when GC analysis revealed 52% conversion, the reaction was stopped and flash chromatography of the product mixture gave the enantiomerically pure *(S)-(−)*-isomer in high yield (97%, 100% ee).

Similarly, we have studied the enzymatic alkoxycarbonylation of enynes **A1–A4** [22]. It was found for the alkoxycarbonylation of the $1\alpha,25-(OH)_2-D_3$ A-ring enynes

CAL is the best enzyme for effecting practical levels of regioselectivity, although CVL is also effective (Scheme 15).

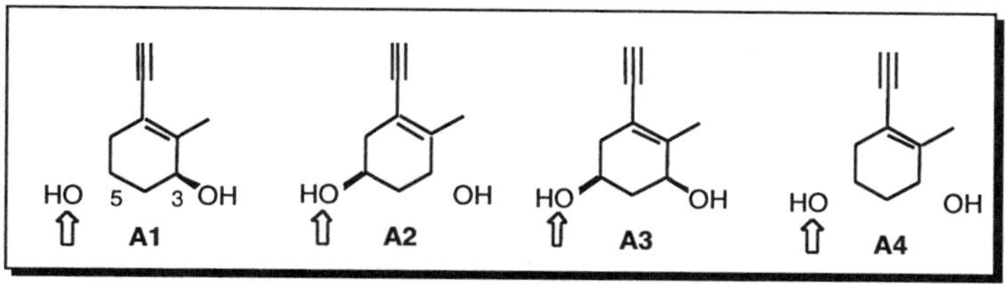

CAL (CVL also shows high regioselectivity),
VCO , toluene

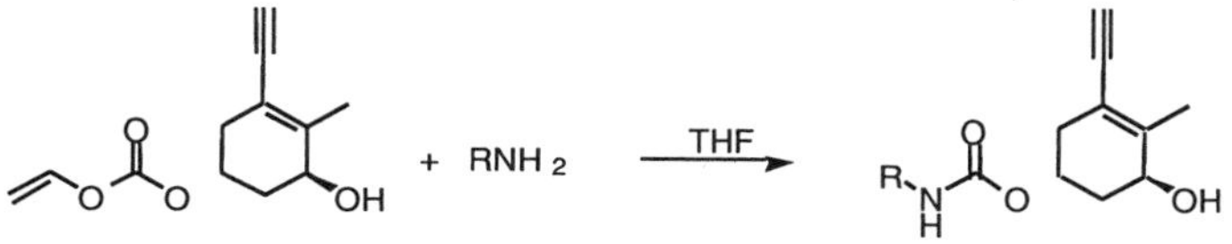

R= H, ^{n}Bu, N H$_2$, HX(CH$_2$)$_n$ [X= O,NH; n= 2,3,6]
or
RN H$_2$= amino acids and spermine

* The process has also been performed for the other three stereoisomers

Scheme 15

In these enzymatic alkoxycarbonylations using VCO (See Scheme 17) as transfer agent, the acetone oxime moiety served as the leaving group, giving rise to regio-selective formation of the corresponding C-5 carbonate derivatives. The synthetic utility of this carbonates have been demonstrated by the preparation of important carbamates (Scheme 15). Thus, the subsequent transformation with ammonia, amines, amino alcohols, diamines, and amino acids produced the corresponding 1α,25-(OH)$_2$-D$_3$ A-ring C-5 carbamates. Special mention should be made of the use of hydrazines and spermine, which permit the preparation of carbazoyl and polyamine derivatives, respectively (Scheme 15) [23].

The difference in behavior between enzymatic acylation and alkoxycarbonylation with these A-ring stereoisomers is worth noting and should be explained by a detailed study of the enzymatic catalysis [24].

5. Conclusion

The enzyme-mediated syntheses covered in this review substantiate the increasing importance of biocatalysis in regioselective processes, such as enzymatic acylation and alkoxycarbonylation reactions in organic solvents. The combination of chemical and enzymatic methods provides efficient routes to bioactive natural (or unnatural) products and analogues, which may be of great interest in view of their potential medical applications.

6. Acknowledgments

The success of these projects is due to the many contributions from a number of co-workers, whose names are mentioned in the references. The work presented here has been generously supported by Spanish Ministry of Education (MEC) (through the projects BIO92-0687 and BIO98-0770). M. Ferrero would like to thank *Fundación Príncipe de Asturias* for financial support for the Vitamin D project, through "Dr. Severo Ochoa" Award. The authors wish to express their cordial thanks to Novo Nordisk Co. and Genzyme Co. for the generous donation of *Candida antarctica* lipase (CAL) and *Chromobacterium viscosum* lipase (CVL), respectively.

7. References

1.   Gotor, V. and Pulido, R. (1991) An Improved Procedure for Regioselective Acylation of Carbohydrates: Novel Enzymatic Acylation of □-D-Glucopyranose and Methyl □-D-Glucopyranoside, *J. Chem. Soc., Perkin Trans. 1*, 491-492.

2.   Pulido, R. Ortiz, F. L. and Gotor, V. (1992) Enzymatic Regioselective Acylation of Hexoses and Pentoses Using Oxime Esters, *J. Chem. Soc., Perkin Trans. 1*, 2891-2898.

3.   Pulido, R. and Gotor, V. (1993) Enzymatic Regioselective Alkoxycarbonylation of Hexoses and Pentoses with Carbonate Oxime Esters, *J. Chem. Soc., Perkin Trans. 1*, 589-592.

4.   Pulido, R. and Gotor, V. (1994) Towards the Selective Acylation of Secondary Hydroxyl Groups of Carbohydrates Using Oxime Esters in an Enzyme-Catalyzed Process, *Carbohydr. Res.* **252**, 55-68.

5.   de la Cruz, A., Elguero, J., Gotor, V., Goya, P., Martinez, A., and Morís, F. (1991) Lipase-Mediated Acylation of Acyclonucleosides. Application to Novel Fluoroquinolone Derivatives, *Synth. Commun.* **21**, 1477-1480.

6.   de la Cruz, A., Elguero, J. , Goya, P., Martinez, A., Gotor, V., Morís, F., and de Clercq, E. (1992) Synthesis and Antiviral Evaluation of New 4-Quinolone Acyclonucleosides, *J. Chem. Res. (S)* , 216-217; *J. Chem. Res. (M)* , 1673-1681.

7. Gotor, V. and Moris, F. (1992) Regioselective Acylation of 2'-Deoxynucleosides Through an Enzymatic Reaction with Oxime Esters, *Synthesis*, 626-628.

8. Moris, F. and Gotor, V. (1993) A Useful and Versatile Procedure for the Acylation of Nucleosides through an Enzymatic Reaction, *J. Org. Chem.* **58**, 653-660.

9. Moris, F. and Gotor, V. (1994) Selective Aminoacylation of Nucleosides through an Enzymatic Reaction with Oxime Aminoacyl Esters, *Tetrahedron* **50**, 6927-6934.

10. Moris, F. and Gotor, V. (1992) A Novel and Convenient Route to 3'-Carbonates from Unprotected 2'-Deoxynucleosides through an Enzymatic Reaction, *J. Org. Chem.* **57**, 2490-2492.

11. Moris, F. and Gotor, V. (1992) Lipase-Mediated Alkoxycarbonylation of Nucleosides with Oxime Carbonates, *Tetrahedron* **48**, 9869-9876.

12. Garcia-Alles, L. F., Moris, F., and Gotor, V. (1993) Chemoenzymatic Synthesis of 2'-Deoxynucleoside Urethanes, *Tetrahedron Lett.* **34**, 6337-6338.

13. Garcia-Alles, L. F. and Gotor, V. (1995) Synthesis of 5'-*O*- and 3'-*O*-Nucleoside Carbamates, *Tetrahedron* **51**, 307-316.

14. Moris, F. and Gotor, V. (1993) Enzymatic Acylation and Alkoxycarbonylation of □-, Xylo-, Anhydro-, and Arabino-Nucleosides, *Tetrahedron* **49**, 10089-10098.

15. Gotor, V., Morís, F., and Garcia-Allés, L. F. (1994) Enzymatic Regioselective Alkoxycarbonylation of Nucleosides and Its Utility in Nucleoside Derivative Synthesis, *Biocatalysis* **10**, 295-305.

16. García-Alles, L. F., Magdalena, J., and Gotor, V. (1996) Synthesis of Purine and Pyrimidine 3'-Amino-3'-deoxy- and 3'-Amino-2',3'-dideoxyxylonucleosides, *J. Org. Chem.* **61**, 6980-6986.

17. Magdalena, J., Fernández, S., Ferrero, M., and Gotor, V. (1998) Chemoenzymatic Synthesis of Novel 3'- and 5'-Carbazoyl Nucleoside Derivatives. Regioselective Preparation of 3'-and 5'-Alkyliden-carbazoyl Nucleosides, *J. Org. Chem.* **63**, 8873-8879.

18. Magdalena, J., Fernández, S., Ferrero, M., and Gotor, V. (1999) Synthesis of Novel Carbazoyl Linked Pyrimidine-Pyrimidine and Pyrimidine-Purine Dinucleotide Analogues, *Tetrahedron Lett.* **40**, 1787-1790.

19. Ferrero, M. and Gotor, V. (1999) Biocatalytic Synthesis of Steroids, in R. N. Patel (ed.), Chapter 20 *Stereoselective Biocatalysis*, Marcel Dekker: New York, pp. 579-631 (in press).

20. Fernández, S., Ferrero, M., Gotor, V., and Okamura, W. H. (1995) Selective Acylation of A-Ring Precursors of Vitamin D Using Enzymes in Organic Solvents, *J. Org. Chem.* **60**, 6057-6061.

21. Okamura, W. H., Elnagar, H. Y., Ruther, M., and Dobreff, S. (1993) Thermal [1,7]-Sigmatropic Shift of Previtamin D_3 to Vitamin D_3: Synthesis and Study of Pentadeuterio Derivatives, *J. Org. Chem.* **58**, 600-610.

22. Ferrero, M., Fernández, S., and Gotor, V. (1997) Selective Alkoxycarbonylation of A-ring Precursors of Vitamin D Using Enzymes in Organic Solvents. Chemoenzymatic Synthesis of 1□,25-Dihydroxyvitamin D_3 C-5 Carbamate Derivatives, *J. Org. Chem.* **62**, 4358-4363.

23. Gotor-Fernández, V., Ferrero, M., Fernández, S. and Gotor, V. (1999) 1□,25-Dihydroxyvitamin D_3 A-Ring Precursors: Studies on Regioselective Enzymatic Alkoxylation Reaction of Their Stereoisomers. Chemoenzymatic Synthesis of A-Ring Synthon Carbamate Derivatives, Including Carbazates and Polyamino Carbamates, submitted.

24. Fernández, S., Díaz, M., Ferrero, M. and Gotor, V. (1997) New and Efficient Enantiospecific Synthesis of (–)-Methyl 5-*epi*-Shikimate and Methyl 5-*epi*-Quinate from (–)-Quinic Acid, *Tetrahedron Lett.* **38**, 5225-5228.

STEREOSELECTIVE BIOCATALYTIC FORMATION OF CYANOHYDRINS, VERSATILE BUILDING BLOCKS FOR ORGANIC SYNTHESIS

ARNE VAN DER GEN AND JOHANNES BRUSSEE

Leiden Institute of Chemistry, Leiden University, Einsteinweg 55, 2333 CC Leiden, The Netherlands

1.   HYDROXYNITRILE LYASES AS ENANTIOSELECTIVE CATALYSTS

1.1.   INTRODUCTION

Cyanohydrins have been shown to be excellent chiral building blocks in organic synthesie. They are expedient starting materials for the preparation of several important classes of compounds such as α-hydroxy acids, acyloins, α -hydroxy aldehydes, vicinal diols, ethanolamines and α- and β-amino acids. This will be illustrated in the second part 2 of this chapter. During the past decades there has been an upsurge of interest in methods for the synthesis and application of chiral cyanohydrins in non-racemic form. It may be expected that in the near future these products will constitute an important addition to the range of commercially available chiral building blocks for the production of pharmaceuticals, pesticides, vitamins, unusual amino acids and resolving agents [1-7]. Cyanohydrins are usually prepared by the addition of hydrogen cyanide (HCN) to aldehydes and ketones or, indirectly, by the addition of trimethylsilyl cyanide followed by acid hydrolysis. When aldehydes or unsymmetrical ketones are used as substrates, a stereogenic center is created. Methods for the synthesis of chiral cyanohydrins in non-racemic form can be divided in chemical methods and enzymatic methods. In general the chemical approaches, using chiral catalysts, have the advantage that a broad range of aldehydes can be used. Up to now major disadvantages are the moderate enantiomeric excess (e.e.) and the problems encountered in separating the catalyst from the product.

The two most important enzymatic approaches to chiral cyanohydrins are enzymatic kinetic resolution by lipases or esterases (whole cells or purified enzymes) and

B. Zwanenburg et al. (eds.), Enzymes in Action, 365–396.
© 2000 *Kluwer Academic Publishers. Printed in the Netherlands.*

enzymatic asymmetric synthesis by R- or S-hydroxynitrile lyases (HNL's). Inherent disadvantages of kinetic resolution are the need of separating the hydrolyzed product from its ester and the maximum theoretical yield of only 50%. Biocatalysis using the R-HNL from almonds, and synthetic applications of chiral cyanohydrins have been thoroughly studied in our laboratories over the past decade.

1.2. CYANOGENESIS

As a means of defence many plants release chemicals to discourage predators from damaging their structure. One such chemical is the commonly found highly toxic hydrogen cyanide (HCN). This compound is mostly present in a chemically masked form in cyanohydrins (α-hydroxynitriles), which in turn are stabilized by β-glycosidic linkages to mono- or oligo-saccharides. If the plant cells are disrupted by herbivoral-, fungal- or mechanical means, catabolic enzymes can act on the cyanogenic glycosides. First, the carbohydrate moiety is cleaved by the action of one or more β-glycosidases. After loss of the carbohydrates, the resulting intermediate cyanohydrins decompose either spontaneously or enzymatically into the corresponding carbonyl compound and HCN. The enzymatic decomposition of cyanohydrins is caused by the presence of a representative of a large family of enzymes called hydroxynitrile lyases (HNL's). The release of HCN as a chemical defence mechanism for either plants or other organisms is called cyanogenesis.

1.3. FIRST APPLICATIONS OF HYDROXYNITRILE LYASE

Rosenthaler first observed the activity of HNL in 1908 [8]. It was discovered that an extract from almonds (the sweet as well as the bitter variety of *Prunus amygdalus*) is capable of metabolizing the cyanogenic glycoside amygdalin (Fig 1). The extract from defatted almond meal, called emulsin, contains glucosidases (amygdalase and prunase) and hydroxynitrile lyases, which, upon tissue damage, convert the physiological substrate amygdalin in a series of steps into glucose, benzaldehyde and HCN.
Like all enzymes, HNL catalyzes the adjustement of a chemical equilibrium, in this case between α-hydroxynitriles (cyanohydrins) and their corresponding aldehydes and HCN. Conversely, in the presence of an excess of HCN, HNL catalyzes the addition of HCN to aldehydes to form cyanohydrins. It is assumed that this addition proceeds with complete enantioselectivity.

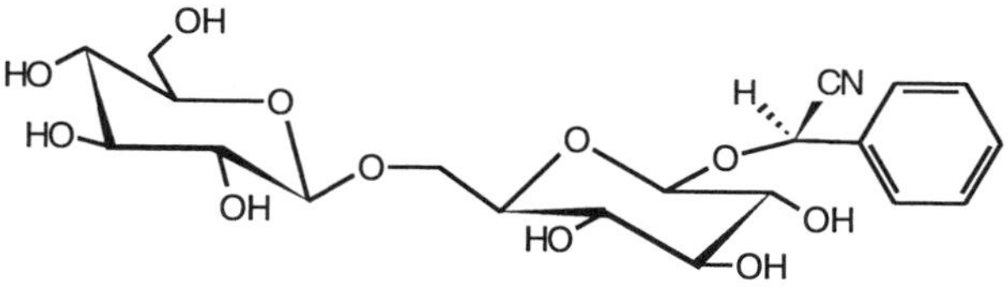

Figure 1. Amygdalin

After the discovery of HNL in almonds, numerous other species were found to catalyze this type of reaction and several studies have appeared on the mechanism and kinetics of either the asymmetric synthesis or the decomposition of cyanohydrins. In contrast, it was more than fifty years after the discovery by Rosenthaler that Becker and Pfeil published the first synthetically useful application [9]. Shortly thereafter the same authors reported a method for continuous synthesis of cyanohydrins in non-racemic form [10]. For that purpose, the hydroxynitrile lyase from almond (E.C. 4.1.2.10) was absorbed to ECTEOLA cellulose. From 1985 on several research groups also began to investigate these synthetic applications [11-20].

1.4. DISTRIBUTION OF HYDROXYNITRILE LYASES IN NATURE

HNL from almonds is yellow-colored, due to the presence of a flavine adenine dinucleotide (FAD) molecule in the glycoprotein (Fig. 2). Multiple forms of FAD-containing HNL's have been isolated from plants, especially from the *Rosaceae* subfamilies *Prunoideae* and *Maloideae*, including cherry, peach, apricot, almond, plum, and apple. Enzymes from these sources were found to be quite similar with respect to their molecular weight, specific activity, natural substrate (mandelonitrile) and the fact that they exist in several iso forms.

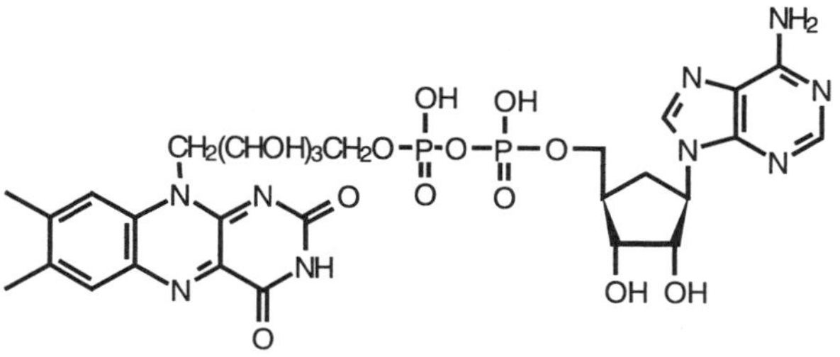

Figure 2. Flavine adenine dinucleotide (FAD)

Other HNL's, isolated from cyanogenic plants like flax, sorghum and the tropical rubber tree, basically all catalyze the same reaction, although they clearly differ in more than one property from those of the *Rosaceae*.

Striking differences are:

- The natural substrates for the different HNL's; e.g.: 4-hydroxybenzaldehyde cyanohydrin in sorghum and acetone cyanohydrin in flax.
- In contrast to HNL's from *Rosaceae*, other HNL's, including those from flax, cassava, sorghum, and ximenia, contain no FAD group.
- Enzymes from *Sorghum bicolor*, *Ximenia americana*, *Sambucus nigra*, and *Hevea brasiliensis*, exclusively catalyze formation and decomposition of *S*-cyanohydrins, whereas the *Rosaceae* HNL's only catalyze the conversion of *R*-cyanohydrins with the configuration shown in Fig. 3.

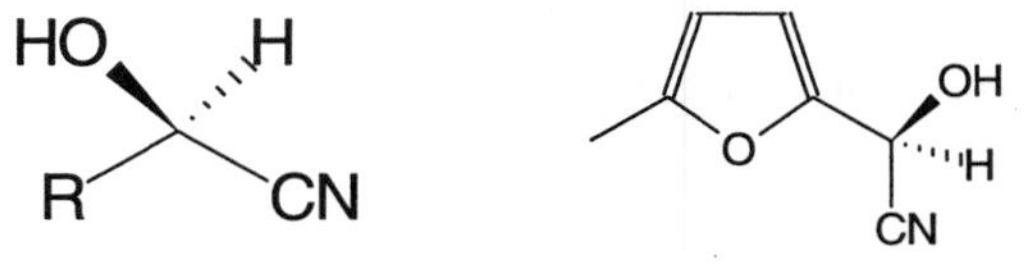

Figure 3. Cyanohydrins from *Rosaceae* HNL's

It should be noted however, that in some instances, e.g.: furfural cyanohydrin (Fig. 3), where the ring oxygen gives the substituent R priority over the nitrile group, the configuration is *S* according to the Cahn-Ingold-Prelog rules.

1.5. ISOLATION AND PURIFICATION OF HYDROXYNITRILE LYASES; ISOENZYMES

Various species in which hydroxynitrile lyase activity has been detected have been described. Sofar, only eleven different HNL's, belonging to six different plant families, have actually been isolated. In 1963 Becker and Pfeil purified the enzyme from bitter almonds for the first time [21]. One year later they reported to have succeeded in crystallizing the enzyme [22]. However, Pfeil communicated in 1980 that they had not been able to reproduce the experiment and at that time it was even uncertain whether the crystals had indeed been from hydroxynitrile lyase [23].

In 1991 Smitskamp-Wilms et al. isolated four isoenzymes from sweet almonds [24]. Two iso forms dominate and constitute over 90% of the HNL content. The four isoenzymes all contain a single polypeptide chain with a molecular weight of approximately 60 kD. More recently, Effenberger isolated also four different iso forms from the almond enzyme and succeeded in obtaining three of them in crystalline form.

Only preliminary X-ray diffraction results, like unit cell parameters, have been published so far [25]. The reason underlying this apparent lack of success in determining the X-ray structure may well be the fact that the enzyme molecules contain a variable number of monosaccharide units attached to the protein core. This severely complicates the interpretation of diffraction data.

The amount of HNL largely varies from one species to another: from less than 1 mg (*Ximenia americana*) to 900 mg (sweet almond) per 100 g of dried or defatted material. R-HNL from almonds [EC 4.1.2.10] and S-HNL from *Sorghum bicolor* [EC 4.1.2.11] are commercially available enzymes and can be used for the laboratory synthesis of the *R*- and *S*-enantiomers of cyanohydrins, respectively. Larger amounts of the S-HNL from *Hevea brasiliensis* [EC 4.1.2.39] have been prepared by over-expression in *Pichia pastoris*. This considerably enhanced the scope of S-HNL's [26,27]. The enzyme was crystallized and its three-dimensional structure was recently elucidated (see section 1.6.2). The very similar S-HNL from the leaves of *Manihot esculenta* [EC 4.1.2.37] was successfully over-expressed in *E. coli* [28].

1.6. THE CATALYSIS MECHANISM OF HYDROXYNITRILE LYASES

Since the discovery of the synthetic capabilities of emulsin by Rosenthaler in 1908, the mechanism underlying the formation of chiral cyanohydrins in non-racemic form has been studied intensively by several research groups. Although substantial progress has been made, it must be admitted that the understanding of the catalysis mechanism is still largely incomplete. As outlined earlier, the HNL's can be divided in FAD-containing and non-FAD-containing enzymes. It must be assumed that fundamental differences exist between the mode of action and/or substrate binding of these two classes. Therefore, the state of affairs will be reviewed separately for these two classes of enzymes.

1.6.1 *FAD-Containing Hydroxynitrile Lyase*
In the sixties and seventies, Pfeil and coworkers published several important contributions to the knowledge of the enzyme. They found that the presence of the tightly but not covalently bound FAD molecule in the enzyme is essential, both for its stability and for its activity. The apo enzyme is catalytically inactive, but activity can be fully regenerated by addition of FAD in its oxidized form [29]. A reaction scheme was proposed for the asymmetric synthesis as depicted in Scheme 1.1, in which the carbonyl compound first forms a complex with the chiral catalyst. This complex then reacts in a

rate-determining step with cyanide ion. In a third, fast, step the complexed cyanohydrin anion is protonated by HCN.

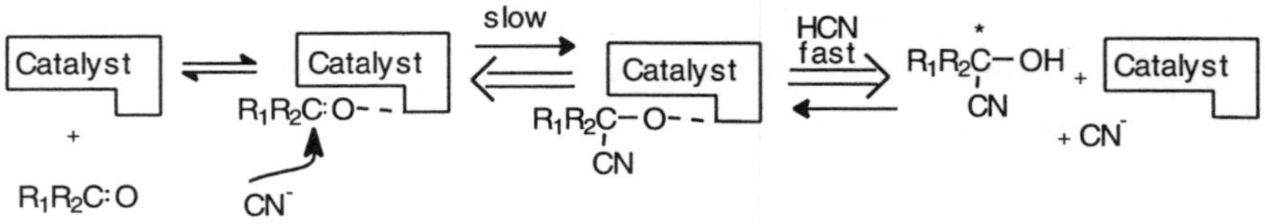

Scheme 1.1

Important contributions to the understanding of the enzyme mechanism were made in the late seventies and early eighties by the group of Lothar Jaenicke at the University of Köln and at that time by the group of Marilyn Schuman Jorns, at Ohio State University.

Kinetic studies by Jorns were consistent with an 'ordered uni-bi mechanism', 'uni-bi' meaning that there is one substrate and two products in the forward (decomposition) reaction and 'ordered' meaning that the products leave in obligatory order (in this case HCN first) [30]. Inhibitors which are reasonable analogs for cyanide ion, such as azide and thiocyanate, do not bind to the enzyme-aldehyde complex, suggesting that the binding species is HCN, rather than cyanide ion. It was also found that replacement of FAD by 5-deaza FAD leads to a large decrease in enzyme activity, indicating that binding to N5 plays an essential role.

It was observed by Bärwald and Jaenicke that no spectral changes occur upon addition of either mandelonitrile, benzaldehyde or HCN to the enzyme. Apparently, binding of substrate and product does not involve the flavine molecule [31]. It was subsequently found that one of the cysteine residues in the enzyme is essential for activity and can be selectively reacted with Michael acceptors. Upon reaction with labeled 3-oxo-3-phenylpropene, followed by reduction (to prevent a retro-Michael reaction) and extensive hydrolysis, L-2-amino-4-thia-DL-7-hydroxy-7-phenylheptanoic acid, the reduced conjugate addition product of cysteine to the inhibitor, was isolated (Scheme 1.2) [32].

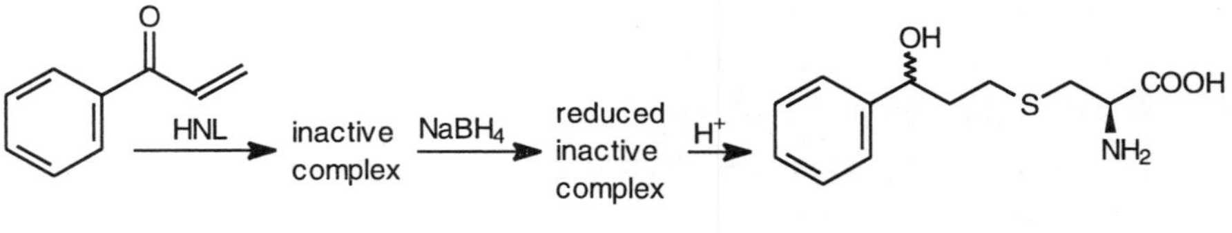

Scheme 1.2

Jaenicke and Preun then postulated a mechanism, in which the essential role of the cysteine residue and of the flavine molecule in the active site was emphasized. As depicted in Scheme 1.3, the proposed reaction mechanism involved addition of cyanide ion to N5, and of the carbonyl carbon to C4a, followed by an unprecedented four-center rearrangement. In view of the results reported earlier by Jorns, it is not surprising that this proposal met with severe criticism [33].

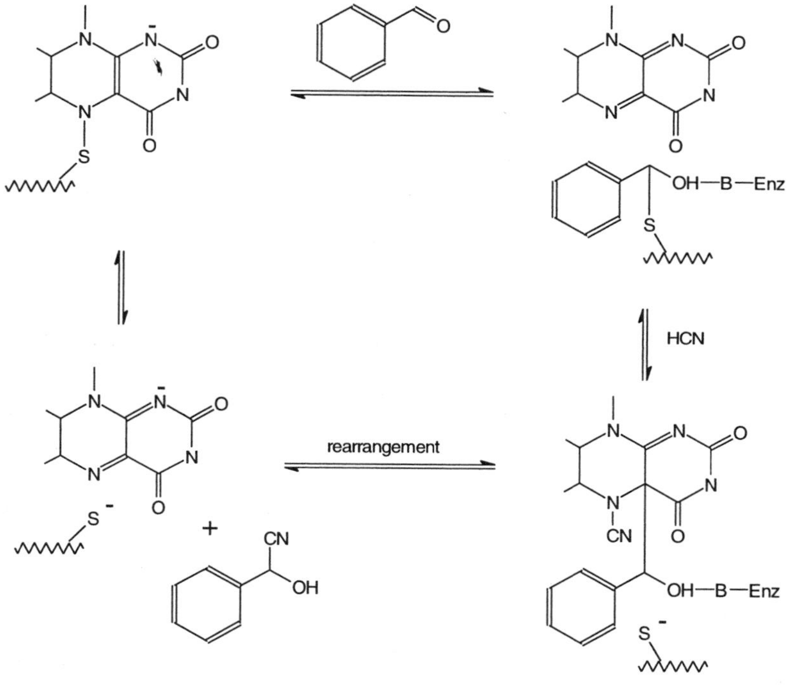

Scheme 1.3

The conversion of aldehydes into cyanohydrins, which can be both acid- and base catalyzed, does not necessitate the intermediacy of a complex transition state. Assuming that the enzymatic conversion involves both base catalysis to provide the nucleophilic cyanide ion and acid catalysis to protonate the incipient oxy-anion, the mechanism can in essence be presented as depicted in Scheme 1.4

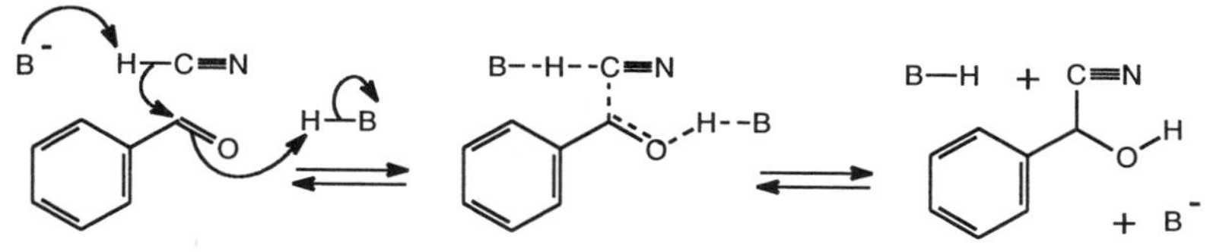

Scheme 1.4

Assuming that the essential cysteine residue plays a role, both in activating the carbonyl group in the first substrate and in deprotonating the HCN, a mechanism neither involving nucleophilic addition to the flavine molecule, nor the presence of intermediates for which there is no evidence, is depicted in Scheme 1.5 [34].

In view of the recently obtained structural information on the non-FAD containing enzyme from *Hevea brasiliensis*, this proposal may have to be modified in the sense that protonation and deprotonation may involve different amino acid residues in the active site. Clarification of this point must await the elucidation of the three-dimensional structure of the almond enzyme.

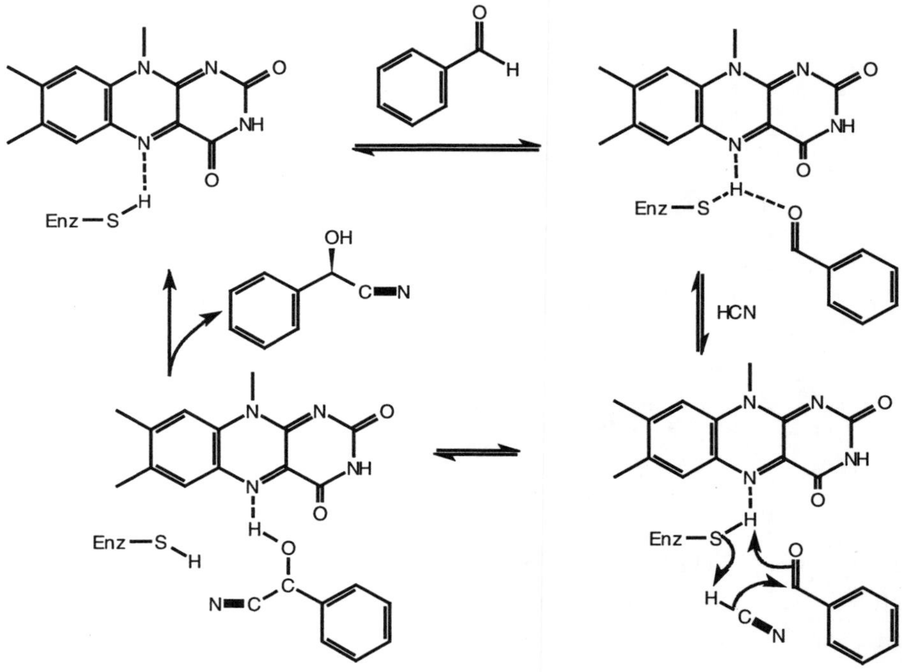

Scheme 1.5

1.6.2 Non-FAD-Containing Hydroxynitrile Lyases

Recently, important contributions to the understanding of the catalysis mechanism of non-FAD-containing HNL's have been made, especially by Herfried Griengl and collaborators at the Technical University Graz. A breakthrough, both with regard to synthetic applications as to the structure determination, was made by the successful overexpression of the HNL from maniok (*Manihot esculenta*) in *E. coli* and of the HNL from the rubber tree (*Hevea brasiliensis*) in *Pichia pastoris* [6]. Both enzymes are *S*-selective and show a wide substrate acceptance. Their molecular weight (30 kD) and amino acid sequence were found to be highly similar. In the case of the Hb-HNL, overexpression led to extremely high levels (up to 60% of total cellular proteins) of soluble HNL, exhibiting a specific activity of 40 U/mg, which is about twice the activity found for the purified enzyme isolated from *Hevea brasiliensis*. This highly purified enzyme was crystallized and for the first time the three-dimensional structure of an enzyme with HNL activity could be determined by X-ray crystallography [35]. The enzyme turned out to be a member of the well-studied α,β-hydrolase fold family. This family comprises enzymes with a wide variety of activities, including acetyl cholinesterase-, hydrolase-, lipase- and carboxypeptidase activity. They nevertheless show large similarities in their catalytic mechanisms. On the basis of the established three-dimensional structure, site directed mutagenisis and inhibition experiments, Griengl and collaborators have formulated a tentative mechanism for the stereoselective formation of cyanohydrins by the Hb-HNL enzyme (Scheme 1.6) [4].

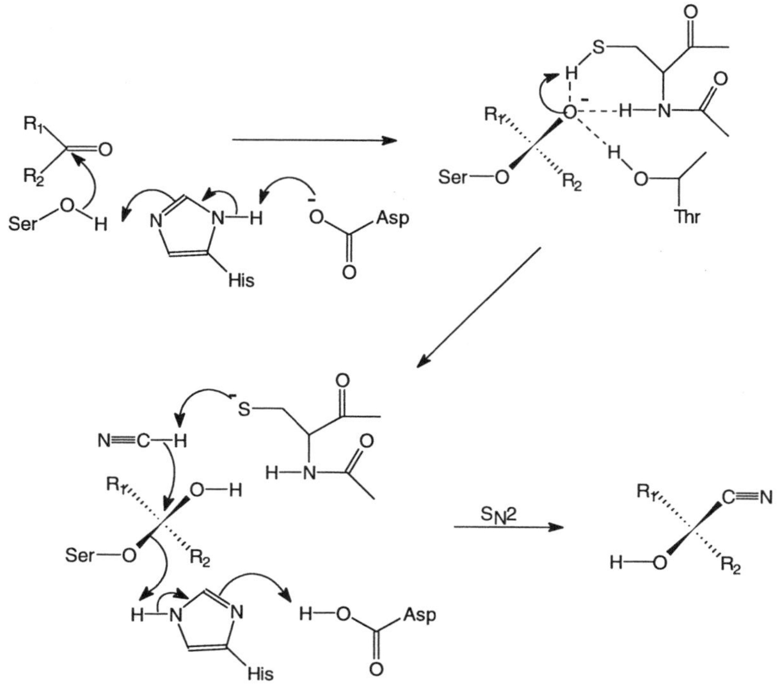

Scheme 1.6

Important contributing factors to this proposal are the following:

- The active site is buried deep inside the protein and is only accessible by a narrow channel. This is consistent with an ordered uni-bi mechanism, as predicted both for the FAD-containing [30] and for the non-FAD-containing [6] HNL's.
- In analogy with other members of the family of α,β-hydrolase fold enzymes, the presence of a catalytic triade in the active site was predicted and could be confirmed by site directed mutagenisis. The triade consists of Ser[80], the oxygen atom of which is rendered nucleophilic by incipient proton abstraction by His[235], which in turn is activated by Asp[207].
- Again in analogy with other enzymes of the family the activated Ser[80] residue is believed to act as a nucleophile and attacks the carbonyl carbon. The incipient negative charge at the carbonyl oxygen is stabilized by a set of three hydrogen bonds, arising from the Thr[11] hydroxyl, the Cys[81] N-H, and the Cys[81] S-H, together forming the 'oxy-anion hole'.
- The oxy-anion of the resulting tetrahedral intermediate is protonated by the Cys[81] thiol group, thereby forming a hemi-acetal.
- The concurrently formed thiolate anion deprotonates HCN to give cyanide ion.
- The cyanide ion attacks the tetrahedral intermediate and displaces the Ser[80] residue in an S_N2 reaction, thereby liberating the S-cyanohydrin.

1.7.   SUBSTRATE ACCEPTANCE OF HYDROXYNITRILE LYASES FROM DIFFERENT SOURCES

The substrate acceptance of the almond enzyme is remarkably wide, including both saturated- and unsaturated-, aliphatic- and aromatic aldehydes, as well as aliphatic methyl ketones. Also heteroaromatic aldehydes (thienyl, furyl) are good substrates. The chain length of aliphatic aldehydes was originally thought to be limited to C_6. Recently it was shown by Kanerva that aldehydes up to C_{10} can be converted by the almond enzyme.

The substrate specificity of the sorghum enzyme appears to be limited to aromatic and heteroaromatic aldehydes. The S-HNL's from *Hevea brasiliensis* and from *Manihot esculenta*, on the other hand, show a wider substrate acceptance. Both aliphatic-, aromatic-, and heteroaromatic aldehydes, saturated as well as α,β-unsaturated, and also methyl ketones are accepted by these enzymes [6,28]. A survey of all substrates that have been reported to be converted into chiral cyanohydrins in at least moderate yield and useful e.e. is presented in Scheme 1.7.

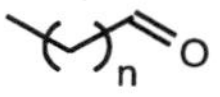

n = 1 to 8

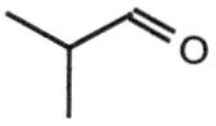

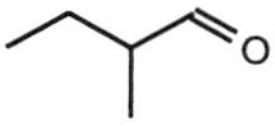

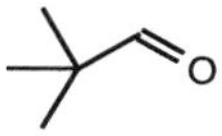

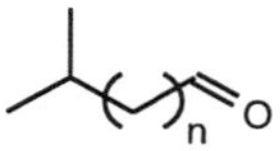

n = 1, 2

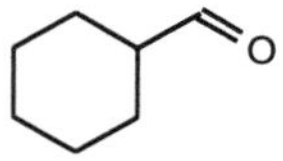

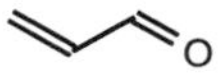

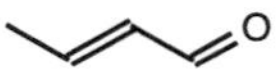

n = 1, n = 3

n = 1 to 3

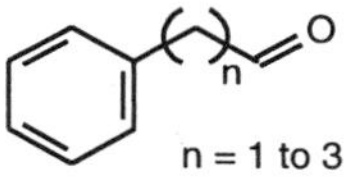

n = 1 to 3

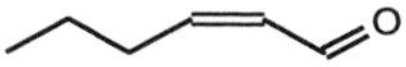

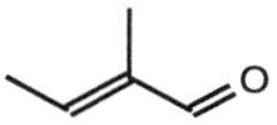

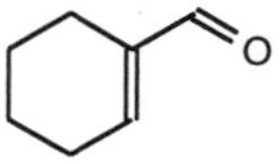

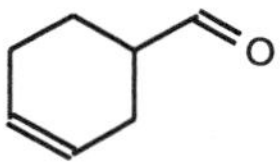

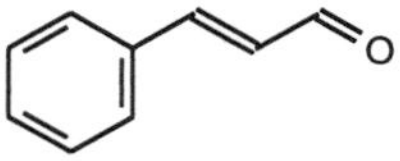

n = 1 , 2

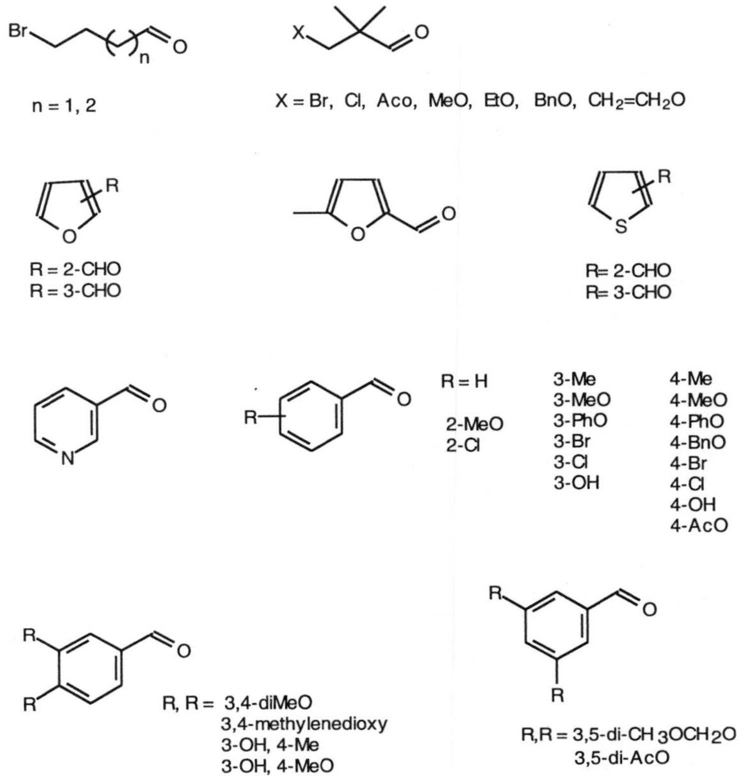

Br⌒⌒⌒(⌒)ₙ⌒CHO

n = 1, 2

X⌒⌒⌒CHO

X = Br, Cl, Aco, MeO, EtO, BnO, CH₂=CH₂O

R = 2-CHO
R = 3-CHO

R= 2-CHO
R= 3-CHO

R = H	3-Me	4-Me
2-MeO	3-MeO	4-MeO
2-Cl	3-PhO	4-PhO
	3-Br	4-BnO
	3-Cl	4-Br
	3-OH	4-Cl
		4-OH
		4-AcO

R, R = 3,4-diMeO
3,4-methylenedioxy
3-OH, 4-Me
3-OH, 4-MeO

R,R = 3,5-di-CH₃OCH₂O
3,5-di-AcO

Scheme 1.7. Aldehydes as substrates for HNL's.

1.8. APPLICATION OF HYDROXYNITRILE LYASES ON A PRACTICAL SCALE

Interesting preparative applications in organic synthesis have been described for only four enzymes, the FAD-containing R-HNL from *Rosaceae*, in particular from almonds *(Prunus amygdalus)*, and the non-FAD-containing S-HNL's from sorghum *(Sorghum bicolor)*, cassava *(Manihot esculenta)*, and from the rubber tree *(Hevea brasiliensis)*.

1.8.1 R-Hydroxynitrile Lyase from Almonds

A.   *Purified HNL, immobilized on ECTEOLA cellulose, in aqueous methanol.*
The purified enzyme combines with cellulose-based ion exchangers to form an active and stable catalyst. (*R*)-(+)-mandelonitrile with an optical purity of 94% could be

obtained on a multi-gram scale, using aqueous ethanol or methanol as the solvent [10]. The small amount of racemic cyanohydrin formed originates from the spontaneous, non-enzymatic reaction, which cannot be completely suppressed in aqueous solvents. The almond enzyme has been shown to consist of several iso-enzymes, differing in the protein composition of the apo-enzyme. For preparative purposes it is not useful to separate these iso-enzymes, as they show very similar substrate specificity [24].

B.   Crude extract from almond meal in aqueous ethanol.

It was shown by Brussee *et al.* that laborious purification of the enzyme is not necessary. Results, identical to those obtained by Becker and Pfeil, could be realized using a crude aqueous extract from defatted, ground almonds. Another improvement, in particular regarding the safety aspects, consisted of the generation of HCN *in situ* from a solution of KCN in an acetate buffer at pH 5.4 [11,19].

C.   Purified HNL, supported on cellulose, using a non-aqueous solvent system.

Effenberger and coworkers showed for the first time that the enzyme catalyzed reaction can proceed in solvents that are not miscible with water [12]. Best results were obtained using ethyl acetate. Non-enzymatic cyanohydrin formation is almost completely suppressed under these conditions and, although longer reaction times are required, the enantiomeric purity of the cyanohydrins obtained from a variety of aldehydes was in fact appreciably better than in the alcohol/water systems. The purified enzyme was bound to crystalline cellulose (Avicel). Other hydrophobic carriers, like Celite, have also been used.

D.   Synthesis of cyanohydrins using almond meal in an organic solvent.

A method that is particularly convenient for the preparation of cyanohydrins on a laboratory scale was published in 1991. Rather than first purifying the enzyme and then absorbing it on a solid support, the unpurified enzyme was used as present on its natural support, almond meal [37,38]. Almond meal is commercially available (Sigma-Aldrich), and is also easy to prepare in the laboratory. By controlling the pH (3.5-5.5) and temperature (0-5 °C.), the non-enzymatic conversion in the minor aqueous phase (4% v/v) is efficiently suppressed and e.e.'s of more than 99 % are regularly obtained. The almond meal method has been used to great advantage in several recent investigations. Under 'micro-aqueous conditions', which can be achieved by not swelling the almond meal in a buffer solution, the enzyme performs well, also at higher temperatures (up to 30 °C.) and can be more easily recovered [39].

E.   Trans-cyanation, using acetone cyanohydrin as the HCN donor.

A different way of suppressing the non-enzymatic reaction is by keeping the concentration of HCN low at all times. This can be achieved by performing a hydroxynitrile lyase mediated trans-cyanation, using acetone cyanohydrin as an HCN source [40]. The e.e. values obtained were in several cases higher than those reported previously. However, the selectivities claimed by the authors could not be reproduced. An inherent disadvantage of this method is that the desired reaction product has to be separated from unreacted acetone cyanohydrin. This can be avoided by carrying out the decomposition of the trans-cyanating agent, either enzyme- or base-catalyzed, in a separate vessel and allow the liberated HCN to diffuse into the reaction mixture [41].

F.   Mass transport limitation as a tool to enhance enantioselectivity.

Recently, it was found in our laboratories that mass transport limitation can be used as an effective tool to enhance enantioselectivity [42]. In an optimized aqueous/organic solvent two-phase system, interesting substrates, previously reported to give poor results, such as acrolein and cinnamic aldehyde, could be successfully transformed into the corresponding cyanohydrins of high enantiomeric purity.

G.   S-cyanohydrins by selective cleavage of the R-form in a racemate.

In principle it should be possible to obtain *S*-cyanohydrins by selective enzymatic decomposition of the *R*-enantiomer in a racemic mixture. This is difficult to achieve on a practical scale, because the HCN that is liberated can engage in both enzyme catalyzed and non-enzymatic reactions by which the *R*-enantiomer is again formed. This problem has been elegantly solved by Gotor and coworkers by combining this approach with that of trans-cyanation [43]. A mixture of aldehyde and racemic cyanohydrin from a methyl ketone was treated with almond meal. The *R*-cyanohydrin of the methyl ketone was selectively cleaved and the *R*-cyanohydrin of the aldehyde was selectively formed by trans-cyanation. The liberated methyl ketone does not compete with the aldehyde for the liberated HCN, because its hydrocyanation proceeds at a much lower rate. After chromatographic separation, the *S*-cyanohydrins of 2-pentanone and 2-hexanone were obtained, as well as the *R*-cyanohydrins of 4-bromobutanal and 5-bromopentanal. The latter compounds are valuable starting materials for heterocyclic synthesis [44].

H.   Purified enzyme in a membrane reactor in a continuous process.

The possibility to produce cyanohydrins of high enantiomeric purity in a continuous process was studied by Kragl, Kula and coworkers [45]. The spontaneous, non-enzymatic reaction was suppressed by working at very low pH (3.75) and by mixing the reactants just before entering the reactor. The enantiomeric excess was further enhanced by making use of the different kinetics of the two reactions. The non-enzymatic reaction is first order in (benz)aldehyde and therefore its rate increases linearly with aldehyde concentration. The enzyme-catalyzed reaction on the other hand obeys typical Michaelis-Menten kinetics and reaches its V_{max} already at low aldehyde concentrations. It is therefore possible to suppress the spontaneous reaction effectively by working with high enzyme and low aldehyde concentrations. The enzyme was contained by an ultrafiltration membrane. Enantiomerically pure (R)-mandelonitrile (e.e. > 99%) was produced with a space-time-yield of up to 2400 g. l^{-1}. day^{-1}. The enzyme consumption was about 17,000 U/kg of product. Because of the easy accessibility of the enzyme, this was not considered to be a limiting factor. [1 kg of almonds provides about 400,000 U, one unit being defined as the quantity that will form 1.0 μmole of benzaldehyde and HCN from (R)-mandelonitrile per min at pH 5.4 and at 25 °C].

I.   Purified HNL in a water / organic solvent biphasic system.

A second procedure aiming at the production of chiral cyanohydrins on an industrial scale was developed by Loos *et al* [46]. Purified enzyme was used in a liquid-liquid biphasic system. Applying a so-called 'D-optimal' design, the process was optimized for a multitude of parameters, using piperonal and anisaldehyde as representative substrates for fast and slow cyanohydrin formation, respectively. On the basis of the results obtained, four critical parameters (pH, concentration of the two reactants and temperature) were optimized in depth with a Full Factorial Design (2^4-FFD). It was shown that (R)-mandelonitrile can be prepared in this two-phase system in excellent yield (98%) and e.e. (98%), using methyl *t*-butyl ether (MTBE) as the solvent and a citrate buffer as the aqueous phase at pH 5.0 at room temperature. The through-put under these conditions was 2.1 $mol.l^{-1}.h^{-1}$ (corresponding to 6700 g. $l^{-1}.day^{-1}$). Notably, the activity of the enzyme had not measurably decreased after re-using the aqueous enzyme-containing layer in five successive experiments.

1.8.2 R-Hydroxynitrile Lyases from Other Rosaceae

With the aim of extending the substrate acceptance, a number of other *Rosaceae* R-hydroxynitrile lyases, namely those from apple, cherry, apricot and plum, were investigated and found to be indeed different [41]. The meal from the seeds of these fruits was used, rather than the purified enzymes. The properties of the enzyme from apple seeds turned out to be the most promising. For the more hindered substrates (3- and/or 4-substituted aromatic-, long chain- and branched aliphatic) this HNL gave improved results, both with regard to reaction rate and to e.e., over those obtained with almond meal. The properties of the other *Rosaceae* HNL's were less interesting. It can be concluded that only in those cases where the almond meal, which is commercially available and easy to prepare, fails to give satisfactory results, it may be worthwhile to try meal from apple seeds as an alternative.

1.8.3 S-Hydroxynitrile Lyases

A.   S-Hydroxynitrile lyase from *Sorghum bicolor*

The first enzyme to be investigated for its potential to catalyze the preparative synthesis of *S*-cyanohydrins was the hydroxynitrile lyase isolated from the seedlings of *Sorghum bicolor* (EC 4.1.2.11) [18]. The first results were rather promising. 4-Hydroxybenzaldehyde could be converted to the natural substrate, 4-hydroxy-mandelonitrile, on the mmole scale in high yield in optically pure form (sodium citrate buffer, pH 3.75, room temperature). Some other aromatic aldehydes also performed well. Larger amounts could be converted in a CST reactor using the enzyme, immobilized on Eupergit C. The results were confirmed and extended by Effenberger and coworkers, who used a suspension of the enzyme, immobilized at pH 5.4 on Avicel-cellulose, in an aprotic organic solvent [diisopropyl ether (DIPE)] [17].

Subsequently, Kiljunen and Kanerva showed that for preparative purposes it is not necessary to use the extensively purified enzyme. Similar results can be obtained using lyophilized and dechlorophylled *Sorghum bicolor* shoots. Despite these promising results, it is unlikely that the *Sorghum bicolor* enzyme will see much future application, for the following reasons:

- Only aromatic aldehydes are accepted by Sb-HNL. This narrow substrate range severely hampers wide application of the enzyme.

- The limited availability of the enzyme. In contrast to *Prunus amygdalus*, which contains about 4 g of HNL per kg of almonds, *Sorghum* holds only about 30 mg of HNL per kg of shoots. Attempts to produce larger amounts by genetic engineering have been frustrated by the complex post-translational processing needed for its expression.

B.   S-Hydroxynitrile lyases from Manihot esculenta and from Hevea brasiliensis

Much wider substrate acceptance, similar to the one shown by the R-HNL from almonds, is shown by the S-HNL's obtained from the leaves of cassava (*Manihot esculenta*, EC 4.1.2.37) [28] and of the rubber tree (*Hevea brasiliensis*, EC 4.1.2.39) [27]. Both enzymes accept aromatic-, heteroaromatic-, aliphatic-, and α,β-unsaturated aldehydes. Also, methyl ketones have been shown to act as substrates. The enzyme from cassava leaves was the first HNL to be made available in larger quantities by genetic engineering. The cloned Me-HNL-gene was overexpressed in *Escherichia coli* cells. The biochemical properties of recombinant Me-HNL and Me-HNL isolated from the leaves of *Manihot esculenta* were identical, but the specific activity of the recombinant enzyme was shown to be twenty-five times higher. Whereas the original experiments obtained with Me-HNL from cassava gave unsatisfactory optical yields, the recombinant Me-HNL produced excellent results, both with regard to yield and e.e. The enzyme was absorbed on nitrocellulose that had been soaked in citrate buffer (pH 3.3) and thoroughly dried. DIPE was again the solvent of choice.

The hydroxynitrile lyase from *Hevea brasiliensis* was first described in 1989. Its potential for the synthesis of *S*-cyanohydrins was extensively studied by Griengl and coworkers [26,27]. Interestingly, this enzyme, which is highly homologous to the HNL from cassava, appeared to perform best in aqueous media. The non-enzymatic reaction was suppressed by working at low pH (citrate buffer, pH 4.0). Although trans-cyanation with acetone cyanohydrin can be applied to keep the HCN concentration down, better results were obtained by using 2 mole equivalents of potassium cyanide at pH 4.0. Of particular interest were the results obtained with 3-phenoxybenzaldehyde, the *S*-cyanohydrin derived therefrom is the alcohol component of some important pyrethroid insecticides [34].

A breakthrough with regard to the availability of this enzyme was achieved by its successful overexpression in several micro-organisms [47]. The most efficacious expression was accomplished in the yeast *Pichea pastoris*. Yields of more than 20 g of pure HNL protein per liter of culture volume could be obtained. It was subsequently shown that the crude enzyme extract (140 IU/ml) can be used for efficient conversion of all substrates. Under standard conditions 100 IU of the enzyme are used per mmole of aldehyde. Substrates that give unsatisfactory e.e.'s under these conditions can be converted with a 20-fold quantity of enzyme at pH 4.5. All reactions were carried out at 0 °C.

Summarizing, it can be concluded that excellent preparative procedures have been described for the use of R-hydroxynitrile lyase from almonds as well as for the overexpressed S-hydroxynitrile lyases from cassava and from the rubber tree. Large scale biocatalytic production of both *R*- and *S*-cyanohydrins in enantiomerically pure form seems feasible. As illustrated in the second part of this chapter, this will provide the fine chemical industry with a host of valuable new chiral building blocks.

382

1.8.4 Safety Aspects

Toxicity of HCN

Hydrocyanic acid is a highly toxic, and easy detectable chemical. Details about the toxicity of HCN (CAS registry No: 74-90-8) and sodium cyanide (CAS registry No: 143-33-9) and about the hazards involved can be found in the regular handbooks [48,49]. Appropriate procedures for the safe use of HCN, and related compounds such as sodium cyanide, are described in literature. HCN detection equipment is commercially available from multiple sources. The TLV (threshold limit value) or MAC-value (maximum allowed concentration) of HCN is 10 ppm, which corresponds to 11 mg/m^3. This concentration should also be regarded as a MAC-C (ceiling) value.

Waste disposal

Small amounts of HCN can be disposed conveniently by conversion into cyanide and oxidation to relatively innocuous cyanate with the aid of sodium hypochlorite (commercial laundry bleach) [50]. The cyanate can be further processed to ammonia and carbon dioxide by addition of sulfuric acid to pH 7, according to:

$HCN + NaOH \rightarrow NaCN + H_2O$

$NaCN + NaOCl \rightarrow NaCl + NaCNO$

$2NaCNO + H_2SO_4 + 2H_2O \rightarrow Na_2SO_4 + 2CO_2 + 2NH_3$

2. CYANOHYDRINS AS MULTIFUNCTIONAL SYNTHONS

Cyanohydrins offer a wide spectrum of opportunities for transformations because of their multifunctional character (Fig 4).

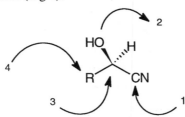

Figure 4. Transformations of cyanohydrins

1.   Reactions at the cyano group (solvolysis, Grignard reaction, hydride addition).

2.   Reactions at the hydroxyl group (protecting groups).

3.   Inversion of the stereogenic center by conversion of the hydroxyl function into a good leaving group, followed by attack of a nucleophile.

4.   Reactions at the substituent R. The presence of functionality (e.g.:unsaturation) in this substituent offers a whole range of possibilities for further conversions.

2.1.   REACTIONS UNDER ACIDIC CONDITIONS:
FORMATION OF α-HYDROXY ACIDS AND ESTERS, AND LACTONES

Any conversion of cyanohydrins under basic conditions is attended with severe decomposition and racemization. Acid-catalyzed hydrolysis of cyanohydrins on the other hand is generally considered to be a simple and straightforward reaction. Nevertheless, appreciable decomposition occurs when a cyanohydrin and concentrated HCl are mixed and directly heated. This is hardly surprising, as most cyanohydrins are unstable at temperatures exceeding 40 °C. Indeed, stirring the cyanohydrin with concentrated HCl at lower temperatures, until all of the starting material is converted to the amide, before raising the temperature in order to complete the hydrolysis, gives much better results [51].

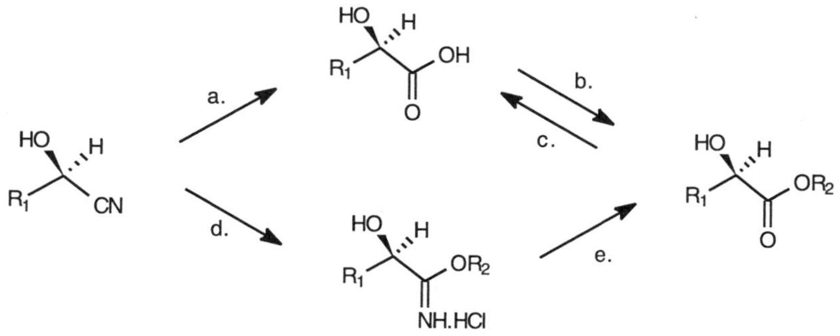

a. HCl / H$_2$O; b. ROH / H$^+$; c. NaOH / dioxan; d. HCl gas, ROH, ether; e. H$_2$O

Scheme 2.1. Products from unprotected cyanohydrins.

α-Hydroxy esters can be obtained from these acids in the usual way. They are also directly accessible from the cyanohydrins by a modified Pinner synthesis: conversion in diethyl ether under optimized conditions, affords the imidate salts (Scheme 2.1). By dissolving this salt in water, it is rapidly and quantitatively converted into the corresponding ester [52].

If R-cyanohydrins of β-substituted pivaldehydes, like β-methoxy- or β-chloropivaldehyde, are hydrolized with concentrated HCl, (R)-pantolactone is obtained in good yield and in high enantiomeric excess (Scheme 2.2) [53].

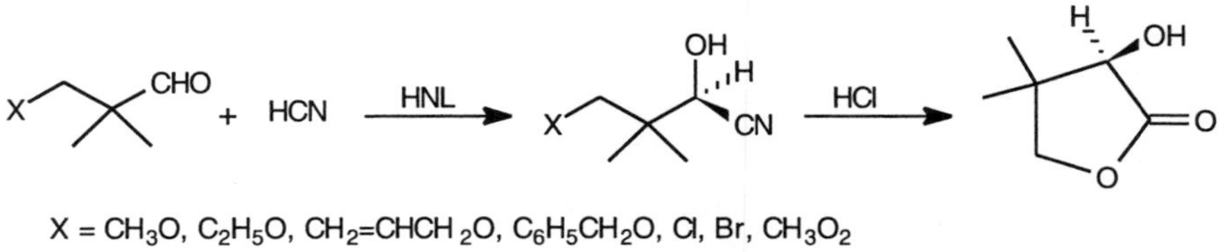

X = CH$_3$O, C$_2$H$_5$O, CH$_2$=CHCH$_2$O, C$_6$H$_5$CH$_2$O, Cl, Br, CH$_3$O$_2$

Scheme 2.2. Synthesis of (R)-pantolactone.

2.2. PROTECTION OF THE HYDROXYL FUNCTION

For application as chiral building blocks, cyanohydrins have the disadvantage of their chemical instability. Furthermore, chemical transformations using basic-, reducing-, or organometallic reagents often result in low yields. These problems can be overcome by introducing a protective group at oxygen. Care has to be taken that this protective group is introduced under sufficiently mild conditions to avoid racemization.

Chiral cyanohydrins have been subjected to mild esterification for e.e. determination purposes. Yet, because of the reactivity of the ester group, cyanohydrin esters are not suitable as starting materials in most subsequent reactions at the nitrile function.

Chiral cyanohydrins have been protected as tetrahydropyranyl (THP) ethers by reaction with 3,4-dihydro-2H-pyran, catalyzed by p-toluenesulfonic acid [54]. Introduction of the THP protecting group is accompanied by formation of a second chiral center and thus a mixture of diastereomers is formed. This can complicate subsequent purification and analytical procedures. To avoid this problem, chiral cyanohydrins have been protected as 2-methoxy-i-propyl (MIP) [55-58] or 2-phenoxy-i-propyl (PIP) ethers [59]. The latter have the additional advantage of being UV-active, thereby facilitating detection of UV-inactive cyanohydrins (e.g. those derived from aliphatic aldehydes). MIP- and PIP protection can be removed by mild acid hydrolysis.

Most frequently, cyanohydrins are protected as silyl ethers. Silylated cyanohydrins can be obtained, without racemization and in high yields, by reaction with a suitable silyl chloride in the presence of two equivalents of imidazole [60.61]. The stability of the silyl-protecting group can be tuned by proper choice of the substituents at silicon. The trimethyl silyl (TMS) group can be removed under mildly acidic conditions. The *t*-butyldimethylsilyl (TBDMS or TBS) group and the UV-active *t*-butyldiphenylsilyl (TBDPS) group are progressively more acid stable and can be removed with fluoride ions (TBAF, HF) or, in special cases, by intramolecular reductive cleavage with LiAlH$_4$. In the latter case, the presence of an atom bearing a proton that reacts with LiAlH$_4$, or a functional group which is reduced by LiAlH$_4$, at the carbon atom neighbouring the one carrying the silyl ether, is a prerequisite. This allows selective cleavage of one TBS ether over another one that lacks such a structural feature [62].

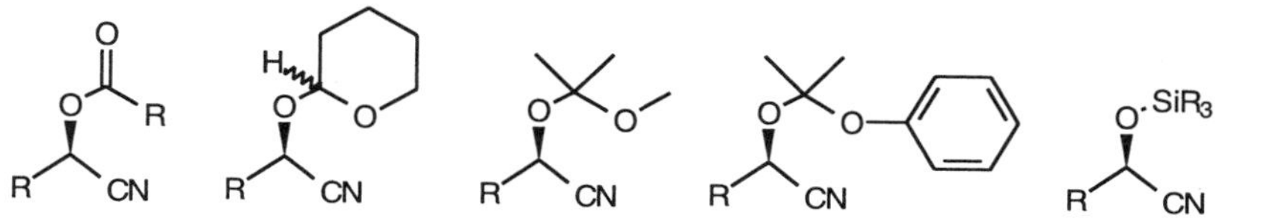

Figure 5. Protected cyanohydrins.

2.3. INVERSION OF THE STEREOGENIC CENTER

Optically active α-sulfonyloxy nitriles, which can be used for inversion of the configuration of cyanohydrins, have been prepared [63]. Aliphatic α-sulfonyloxy nitriles are stable at room temperature and react with a series of reagents (Scheme 2.3). Aromatic α-sulfonyloxy nitriles are too unstable to be synthetically useful.

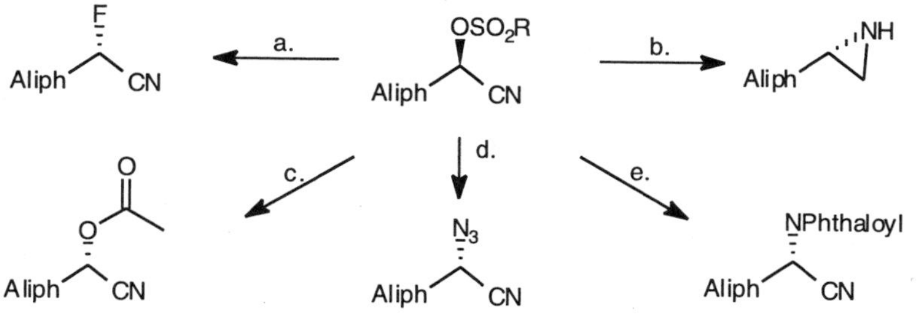

a. Amberlyst/F; b. LiAlH$_4$; c. CH$_3$CO$_2$K/DMF; d. KN$_3$/DMF; e. potassium phtalimide

Scheme 2.3. Substitution products from α-sulfonyloxy nitriles

Allylic- and benzylic cyanohydrins have been transformed into cyanohydrin esters of opposite configuration under Mitsunobu conditions (Scheme 2.4) [64]. Cyanohydrins containing strongly electron-donating substituents (e.g. 4-methoxymandelonitrile) gave extensive racemization, while saturated aliphatic cyanohydrins afforded esters in which the original configuration was retained.

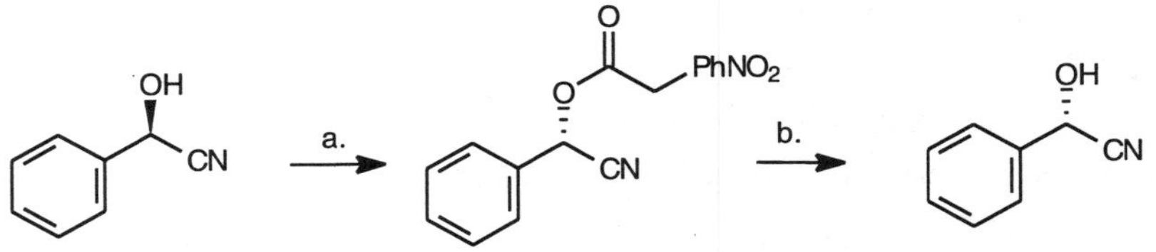

a. triphenylphosphine, diethyl azocarboxylate, p.nitro phenylacetic acid
b. methylsulfonic acid, methanol

Scheme 2.4. Inversion under Mitsunobu conditions

As the Mitsunobu inversion gave excellent results with two important categories of cyanohydrins that did not behave satisfactorily via the sulfonyloxy route, namely aromatic cyanohydrins and α,β-unsaturated cyanohydrins, the two approaches are nicely complementary.

Using an appropriately protected amine as the Mitsunobu nucleophile it is possible to convert aliphatic cyanohydrins into protected α-aminonitriles, which are precursors to chiral amino acids [65].

2.4. NUCLEOPHILIC ADDITION TO THE NITRILE FUNCTION

2.4.1 DIBAL Reduction/Hydrolysis: Formation of α-Hydroxy Aldehydes

DIBAL is commonly used for the partial reduction of esters and nitriles. Addition of DIBAL to a nitrile results in the formation of an imine-aluminum complex that is decomposed, during work-up, to the aldehyde (Scheme 2.5). Protected cyanohydrins have to be used, because dimerization and isomerization can take place. Even in the protected form special care has to be taken during synthesis (low temperatures) and work-up [56].

2.4.2 LiAlH₄ or DIBAL/NaBH₄ Reduction: Formation of Ethanolamines

In 1965 the asymmetric synthesis of ethanolamines via enzyme-catalyzed addition of hydrogen cyanide to aldehydes, followed by reduction with LiAlH₄ was first described

[66]. Subsequently, TBS-protected cyanohydrins were used in order to avoid decomposition and racemization [60]. Surprisingly, quantitative deprotection by an intramolecular reductive cleavage occurred and the free ethanolamines were obtained in high yields [60,62]. TBS-protected ethanolamines (with one chiral center) could be obtained by DIBAL reduction at low temperature, followed by $NaBH_4$ reduction. (Scheme 2.5) [56].

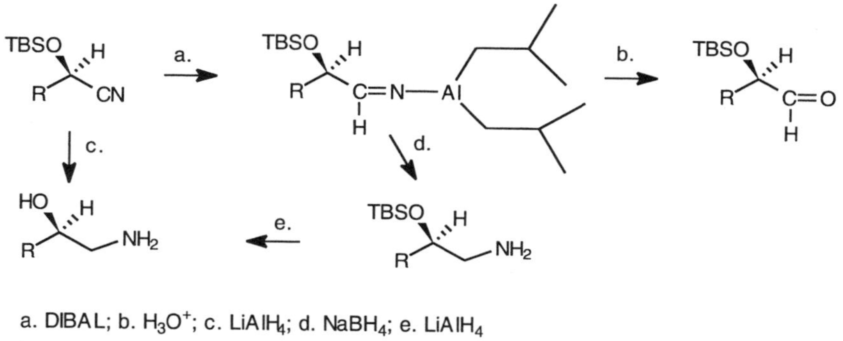
TBSO
H
R
CN
a.
TBSO
H
R
C=N—Al
H
b.
TBSO
H
R
C=O
H
c.
d.
HO
H
R
NH2
e.
TBSO
H
R
NH2
a. DIBAL; b. H3O+; c. LiAlH4; d. NaBH4; e. LiAlH4

Scheme 2.5. Hydride reduction of O-protected cyanohydrins

2.4.3 Grignard Reaction/Hydrolysis: Formation of Acyloins

α-Hydroxyketones (acyloins) were prepared by a Grignard reaction on TMS- or TBS-protected mandelonitriles (Scheme 2.6) [14]. Because of the acid-lability of the TMS group, in the former case the free acyloins were isolated after acidic work-up. In the latter case the TBS-protected derivatives were isolated. In both cases high e.e. values were recorded. The protected acyloins are excellent starting materials for further stereoselective transformations. E.g.: reductive amination in the presence of magnesium salts afforded ethanolamines with high diastereoselectivity [61].

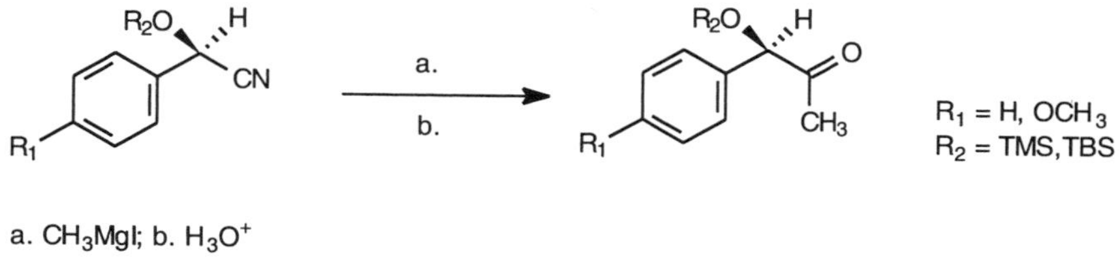
R2O
H
CN
R1
a.
b.
R2O
H
O
CH3
R1
R1 = H, OCH3
R2 = TMS,TBS
a. CH3MgI; b. H3O+

Scheme 2.6. Synthesis of O-protected acyloins.

2.4.4 Grignard Reaction/Hydride Reduction: Introduction of a Second Stereogenic Center

In the previous paragraph it was shown that Grignard reagents add smoothly to silyl protected cyanohydrins. Reduction with $NaBH_4$ of the imine intermediates, instead of acidic work-up, affords ethanolamines with two stereogenic centers (Scheme 2.7). Most likely, steric control during formation of the second stereogenic center is enhanced by coordination of magnesium-(II)-ions to both the nitrogen- and the oxygen moiety. Erythro/threo ratios of the crude products varied between 85/15 and 91/9. Pure erythro ethanolamines were obtained after one crystallization of their HCl-salts [60,67].

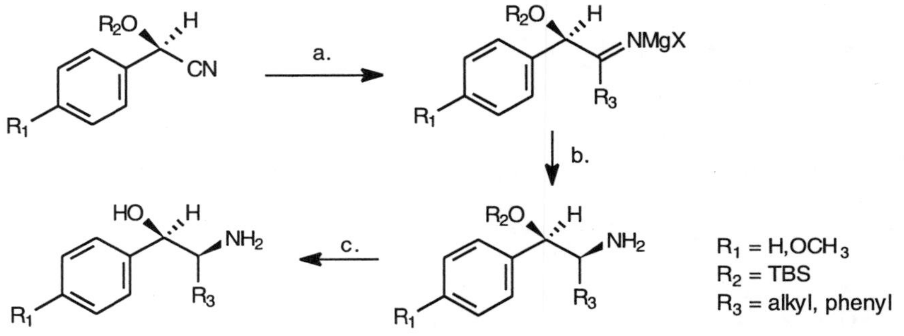

a. R_3MgX; b. $NaBH_4$; c. HF or $LiAlH_4$

Scheme 2.7. Synthesis of ethanolamines with two stereogenic centers

2.4.5 Addition of Vinylmagnesium Bromide: Formation of 2-Amino-1,3-diols

Attempts by Effenberger et al. to prepare 2-amino-1,3-diols using functionalized organometallic reagents like Grignard compounds derived from halo ethers or silyl methyl chlorides resulted in racemization and decomposition of the starting material. An alternative route to 2-amino-1,3-diols was found by introducing a vinyl group via Grignard addition to the nitrile function and subsequent ozonolysis, followed by reductive work-up (Scheme 2.8) [68].

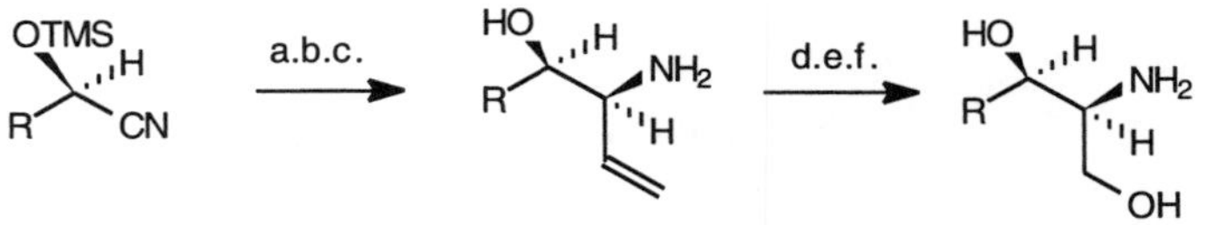

a. vinylmagnesium bromide, b. $NaBH_4$, c. H_3O^+, d. AcO_2/py; e. O_3, f. NaBH

Scheme 2.8. Synthesis of 2-amino-1,3-diols

2.4.6 Blaise Reaction: Formation of Tetronic Acids

Tetronic acids (Scheme 2.9) are of great interest because their derivatives are foound widespread in nature. The Blaise reaction (addition of the zinc derivative of an α-bromo ester to a nitrile) has been applied to O-protected cyanohydrins. Isolation of the tetronic acids was achieved, in moderate yields, by acid hydrolysis and cyclization of the intermediate imines (94). Better results were obtained in an optimized reaction sequence (Scheme 2.9) [69].

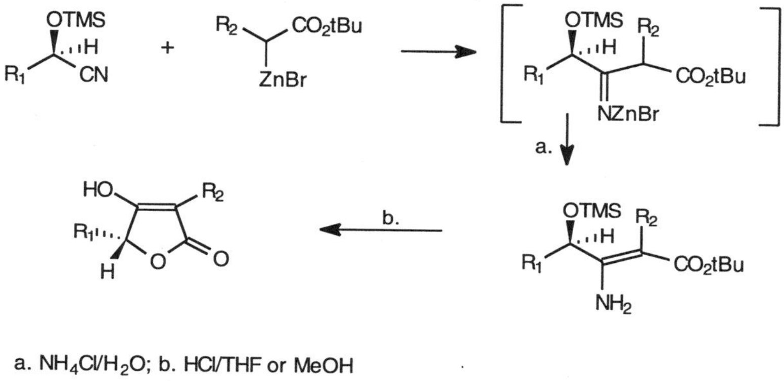

a. NH_4Cl/H_2O; b. HCl/THF or MeOH

Scheme 2.9. Synthesis of tetronic acids

2.5. TRANSIMINATION

Transimination (interconversion of Schiff bases), also called transaldimination or trans-Schiffization, denotes the reaction of an imine with a primary amine so that the original imine is converted to a new imine and a primary amine or ammonia is liberated (Scheme 2.10). Chemically it is a symmetric and reversible process.

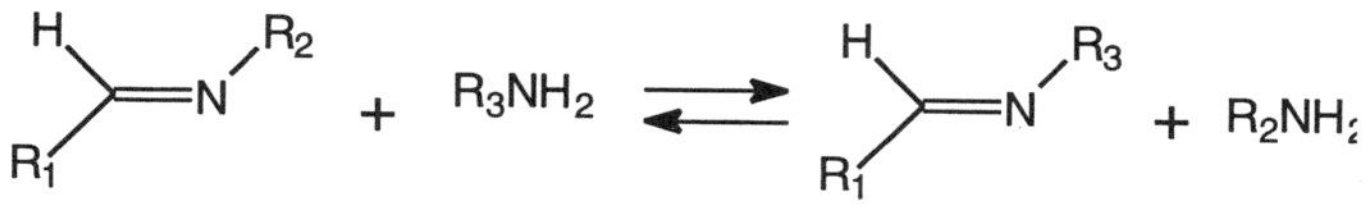

Scheme 2.10. Transimination

The structures of the participating reactants strongly affect the rate of the reaction and the position of the equilibrium. The thermodynamic stability of -C=N- linkages increases with the type of amine used, in the order NH_3 << aliphatic amines < aromatic amines < amines with an adjacent electronegative atom bearing a free electron pair (e.g. H_2N-OH). Consequently, when an *in situ* prepared primary imine is allowed to react with an alkylamine under equilibrating conditions, transimination takes place and the more stable secondary imine is readily formed.

By partial hydrogenation with DIBAL, or reaction with RMgX, of O-protected cyanohydrins, metallated primary imines are formed initially. Upon protonation, the free imines can be subjected to transimination to furnish biologically interesting N-substituted amino alcohols, e.g.: ephedrine, tembamide, aegeline, denopamine, and analogues of salbutamol [56,67,70-72].

2.5.1 Grignard Reaction / Transimination / Hydride Reduction:
Synthesis of N-Substituted Ethanolamines

N-substituted ethanolamines with two stereogenic centers have been prepared in a one-pot procedure in which transimination is followed by NaBH$_4$ reduction (Scheme 2.11) [73]. Transimination is not restricted to primary amines, but also works well with amino acid esters [54] and with ethanolamines [74]. In the latter case, diethanolamines are formed.

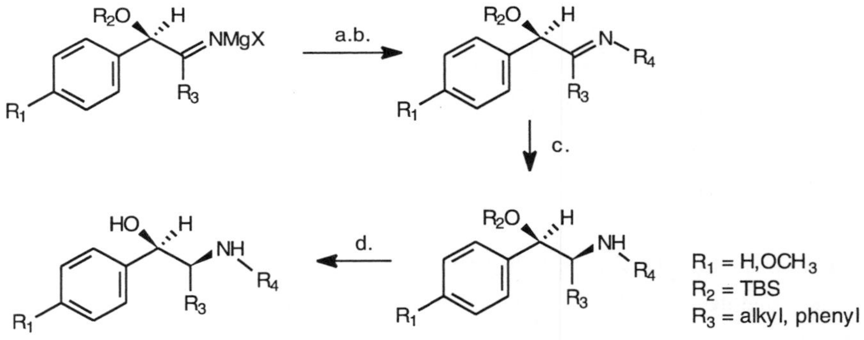

a. MeOH, b. R$_4$NH$_2$, c. NaBH$_4$, d. LiAlH$_4$

Scheme 2.11. Synthesis of N-substituted ethanolamines.

2.5.2 DIBAL Reduction/Transimination/HCN Addition/Hydrolysis:
Synthesis of □-Hydroxy-□-Amino Acids

MIP-protected (R)-mandelonitrile was reduced by DIBAL at low temperature (-70 °C). Quenching with ammonium bromide in methanol afforded the free primary imine which was transiminated with a primary amine to introduce the N-alkyl group. This secondary imine was directly converted to the α-cyano amine by addition of HCN, creating a second stereogenic center. During acidic work-up the MIP group is removed, affording the HCl salt of the β-hydroxy-α-cyanoamine. After conversion to the oxazolidinone derivative, mild hydrolysis (solvolysis) of the nitrile group was accomplished with K_2CO_3 in ethanol. Saponification of the ester and oxazolidinone moiety afforded $(2S,3R)$-N-alkyl-β-hydroxy-α-amino acids in high overall yields (Scheme 2.12) [58].

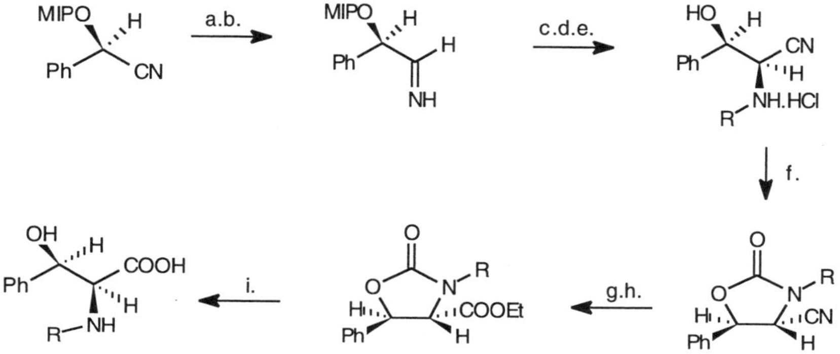

a. DIBAL; b. MeOH; c. RNH_2; d. HCN; e. HCl; f. $CO(Im)_2$; g. K_2CO_3, EtOH; h. HCl; i. KOH/H

Scheme 2.12. Synthesis of β-hydroxy-α-amino acids

2.5.3 Transimination with N-Substituted Hydroxylamines:
Chiral Aldo- and Keto Nitrones

Nitrones are usually prepared by condensation of aldehydes with hydroxylamines. Cyanohydrins can be converted to α-hydroxy aldehydes (partial reduction with DIBAL, followed by hydrolysis), which in turn can be used for the synthesis of nitrones. However, a particularly attractive method is the direct conversion of chiral cyanohydrins into chiral nitrones via transimination (Scheme 2.13).

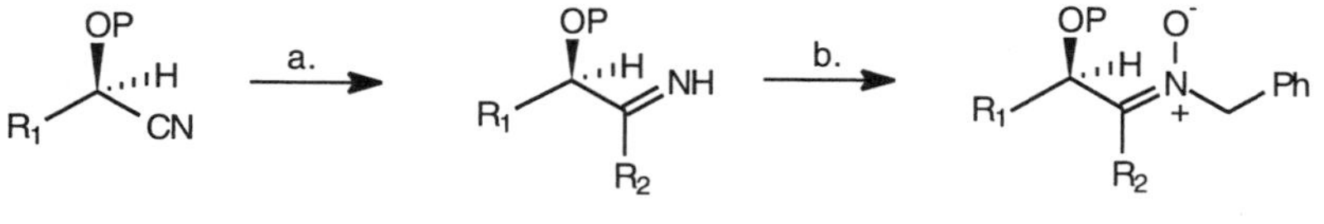

a. DIBAL or RMgX; b. Benzylhydroxylamine; P = MIP, TBS. R_1, R_2 = alkyl, alkenyl, ar

Scheme 2.13. One-pot synthesis of aldo- and keto nitrones

Conversion of the nitrile to a primary imine, either by DIBAL reduction or Grignard addition, is followed by transimination with N-benzylhydroxylamine. Work-up after 4 hours provides aldo- and keto nitrones in high yields and of high enantiomeric purity

[55]. These chiral nitrones, in turn, are valuable starting materials for intra- and intermolecular 1,3-dipolar cycloadditions [75].

2.6. CYANOHYDRINS FROM α,β-UNSATURATED ALDEHYDES

Croton aldehyde (2-butenal), like many other α,β-unsaturated aldehydes, is a good substrate for almond hydroxynitrile lyase. The resulting unsaturated cyanohydrin (*R,E*)-2-hydroxy-3-pentenenitrile has been used, after protection of the hydroxyl group, as a chiral C_3-synthon, in which the double bond serves as a masked oxygen functionality.

2.6.1 Ozonolysis/Reduction: Synthesis of β-Blockers

Ozonolysis allows several modes of work-up, thus providing routes for converting the olefin into an alcohol, an aldehyde, or an acid. A method was developed for the preparation of the pure enantiomers of 5-(hydroxymethyl)-3-isopropyloxazolidin-2-one (Scheme 2.14) [56]. These compounds are excellent building blocks for the preparation of chiral β-blockers (β-adrenoceptor antagonists) which are among the world's most applied pharmaceuticals.

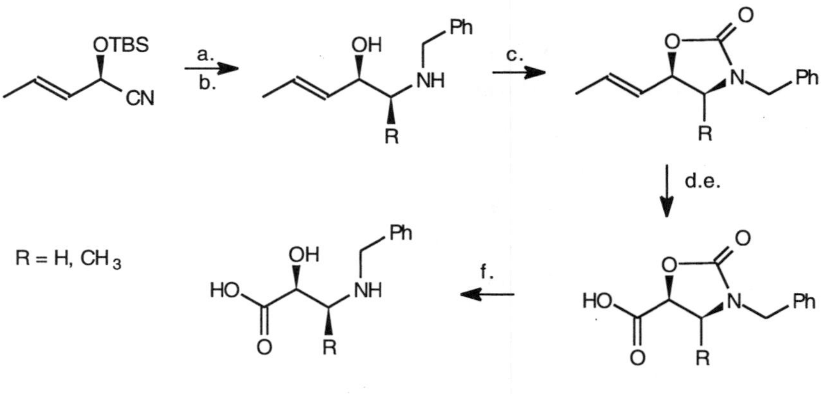

a. DIBAL or MeMgI, MeOH, benzylamine, NaBH$_4$; b. LiAlH$_4$; c. CO(Im)$_2$; d. O$_3$
e. Jones' ox.; f .2 N KOH, H$_3$O$^+$

Scheme 2.14, Synthesis of both enantiomers of 5-hydroxymethyl-3-*i*-propyloxazolidin-2-one

Most β-blockers presently marketed are racemic 1-alkylamino-3-aryloxy-2-propanols, in which the aryl group can be one of a large variety of aromatic substituents and the N-alkyl substituent is generally *i*-propyl or *t*-butyl. The stereoselective approach depicted in Scheme 2.15 allows the introduction of a large variety of substituents at nitrogen, since the transimination step proceeds well with most primary amines, even with strongly sterically hindered ones like *t*-butylamine. Inversion of the chiral center was accomplished by treating the *N*-benzyloxycarbonyl derivative with thionyl chloride. Ozonolysis is carried out at a stage where both the hydroxyl- and the amino substituent are protected as an oxazolidinone.

2.6.2 Ozonolysis/Oxidation: Synthesis of α-Hydroxy-β-Amino Acids

(*R,E*)-2-hydroxy-3-pentenenitrile is also an excellent chiral starting material for the synthesis of α-hydroxy-β-amino acids (Scheme 2.15) [73]. In this case the first step can be a DIBAL reduction (R = H) or a Grignard addition (R = alkyl or aryl) and the ozonolysis is directly followed by oxidative work-up (Jones' reagent). Hydrolysis of the oxazolidinones provided the desired N-protected α-hydroxy-β-amino acids. of high enantiomeric and diastereomeric purity.

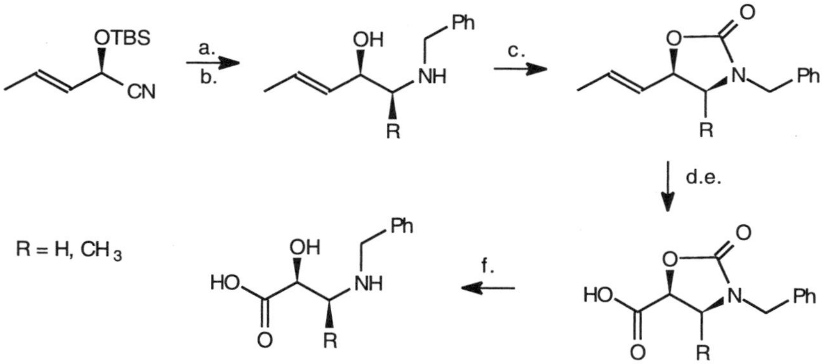

a. DIBAL or MeMgI, MeOH, benzylamine, NaBH$_4$; b. LiAlH$_4$; c. CO(Im)$_2$; d. O$_3$
e. Jones' ox.; f .2 N KOH, H$_3$O$^+$

Scheme 2.15. Synthesis of α-hydroxy-β-amino acids

3. REFERENCES

1. Effenberger, F. (1994) *Angew Chem Int Ed Engl.* **33**, 1555-1564.
2. Kanerva, L.T. (1996) *Acta Chem Scand* **50**, 234-242.
3. Effenberger, F. (1996) *Enantiomer* **1**, 359-363.
4. Hickel, A., Hasslacher, M., Griengl, H. (1996) *Physiol Plant* **98**, 891-898.
5. CG Kruse, C.G. (1992) In: *Chirality in Industry*, Eds Collins, A.N., Sheldrake, G.N., Crosby, J., John

394

Wiley, New York, p 279-299.

6. Griengl,H., Hickel, A., Johnson, D.V., Kratky, C., Schmidt, M., Schwab, H. (1997) *Chem Commun* 1933-1940.

7. North, M. (1993) *Synlett* 807-820.

8. Rosenthaler, L., (1908) *Biochem Z* **14**, 238-253.

9. Becker, W., Freund, H., Pfeil, E. (1965) *Angew Chem Int Ed Eng* **4**, 1079.

10. Becker, W., Pfeil, E. (1966) *J Am Chem Soc* **88**, 4299-4300.

11. Brussee, J., Jansen, A.C.A., Kühn, A. (1985) *Proc Symp Med Chem, Acta Pharm Suecica Suppl* **2**, 479.

12. Effenberger, F., Ziegler, T., Förster, S. (1987) *Angew Chem Int Ed Engl* **26**, 458-459.

13. Effenberger, F., Ziegler, T., Förster, S. (1987) *Eur Pat 0 276 375 A2*, to Degussa A.G.

14. Brussee, J., Roos, E.C., Van der Gen, A. (1988) *Tetrahedron Lett* **29**, 4485-4488.

15. Brussee, J., Van der Gen, A. (1988) *Eur Patent 322, 973* to Duphar.

16. Kula, M-R., Niedermeyer, U., Stürtz, I.M. (1989) *Eur Pat Appl 0 350 908 A2*, to Degussa AG.

17. Effenberger, F., Hörsch, B., Förster,S., Ziegler, T. (1990) *Tetrahedron Lett* **31**, 1249-1252.

18. Niedermeyer, U., Kula, M-R. (1990) *Angew Chem Int Ed Engl* **29**, 386-387.

19. Brussee, J., Loos, W.T., Kruse, C.G., Van der Gen, A. (1990) *Tetrahedron* **46**, 979-986.

20. Ziegler, T., Hörsch, B., Effenberger, F. (1990) *Synthesis* 575-578.

21. Becker, W., Benthin, U., Eschenhof, E., Pfeil, E. (1963) *Angew Chem* **75**, 93.

22. Becker, W., Pfeil, E. (1964) *Naturwissenschaften* **51**, 193.

23. Pfeil, E. (1980) personal communication.

24. Smitskamp-Wilms, E., Brussee, J., Van der Gen, A., Van Scharrenburg, G.J.M., Sloothaak, J.B. (1991) *Recl Trav Chim Pays-Bas* **110**, 209-215.

25. Lauble, H., Müller, K., Schindelin, H., Förster, S., Effenberger, F. (1994) *Proteins: Structure, Functions and Genetics* **19**, 343-347.

26. Klempier, N., Pichler, U., Griengl, H. (1995) *Tetrahedron: Asymmetry* **6**, 845-848.

27. Schmidt, M., Hervé, S., Klempier, N., Griengl, H. (1996) *Tetrahedron* **52**, 7833-7840.

28. Förster, S., Roos, J., Effenberger, F., Wajant, H., Sprauer, A. (1996) *Angew Chem Int Ed Engl* **35**, 437-439.

29. Becker, W., Benthin, U., Eschenhof, E., Pfeil, E. (1963) *Biochem Z* **337**, 156-166.

30. Jorns, M.S. (1980) *Biochim Biophys Acta* **613**, 203-209.

31. Bärwald, K.R., Jaenicke, L. (1978) *FEBS Lett* **90**, 255-260.

32. Jaenicke, L., Preun, J. (1984) *Eur J Biochem* **138**, 319-325.

33. Jorns, M.S. (1985) *Eur J Biochem* **146**, 481-482.

34. Brussee, J., Van der Gen, A. unpublished results.

35. Wagner, U., Hasslacher, M., Griengl, H., Schwab, H., Kratky, C. (1996) *Structure* **4**, 811-822.

36. Huuhtanen, T.T., Kanerva, L.T. (1992) *Tetrahedron: Asymmetry* **3**, 1223-1226.

37. Zandbergen, P., Van der Linden, J., Brussee, J., Van der Gen,A. (1995) *Preparative Biotransformations. Whole cell and isolated enzymes in organic synthesis.* Update 5, 4:5.1. Wiley Looseleaf Publications, Ed. Roberts, S.M.

38. Zandbergen, P., Van der Linden, J., Brussee, J., Van der Gen, A. (1991) *Synth Commun* **21**, 1387-1391.

39. Han, S., Lin, G., Li, Z. (1998) *Tetrahedron: Asymmetry* **9**, 1835-1838.

40. Ognyanov, V.I., Datcheva, V.K., Kyler, K.S. (1991) *J Am Chem Soc* **113**, 6992-6996.

41. Kiljunen, E., Kanerva, L.T. (1997) *Tetrahedron: Asymmetry* **8**, 1225-1234.

42. Gerrits, P.J., Brussee, J., Van der Gen, A., Straathof, A.J.J., Heijnen, J.J. Submitted for publication.

43. Menéndez, E., Brieva, R., Rebolledo, F., Gotor, V. (1995) *J Chem Soc Chem Commun* 989-990.

44. Nazabadioko, S., Pérez, R., Brieva, R., Gotor, V. (1998) *Tetrahedron: Asymmetry* **9**, 1597-1604.

45. Kragl, U., Niedermeyer, U., Kula, M-R., Wandrey, C. (1990) *Ann New York Acad Sc* 167-175.

46. Loos, W.T., Geluk, H.W., Ruijken, M.M.A., Kruse, C.G., Brussee, J., Van der Gen, A. (1995) *Biocatalysis and Biotransformation* **12**, 255-266.

47. Hasslacher, M., Schall, M., Hayn, M., Griengl, H., Kohlwein, S.D., Schwab, H. (1996) *J Biol Chem* **271**, 5884

48. Lewis, R. J., Sr (1992) *Sax's dangerous properties of industrial materials*. 8th ed. Van Nostrand Reinhold. New York.

49. National Research Council. (1981) *Prudent Practices for Handling Hazardous Chemicals in Laboratories*. National Academy Press. Washington, D.C.

50. National Research Council. (1983) *Prudent Practices for Disposal of Chemicals from Laboratories*. National Academy Press. Washington, D.C.

51. Effenberger, F., Hörsch, B., Weingart, F., Ziegler, T., Kühner, S. (1991) *Tetrahedron Lett* **32**, 2605-2608.

52. Warmerdam, E.G.J.C., Van den Nieuwendijk, A.M.C.H., Kruse, C.G., Brussee, J., Van der Gen, A. (1996) *Recl Trav Chim Pays-Bas* **115**, 20-24.

53. Effenberger, F., Eichhorn, J., Roos, J. (1995) *Tetrahedron: Asymmetry* **6**, 271-282.

54. Van den Nieuwendijk, A.M.C.H., Warmerdam, E.G.J.C., Brussee, J., Van der Gen, A (1995) *Tetrahedron: Asymmetry* **6**, 801-806.

55. Hulsbos, E., Marcus, J., Brussee, J., Van der Gen, A. (1997) *Tetrahedron: Asymmetry* **8**, 1061-1067.

56. Zandbergen, P., Van den Nieuwendijk, A.M.C.H., Brussee, J., Van der Gen, A., Kruse, C.G. (1992) *Tetrahedron* **48**, 3977-3982.

57. Warmerdam, E.G.J.C., Brussee, J., Kruse, C.G., Van der Gen, A. (1994) *Helv Chim Acta* **77**, 252-256.

58. Zandbergen, P., Brussee, J., Van der Gen, A., Kruse, C.G. (1992) *Tetrahedron: Asymmetry* **3**, 769-774.

59. Zandbergen, P., Willems, H.M.G, Van der Marel, G.A., Brussee, J., Van der Gen, A. (1992) *Synth Commun* **22**, 2781-2787.

60. Brussee, J., Dofferhoff, F., Kruse, C.G., Van der Gen, A. (1990) *Tetrahedron* **46**, 1653-1658.

61. Brussee, J., Van Benthem, R.A.T.M., Van der Gen, A. (1990) *Tetrahedron: Asymmetry* **1**, 163-166.

62. De Vries, E.F.J., Brussee, J., Van der Gen, A. (1994) *J Org Chem* **59**, 7133-7137.

63. Effenberger, F., Stelzer, U. (1991) *Angew Chem Int Ed Engl* **30**, 873-874.

64. Warmerdam, E.G.J.C., Brussee, J., Kruse, C.G., Van der Gen, A. (1993) *Tetrahedron* **49**, 1063-1070.

65. Decicco, C.P., Grover, P. (1997) *Synlett* 529-530.

66. Becker, W., Freund, H., Pfeil, E. (1965) *Angew Chem* **77**, 1139.

67. Effenberger, F., Jäger, J. (1997) *Chem Eur J* **3**, 1370-1374.

68. Effenberger, F., Gutterer, B., Syed, J. (1995) *Tetrahedron: Asymmetry* **6**, 2933-2943.

69. Effenberger, F., Syed, J. (1998) *Tetrahedron: Asymmetry* **9**, 817-825.

70. Effenberger, F., Jäger, J. (1997) *J Org Chem* **62**, 3867-3873.

71. Brussee, J., Van der Gen, A. (1991) *Recl Trav Chim Pays-Bas* **110**, 25-26.

72. Brown, R.F.C., Donohue, A.C., Jackson, W.R., McCarthy, T.D. (1994) *Tetrahedron* **50**, 13793-13752.

73. Warmerdam, E.G.J.C., van Rijn, R.D., Brussee, J., Kruse, C.G., Van der Gen, A. (1996) *Tetrahedron: Asymmetry* **7**, 1723-1732.

74. De Vries, E.F.J., Steenwinkel, P., Brussee, J., Kruse, C.G., Van der Gen, A. (1993) *J Org Chem* **58**, 4315-4325.

75. Marcus, J., Brussee, J., Van der Gen, A. (1998) *Eur J Org Chem* 2513-2517.

ENZYMES IN POLYMERS AND POLYMERS FROM ENZYMES

B.J. KLINE, G. DREVON, AND A.J. RUSSELL*
Department of Chemical and Petroleum Engineering and Center for Biotechnology and Bioengineering, University of Pittsburgh, Pittsburgh, Pennsylvania 15261 USA

1. Introduction

The interface of biology and materials encompasses a wide array of topics. The two main categories inclusive to this subject are materials applied in biological systems and materials produced through biological means. Materials associated with the former topic can include medical implants, polymeric drug carriers, and surgical glues while the latter category can include synthetic materials usually formed chemically. In this chapter, two areas of increasing interest will be reviewed: first, the enzymatic synthesis of polymers, and second, the incorporation of enzymes into polymers.

2. Polymers from Enzymes

Growing concern for the environment and depletion of fossil fuels has led to initiatives for the development of better synthetic processes through green chemistry. This involves minimizing the energy required for a particular chemical process. The ability of enzymes to function under conditions of ambient temperature and pressure and to catalyze highly selective transformations renders them ideal for the development of these green processes.

B. Zwanenburg et al. (eds.), Enzymes in Action, 397–431.
© 2000 *Kluwer Academic Publishers. Printed in the Netherlands.*

The application of enzymes to industrial processes has recently become more feasible since the discovery that most enzymes retain some activity in organic solvents [1]. One of the most interesting aspects of enzymatic reactions in unconventional milieu is the reversal of thermodynamic equilibria toward synthesis rather hydrolysis. For example, lipases suspended in organic solvents can catalyze reactions such as transesterification, esterification, and aminolysis that are usually suppressed by hydrolysis when conducted in water. The following sections review the research that has been performed concerning enzymatic polymerizations in nonaqueous environments.

2.1.   PREPARATION OF MONOMERS USING ENZYMES

2.1.1 Acrylate and Methacrylate Esters
Polyacrylates are highly stable polymers used for a wide variety of applications including contact lenses, coatings, and Plexiglass®. The synthesis of poly(acrylates), poly(methacrylates) and their copolymers is carried out through polymerization of acrylic and methacrylic acid esters.   Different structures of the ester can result depending on the method of synthesis that is employed. Generally, (meth)acrylic acid can react with an alcohol in the presence of sulfuric acid at elevated temperatures to yield the corresponding ester and water [2]. The product must endure a thorough purification process to remove alcohol, residual monomer and the water byproduct. The presence of another functional group (i.e. the use of a diol) can lead to unwanted products such as diester or oligomers. The current method to synthesize functionalized acrylates such as 2-hydroxyethyl (meth)acrylate involves the reaction of the acid with ethylene oxide [2].

Enzymatic synthesis of acrylates was suggested as an alternative to traditional synthesis because it would eliminate the need for acid use and could minimize the occurrence of side reactions that form diesters or oligomeric products. Additionally, because of the specificity of the lipase, interesting properties such as functionality and chirality could be incorporated into the subsequent polymer. Figure 1 shows the reaction scheme for the chemoenzymatic synthesis of polyacrylates. The enzymatic step (1) forms the acrylate ester from acrylic acid (or its derivatives) and a mono-functional alcohol. The chemical step (2) shown in figure 1 represents a radical polymerization of the monomer using an azo compound as an initiator. In Figure 1, R_1 can either be H or CH_3 for the acrylate or the methacrylate to be formed, respectively. R_2 can represent various alkyl groups. R_3 can represent alkyl, vinyl or oxime groups. When R_3 possesses a functional group such as –(CH$_2$)$_4$OH, the resulting polyacrylate will possess a pendant hydroxyl group.

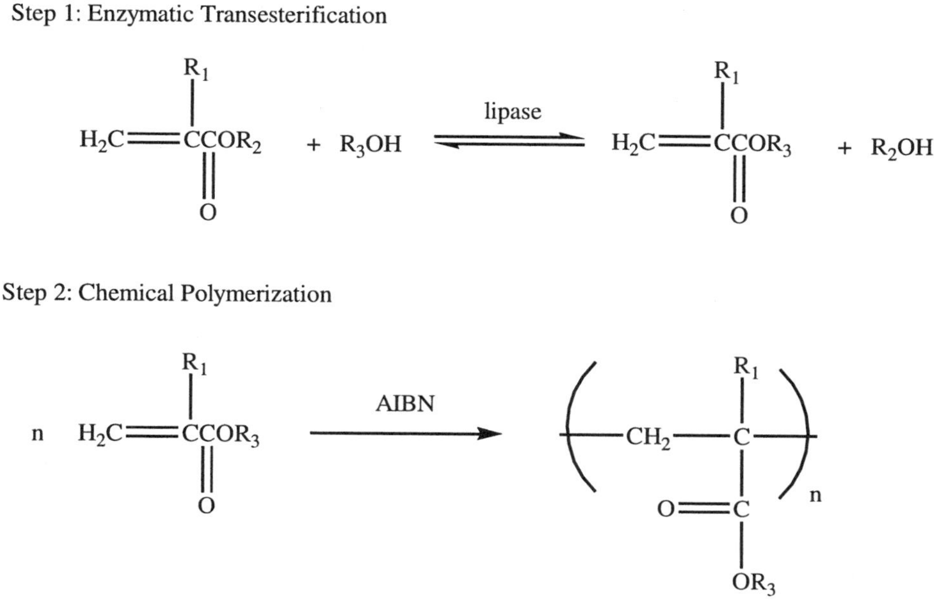

Step 1: Enzymatic Transesterification

Step 2: Chemical Polymerization

Figure 1: Chemoenzymatic synthesis of polyacrylates and polymethacrylates

Hydroxyalkyl esters were synthesized by Tor *et al.* and Hajjar *et al.* by reacting diols or triols with various (meth)acrylates [3,4]. The first mentioned authors used a prestabilized and immobilized pig liver esterase in an aqueous organic solvent composed of ethylene glycol and phosphate buffer (pH 7.5) at 37°C [3]. By doing so, the ethylene glycol acts as both the solvent and acyl agent. By investigating various ratios of the water and organic components, it was found that a maximum rate of transesterification was achieved at 40% glycol; however, hydrolysis also occurred at this concentration. At a concentration of 70% organic, no hydrolysis occurred [3].

To eliminate hydrolysis, Hajjar *et al.* used a nearly anhydrous organic solvent (ethyl acrylate) [4]. The lipase from *Chromobacterium viscosum* was found to possess the best activity of the enzymes tested and no stabilization or immobilization was necessary prior to the reaction. Linear and branched diols were reacted with ethyl acrylate to investigate the kinetics and formation of side products. While the reaction with linear diols reached its peak within 2 to 3 days, a significant amount of diester was formed. Branched diols resulted in slower rates (maximum formation in 4 to 7 days), but diester formation could be eliminated [4].

Pavel and Ritter formed a carboxyalkyl ester by the reaction of 12-hydroxylauric acid with 11-methacroylaminoundecanoic acid in the presence of lipase from *Candida rugosa* [5]. Subsequent radical polymerization using 2,2-azobisiso-butyronitrile (AIBN) as the initiator resulted in a carboxy-functional comb polymer. Fukusaki *et al.* formed a carboxyalkyl acrylate through a two-stage enzymatic process [6]. In the first step, vinyl acrylate and a hydroxyalkanoic acid methyl ester were reacted with lipase from *Pseudomonas* sp. in isopropyl ether at 60°C to yield the methoxycarbonylalkyl acrylate. The second step employed the lipase from *Candida rugosa* to hydrolyze the methyl group to form the carboxy-functional acrylate [6]. Finally, Derango *et al.* was able to synthesize an amino-functional methacrylate through the reaction of 2-hydroxyethyl carbamate with vinyl methacrylate at 50°C using *Pseudomonas* sp. lipase [7]. Various solvents and solvent mixtures were studied and the highest product formation resulted when using a 3:1 mixture of toluene and tetrahydrofuran (THF) [7].

The synthesis of alkyl esters (without functional groups) has also been investigated. Pavel and Ritter reacted 11-methacroylaminoundecanoic acid with various alcohols including isobutyl alcohol, cyclohexanol, menthol, cholesterol, and testosterone using lipase from *C. rugosa* in cyclohexane and water at 42°C [5]. They observed that the rate of esterification decreased as the steric hindrance of the hydroxyl functionality increased. Ikeda *et al.* used vinyl acrylate with alcohol to synthesize the ester in isooctane at 37°C with *C. rugosa* lipase [8]. Alcohols with various chain lengths and substituted structures were studied and it was found that n-hexyl acrylate and β-phenethyl acrylate were produced in the highest yield, 40 and 66%, respectively [8]. While the enzyme-catalyzed synthesis of acrylate esters does possess some disadvantages including long reaction times and low yields, Warwel and coworkers were able to produce undecyl methacrylate in 1.5 hours and 90% yield by using Novozym® 435 at 30°C [9].

Russell and coworkers have published a series of reports on the synthesis of 2-ethylhexyl methacrylate using organic solvents and supercritical fluids with lipase from *C. rugosa* [10-12]. In the first paper, the enzyme activity in supercritical carbon dioxide (CO_2), ethane, ethylene, sulfur hexafluoride (SF_6), and fluoroform was investigated and compared to that in hexane [10]. The reaction rates in all the supercritical fluids (SCFs) except CO_2 were greater than in most organic solvents and SF_6 was faster by an order of magnitude. The effect of temperature (40 to 50°C) and pressure (atmospheric pressure to 3000 psi) was also determined in ethane. While increasing the temperature from 40 to 45°C enhanced the rate, the activity decreased by 25% when the pressure was increased from 1600 to 3000 psi [10].

These results led to a more in-depth study concerning how the pressure and density changes of the SCFs affect their properties, especially the dielectric constant (ε) and Hildebrand solubility parameter (δ) [11]. Equations of state were used to predict the density at different pressures of the SCFs. Equations were also developed to relate the ε and δ to either density and temperature or pressure and temperature. From these studies, Kamat et al. were able to conclude that the activity of the enzyme is more dependent on the dielectric constant than the Hildebrand solubility parameter [11]. In fluoroform, changes in pressure from 850 to 4000 psi altered ε from 1.5 to 7. This increase in pressure (and ε) led to a decreased enzyme activity in fluoroform [11].

Kamat et al. found that CO_2 is not a good solvent for the transesterification of acrylate esters [10]. However, because there are many advantages for using supercritical CO_2 including low cost, availability, and low toxicity, the bulk of enzymatic reactions in SCFs are performed in CO_2. It was thought that by divining the reasons that enzymes have low activity in CO_2 would be beneficial and perhaps lead to methods or conditions in which increased activity is observed. Two possible explanations for the diminished capacity of enzymes in CO_2 were presented. First, the pH of the water surrounding the enzyme could be altered by dissolved CO_2; however, it was found that the change in pH before and after CO_2 addition was negligible [12]. Second, the formation of carbamates between CO_2 and the lysine residues on the enzyme surface could cause low reaction rates. Kamat et al. varied the temperature of the reaction between 40 to 70°C because high temperatures have been shown to decrease carbamate stability. As a result, the rate of transesterification increased significantly above 55°C [12]. The presence of carbamate structures in the enzymes was also confirmed by matrix-assisted laser desorption/ionization – time of flight (MALDI-TOF) mass spectrometry [12].

Finally, enzymes have been used to introduce optical activity to the acrylate esters and polymers. Ghogare and Sudesh Kumar reacted a chiral alcohol, 2-ethylhexanol with O-acryloyl oximes at 35°C using porcine pancreatic lipase (PPL). The S-enantiomer of the acrylate ester was formed [13]. Margolin et al. reacted vinyl acrylate or trifluoroethyl (meth)acrylate with various chiral alcohols or amines using lipase from Pseudomonas cepacia or subtilisin Carlsberg. The enantiomeric form (R or S) of the ester depended on the individual alcohol or amine and enantiomeric excesses (ee) ranged between 77-99% [14]. Athawale and Gaonkar recently published a report on the reaction of chiral alcohols with 2,3-butanedione monoxime acrylate in the presence of immobilized *Mucor miehei* lipase. Complete conversion was obtained in 4 to 6 hours. The R-configuration of the ester was produced with only 25-30% ee [15].

2.1.2 Monomeric Sugars

Incorporation of sugars into polymers changes their nature so that they are more hydrophilic. Rethwisch and coworkers have developed a chemoenzymatic method to synthesize polyacrylates with sugar moieties attached as pendant groups [16,17]. They reacted vinyl acrylate and monosaccharides with various lipases and proteases in pyridine to yield 6-acryloyl esters. Enzyme regioselectivity in which the only primary hydroxyl groups are acylated, was first reported by Theridsod and Klibanov [18]. This finding was significant since it allowed regioselective modification of unprotected saccharides. Chemical methods usually require the blockage of free hydroxyls through the use of protecting groups [19]. By investigating conditions for both the enzymatic reaction and the chemical polymerization, rates were dramatically increased and molecular weights (M_n) for the water-soluble polymers as high as 240000 Da have been obtained [17]. The addition of cross-linkers during polymerization produced polyacrylates that are gel-like in consistency and swellable when placed in water [20,21].

Shibatani *et al.* and Kitagawa and Tokiwa have produced sugar esters through the reaction of divinyl esters with glucose in N,N-dimethylformamide (DMF) using either the alkaline protease from *Streptomyces* sp. or the lipase from *Alcaligenes* sp. [22,23]. The monomer 6-O-vinylsebacyl-D-glucose was then polymerized using AIBN as the initiator to give a molecular weight of 11000 Da (polydispersity index (PDI) or M_w/M_n = 1.5) [23]. Because sugars are only soluble in hydrophilic solvents such as DMF or pyridine, which have been shown to have denaturing effects on some enzymes, Ikeda and Klibanov have treated glucose with a hydrophobic acid, phenyl boronic acid [24]. The modified glucose can then be solubilized in more hydrophobic solvents, enzymatically reacted with vinyl acrylate, and polymerized. However, the report did not mention if the attached phenyl boronic acid could be easily removed or if the resulting polymer with the modified sugar was still water-soluble.

Blinkovsky and Dordick have also synthesized polyethylene and polyacetylene with pendant sugar groups by first, reacting disaccharides such as lactose, maltose or cellobiose with allyl or propargyl alcohol in the presence of glycosidases [25]. The reaction of the disaccharides with the propargyl or allyl alcohol took place at the glycosidic bond in 42% and 13% yield, respectively. While the propargyl group was more reactive than the allyl alcohol, polymerization using AlBr$_3$ only produced polyacetylene oligomers of 1300 Da (M$_w$). Radical polymerization did not take place to produce polyacetylene, while polyethylene was synthesized in DMF or water using AIBN producing molecular weights of 31000 to 36000 Da [25].

2.2. POLYPHENOLS

Phenol-formaldehyde resins are known for their mechanical toughness and temperature resistance and are used in many applications. They are synthesized by curing either novolak or resol oligomers which consist of condensed phenol and formaldehyde residues [26]. Catalytic polymerization of phenols was originally suggested as an alternative for phenol-formaldehyde resins due to concerns about the toxicity of formaldehyde and the synthesis conditions. However, poly(phenylene oxide) produced with inorganic catalysts are constrained to the use of 2,6-disubstituted phenols [27].

The use of enzymes in the synthesis of polyphenol is not straightforward due to the poor solubility of phenol in water. This necessitates the use of very small substrate concentrations or water-miscible organic solvents. Because the dimers and trimers of phenol are even less soluble, precipitation occurs soon after their formation preventing further polymerization. Schwartz and Hutchinson did attempt to couple phenol in phosphate buffer (pH 6.5) using horseradish peroxidase as the catalyst and peroxide as the oxidizing agent [28]. Horseradish peroxidase (HRP) catalyzes the one-electron oxidation of phenol compounds in the presence of peroxide. They observed that phenol could be coupled, but were somewhat surprised to learn that 4,4'-biphenol (dimer) reacted as well [28]. The formation of brown precipitate complicated their analysis.

To avoid the problems associated with the solubility in water, Dordick *et al.* performed the polymerization of p-phenylphenol in various solutions of 1,4-dioxane and acetate buffer (pH 5.0) [29]. Polymers formed in solvent with greater than 60 vol% dioxane. The polymers that were produced using 80 to 90 vol% dioxane were only partly soluble in DMF. The molecular weight was obtained by analyzing the DMF-soluble portion of the polymer using gel permeation chromatography with DMF and methanol (4:1) as eluent. In 1,4-dioxane and water (85:15), poly(p-phenylphenol) with a molecular weight of approximately 26,000 Da was synthesized [29]. The authors also investigated the polymerization of other phenol compounds employing the same reaction conditions. Most of these compounds, such as phenol, p-cresol, and aniline produced oligomers with an average molecular weight between 1400-2000 Da. Only the polymer derived from 2-naphthol was completely insoluble in 1,4-dioxane and DMF implying a very high molecular weight [29].

Since then, there has been a significant amount of work performed using 1,4-dioxane and other water-miscible organic solvents to polymerize different types of phenols such as o-, m-, and p-cresol, phenol, and various p-alkylphenols in the presence of either horseradish peroxidase or soybean peroxidase [30-38]. One of the most striking features in all of the reports was the difficulty in analyzing these polymers. First, in most cases, polyphenols are not soluble in most common organic solvents, and they are only slightly soluble in highly polar solvents like DMF and dimethyl sulfoxide (DMSO). This trait leads to complications when determining the molecular weight; only the DMF-soluble part can be characterized. Uyama *et al.* observed, after analyzing the DMF-soluble and insoluble parts of polyphenol by IR spectroscopy, that the spectra were similar [34]. Therefore, the DMF-insoluble portion is structurally similar, but higher in molecular weight than the soluble part. Recently, Tonami *et al.* and Kobayashi *et al.* have synthesized soluble polyphenols derived from m-cresol and bisphenol A, respectively [37,38]

Second, molecular weight determination of the DMF-soluble portion can, at best, be termed challenging. Akkara and coworkers claimed to have enzymatically synthesized poly(p-phenylphenol) with molecular weights of 400,000 Da in 1,4-dioxane and water [30]. They employed the same solvent system in their molecular weight analysis as Dordick *et al.* [29]. In subsequent papers, the same group reported that using only DMF or DMF and methanol (4:1) yields higher values of the molecular weight due to inter- and intramolecular interactions of the polymer chains. Polyphenols form aggregates, which elute early when using GPC [32]. It was found that addition of LiBr or LiCl breaks up the aggregates sufficiently and affords better estimates of the molecular weight [32,33].

Figure 2 shows the reaction of para-substituted phenol to form polyphenol. Two groups have offered structural characterization of the enzymatically synthesized polyphenols. Kaplan and coworkers have synthesized poly(p-phenylphenol) [30], polybenzidine [30], and poly(p-ethylphenol) [32]. Poly(p-phenylphenol) was analyzed by FTIR, and ^{13}C NMR. No phenoxy ether linkages were observed in the FTIR spectrum while ^{13}C NMR indicated that the polymer consisted predominantly of ortho-substitutions on the ring structure [30]. In Figure 2, this translates to x=n, and y=0. Conversely, ^{13}C NMR and proton NMR showed that polybenzidine was made up of linkages through both the ortho-carbon on the ring and the amine group [30]. This structure is not represented in Figure 2. The ^{13}C NMR analysis of poly(p-ethylphenol) supported the absence of phenoxy ether bonds and the existence of ortho linkages [32]. Kobayashi and coworkers have also enzymatically polymerized a number of different phenol compounds including cresols, p-alkylphenols, and phenol [31,32,34,38]. They used IR and proton NMR spectroscopy to analyze the polymers. In direct opposition to Kaplan and colleagues, they claim that a peak at 1187-1235 cm^{-1} in the IR spectra represents overlapping vibration for C-O-C and C-OH bonds. At times, these peaks are separated. Furthermore, they calculated the integrated ratio of peaks in the proton NMR spectra representing aromatic protons and hydroxyl protons and found that both phenylene (50%) and oxyphenylene (50%) repeat units make up the polymer. Therefore, x=0.5n and y=0.5n in Figure 2. This discrepancy in the structural characterization between the two groups has never been explained.

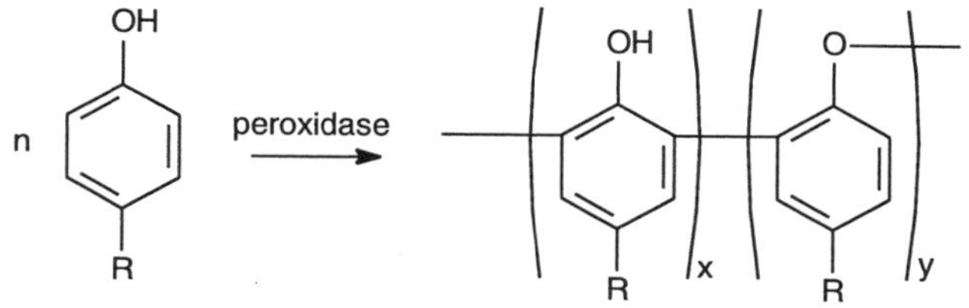

Figure 2: Peroxidase-catalyzed synthesis of polyphenol. The reaction scheme is derived from a figure in Reference 31.

The research concerning enzymatic phenol polymerization in aqueous organic solvents involved various reaction parameters such as organic solvent content, buffer salt type, pH, temperature, enzyme and phenol type to evaluate their influence on polymer molecular weight and yield. In general, it has been found that by increasing organic solvent content to approximately 85%, the highest molecular weight is achieved because solubility conditions for the monomer and polymer are substantially improved. For the most part, however, molecular weights (M_n) have not exceeded 5000 Da. Polymer yield is more difficult to generalize and results depend on the individual set of reaction conditions.

Kaplan and coworkers have attempted to orient the phenol derivatives in a certain direction so that their synthesis will occur at certain locations [32,39,40]. This is accomplished by the use of reversed micelles, which are formed from water, isooctane and surfactant (dioctyl sodium sulfosuccinate or AOT). By forming these water-in-oil microemulsions, the polymerization could be controlled to a higher degree because the phenols might order themselves along the oil-water interface with the hydroxyl group facing toward the micelle center. The synthesis of poly(p-ethylphenol) [32,39] and fluorescent poly(2-naphthol) [40] has been achieved in this manner producing peak molecular weights between 800-1400 Da. Interestingly, the polymerization of phenols in certain reversed micellar systems forms particles with spherical morphology. The reaction environment was investigated to elucidate its role in the formation of the particle and it was observed that the beads appeared when high ratios of surfactant to monomer (> 3:1) were used [41].

Another polymer involving phenol derivatives is poly(phenylene oxide) which can be produced enzymatically through two methods. Similarly the chemical catalytic processes, the enzymatic oxidative polymerization of 2,6-dimethylphenol can be carried out using peroxidase or laccase to yield polymer with molecular weights between 2700-6900 Da (PDI = 1.2-2.1) [42]. Alternatively, the oxidative polymerization of syringic acid (3,5-dimethoxy-4-hydroxybenzoic acid) and other related monomers has been accomplished by peroxidase and laccase with the elimination of carbon dioxide and hydrogen from the monomer [43,44]. Molecular weights as high as 18000 Da have been observed [43]. Other polymerizations involving phenol include the incorporation of its derivatives into lignin, a naturally occurring phenolic resin [45,46] and the formation of poly(hydroquinone) [47] or nucleoside-containing polyphenol [48] using a two-step enzymatic synthesis.

2.3. POLYESTERS

The synthesis of polyesters by biocatalytic means is the most widely studied among all enzymatic polymerizations. The many and diverse applications of polyesters result by varying the backbone structure, molecular weight, and end group functionality. The structure of the backbone is a key factor in determining the melting point of the polyester as well as its sensitivity to hydrolysis. Aliphatic polyesters have been used commercially as plasticizers, intermediates in polyurethane synthesis, and bioresorbable surgical sutures [49]. Aromatic polyesters are used as molding plastics, fibers, and films [49]. Poly(ethylene terephthalate) (PET) is one of the most widely known polymers as it is used for recyclable drink containers.

Although there are three methods to produce polyester, traditional synthesis usually occurs at high temperature using dicarboxylic acids or their derivatives and diols [49]. The other methods are self-condensation of hydroxyacids and ring-opening polymerization of lactones. All three types have been attempted enzymatically and each will be covered separately.

2.3.1 [AA-BB]x-Type Polyesters

The first attempt to synthesize polyesters enzymatically was published by Okumura *et al.* in 1984 [50]. Lipase from *Aspergillus niger* was used in reactions of dicarboxylic acids with varying lengths and glycols to produce low molecular weight oligomers. All oligomers were hydroxy-terminated, as a 80-fold excess of diol was used as both solvent and substrate. The species present in the greatest amount was made up of five condensed monomers [50].

Binns *et al.* then studied the reaction of adipic acid and 1,4-butanediol (BD) in diisopropyl ether with Lipozyme® IM (lipase from *Mucor miehei* immobilized on an anion exchange resin) as catalyst [51]. Reactions took place at 40 to 44°C for 70 to 140 hours producing hydroxy-terminated polyester with weight average molecular weights of 4650 Da and dispersities of 1.11. Recently, Binns *et al.* have reported the polymerization of the same reaction system in bulk using immobilized lipase from *Candida antarctica* [52]. They suggest that the formation of the condensed diol and acid, called an AB synthon, is essential for polymerization. In their proposed mechanism, the AB synthon is synthesized and reacts with other like compounds to form the polymer.

Linko and coworkers have also reacted a dicarboxylic acid (sebacic acid) with 1,4-butanediol in diphenyl ether in the presence of *Rhizomucor miehei* lipase at 37°C [53,54]. They observed an immediate drop in activity. However, by using intermittent periods of reduced pressure to remove the byproduct, the molecular weight reached 56000 Da in 5 days.

Because the biocatalytic reaction of dicarboxylic acids and diols is very slow and usually produces low molecular weight polyesters, several groups have used activated diesters in place of the diacids. Higher molecular weight polyester was achieved by Wallace and Morrow through the reaction of bis(2,2,2-trichloroethyl) alkanedioates with diols in the presence of PPL [55]. These diesters can be considered activated because the presence of the halogen atoms limits the reverse reaction (toward polymer degradation) thereby shifting the reaction in the forward direction. Molecular weights varied between 1300 to 8200 Da (M_n) depending on the substrate, solvent, and reaction time (74 to 135 hours) [55]. The use of aromatic diesters resulted in no polymerization.

A subsequent paper by Morrow and coworkers was published in which bis(2,2,2-trifluoroethyl) glutarate and BD were reacted with PPL to investigate those parameters that limit molecular weight. It was found that limiting the hydrolysis cause by enzyme-bound water and driving the reaction in the forward direction by removing trifluoroethanol resulted in high molecular weight polyesters of 40000 Da after 24 to 48 hours of reaction [56]. Linko *et al.* have studied the biocatalytic polymerization of aliphatic diols with bis(2,2,2-trifluoroethyl) sebacate. The reaction with immobilized *M. miehei* lipase in diphenyl ether at 37°C produced high molecular weights of 46400 Da (M_w) in 7 days [57].

A new diester, viz. bis(2,3-butanedione monoxime) alkanedioate, was introduced by Athawale and Gaonkar [58]. They attempted to use this diester as a monomer for enzymatic polymerization with either PPL or immobilized *M. miehei* lipase because it had previously been demonstrated that oxime esters possessed higher transesterification rates than both alkyl and vinyl esters. Various diols and solvents were reacted with three different oxime diesters producing polymers with weight average molecular weights ranging between 3100 to 9200 Da (PDI = 1.3-1.8) [58].

Uyama and Kobayahsi first reported the use of divinyl adipate (DVA) as a monomer in the polyesterification with various diols [59]. The reaction took place in i-propyl ether with *Pseudomonas fluorescens* lipase at 45°C for 48 hours. The highest molecular weights were obtained using BD (M_n 6700 Da, PDI=1.9) [59]. Figure 3 shows the reaction scheme. Since then, our group has studied this reaction in detail using the immobilized lipase from *Candida antarctica* (Novozym® 435).

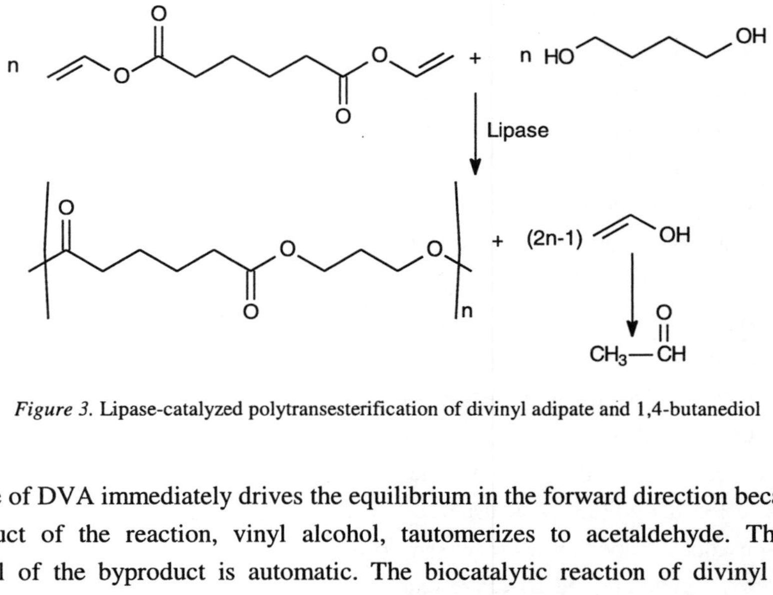

Figure 3. Lipase-catalyzed polytransesterification of divinyl adipate and 1,4-butanediol

The use of DVA immediately drives the equilibrium in the forward direction because the byproduct of the reaction, vinyl alcohol, tautomerizes to acetaldehyde. Therefore, removal of the byproduct is automatic. The biocatalytic reaction of divinyl adipate (DVA) and 1,4-butanediol (BD) is governed by the competing processes of transesterification and hydrolysis, the latter of which is partly accountable for limiting the molecular weight of the resulting polyester [60]. The reaction of vinyl ester end group functionality with water results in the formation of acid-terminated polyesters.

These acid end groups essentially stop further polymerization. The functionality of polyester is dependent on the enzyme concentration, water content, enzyme specificity, and initial stoichiometric ratio. It is interesting that the specificity of the enzyme leads to preferential reaction with vinyl ester-terminated oligomers [60]. These findings were used as the basis for the development of a non-equal reactivity model for biocatalytic polytransesterification [61]. Our knowledge of initial kinetic parameters and enzyme specificity were used to develop relationships that predict the composition, in terms of end group functionality, for polyester chains.

Chaudhary *et al.* also produced polyester in a solvent-free environment using DVA and BD with Novozym® 435 at 50°C [62]. This reaction represented the most efficient biocatalytic synthesis of polyester with regard to productivity up to that time. High molecular weight polyester, greater than 20000 Da (M_n) was produced with very low enzyme concentration and short reaction times of 4 hours [62]. Additionally, the enzyme was recycled and reused 5 times after its initial use, and still produced polyester with molecular weights in the range of 10000 Da [62].

The reactions were performed in vials using an incubator/shaker to maintain the temperature. However, because the reaction is exothermic, the reaction temperature increased by as much as 25°C depending on the substrate and enzyme concentrations. As a result, the kinetics of the initial part of the polymerization are difficult to model. Reaction calorimetry was used in order to develop a relationship between the temperature and the polymerization extent [63]. Using these techniques, the apparent heat of reaction and apparent activation energy for this polytransesterification are –30 kJ/mol and 15.1 kcal/mol [63].

A batch stirred reactor was then fabricated in an attempt to optimize the reaction conditions even further [64]. The reactor was equipped with an overhead stirrer so that constant agitation could be provided even as the viscosity of the reaction medium increased. Parameters such as enzyme deactivation, water content, temperature, enzyme concentration and enzyme particle size were investigated in an effort to obtain maximal molecular weights. By crushing the enzyme particles, polyester was formed with a molecular weight (M_w) of 23400 Da (PDI=1.7) in only 1 hour at 60°C [64]. Because the previous findings led us to believe that mass transfer limitations were affecting the reaction, we performed an analysis to better understand the role of diffusion in biocatalytic polytransesterification [65]. Factors such as molecular weight, exo-thermicity, decreased enzyme specificity with increasing chain length, enzyme deactivation, and vinyl hydrolysis were taken into account in our analysis. Initially, both bulk and pore diffusion affect the reaction, but as the molecular weight increases, the role of pore diffusion diminishes. As a result, the dispersity of the polymer as a function of the molecular weight proceeds through a maximum at approximately 5000 Da [65].

Investigation of enzymatic reactions forming aromatic polyesters is limited because aromatic monomers are bulky. Unless the lipase possesses broad substrate specificity, the aromatic monomer may be difficult to accommodate in its active site. Mezoul et al. did attempt to react 1,6-hexanediol with various aromatic diesters including dimethyl o-phthalate, dimethyl isophthalate, and dimethyl terephthalate using Novozym® 435 [66]. While dimethyl o-phthalate did not react, poly(1,6-hexanediyl terephthalate) and poly(1,6-hexanediyl isophthalate were produced with molecular weights (M_w) of 2790 Da and 29900 Da, respectively. The former polyester was produced with low molecular weights due to low reaction extents. The latter polyester, while possessing higher molecular weights, also had a very broad distribution (PDI =7.3) [66].

Recently, Rodney et al. reported the synthesis of aromatic polyesters by reacting aromatic diols with divinyl adipate in the presence of immobilized Candida antarctica lipase B at 50°C. Polyesters with molecular weights of 3500 Da (M_w) were generated [67]. Conversely, Uyama et al. reacted glycols with various aromatic divinyl esters using the same lipase. Molecular weights as high as 5500-6000 Da (PDI =1.6-1.8) were obtained [68].

Geresh and Gilboa attempted to enzymatically synthesize unsaturated alkyds through the reaction of BD with unsaturated esters (fumarate ester) [69]. Various enzymes were screened and it was found that PPL, Lipozyme® IM and lipases from Candida rugosa and Pseudomonas fluorescens catalyzed the polymerization to varying extents. The molecular weight of the oligomers depended on the solvent with reactions in acetonitrile producing polyester of 1250 Da (M_w) while those in THF produced oligomers of 600 to 800 Da [69]. A later paper by the same group investigated the feasibility of incorporating aromatic groups as well as double bonds into the backbone structure. P. fluorescens and immobilized Mucor miehei lipases were used to catalyze reactions with various aromatic diesters with aromatic, unsaturated, and aliphatic diols [70]. The best case produced low molecular weight polymers of approximately 2200 Da [70]. Most recently, Mezoul et al. synthesized copolyesters of 1,6-hexanediol, dimethyl maleate, and dimethyl fumarate. It was observed that by varying the initial ratio of maleate and fumarate monomers, the resulting composition could be controlled [71].

Our group has investigated the potential for polyester syntheses in SCFs. Polyesters were produced biocatalytically using PPL in supercritical fluoroform from the reaction of bis(2,2,2-trichloroethyl) adipate and BD. It was observed that the molecular weight of the polyester could be controlled by varying the pressure of the supercritical fluid [72]. As discussed previously, supercritical CO_2 is an ideal solvent for industrial processes with regard to environmental concerns and cost. However, CO_2 possesses less than ideal solvating power and many monomers and polymers have very low solubility in supercritical CO_2. It has been shown that fluorinated compounds are more soluble in CO_2 than most compounds. Therefore, Mesiano et al. investigated the phase behavior of fluorinated diols of varying length in supercritical CO_2 as well as the reaction of these diols with DVA in organic solvents [73,74]. The ultimate goal is to perform polymerizations using this reaction system and supercritical CO_2 as the solvent.

Klibanov and coworkers attempted a stereoselective polymerization utilizing either a racemic diester with a diol or a racemic diol with a diester [75]. Nine lipases were screened with lipases from *Aspergillus niger* and *Chromobacterium* sp. exhibiting the greatest activity. The reactions took place at 45°C in toluene with large amounts of enzyme (240 to 380 mg/mL) for 7 to 14 days. In all cases, optically active trimer and pentamer were formed [75].

Incorporation of reactive functionality into polyesters would widen the pool of potential applications for polyesters and could possibly change their nature so that they are less hydrophobic. Wallace and Morrow first introduced the idea that, by using enzymes for polymerization, reactive functionality could be incorporated into the polymer that would otherwise be destroyed. They attempted to react bis(2,2,2-trichloroethyl) *trans*-3,4-epoxyadipate with BD in anhydrous ether in the presence of PPL at ambient temperature for 3.5 days [76]. The epoxy-substituted polyester produced by this reaction was optically active with molecular weights of 5300 Da (M_n). If the epoxy groups were then reduced, hydroxyl-substituted polyester would result. Dordick and coworkers used sucrose as a comonomer with bis(2,2,2-trifluoroethyl) adipate to produce linear hydroxyl-functional polyester [77]. The reaction was catalyzed by Proleather, an alkaline protease. Polyester was produced with molecular weights as high as 12500 Da (M_w) after 25 days [77].

Kline et al. were able to enzymatically produce polyesters with hydroxyl pendant groups in a one-step synthesis [78]. Reactions took place at 50°C between DVA and triols such as glycerol, 1,2,4-butanetriol, and 1,2,6-hexanetriol in the presence of Novozym® 435. Molecular weights (M_w) varied between 4000 to 14000 Da (PDI = 2.2-3.0) depending on the triol type. The presence of the hydroxyl pendant group was confirmed by MALDI-TOF mass spectrometry and titrimetric techniques. The enzyme was regioselective, polymerizing predominantly through the primary hydroxyl groups. When reactions were performed with divinyl adipate and both glycerol and 1,4-butanediol (BD), copolymers were formed [78]. The number of pendant hydroxyl groups could be controlled by varying the initial diol/triol ratio. Recently, Uyama *et al.* performed the same reaction with divinyl sebacate [79]. While a molecular weight (M_w) as high as 27000 Da were obtained, very high dispersity of 6.7 also resulted.

2.3.2 [A-B]$_x$- Type Polyesters

The first attempt to enzymatically synthesize an [A-B]$_x$-type polyester was made by Ajima *et al.* in which 10-hydroxydecanoic acid was reacted in the presence of *Pseudomonas fluorescens* lipase modified by poly(ethylene glycol) (PEG) [80]. After 44 hours, the substrate peak on the GPC trace disappeared while several small peaks appeared. These were attributed to low molecular weight oligomers. However, a dominant peak formed at a retention time longer than that of the substrate [80]. The authors suggest that it is high molecular weight linear polyester, but it is probable that the peak could have been representative of lactone formation.

Matsumura and Takahashi investigated the polymerization of various hydroxyacids possessing either primary or secondary hydroxyl groups using lipases from *Candida rugosa* or *Chromobacterium viscosum*. While mostly trimers or tetramers were formed, primary hydroxyacids produced a much narrower molecular weight distribution than those with secondary groups [81]. Morrow *et al.* attempted the enzymatic polymerization of 2,2,2-trihaloethyl ω-hydroxyalkanoates in an effort to produce improved molecular weights than produced when using hydroxyacids. However, polyester with molecular weights ranging between 272 to 1975 Da (M_n) were formed depending on the monomer type [82].

Knani *et al.* synthesized polyester from aliphatic hydroxyesters such as methyl 6-hydroxyhexanoate. They found that aromatic hydroxyesters did not produce polymers. Various reaction parameters were studied such as enzyme type, solvent, substrate concentration, reaction temperature, reaction time, and agitation. While polyester with degrees of polymerization up to 100 could be synthesized in n-hexane using PPL at 69°C, reaction times were very long, lasting between 14 to 58 days [83]. In a subsequent paper, the authors synthesized substituted polyester using the same reaction conditions. The substituents included alkyl and aromatic functionalities and the molecular weight of the resulting polymer decreased as the size of the substituent increased [84].

In an effort to increase the efficiency of the [A-B]$_x$-type polymerizations, O'Hagan and Zaidi studied the reaction of 10-hydroxydecanoic acid [85] and 11-hydroxyundecanoic acid [86] using *Candida rugosa* lipase in n-hexane at 48 hours. By increasing the temperature to 55°C and adding molecular sieves, polyesters with molecular weights of 12065 Da (M_w) (PDI = 1.29) were achieved using the former hydroxyacid [85]. Other hydroxyacids, both longer and shorter in length, did not produce significant amounts of polymer.

The reason that enzyme-catalysed polymerizations of hydroxyacids have not been as successful as [AA-BB]$_x$-type polymerizations is still unexplained. Most papers have reported very long reaction times and low molecular weights. Theoretically, these reactions possess an advantage because stoichiometry is not an issue. Nevertheless, the use of an activated ester group in place of the acid group on the hydroxyacid did not improve the polyester molecular weight [82]. Two reports by Gutman *et al.* have studied the conditions in which lactones are formed from hydroxyacids instead of polymers. They found that, in most cases, the formation of oligomers was preferred [86,87].

2.3.3 Ring-Opening Polymerization of Lactones

Ring-opening polymerization of lactones also has the same advantage that the former polymerizations possessed in that stoichiometry (initial ratio of hydroxyl and carboxyl groups) is automatically equal. This type of polyesterification can essentially be split into two groups based on the ring-size of the lactones. The small to medium-sized lactones that have been studied range from 4 to 7 member ring structures and include β-propiolactone, γ-butyrolactone, and ε-caprolactone, while the macrolactones include 11-undecalactone, 12-dodecalactone, and 15-pentadecalactone.

Uyama and Kobayashi first accomplished the ring-opening polymerization of ε-caprolactone in bulk using *Pseudomonas fluorescens* lipase at 60°C. The influence of reaction temperature was investigated and at 75°C, the best conversion (92%) and highest molecular weight (M_n 7700 Da, PDI = 1.24) were achieved [88]. However, long reaction times of 10 days were employed. Knani *et al.* also investigated the same reaction with PPL in n-hexane using methanol or methyl 6-hydroxyhexanoate as the initiator for ring opening. Substrate/initiator ratios were varied to optimize the molecular weight, but again long reaction times of 96 to 624 hours were needed to produce molecular weights between 300 to 2000 Da [83].

Since then, other groups have attempted the ring-opening polymerization of β-propiolactone [90], racemic α-methyl-β-propiolactone [91], β-butyrolactone [92-94], γ-butyrolactone [93], and ε-caprolactone (ε-CL) [94-99] using various initiators including water [89,90], butanol [95,96], butylamine [96], and ethyl glucoside [98]. Most papers either reported the production of low molecular weight polymers or used long reaction times to obtain higher molecular weights. For example, the highest molecular weight reported for the enzymatic ring-opening polymerization of ε-CL was 14500 Da (M_w) (PDI = 1.23), which was obtained after a 10-day reaction [99]. It is interesting to note that the polymerization of racemic α-methyl-β-propiolactone in heptane using lipase from *P. fluorescens* resulted in an optically active polymer made up of S enantiomer preferentially [91].

The success of ring-opening polymerizations using chemical means is based on the ring strain of the lactones. Smaller ring structures such as those just described, are under more strain than larger macrolactones. Therefore, it was surprising to find that enzyme-catalyzed ring-opening polymerization of macrolactones usually proceed at a faster rate and produce higher molecular weights than that of ε-CL. Uyama *et al.* first studied the poymerization of 11-undecalactone and 15-pentadecalactone using *P. fluorescens* and *C. rugosa* lipases at 60°C for 120 hours. The former lipase produced poly(undecalactone) with high yields (98%) and molecular weights (M_n 8400 Da, PDI = 2.5) [100]. They later studied the effects of reaction temperature and enzyme immobilization of *Pseudomonas* sp. on the bulk polymerization of 12-dodecalactone and were able to produce molecular weights as high as 25000 Da (M_n) [101].

Uyama *et al.* and Bisht *et al.* have also investigated the polymerization of 15-pentadecalactone [100,102,103]. The latter group performed the reaction at 80°C in bulk for 24 or 72 hours. By screening enzymes, they found that *Pseudomonas* sp. lipases as well as Lipozyme® IM and Novozym® 435 yielded high monomer conversions (80-100%) and molecular weights (M_n 15000-34400 Da) [103]. Control of the water content led to further increases in the molecular weight to 62000 Da (PDI = 1.9) [103].

Mechanistic studies for macrolactone ring-opening polymerization by both groups suggest that the initiation step is slower than chain propagation [100,103,104]. Specifically, Uyama *et al.* believed that the formation of an acyl-enzyme intermediate (before initiation) was the rate-determining step in the overall reaction. Further, they proposed the same mechanism for the ring-opening polymerization of both small and large ring structures [104]. However, Gross and coworkers also studied the mechanism for polymerization of ε-CL [95,96]. Conversely to the polymerization of macrolactones, they observed that initiation was rapid and followed by slow propagation. Chain transfer and termination also did not take place suggests that ring opening polymerization of ε-CL proceeds by a living mechanism [96].

2.4.   POLYCARBONATES

Polycarbonates possess the same diversity in terms of structure, properties, and applications as polyesters.   However, the carbonate linkage can provide some advantages over the ester bond including hydrolytic and mechanical stability. Aliphatic polycarbonates in the molecular weight range of 500 to 5000 have been used industrially for the same applications as low molecular weight aliphatic polyesters such as intermediates in the synthesis of polyurethane elastomers and plasticizers for poly(vinyl chloride) [105].   Because they are biodegradable, aliphatic polycarbonates have also been suggested for application as bioresorbable biomaterials [105].   Aromatic polycarbonates, especially those derived from bisphenol A, are widely used high performance plastics [105].

Research for enzymatic polymerization of carbonates has been limited and all work has appeared in the last five years with one exception.   Table 1 provides an overview of the biocatalytic synthesis of polycarbonates [67,106-113].   In 1989, Abramowicz and Keese published a report mainly concerned with the enzymatic transesterification of carbonates with mono-functional alcohols.   However, they observed that reaction of diphenyl carbonate (DPC) with bisphenol A could potentially form polycarbonates and in fact, a molecular weight of approximately 900 was obtained [106].   While this molecular weight represents the formation of predominantly dimer, both monomers possessed aromatic structures, which have traditionally been difficult for lipases to accommodate.

Two major types of synthesis exist for the polymerization of carbonates: condensation and ring-opening polymerization. The former method is most widely used for industrial production. Unfortunately, condensation requires the use of phosgene either for synthesis of a more active carbonate monomer such as DPC or for direct polymerization with a bifunctional alcohol.   Therefore, the use of enzymes in the synthesis of polycarbonates could potentially improve the environmental impact of the process.

TABLE 1. Overview of polycarbonate synthesis using enzymes

Polymerization Type	Chain Structure	Enzyme (Conc.)	Substrate(s) (Solvent)	Time	Temp. (°C)	M_w (PDI)	Ref. #
Condensation	Aromatic	CRL (1 mg/ml in PBS)	DPC with Bisphenol A (75:25 PBS:acetone)	-	-	Mostly Dimer 900	106
Ring-Opening	Aliphatic	PPL (0.1 wt%)	1,3-Dioxan-2-one (bulk)	24 hrs	100	169000 (3.5)	107
Ring-Opening	Aliphatic	CAL (50 wt%)	1,3-Dioxan-2-one (bulk)	72-120 hrs	60-75	6900-8500 (3.0-3.4)	108
Ring-Opening	Aliphatic	CAL (25 wt%)	1,3-Dioxan-2-one (bulk)	48 hrs	55-85	32500-61000 (2.4-2.6)	109 110
Ring-Opening	Aliphatic	PPL, PCL (150 mg/ml)	1,3-Dioxan-2-one (dioxane or toluene)	-	65	Oligomers DP = 3-9	109 110
Condensation	Aliphatic	CAL (20 wt%)	Activated Carbonates with 1,3-Propanediol (bulk)	72 hrs	40-50	6500-8500 (-)	111
Condensation	Aliphatic (OH-Pendant)	CAL (10 wt%)	Activated Carbonates with Triols (bulk)	72 hrs	50	2500-5300 (-)	111
Condensation	Aromatic	CAL (10 wt%)	Activated Carbonates with Aromatic Diols (bulk)	24 hrs	70	Up to 5200 (-)	67
Ring-Opening	Aliphatic (COOH-Pendant)	PFL (50 wt%)	5-Methyl-5-benzyloxycarbonyl-1,3-dioxan-2-one (bulk)	72 hrs	80	10100 (1.65)	112
Condensation	Aliphatic	CAL (18 wt%)	DEC with 1,4-Butanediol (bulk)	31-40 hrs (2 stages)	70-75	24400-40000 (2.2-2.5)	113

Nomenclature: M_w – weight average molecular weight, PDI – polydispersity index (M_w/M_n), CRL – *Candida rugosa* lipase, DPC – diphenyl carbonate, PPL – porcine pancreatic lipase, CAL – *Candida antarctica* lipase, PCL – *Pseudomonas cepacia* lipase, DP – degree of polymerization, PFL – *Pseudomonas fluorescens* lipase

Rodney and coworkers have attempted to improve polycarbonate production by first, synthesizing DPC enzymatically and second, directly polymerizing activated carbonates with diol [111]. As mentioned previously, DPC is traditionally produced through the reaction of phenol with phosgene. Alternatively, the reaction of dimethyl carbonate (DMC) with phenol was performed at 40°C using several different lipases. However, this process was complicated by many factors including side reactions, enzyme specificity and the poor nucleophilic character of phenol. As a result, DPC was produced after long reaction times in the order of weeks and with low yield (maximum 4.2%) [111].

Since our group has achieved much in the area of enzymatic polyester synthesis using activated diesters, the same method was used for the production of polycarbonates by synthesizing dicarbonates with vinyl end groups. As discussed previously, when the reaction of these activated compounds takes place, the leaving group is in the form of an unstable vinyl alcohol which tautomerizes to acetaldehyde and drives the reaction in the forward direction toward polymerization. The activated carbonates were subsequently reacted with diols in the presence of Novozym® 435. Various reaction conditions including temperature and enzyme concentration were investigated, and by doing so, the molecular weight of the polycarbonate was increased from 1300 to 8500 Da [111].

Recently, Matsumura *et al.* has also reported the synthesis of an aliphatic polycarbonate through the direct, two-stage synthesis of diethyl carbonate (DEC) and 1,4-butanediol (BD) [113]. The purpose of the first 24-hour stage was to produce oligomeric products of DEC and BD, which were then polymerized under reduced pressure. Because of the formation of the dimers and trimers in the first stage, loss of significant monomer was avoided in the second stage. By varying the stoichiometry of the monomers and the polymerization time, the molecular weight and end group functionality could be effectively controlled [113].

The second process to produce polycarbonates is through ring-opening polymerization, which was first accomplished by Carothers by reacting a six-membered ring structure at elevated temperatures [114]. Since then, the reaction of cyclic carbonates has been carried out by anionic [115] and cationic [116,117] polymerization. Some decarboxylation with the concomitant formation of carbon dioxide and ether linkages in the polymer has been observed using the latter type of ring-opening polymerization.

Reports by three different groups were released in a very short span of time detailing the enzymatic ring-opening polymerization of 1,3-dioxan-2-one (or trimethylene carbonate) (TMC) in bulk [107-110]. It is clear from Table 1 that research conducted by each group produced highly dissimilar results. Matsumura *et al.* reported the formation of polymer with extremely high molecular weights using low enzyme concentration and elevated temperatures [107]. Conversely, Kobayashi *et al.* produced polymers with low molecular weights using high enzyme concentrations [108]. Finally, Bisht *et al.* investigated various reaction parameters of the ring-opening polymerization of TMC and were able to obtain polymers with molecular weights of 61000 Da (PDI = 2.4-2.6) [109,110].

Bisht and coworkers did observe the conflicting data and attempted to rationalize it with the following explanations [109]. First, the discrepancy between Kobayashi *et al.* and Bisht *et al.* could be due to reaction preparation and conditions, especially in terms of water content. Second, all three groups obtained no conversion of the monomer using *Candida antarctica* lipase B at 100°C. However, no explanation was given concerning the high molecular weight obtained by Matsumura *et al.* using PPL at 100°C. Kobayashi *et al.* reported that PPL still produced polymer after thermal treatment at 100°C [108]. Therefore, either polymerization by PPL can still take place at that temperature or impurities in the crude enzyme preparation are responsible for the reaction.

As mentioned previously, enzymatic condensation of aromatic monomers is difficult and usually produces very low molecular weight products, further confirmed by the results of Abramowicz and Keese [106]. Recently, Rodney *et al.* reported the synthesis of aromatic polycarbonates by the reaction of activated dicarbonates with six different aromatic diols [67]. By varying the reaction temperature in bulk between 50 to 110°C, molecular weights up to 5200 Da were produced at an optimum temperature of 70°C. Furthermore, the immobilized *Candida antarctica* lipase exhibited regioselectivity at this temperature for the para-position over the ortho- and meta-positions of benzene dimethanol [67].

Finally, the synthesis of functional polycarbonates has been of recent interest for use in the biomedical field. A carboxyl-pendant polymer was produced by Al-Azemi and Bisht through the ring-opening polymerization of a protected TMC monomer by lipase from *Pseudomonas fluorescens* at 80°C [112]. The synthesis of the monomer occurred with high yield (89%). After deprotection of the carboxyl group, the resulting polymer possessed a molecular weight of 10100 Da (PDI = 1.65). Rodney *et al.* also reported the production of polycarbonates with hydroxyl pendant groups through the direct synthesis of activated carbonates with three multi-hydroxyl compounds. Investigation of enzyme concentration resulted in hydroxy-functional polycarbonates of approximately 5300 Da with 10% enzyme concentration [111].

3.   Enzymes in Polymers

The motivation to incorporate enzymes into polymers is highly distinct from the synthesis of polymers using enzymes. Bestowing catalytic activity on a polymer can be useful for applications concerning biosensors, nerve-agent degradation, and separation processes. The fixed enzyme is rendered more stable because its conformation is less mobile.

3.1.   TRADITIONAL IMMOBILIZATION

Biocatalysts offer numerous advantages such as high catalytic efficiency and selectivity over conventional chemical catalysts. However, they often display instability under the operational conditions of industrial processes. Therefore, immobilization constitutes a strategy to overcome this problem as well as a means to facilitate the enzyme separation from reactants and products [118].

Many techniques involving the use of a wide range of materials as a support have been developed to perform enzyme immobilization [119,120]. Polymers have been extensively employed as matrices due to their attractive thermo-mechanical and chemical properties. Synthetic polymers are usually selected for enzyme immobilization [121], though various examples involving the use of natural polymers such as cellulose are described in the literature [122,123].

Traditional immobilization of enzyme can be achieved by entrapment within the polymer during polymerization [124,125], by selective or non-selective adsorption on a polymeric material [126], and by covalent attachment to a support. The last strategy often necessitates either the activation of the polymer surface or the modification of the enzyme to enable a covalent enzyme-support connection [127].

3.2.   ACTIVITY AND STABILITY OF ENZYME MODIFIED VIA TRADITONAL IMMOBILIZATION

Immobilization affects the enzyme microenvironment due to the physical and chemical characteristics of the polymeric support resulting in modifications in the biocatalyst structure, thereby influencing its biological properties. As adsorption and entrapment constitute mild immobilization processes with respect to covalent linkage, they are expected to have a less dramatic effect on the enzyme. Covalent immobilization usually leads to an increase in the intrinsic K_M combined with a decrease in k_{cat} [128]. Multiple point covalent immobilization may result in even more severe changes in the enzyme structure and more pronounced alteration of biocatalytic kinetic characteristics than single point covalent attachment, unless one can learn how to maintain structure during preparation [129].

The effect of traditional immobilization on enzyme stability is usually exhibited by the enhancement of thermostability in aqueous media [130]. Alteration in pH optima and dependence of enzymatic activity on pH is also often associated with immobilization [131-133]. The polymeric microenvironment and the non-covalent or covalent interactions connecting the biocatalyst to the support lead to a reduction in the enzyme mobility, thereby preventing denaturation that results from conformational changes. Multiple point covalent attachment enhances conformational stability [129,134].

3.3.   BIOPLASTICS

Enzyme modification through the formation of a bioplastic constitutes an original concept for immobilization. Bioplastics differentiate themselves from traditional enzyme immobilization on a polymeric support in the sense that the enzyme participates in the polymer synthesis. As the enzyme acts as a monomer during the polymerization, its immobilization is performed within the polymeric matrix by covalent linkage. Moreover, this type of incorporation into a polymer relies on the chemical reaction of polymeric monomers with specific functionalities on the enzyme surface, enabling a multiple-point covalent immobilization. The enzyme-containing polymer is generated in a single reaction step.

3.3.1 Bioplastics Synthesis in Organic Solvents

Bioplastic synthesis represents a processing strategy designed to obtain modified enzymes, which are active in both aqueous and organic media. The resulting enzymes should ideally exhibit increased chemical resistance and thermostability in both environments. Several examples involving bioplastic synthesis are described in the literature. Interestingly the bioplastics have themselves been synthesized in either aqueous or non-aqueous medium [135-138].

Incorporation of Subtilisin into Acrylic Polymers.   The choice of the medium for bioplastic synthesis may influence the amount of enzyme that can be incorporated into the polymer, and hence its activity retention and stability once immobilized. Yang *et al.* proposed that the formation of protein-containing polymers in the presence of a non-conventional medium would be beneficial in obtaining a viable enzyme-polymer system for utility in organic media [139].

To examine their hypothesis they synthesized subtilisin-containing polyacrylate in toluene. Subtilisin was modified with polyethylene glycol aldehyde acrylate (PEG). The PEG-modification involved the reaction of the PEG aldehyde functionality with the ε-amino groups of lysine residues on the enzyme surface. This procedure ensured that the subtilisin was soluble in organic solvents. After modification, subtilisin could participate in the polyacrylate synthesis via its own acrylate groups.

As expected, the activity of biopolymer in aqueous media was considerably reduced with respect to the activity of native subtilisin, but was comparable with respect to the activity of subtilisin traditionally modified by single-point covalent bonding to oxirane activated macroporous acrylic beads. The effect of bioplastic immobilization was further investigated by comparing the activity of biopolymer in the aprotic, apolar solvent dioxane with the activity of native subtilisin, PEG modified subtilisin and traditionally immobilized subtilisin. Dioxane is known as a strong inhibitor of α-chymotrypsin catalyzed hydrolysis. Moreover, its presence brings about electrostatic changes in the enzyme environment and provokes an augmentation in the repulsion between substrate and enzyme. As a results, K_M value increases, while k_{cat} remains unchanged, leading to a reduced k_{cat}/K_M ratio. The analysis of the dioxane effect on the different subtilisin systems indicated that the solvent dramatically alters the activity of all systems except the biopolymer. The increased resistance of the biopolymer to dioxane thus results from the smaller electrostatic influence of solvent on the hydrophobic environment of subtilisin. The enzyme rigidity inside the biopolymer may also improve its resistance to denaturing effects of solvent.

Yang *et al.* showed that the increase in thermostability of modified subtilisin in aqueous media with respect to the native subilitisin depends on the number of covalent bonds between enzyme and immobilization support [139]. The biopolymer displays the highest thermostability as a consequence of multiple covalent attachments of subtilisin to the matrix. As expected, the biopolymer exhibited higher thermostability and recyclability than both traditionally immobilized and native subtilisin in organic media.

Incorporation of α-Chymotrypsin into Polymers. Wang *et al.* incorporated α-chymotrypsin into various polymers (polystyrene, poly(vinyl acetate), poly(ethyl vinyl ether), poly(methyl methacrylate)) [140]. α-Chymotrypsin activation for polymerization was obtained through acryloylation, while hydrophobic ion pairing enabled solubilization of modified α-chymotrypsin in the organic medium used for biopolymer synthesis. Diffusional limitations were small enough to enable the investigation of intrinsic kinetic characteristics of the bioplastic. It was suggested that α-chymotrypsin-containing biopolymers and ion-paired soluble enzyme exhibit comparable reactivity in solvents like hexane and toluene. However, differences in the kinetic behavior of ion-paired α-chymotrypsin and α-chymotrypsin incorporated into polyacrylates are observed when used in polar solvents. Indeed, the bioplastic displayed significantly higher activity. Analysis in organic solvents such as hexane and THF suggested that the thermostability of α-chymotrypsin was enhanced by the immobilization into plastics. High activity of bioplastics in non-conventional media combined with increased thermostability leads to the widening of the potential applicability of enzymes in industrial processes.

3.3.2 Bioplastics Synthesized in Aqueous Media

A number of researchers, including our own laboratory, have assessed the properties of enzyme-polyurethanes. We have focused much of our attention on the properties of organophosphorus hydrolase (OPH)-containing polyurethane polymers. The bioplastics were prepared by reacting an aqueous solution with a polyurethane prepolymer containing multiple isocyanate functionalities [141-143]. During the foaming process, isocyanates react with water, initiating the formation of carbamic acid, which degrades to amine and evolves carbon dioxide. Carbon dioxide bubbles through the reacting polymer generating a porous foam structure. Amines further react with isocyanate groups resulting in the formation of a crosslinked polyurethane matrix. Enzyme added to the aqueous solution can participate in the polymeric synthesis via their lysine residues, creating an enzyme-polymer network with multiple-point attachments.

No internal or external diffusional limitations were detected during activity measurements in aqueous media. The covalent incorporation of OPH into sponge-like foams leads to more than 50% of activity retention with a 3-fold increase in K_M value. Interestingly, no essential change in the inhibition effect of chemicals like DMF, dioxane and phenol on OPH was observed as a result of immobilization process. These results, combined with the kinetic analysis suggest that the immobilization process does not significantly alter OPH intrinsic reactivity. The thermostability of OPH-containing bioplastics was examined at 25 and 50°C in the absence and presence of 20% DMSO. OPH incorporation into polyurethane foams dramatically stabilized their half-lives, which changed from days to years under ambient conditions in aqueous media. Interestingly, the addition of DMSO to aqueous media enhanced the thermostability of native and immobilized enzyme in a similar manner. Since hypochlorite bleach constitutes an important means of nerve agent decontamination, the effect of oxidizers on OPH-containing biopolymers was also examined in aqueous media. We demonstrated that OPH incorporation into polymers results in a viable biopolymer in buffered bleach solutions whereas the native enzyme is immediately denatured. Similar results are obtained for stability against pH fluctuations.

4.   Summary and Conclusions

Research combining the multidisciplinary fields of biotechnology and polymeric materials will continue to grow as the use of enzymes becomes more accepted as an industrial practice. Most of the work presented in this review represents preliminary data and analyses needed to develop predictable biological processes. It is clear that enzymatic polymerizations are becoming more efficient in terms of reaction time and conditions. Furthermore, methods to produce novel polymers with optical activity and/or functionality are much more feasible using enzymes. Enzymes incorporated into polymers possess dramatically enhanced stability when compared to regular enzymes. As a result, they should be capable of withstanding harsh conditions sometimes necessary in industrial processes. Our group is continuing research in the area of enzymatic polyester synthesis and enzyme stability, specifically concerning DFPase, in polyurethane foams [144].

5.   References

1.   Zaks, A., and Klibanov, A.M. (1985) Enzyme-catalyzed processes in organic solvents, *Proc. Natl. Acad. Sci. USA* **82**, 3192-3196.
2.   Kine, R.B., and Novak, R.W. (1985) Acrylic and methacrylic ester polymers, in H.F. Mark, N.M. Bikales, C.G. Overberger, and C. Menges (eds.), *Polymer Science and Engineering*, 2nd edition, Vol. 1, John Wiley and Sons, Inc., New York, pp. 234-299.
3.   Tor, R., Dror, Y., and Freeman, A. (1990) Enzymatically catalysed transesterifications of acryl and methacryl monomeric esters, *Enzyme Microb. Technol.* **12**, 299-304.
4.   Hajjar, A.B., Nicks, P.F., and Knowles, C.J. (1990) Preparation of monomeric acrylic ester intermediates using lipase catalysed transesterifications in organic solvents, *Biotechnol. Lett.* **12**, 825-830.
5.   Pavel, K., and Ritter, H. (1991) Radical polymerization of different 11-meth-acryloylaminoundecanoic acid esters and oligoesters esterified by lipases, *Makromol. Chem.* **192**, 1941-1949.

424

6. Fukusaki, E., Senda, S., Nakazono, Y., Yuasa, H., and Omata, T. (1992) Preparation of carboxyalkyl acrylate by lipase-catalyzed regioselective hydrolysis of corresponding methyl ester, *Bioorg. Med. Chem. Lett.* **2**, 411-414.

7. Derango, R., Wang, Y.-F., Dowbenko, R., and Chiang, L. (1994) The lipase-catalyzed synthesis of carbamoyloxyethyl methacrylate, *Biotechnol. Lett.* **16**, 241-246.

8. Ikeda, I., Tanaka, J., and Suzuki, K. (1991) Synthesis of acrylic esters by lipase, *Tetrahedron Lett.* **32**, 6865-6866.

9. Warwel, S., Steinke, G., and Rüsch gen. Klaas, M. (1996) An efficient method for lipase-catalysed preparation of acrylic and methacrylic acid esters, *Biotechnol. Techniques* **10**, 283-286.

10. Kamat, S., Barrera, J., Beckman, E.J., and Russell, A.J. (1992) Biocatalytic synthesis of acrylates in organic solvents and supercritical fluids: I. Optimization of the enzyme environment, *Biotechnol. Bioeng.* **40**, 158-166.

11. Kamat, S.V., Iwaskewycz, B., Beckman, E.J., and Russell, A.J. (1993) Biocatalytic synthesis of acrylates in supercritical fluids: II. Tuning enzyme activity by changing pressure, *Proc. Natl. Acad. Sci. USA* **90**, 2940-2944.

12. Kamat, S., Critchley, G., Beckman, E.J., and Russell, A.J. (1995) Biocatalytic synthesis of acrylates in organic solvents and supercritical fluids: III. Does carbon dioxide covalently modify enzymes?, *Biotechnol. Bioeng.* **46**, 610-620.

13. Ghogare, H. and Sudesh Kumar, G. (1990) Novel route to chiral polymers involving biocatalytic transesterification of O-acryloyl oximes, *J. Chem. Soc., Chem. Commun.*, 134-135.

14. Margolin, A.L., Fitzpatrick, P.A., Dubin, P.L., and Klibanov, A.M. (1991) Chemoenzymatic synthesis of optically active (meth)acrylic polymer, *J. Am. Chem. Soc.* **113**, 4693-4694.

15. Athawale, V.D., and Gaonkar, S.R. (1999) Chemoenzymatic syntheses of optically active polyacrylates, *Macromolecules* **32**, 6065-6068.

16. Patil, D.R., Dordick, J.S., and Rethwisch, D.G. (1991) Chemoenzymatic synthesis of novel sucrose-containing polymers, *Macromolecules* **24**, 3462-3463.

17. Chen, X.M., Johnson, A., Dordick, J.S., and Rethwisch, D.G. (1994) Chemoenzymatic synthesis of linear poly(sucrose acrylate) – optimization of enzyme activity and polymerization conditions, *Macromol. Chem. Phys.* **195**, 3567-3578.

18. Therisod, M., and Klibanov, A.M. (1986) Facile enzymatic preparation of monoacylated sugars in pyridine, *J. Am. Chem. Soc.* **108**, 5638-5640.

19. Wulff, G., Schmid, J., and Venhoff, T. (1996) The synthesis of polymerizable vinyl sugars, *Macromol. Chem. Phys.* **197**, 259-274.

20. Martin, B.D., Ampofo, S.A., Linhardt, R.J., and Dordick, J.S. (1992) Biocatalytic synthesis of sugar-containing poly(acrylate)-based hydrogels, *Macromolecules* **25**, 7081-7085.

21. Chen., X.M., Dordick, J.S, and Rethwisch, D.G. (1995) Chemoenzymatic synthesis and characterization of poly(α-methyl galactoside 6-acrylate) hydrogels, *Macromolecules* **28**, 6014-6019.

22. Shibatani, S., Kitagawa, M., and Tokiwa, Y. (1997) Enzymatic synthesis of vinyl sugar ester in dimethylformamide, *Biotechnol. Lett.* **19**, 511-514.

23. Kitagawa, M., and Tokiwa, Y. (1998) Synthesis of polymerizable sugar ester possessing long spacer catalyzed by lipase from *Alcaligenes* sp. and its chemical polymerization, *Biotechnol. Lett.* **20**, 627-630.

24. Ikeda, I., and Klibanov, A.M. (1993) Lipase-catalyzed acylation of sugars solubilized in hydrophobic solvents by complexation, *Biotechnol. Bioeng.* **42**, 788-791.

25. Blinkovsky, A.M., and Dordick, J.S. (1993) Enzymatic derivatization of saccharides and their chemical polymerization, *Tetrahedron: Asymm.* **4**, 1221-1228.

26. Kopf, P.W. (1985) Phenolic resins, in H.F. Mark, N.M. Bikales, C.G. Overberger, and C. Menges (eds.), *Polymer Science and Engineering*, 2nd edition, Vol. 11, John Wiley and Sons, Inc., New York, pp. 45-95.

27. Aycock, D., Abolins, V., and White, D.M. (1985) Poly(phenylene ether), in H.F. Mark, N.M. Bikales, C.G. Overberger, and C. Menges (eds.), *Polymer Science and Engineering*, 2nd edition, Vol. 13, John Wiley and Sons, Inc., New York, pp. 1-30.

28. Schwartz, R.D., and Hutchinson, D.B. (1981) Microbial and enzymatic production of 4,4'-dihydroxybiphenyl via phenol coupling, *Enzyme Microb. Technol.* **3**, 361-363.

29. Dordick, J.S., Marletta, M.A., and Klibanov, A.M. (1987) Polymerization of phenols catalyzed by peroxidase in nonaqueous media, *Biotechnol. Bioeng.* **30**, 31-36.

30. Akkara, J.A., Senecal, K.J., and Kaplan, D.L. (1991) Synthesis and characterization of polymers produced by horseradish peroxidase in dioxane, *J. Polym. Sci. Part A: Polym. Chem.* **29**, 1561-1574.

31. Uyama, H., Kurioka, J., Kaneko, I., and Kobayashi, S. (1994) Synthesis of a new family of phenol resin by enzymatic oxidative polymerization, *Chem. Lett.*, 423-426.

32. Ayyagari, M.S., Marx, K.A., Tripathy, S.K., Akkara, J.A., and Kaplan, D.L. (1995) Controlled free-radical polymerization of phenol derivatives by enzyme-catalyzed reactions in organic solvents, *Macromolecules* **28**, 5192-5197.

33. Uyama, H., Kurioka, H., Sugihara, J., Komatsu, I., and Kobayashi, S. (1995) Peroxidase-catalyzed oxidative polymerization of cresols to a new family of polyphenols, *Bull. Chem. Soc. Jpn.* **68**, 3209-3214.

34. Uyama, H. Kurioka, H., Sugihara, J., and Kobayashi, S. (1996) Enzymatic synthesis and thermal properties of a new class of polyphenol, *Bull. Chem. Soc. Jpn.* **69**, 189-193.

35. Uyama, H., Kurioka, H., Sugihara, J., Komatsu, I., and Kobayashi, S. (1997) Oxidative polymerization of p-alkylphenols catalyzed by horseradish peroxidase, *J. Polym. Sci. Part A: Polym. Chem.* **35**, 1453-1459.

36. Uyama, H., Lohavisavapanich, C., Ikeda, R., and Kobayashi, S. (1998) Chemoselective polymerization of a phenol derivative having a methacryl group by peroxidase catalyst, *Macromolecules* **31**, 554-556.

37. Kobayashi, S., Uyama, H., Ushiwata, T., Uchiyama, T., Sugihara, J., and Kurioka, H. (1999) Enzymatic oxidative polymerization of bisphenol A to a new class of soluble polyphenol, *Macromol. Chem. Phys.* **199**, 777-782.

38. Tonami, H. Uyama, H., Kobayashi, S., and Kubota, M. (1999) Peroxidase-catalyzed oxidative polymerization of m-substituted phenol derivatives, *Macromol. Chem. Phys.* **200**, 2365-2371.

39. Madhusudhan, A., John, V.T., Gonzalez, R.D., Akkara, J.A., and Kaplan, D.L. (1993) Catalytic and interfacial aspects of enzymatic polymer synthesis in reversed micellar systems, *Biotechnol. Bioeng.* **41**, 531-540.

40. Premachandran, R.S., Banerjee, S., Wu, X.-K., John, V.T., McPherson, G.L., Akkara, J., Ayyagari, M., and Kaplan, D. (1996) Enzymatic synthesis of fluorescent naphthol-based polymers, *Macromolecules* **29**, 6452-6460.

41. Karayigitoglu, C.F., Kommareddi, N., Gonzalez, R.D., John, V.T., McPherson, G.L, Akkara, J.A., and Kaplan, D.L. (1995) The morphology of phenolic polymers enzymatically synthesized in surfactant microstructures, *Mater. Sci. Eng.* **C2**, 165-171.

42. Ikeda, R., Sugihara, J., Uyama, H., and Kobayashi, S. (1996) Enzymatic oxidative polymerization of 2,6-dimethylphenol, *Macromolecules* **29**, 8702-8705.

43. Ikeda, R., Uyama, H., and Kobayashi, S. (1996) Novel synthetic pathway to a poly(phenylene oxide). Laccase-catalyzed oxidative polymerization of syringic acid, *Macromolecules* **29**, 3053-3054.

44. Ikeda, R., Sugihara, J., Uyama, H., and Kobayashi, S. (1998) Enzymatic oxidative polymerization of 4-hydroxybenzoic acid derivatives to poly(phenylene oxide)s, *Polym. Int.* **47**, 295-301.

45. Popp, J.L., Kirk, T.K., and Dordick, J.S. (1991) Incorporation of p-cresol into lignins via peroxidase-catalysed copolymerization in nonaqueous media, *Enzyme Microb. Technol.* **13**, 964-968.

46. Blinkovsky, A.M., and Dordick, J.S. (1993) Peroxidase-catalyzed synthesis of lignin-phenol copolymers, *J. Polym. Sci. Part A: Polym. Chem.* **31**, 1839-1846.

47. Wang, P., Martin, B.D, Parida, S., Rethwisch, D.G., and Dordick, J.S. (1995) Multienzymic synthesis of poly(hydroquinone) for use as a redox polymer, *J. Am. Chem. Soc.* **117**, 12885-12886.

48. Wang, P., and Dordick, J.S. (1998) Enzymatic synthesis of unique thymidine-containing polyphenols, *Macromolecules* **31**, 941-943.

49. Goodman, I. (1985) Polyesters, in H.F. Mark, N.M. Bikales, C.G. Overberger, and C. Menges (eds.), Polymer Science and Engineering, 2nd edition, Vol. 12, John Wiley and Sons, Inc., New York, pp. 1-75.

50. Okumura, S., Iwai, M., and Tominaga, Y. (1984) Synthesis of ester oligomer by Aspergillus niger lipase, *Agric. Biol. Chem.* **48**, 2805-2808.

51. Binns, F., Roberts, S.M., Taylor, A., and Williams, C.F. (1993) Enzymic polymerization of an unactivated diol/diacids system, *J. Chem. Soc. Perkins Trans. I*, 899-904.

52. Binns, F., Harffey, P., Roberts, S.M., and Taylor, A. (1998) Studies of lipase-catalyzed polyesterification of an unactivated diacids/diol system, *J. Polym. Sci. Part A: Polym. Chem.* **36**, 2069-2080.

53. Linko, Y.-Y., Wang, Z.-L., and Seppälä, J. (1995) Lipase-catalyzed synthesis of poly(1,4-butyl sebacate) from sebacic acid or its derivatives with 1,4-butanediol, *J. Biotechnol.* **40**, 133-138.

54. Wu, X.Y., Seppälä, J., and Linko, Y.-Y. (1996) Lipase-catalyzed polyester synthesis, *Biotechnol. Techniques* **10**, 793-798.

55. Wallace, J.S., and Morrow, C.J. (1989) Biocatalytic synthesis of polymers. II. Preparation of [AA-BB]$_x$ polyesters by porcine pancreatic lipase catalyzed transesterification in anhydrous, low polarity organic solvents, *J. Polym. Sci. Part A: Polym. Chem.* **27**, 3271-3284.

56. Brazwell, E.M., Filos, D.Y., and Morrow, C.J. (1995) Biocatalytic synthesis of polymers. III. Formation of a high-molecular weight polyester through limitation of hydrolysis by enzyme-bound water and through equilibrium control, *J. Polym. Sci. Part A: Polym. Chem.* **33**, 89-95.

57. Linko, Y.-Y., Wang, Z.-L., and Seppälä, J. (1995) Lipase-catalyzed linear aliphatic polyester synthesis in organic solvent, *Enzyme Microb. Technol.* **17**, 506-511.

58. Athawale, V.D., and Gaonkar, S.R. (1994) Enzymatic synthesis of polyesters by lipase catalysed polytransesterification, *Biotechnol. Lett.* **16**, 149-154.

59. Uyama, H., and Kobayashi, S. (1994) Lipase-catalyzed polymerization of divinyl adipate with glycols to polyesters, *Chem. Lett.*, 1687-1690.

60. Chaudhary, A.K., Beckman, E.J., and Russell, A.J. (1997) Biocatalytic polyester synthesis: Analysis of the evolution of molecular weight and end group functionality, *Biotechnol. Bioeng.* **55**, 227-239.

61. Chaudhary, A.K., Beckman, E.J., and Russell, A.J. (1998) Nonequal reactivity model for biocatalytic polytransesterification, *AIChE J.* **44**, 753-764.

62. Chaudhary, A.K., Lopez, J., Beckman, E.J., and Russell, A.J. (1997) Biocatalytic solvent-free polymerization to produce high molecular weight polyesters, *Biotechnol. Prog.* **13**, 318-325.

63. Chaudhary, A.K., Beckman, E.J., and Russell, A.J. (1998) Rapid biocatalytic polytransesterification: Reaction kinetics in an exothermic reaction, *Biotechnol. Bioeng.* **59**, 428-437.

64. Kline, B.J., Lele, S.S., Lenart, P.J., Beckman, E.J., and Russell, A.J. (in press) Use of a batch stirred reactor to rationally tailor biocatalytic polytransesterification, *Biotechnol. Bioeng.*

65. Kline, B.J., Lele, S.S., Beckman, E.J., and Russell, A.J. (submitted) Role of diffusion in biocatalytic polytransesterification, *AIChE J.*

66. Mezoul, G., Lalot, T., Brigodiot, M., and Maréchal, E. (1996) Enzyme-catalyzed syntheses of poly(1,6-hexanediyl isophthalate) and poly(1,6-hexanediyl terephthalate) in organic medium, *Polym. Bull.* **36**, 541-548.

67. Rodney, R.L., Allinson, B.T., Beckman, E.J., and Russell, A.J. (1999) Enzyme-catalyzed polycondensation reactions for the synthesis of aromatic polycarbonates and polyesters, *Biotechnol. Bioeng.* **65**, 485-489.

68. Uyama, H., Yaguchi, S., and Kobayashi, S. (1999) Enzymatic synthesis of aromatic polyesters by lipase-catalyzed polymerization of dicarboxylic acid divinyl esters and glycols, *Polym. J.* **31**, 380-383.

69. Geresh, S., and Gilboa, Y. (1990) Enzymatic synthesis of alkyds, *Biotechnol. Bioeng.* **36**, 270-274.

70. Geresh, S., and Gilboa, Y. (1991) Enzymatic syntheses of alkyds. II. Lipase-catalyzed polytransesterification of dichloroethyl fumarate with aliphatic and aromatic diols, *Biotechnol. Bioeng.* **37**, 883-888.

71. Mezoul, G., Lalot, T., Brigodiot, M., and Maréchal, E. (1996) Enzyme-catalyzed synthesis of poly(1,6-hexanediyl maleate-co-fumarate) in organic medium. Study of macrolactone formation in relation to the composition of initial monomer mixture, *Macromol. Chem. Phys.* **197**, 3581-3592.

72. Chaudhary, A.K., Beckman, E.J., and Russell, A.J. (1995) Rational control of polymer molecular weight and dispersity during enzyme-catalyzed polyester synthesis in supercritical fluids, *J. Am. Chem. Soc.* **117**, 3728-3733.

73. Mesiano, A.J., Enick, R.M., Beckman, E.J., and Russell, A.J. (submitted) The phase behavior of fluorinated diols, divinyl adipate, and a fluorinated polyester in supercritical carbon dioxide, *Fluid Phase Equilibria.*

74. Mesiano, A.J., Beckman, E.J., and Russell, A.J. (in press) Biocatalytic synthesis of fluorinated polyesters, *Biotechnol. Prog.*

75. Margolin, A.L., Crenne, J.-Y., and Klibanov, A.M. (1987) Stereoselective oligomerizations catalyzed by lipases in organic solvents, *Tetrahedron Lett.* **28**, 1607-1610.

76. Wallace, J.S., and Morrow, C.J. (1989) Biocatalytic synthesis of polymers. Synthesis of an optically active, epoxy-substituted polyester by lipase-catalyzed polymerization, *J. Polym. Sci. Part A: Polym. Chem.* **27**, 2553-2567.

77. Patil, D.R., Rethwisch, D.G., and Dordick, J.S. (1991) Enzymatic synthesis of a sucrose-containing linear polyester in nearly anhydrous organic media, *Biotechnol. Bioeng.* **37**, 639-646.

78. Kline, B.J., Beckman, E.J., and Russell, A.J. (1998) One-step biocatalytic synthesis of linear polyesters with pendant hydroxyl groups, *J. Am. Chem. Soc.* **120**, 9475-9480.

79. Uyama, H., Inada, K., and Kobayashi, S. (1999) Regioselective polymerization of divinyl sebacate and triols using lipase catalyst, *Macromol. Rapid Commun.* **20**, 171-174.

80. Ajima, A., Yoshimoto, T., Takahashi, K., Tamaura, Y., Saito, Y., and Inada, Y. (1985) Polymerization of 10-hydroxydecanoic acid in benzene with polyethylene glycol – modified lipase, *Biotechnol. Lett.* **7**, 303-306.

81. Matsumura, S., and Takahashi, J. (1986) Enzymatic synthesis of functional oligomers 1. Lipase catalyzed polymerization of hydroxy acids, *Makromol. Chem., Rapid Commun.* **7**, 369-373.

82. Morrow, C.J., Wallace, J.S., Bybee, G.M., Reda, K.B., and Williams, E. (1990) Biocatalytic synthesis of polyesters by enzyme-catalyzed transesterification in organic media, *Mater. Res. Soc. Symp. Proc.* **174**, 197-208.

83. Knani, D., Gutman, A.L., and Kohn, D.H. (1993) Enzymatic polyesterification in organic media. Enzyme-catalyzed synthesis of linear polyesters. I. Condensation polymerization of linear hydroxyesters. II. Ring-opening polymerization of ε-caprolactone, *J. Polym. Sci. Part A: Polym. Chem.* **31**, 1221-1232.

84. Knani, D., and Kohn, D.H. (1993) Enzymatic polyesterification in organic media. II. Enzyme-catalyzed synthesis of lateral-substituted aliphatic polyesters and copolyesters, *J. Polym. Sci. Part A: Polym. Chem.* **31**, 2887-2897.

85. O'Hagan, D., and Zaidi, N.A. (1993) Polymerisation of 10-hydroxydecanoic acid with the lipase from *Candida cylindracea, J. Chem. Soc. Perkin Trans. I*, 2389-2390.

86. O'Hagan, D., and Zaidi, N.A. (1994) Enzyme-catalysed condensation polymerization of 11-hydroxyundecanoic acid with lipase from *Candida cylindracea, Polymer* **35**, 3576-3578.

87. Gutman, A.L., Oren, D., Boltanski, A., and Bravdo, T. (1987) Enzymatic oligomerisation versus lactonisation of ω-hydroxyesters, *Tetrahedron Lett.* **28**, 5367-5368.

88. Gutman, A.L. Knani, D., and Bravdo, T. (1997) Enzymatic condensation and polycondensation reactions in organic solvents, *Macromol. Symp.* **122**, 39-44.

89. Uyama, H., and Kobayashi, S. (1993) Enzymatic ring-opening polymerization of lactones catalyzed by lipase, *Chem. Lett.*, 1149-1150.

90. Namekawa, S., Uyama, H., and Kobayashi, S. (1996) Lipase-catalyzed ring-opening polymerization and copolymerization of β-propiolactone, *Polym. J.* **28**, 730-731.

91. Svirkin, Y.Y., Xu, J., Gross, R.A., Kaplan, D.L., and Swift, G. (1996) Enzyme-catalyzed stereoselective ring-opening polymerization of α-methyl-β-propiolactone, *Macromolecules* **29**, 4591-4597.

92. Nobes, G.A.R., Kazlauskas, R.J., and Marchessault, R.H. (1996) Lipase-catalyzed ring-opening polymerization of lactones: A novel route to poly(hydroxyalkanoate)s, *Macromolecules* **29**, 4829-4833.

93. Xie, W.H., Li, J., Chen, D.P., and Wang, P.G. (1997) Ring-opening polymerization of β-butyrolactone by thermophilic lipases, *Macromolecules* **30**, 6997-6998.

94. Dong, H., Wang, Z., Li, Z.-Q., You, D.-L., Han, S.-P., Cao, S.-G., and Shen, J.-C. (1998) Enzyme-catalyzed synthesis of polyester: Effect of the microenvironment of enzyme on polyesterification, *Ann. N.Y. Acad. Sci.* **864**, 263-266.

95. MacDonald, R.T., Pulapura, S.K., Svirkin, Y.Y., Gross, R.A., Kaplan, D.L., Akkara, J., Swift, G., and Wolk, S. (1995) Enzyme-catalyzed ε-caprolactone ring-opening polymerization, *Macromolecules* **28**, 73-78.

96. Henderson, L.A., Svirkin, Y.Y., Gross, R.A., Kaplan, D.L., and Swift, G. (1996) Enzyme-catalyzed polymerizations of ε-caprolactone: Effect of initiator on product structure, propagation kinetics, and mechanism, *Macromolecules* **29**, 7759-7766.

97. Dong, H., Cao, S.-G., Wang, H.-D., Han, S.-P., Zhang, N.-X., and Yang, H. (1996) Lipase-catalyzed synthesis of lactone and biodegradable polyester in organic solvent, *Ann. N.Y. Acad. Sci.* **799**, 588-594.

98. Bisht, K.S., Deng, G., Gross, R.A., Kaplan, D.L., Swift, G. (1998) Ethyl glucoside as a multifunctional initiator for enzyme-catalyzed regioselective lactone ring-opening polymerization, *J. Am. Chem. Soc.* **120**, 1363-1367.

99. Dong, H., Cao, S.-G., Li, Z.-Q., Han, S.-P., You, D.-L., and Shen, J.-C. (1999) Study on the enzymatic polymerization mechanism of lactone and the strategy for improving the degree of polymerization, *J. Polym. Sci. Part A: Polym. Chem.* **37**, 1265-1275.

100. Uyama, H., Takeya, K., and Kobayashi, S. (1995) Enzymatic ring-opening polymerization of lactones to polyesters by lipase catalyst: Unusually high reactivity of macrolides, *Bull. Chem. Soc. Jpn.* **68**, 56-61.

101. Uyama, H., Takeya, K., Hoshi, N., and Kobayashi, S. (1995) Lipase-catalyzed ring-opening polymerization of 12-dodecanolide, *Macromolecules* **28**, 7046-7050.

102. Uyama, H., Kikuchi, H., Takeya, K., and Kobayashi, S. (1996) Lipase-catalyzed ring-opening polymerization and copolymerization of 15-pentadecanolide, *Acta Polym.* **47**, 357-360.

103. Bisht, K.S., Henderson, L.A., Gross, R.A., Kaplan, D.L., and Swift, G. (1997) Enzyme-catalyzed ring-opening polymerization of ω-pentadecalactone, *Macromolecules* **30**, 2705-2711.

104. Uyama, H., Namekawa, S., and Kobayashi, S. (1997) Mechanistic studies on the lipase-catalyzed ring-opening polymerization of lactones, *Polym. J.* **29**, 299-301.

105. Freitag, D., Grigo, U., Müller, P.R., and Nouverné, W. (1985) Polycarbonates, in H.F. Mark, N.M. Bikales, C.G. Overberger, and C. Menges (eds.), *Polymer Science and Engineering*, 2nd edition, Vol. 11, John Wiley and Sons, Inc., New York, pp. 648-718.

106. Abramowicz, D.A., and Keese, C.R. (1989) Enzymatic transesterifications of carbonates in water-restricted environments, *Biotechnol. Bioeng.* **33**, 149-156.

107. Matsumura, S., Tsukada, K., and Toshima, K. (1997) Enzyme-catalyzed ring-opening polymerization of 1,3-dioxan-2-one to poly(trimethylene carbonate), *Macromolecules* **30**, 3122-3124.

108. Kobayashi, S., Kikuchi, H., and Uyama, H. (1997) Lipase-catalyzed ring-opening polymerization of 1,3-dioxan-2-one, *Macromol. Rapid Commun.* **18**, 575-579.

109. Bisht, K.S., Svirkin, Y.Y., Gross, R.A., Kaplan, D.L., and Swift, G. (1997) Lipase-catalyzed ring-opening polymerization of cyclic carbonates, *PMSE Prepr.* **76**, 421-422.

110. Bisht, K.S., Svirkin, Y.Y., Henderson, L.A., Gross, R.A., Kaplan, D.L, and Swift, G. (1997) Lipase-catalyzed ring-opening polymerization of trimethylene carbonate, *Macromolecules* **30**, 7735-7742.

111. Rodney, R.L., Stagno, J.L., Beckman, E.J., and Russell, A.J. (1999) Enzymatic synthesis of carbonate monomers and polymers, *Biotechnol. Bioeng.* **62**, 259-266.

112. Al-Azemi, T.F., and Bisht, K.S. (1999) Novel functional polycarbonate by lipase-catalyzed ring-opening polymerization of 5-methyl-5-benzyloxycarbonyl-1,3-dioxan-2-one, *Macromolecules* **32**, 6536-6540.

113. Matsumura, S., Harai, S., Toshima, K. (1999) Enzymatic synthesis of poly(tetramethylene carbonate) from diethyl carbonate and 1,4-butanediol, *Proc. Japan Acad. Series B* **75**, 117-121.

114. Carothers, W.H., and Van Natta, F.J. (1930) Studies on polymerization and ring formation. III. Glycol esters of carbonic acid, *J. Am. Chem. Soc.* **52**, 314-326.

115. Höcker, H., Keul, H., Kühling, S., Hovestadt, W., Müller, A.J. (1991) The anionic ring-opening polymerization and copolymerization of cyclic carbonates, *Makromol. Chem., Macromol. Symp.* **44**, 239-245.

116. Kricheldorf, H.R., Dunsing, R., and Serra i Albet A. (1987) Polylactones. 12. Cationic polymerization of 2,2-dimethyltrimethylene carbonate, *Makromol. Chem.* **188**, 2453-2466.

117. Ariga, T., Takata, T., Endo, T. (1993) Alkyl halide-initiated cationic polymerization of cyclic carbonate, *J. Polym. Sci. Part A: Polym. Chem.* **31**, 581-584.

118. Katchalski-Katzir E. (1993) Immobilized enzymes - learning from past successes and failures, *Trends Biotechnol.* **11**, 471-478.

119. Munnecke D.M. (1979) Hydrolysis of organophosphate insecticides by an immobilized-enzyme system, *Biotechnol. Bioeng.* **21**, 2247-2261.

120. Vasudevan, P.T., and Thakur, D.S. (1994) Soluble and immobilized catalase, *Appl. Biochem. Biotechnol.* **49**, 173-189.

121. Chiaratini L., Magnani M. (1997) Immobilization of enzymes and proteins on red blood cells, in G.F Bickerstaff (ed.), *Immobilization of Enzymes and Cells*, Humana Press, Totowa, N.J., pp 143-152.

122. Paterson M., Kennedy J.F. (1997) Cellulose paper support for immobilization, in G.F Bickerstaff (ed.), *Immobilization of Enzymes and Cells*, Humana Press, Totowa, N.J., pp 153-165.

123. Zhang M., Desai T., and Ferrari M. (1998) Proteins and cells on PEG immobilized silicon surfaces, *Biomaterials* **19**, 953-960.

124. Almeida N.F., Beckman E.J., and Ataai M.M. (1993) Immobilization of glucose oxidase in thin polypyrrole films: Influence of polymerization conditions and film thickness on the activity and stability of the immobilized enzyme, *Biotechnol. Bioeng.* **42**, 1037-1045.

125. Selampinar F., Akbulut U., Özden M.Y., and Toppare L. (1997) Immobilization of invertase in conducting polymer matrices, *Biomaterials* **18**, 1163-1168.

126. Montero S., Blanco A., Virto M.D., Landeta C., Agud I., Solozabal R., Lascaray J.M., Renobales M., Llama M.J., and Serra J.L. (1993) Immobilization of *Candida rugosa* lipase and some properties of the immobilized enzyme, *Enzyme Microb. Technol.* **15**, 239-247.

127. Caldwell S.R, and Raushel F.M. (1991) Detoxification of organophosphate pesticides using a nylon based immobilized phosphotriesterase from *Pseudomonas diminuta*, *Appl. Biochem. Biotechnol.* **44**, 1018-1022.

128. Li Z.F., Kang E.T., Neoh K.G., and Tan K.L. (1997) Covalent immobilization of glucose oxidase on the surface of polyaniline films graft copolymerized with acrylic acid, *Biomaterials* **19**, 45-53.

129. Yang Z., Williams D., and Russell A.J. (1995) Synthesis of protein-containing polymers in organic solvents, *Biotechnol. Bioeng.* **45**, 10-17.

130. Yang Z., Domach M, Auger R., Yang F.X., and Russell A.J. (1996) Polyethylene glycol-induced stabilization of subtilisin, *Enzyme Microb. Technol.* **18**, 82-89.

131. Ahulekar R.V., Prabhune A.A., Raman S.H., and Ponratham S. (1993) Immobilization of penicillin G acylase on functionalized macroporous beads, *Polymer* **34**, 163-166.

132. Kumakura M. (1995) Effect of heat treatment on enzymes entrapped into polymer gels, *J. Molec. Catal. B: Enzymatic* **1**, L1-L6.

133. Rejikumar, S. and Devi, S. (1995) Immobilization of β-galactosidase onto polymeric supports, *J. Appl. Polym. Sci.* **55**, 871-878.

134. Klibanov A.M. (1979) Enzyme stabilization by immobilization, Anal. Biochem. **93**, 1-25.

135. Fulcrand V., Jacquier R., Lazaro R., and Viallefont P. (1990) Enzymatic peptide synthesis in organic solvent mediated by gels of copolymerized acrylic derivatives of α-chymotrypsin and polyoxyethylene, *J. Peptide Protein Res.* **38**, 273-277.

136. Storey K.B., Duncan J.A., and Chakrabarti A.C. (1990) Immobilization of amyglucosidase using two forms of polyurethane polymer, *Appl. Biochem. Biotechnol.* **23**, 221-236.

137. Braatz J.A., Yasuda Y., Olden K., Yamada K.M., and Heifetz A.H. (1993) Functional peptide-polyurethane conjugates with extended circulatory half-lives, *Bioconjugate Chem.* **4**, 262-267.

138. Dias S.F., Vilas-Boas L., Cabral J.M.S., and Fonseca M.M.R. (1991) Production of ethyl butyrate by *Candida rugosa* lipase immobilized in polyurethane, *Biocatalysis* **5**, 21-34.

139. Yang Z., Mesiano A.J., Venkatasubramanian, Gross S.H., Harris J.M., and Russell A.J. (1995) Activity and stability of enzymes incorporated into acrylic polymers, *J. Am. Chem. Soc.* **117**, 4843-4850.

140. Wang P., Sergeeva M.V., Lim L., and Dordick J.S. (1997) Biocatalytic plastics as active and stable materials for biotransformations, *Nature Biotechnol.* **15**, 789-793.

141. LeJeune K.E., and Russell A.J. (1996) Covalent binding of a nerve agent hydrolyzing enzyme within polyurethane foams, *Biotechnol. Bioeng.* **51**, 450-457.

142. LeJeune K.E., Mesiano A.J., Bower S.B., Grimsley J.K., Wild J.R., and Russell A.J. (1996) Dramatically stabilized phosphotriesterase-polymers for nerve agent degradation, *Biotechnol. Bioeng.* **54**, 105-113.

143. LeJeune K.E., Jeffrey S.S., Hetro A.D., Donahey G.P., and Russell A.J. (1998) Increasing the tolerance of organophosphorus hydrolase to bleach, *Biotechnol. Bioeng.* **64**, 250-250.

144. Drevon, G., and Russell, A.J., unpublished data.

APPLICATION OF ENZYMES IN SYNTHETIC STRATEGY

B. ZWANENBURG* AND A.J.H. KLUNDER

Department of Organic Chemistry, NSR Institute for Molecular Structure, Design and Synthesis, University of Nymegen, Toernooiveld, 6525 ED Nymegen, The Netherlands

1.   Introduction

The use of enzymes in organic synthesis is now commonly accepted. The earlier reluctance that these sensitive bioreagents would not withstand the reaction conditions normally employed in organic synthesis has fainted. The organic chemists amazingly quickly learned how to adjust the experimental conditions of synthetic operations allowing the fruitful use of enzymes. In addition, it was shown that enzymes can be used in organic solvents as well. Furthermore, many new methodologies have been developed for the directed use of enzymes in synthetic conversions.

Many useful reviews [1] on the use of enzymes in organic syntheis have appeared in recent years and in fact, enzymes are now common synthetic tools in the hands of organic chemists.

In this chapter various examples of the strategic use of enzymes in solving synthetic problems are presented. The main focus will be on the production of optically active materials using enzyme-mediated kinetic resolution of racemic compounds. However, it will also be shown that enzymes can successfully be used in solving seemingly simple synthetic problems, such as the chemical differentiation between two isomeric compounds. The highest benefit of enzymes can be achieved when combined with other synthetic methodology. Several examples of such synthetic strategies will be presented.

B. Zwanenburg et al. (eds.), Enzymes in Action, 433–448.
© 2000 *Kluwer Academic Publishers. Printed in the Netherlands.*

2. Chemo- and Regioselective Conversions

The mono-hydrolysis of a diester may constitute a rather difficult problem. The kinetic difference between two ester functions in alkaline saponifications is often marginal or even absent. Enzymes, however, can serve as effective catalysts in the mono-hydrolysis of diesters, as is exemplified by the formation of the monoester of fumaric acid,[2] maleic acid [2] and cubanedicarboxylic acid [3] (Scheme 1). The last-mentioned case is particularly worth mentioning as all attempts to achieve a mono-hydrolysis by conventional methods failed completely. Moreover, cubanedicarboxylic esters are nonnatural substrates beyond any doubt. Nevertheless, these esters are accepted as substrates by the enzyme to give monoester in high yield. This monoester is of great importance for the synthesis of various substituted cubanecarboxylic acids and disubstituted cubanes.

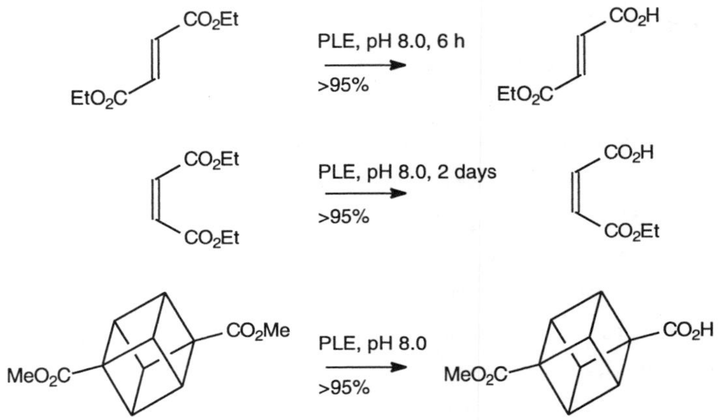

Scheme 1

In our synthetic approach to macrocylic dilactones, such as pyrenophorin and colletallol, the hydrolysis of functionalised α,β-unsaturated esters was required. Conventional methods all failed and led to either no reaction or decomposition of the products. Mild hydrolysis employing Pig Liver Esterase (PLE) as the catalyst smoothly produced the desired unsaturated acids in excellent yields [4], which then in subsequent steps were converted into the dilactones. It should be noted that the saponification of unsaturated esters by aqueous base requires more harsh conditions than needed for saturated esters.

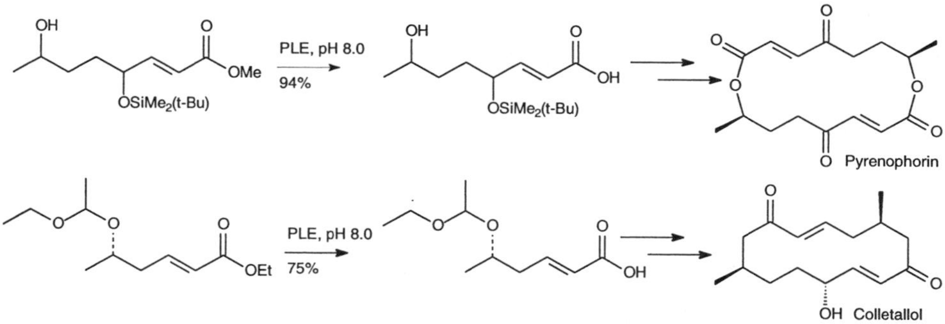

Scheme 2

The third example refers to the separation of *cis* and *trans* oxiranecarboxylic esters which were obtained as a mixture in the Darzens condensation. Enzymatic hydrolysis of this isomeric mixture led to a completely selective hydrolysis of the the *trans* ester, while the *cis* ester remained unchanged (Scheme 3). The separation of the *trans* acid and the *cis* ester caused no problems.

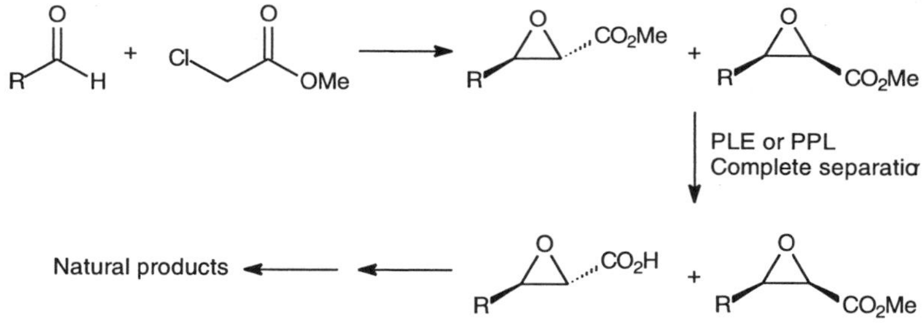

Scheme 3

3. Strategic Use of Enzymes in the Synthesis of Cyclopentenoid Natural Products

Oxygenated cyclopentenones constitue a group of natural products with interesting biological activities. Typical examples are kjellmanianone, pentenomycin, and marine prostanoids, such as clavulones and halovulones.

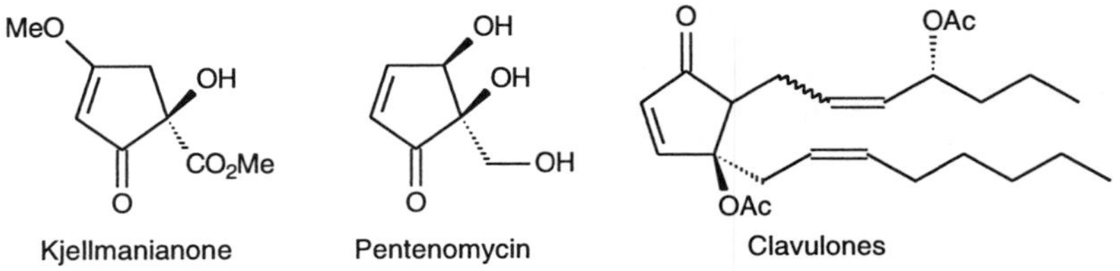

MeO
OH
CO2Me
O
Kjellmanianone
OH
OH
OH
O
Pentenomycin
O
OAc
OAc
Clavulones

Scheme 4

A most direct approach to the synthesis of these cyclopentenoids would be chemical elaboration of cyclopentadienone, e.g. by conjugate addition of suitable nucleophiles followed by electrphilic trapping of the enolate. This approach, however, is not feasible as cyclopentadienone dimerizes at temperatures as low as –100°C. Therefore, the *endo*-tricyclo[5.2.1.0^{2,6}]decadienone system was explored as a synthetic equivalent of cyclopentadienone. These tricyclodecadienones, which essentially are the Diels-Alder adducts of cyclopentadienone and cyclopentadiene, can be considered as cyclopentadienone in which one of the double bonds is protected. Chemical transformation of the remaining enone system, e.g. by nucleophilic addition and/or electrophilic substitution, followed by a retro Diels-Alder reaction induced by either Lewis acid or thermolysis using flash vacuum thermolysis to regenerate the masked enone unit, ultimately yields functionalized cyclopentenones (Scheme 5) [6,7].

The Concept of Transient Chirality
R
chiral
O
introduction of the
stereogenic element
R
+
O
achiral
1. Nu⊖
2. E⊕
regio- and stereoselective
transformations
1. Nu⊖
2. E⊕
R
Nu
chiral
O
E
[4+2] cycloreversion
(FVT)
removal of the
stereogenic element
R
Nu
E
O
chiral

Scheme 5. The concept of transient chirality.

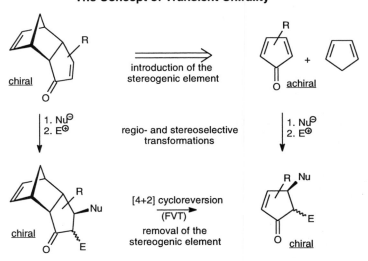

This synthetic strategy not only offers a unique solution to the problem of the intrinsic instability of cyclopentadienones by temporally protecting one of the double bonds, but such a protection simultaneously transforms the nonchiral flat cyclopentadienone into a chiral rigid structure with a pronounced difference in stereoheterotopic faces. It is therefore conceivable that reactions at the enone moiety can be conducted in a stereoselective manner. Cycloreversion removes the protecting ring system, thereby reforming the original double bond, but also eliminating the stereogenic element that initiated the stereoselective introduction of subsequent (new) stereogenic centers. For this introduction and removal of a temporal auxiliary stereogenic center we proposed the term *transient chirality* [8]. This scheme becomes strategically optimal when the key intermediate, i.e. the tricyclodecadienone that serves as synthetic equivalent for cyclopentadienone, can be obtained enantiomerically pure. Assuming that all reactions of the key intermediate proceed with full stereocontrol, the ultimate result will be the synthsis of enantiopure cyclopentenoid products [8].

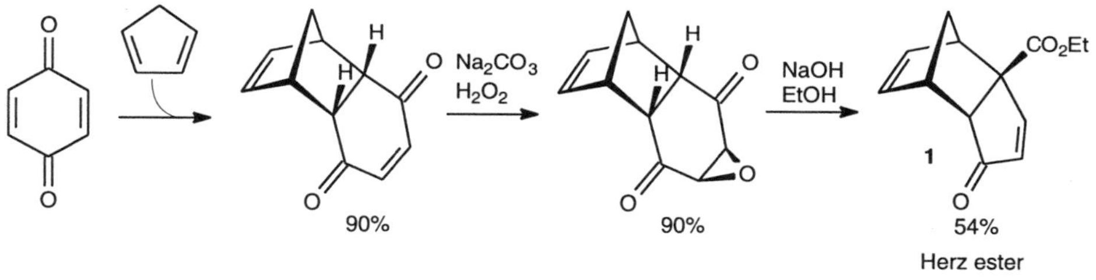

Scheme 6

This concept of transient chirality has been employed by us for the enantiocontrolled synthesis of several biologically important cyclopentenoids [8,9]. For the discussin in this chapter the focus will be on the use of enzymes in the preparation of enantiopure key intermediates of Scheme 5. A versatile synthetic equivalent of a cyclopentadienone is the tricyclic carboxylic ester 1, the so-called Herz ester. This compound can be readily prepared, as is shown in Scheme 6, starting from the Diels-Alder adduct of benzoquinone and cyclopentadiene by epoxidation, followed by a Favorskii-type ring contraction [10]. The thus obtained Herz ester can be kinetically resolved by using PLE with acetonitrile as co-solvent (Scheme 7) [11]. Using this enzymatic resolution in concert with flash vacuum thermolysis allows the synthesis of enantiopure cyclopentenones [12].

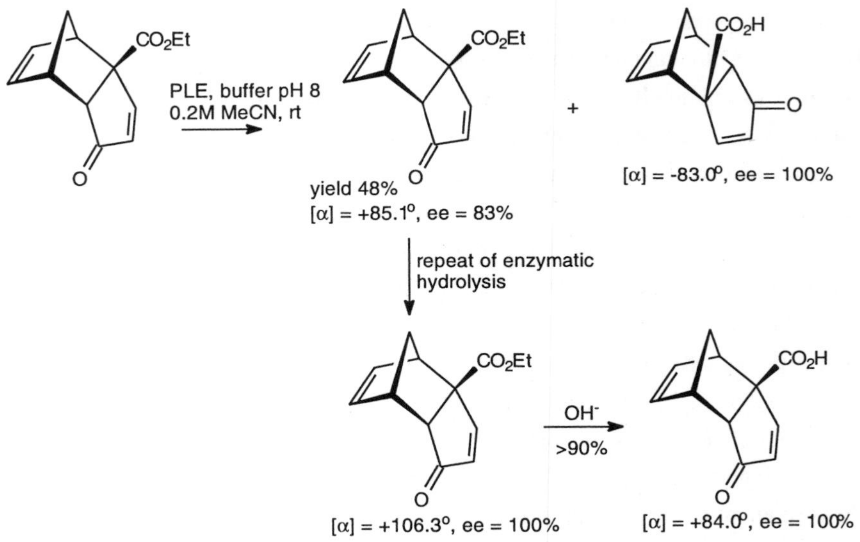
CO2Et
PLE, buffer pH 8
0.2M MeCN, rt
O
CO2Et
yield 48%
[α] = +85.1°, ee = 83%
+
CO2H
O
[α] = -83.0°, ee = 100%
repeat of enzymatic
hydrolysis
CO2Et
OH-
>90%
CO2H
[α] = +106.3°, ee = 100%
[α] = +84.0°, ee = 100%

Scheme 7

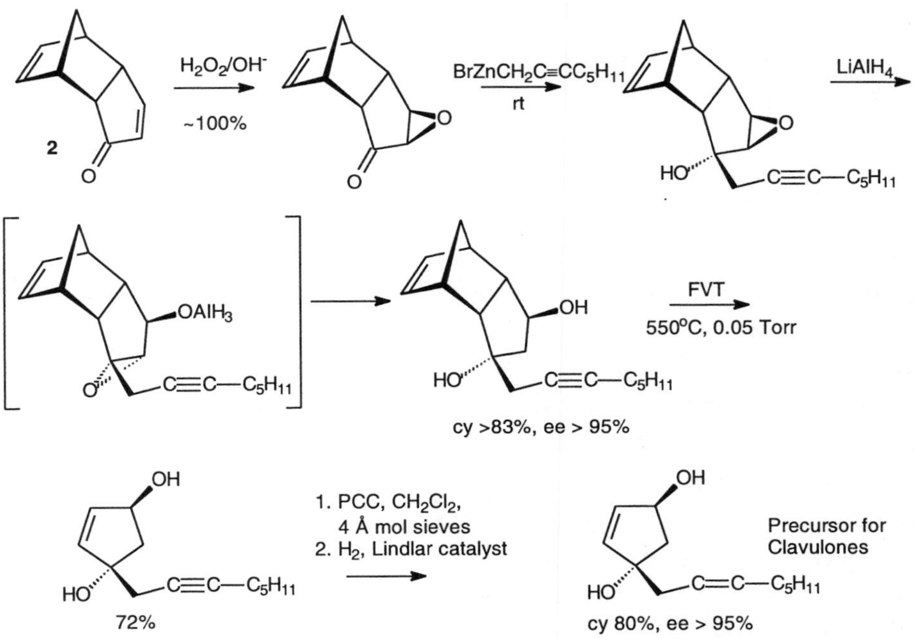
2
H2O2/OH-
~100%
O
BrZnCH2C≡CC5H11
rt
HO
C≡C—C5H11
LiAlH4
OAlH3
OH
FVT
550°C, 0.05 Torr
cy >83%, ee > 95%
72%
1. PCC, CH2Cl2,
4 Å mol sieves
2. H2, Lindlar catalyst
C=C—C5H11
Precursor for
Clavulones
cy 80%, ee > 95%

Scheme 8

The enantiopure tricyclodecadienone **2** was also the chiral synthon for the preparation of the marine prostanoid clavulone (Scheme 4). The essential strategy for the key precursor of these compounds involves epoxidation of the enone, reaction with an appropriate organozinc reagent, followed by reductive opening of the epoxide, then flash vacuum thermolysis and finally oxidation to the ketone and reduction of the triple bond (Scheme 8) [13].

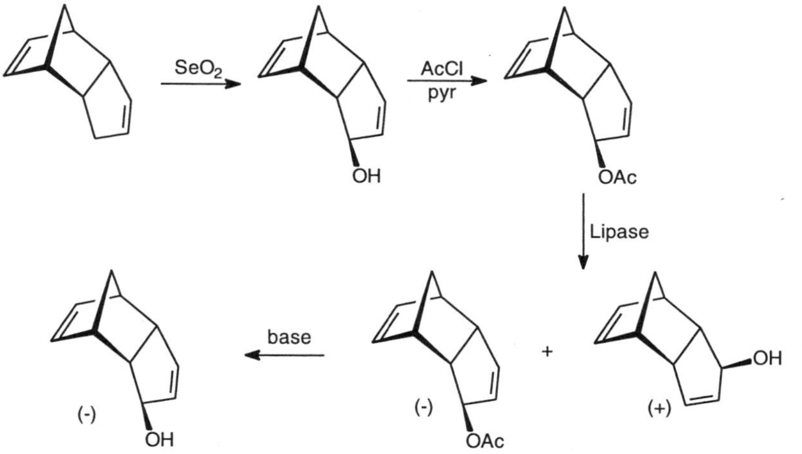

SeO$_2$

AcCl
pyr

Lipase

base

+

(-)   OH

(-)   OAc

(+)   OH

OH

OAc

Scheme 9 [14]

1. Epoxidation
2. LiAlH$_4$

CCL
vinyl acetate
81%

3

OH

OH

4   (+)
OAc

OH

91%

PCC (oxidation)

OSi

(-)   O

O

(+)   O

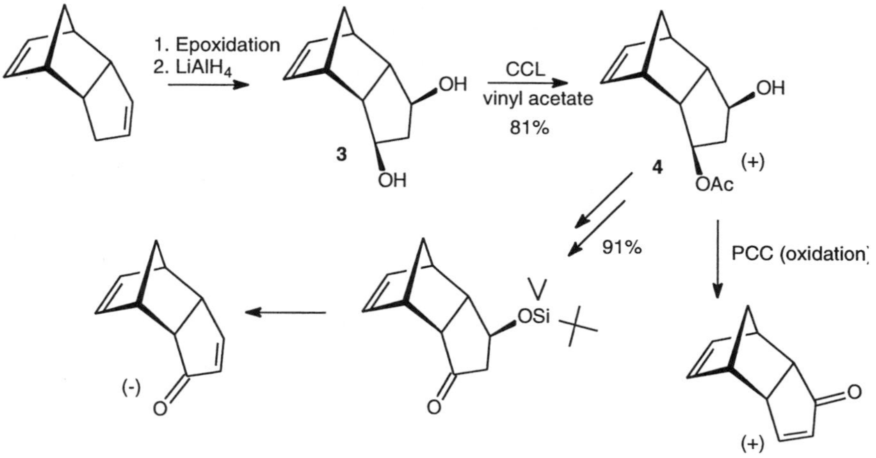

Scheme 10 [15]

Takano et al. [14] used also a kinetic enzymatic resolution to obtain enantiopure tricyclodecadienols (Scheme 9). An enzymatic esterification using vinyl acetate as the acyl donor was described by Liu et al. [15] (Scheme 10). In this case the meso diol **3** was subjected to enzymatic desymmetrization, successfully producing enantiopure monoacetate **4**.

We also attempted the dynamic kinetic resolution of the fast equilibrating enol ketone, shown in Scheme 11; however, without any result due to lack of reactivity of these substrates under enzymatic conditions [16].

The asymmetric desymmetrization of these *pseudo-meso* enol ketones was accomplished in a pure chemical manner, namely by reaction with chiral amines, such as proline methyl ester [16].

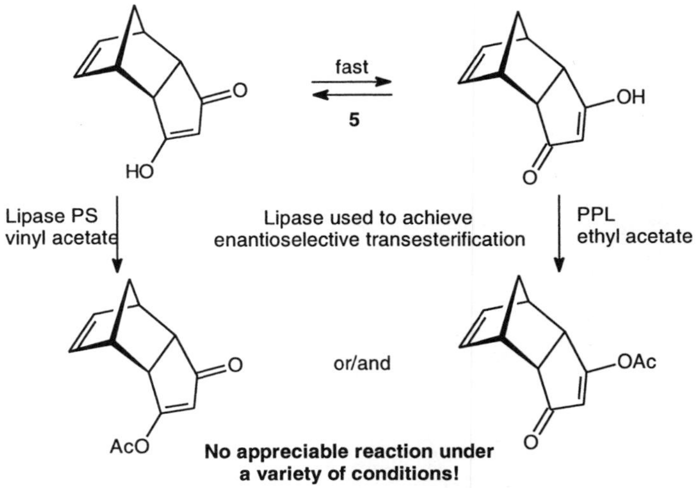

Scheme 11

3.1. SUBSTRATE MODEL FOR BI- AND TRICYCLIC NORBORNANE TYPE ESTERS

The kinetic resolution of the Herz ester **1** as shown in Scheme 7 was in fact a fortuitous finding. Several active-site models have been developed to understand the relationship between the molecular structure of the substrates and their acceptance by enzymes. For this purpose Tamm et al. [17] followed an indirect approach by designing a model substrate in which the ideal structural and stereochemical requirements that a substrate must meet to ensure optimal enantioselectivity and acceptance by the enzyme, are included. The Herz ester **1** could be fitted with this model substrate (see Scheme 12), whereby the polar carbonyl group has a *trans* position with respect to the *exo*-ester moiety. Appropriate hydrogen binding of this ketone function with an amino acid residue near the active site of the enzyme then provides the necessary fixation of the substrate in the enzyme pocket.

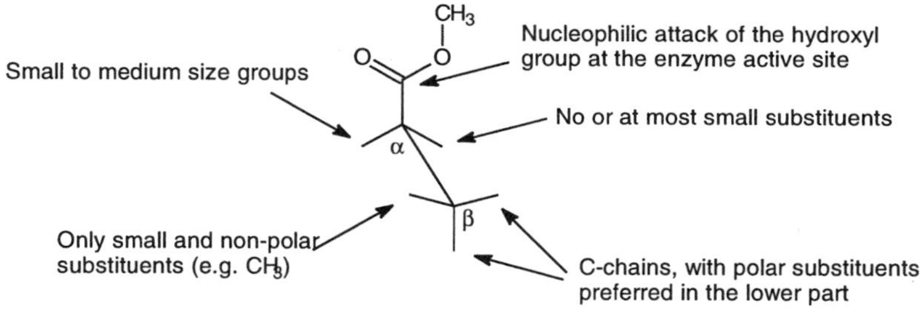
CH3
O
O
Nucleophilic attack of the hydroxyl group at the enzyme active site
Small to medium size groups
No or at most small substituents
α
β
Only small and non-polar substituents (e.g. CH3)
C-chains, with polar substituents preferred in the lower part

Scheme 12. Tamm's substrate model for PLE [17]

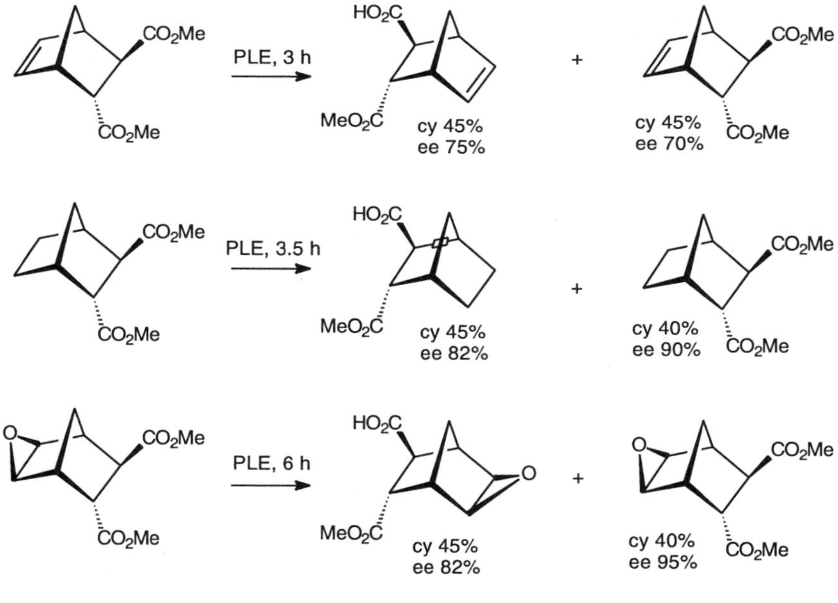
CO2Me
CO2Me
PLE, 3 h
HO2C
MeO2C
cy 45%
ee 75%
+
CO2Me
cy 45%
ee 70%
CO2Me
CO2Me
CO2Me
PLE, 3.5 h
HO2C
MeO2C
cy 45%
ee 82%
+
CO2Me
cy 40%
ee 90%
CO2Me
O
CO2Me
CO2Me
PLE, 6 h
HO2C
O
MeO2C
cy 45%
ee 82%
+
O
CO2Me
cy 40%
ee 95%
CO2Me

Scheme 13. PLE catalysed hydrolysis of norbornane *trans*-diesters

To further substantiate this proposal, a series of rigid bicyclic norbornane type esters was subjected to hydrolysis catalysed by PLE. The results obtained for a series *trans*-norbornane diesters are depicted in Scheme 13. In all three cases the same ester function was preferentially hydrolysed [18] with a high enantioselectivity (Scheme 13). This study [18] was completed by investigating *cis*-diesters and monoesters as well. These results are summarized in Scheme 14. These studies clearly confirm the proposal that according to Tamm's model a polar carbonyl containing group *trans* with respect to the preferentially hydrolysed ester function is necessary for an efficient enantioselective reaction.

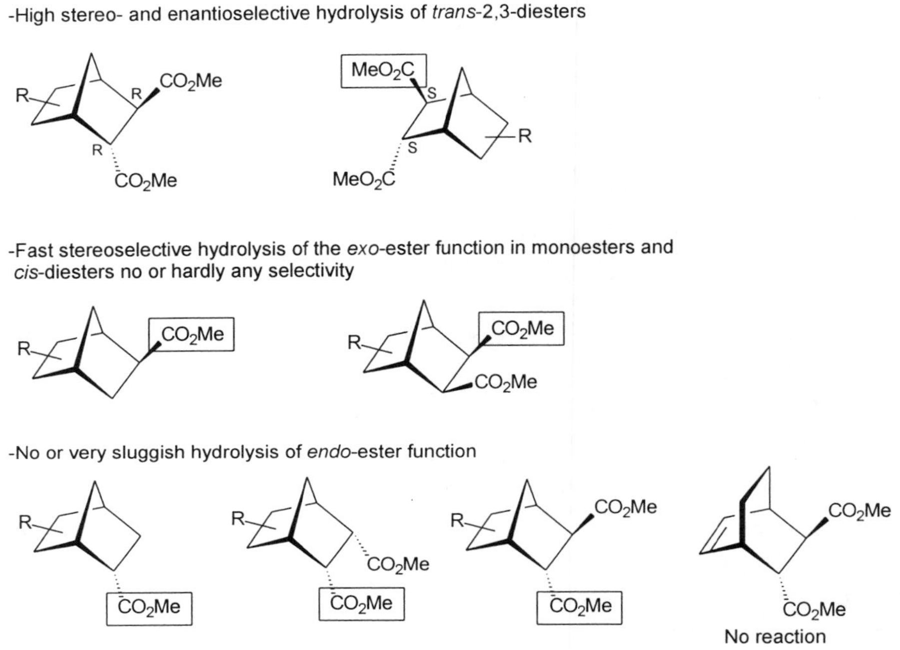

-High stereo- and enantioselective hydrolysis of *trans*-2,3-diesters

R— R CO$_2$Me R CO$_2$Me

MeO$_2$C S R MeO$_2$C S

-Fast stereoselective hydrolysis of the *exo*-ester function in monoesters and *cis*-diesters no or hardly any selectivity

R— CO$_2$Me   R— CO$_2$Me CO$_2$Me

-No or very sluggish hydrolysis of *endo*-ester function

R— CO$_2$Me   R— CO$_2$Me   R— CO$_2$Me CO$_2$Me   CO$_2$Me CO$_2$Me

No reaction

Scheme 14. PLE catalysed hydrolysis of norbornane esters.
(The ester that is hydrolysed by preference is shown in a box)

A conceivable complexation in the enzyme pocket can now be pictured as is shown in Scheme 15 for the *trans*-diesters. It should be noted that the differentiation is also directed by the position of the ethylene bridge, which in the unfavourable complexation cannot adequately be accommodated. This is nicely demonstrated by the observation that the bicyclo[2,2,2]octanedicarboxylic esters are not accepted by PLE.

The enantiodifferentiation by amylase turned out to be opposite to that of PLE, as is shown in Scheme 16. This complimentary behavior of these enzymes has some practical relevance, when an acid with a specific absolute configuration is needed [2]. An example of the strategic use norbornene diester is shown in Scheme 17. The enantiopure monoester is converted to an optically pure lactone with a high degree of efficiency [2]. Note that a base-induced inversion of a stereogenic center is included in this sequence.

-High stereo- and enantioselective hydrolysis of *trans*-2,3-diesters

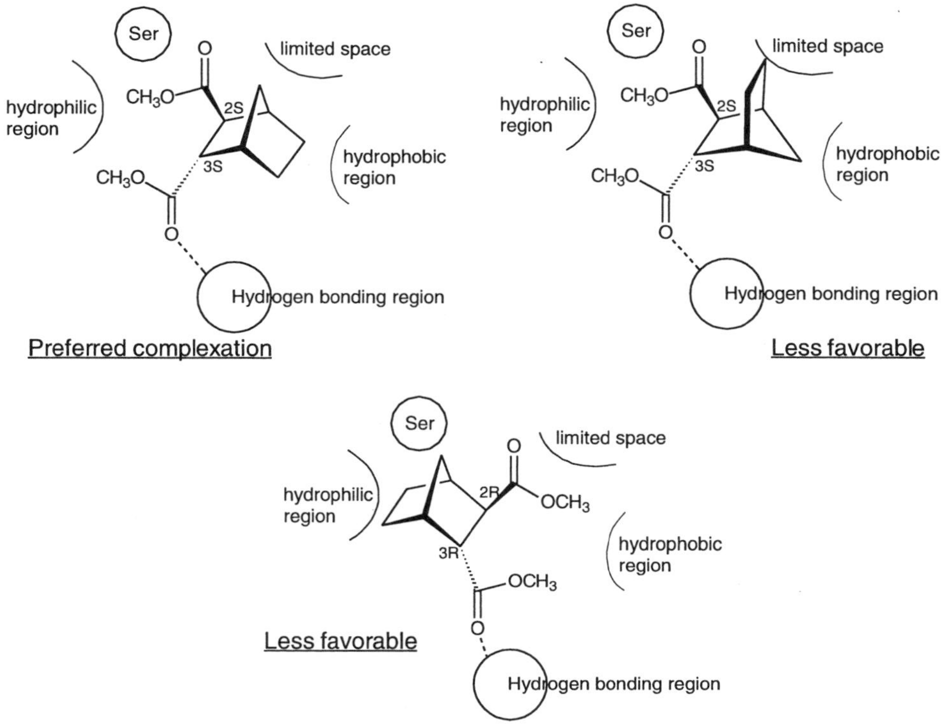

Ser

O

\ limited space

hydrophilic region

CH₃O

2S

hydrophobic region

CH₃O

3S

O

Hydrogen bonding region

Preferred complexation

Ser

O

\ limited space

hydrophilic region

CH₃O

2S

hydrophobic region

CH₃O

3S

O

Hydrogen bonding region

Less favorable

Ser

O

limited space

hydrophilic region

2R

OCH₃

hydrophobic region

3R

OCH₃

O

Less favorable

Hydrogen bonding region

Scheme 15. Conceivable PLE-substrate complexations of *trans*-norbornane diesters

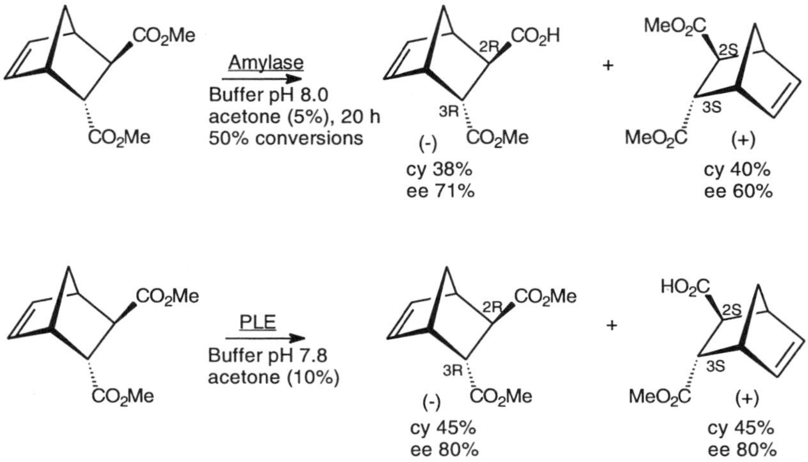

CO₂Me

CO₂Me

Amylase
Buffer pH 8.0
acetone (5%), 20 h
50% conversions

2R CO₂H

3R CO₂Me

(-)
cy 38%
ee 71%

+

MeO₂C

2S

3S

MeO₂C

(+)
cy 40%
ee 60%

CO₂Me

CO₂Me

PLE
Buffer pH 7.8
acetone (10%)

2R CO₂Me

3R CO₂Me

(-)
cy 45%
ee 80%

+

HO₂C

2S

3S

MeO₂C

(+)
cy 45%
ee 80%

Scheme 16. Comparison of PLE and Amylase

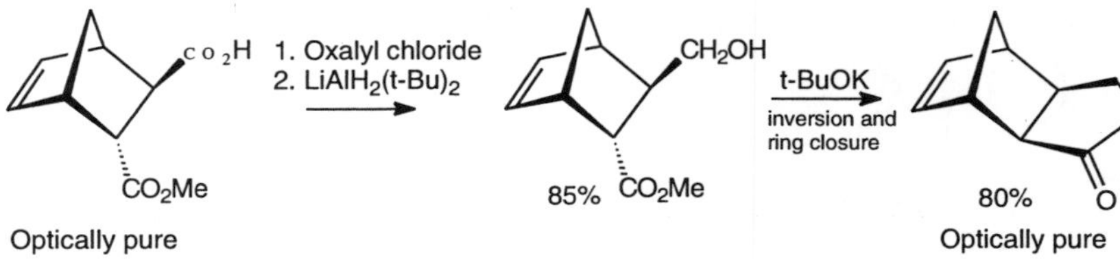

Scheme 17

4. Strategic Use of Enzymes in the Synthesis of Germination Stimulants

The root parasitic plants *Striga* and *Orobanche* spp. have an extremely devastating impact on their hosts, which mostly are important food crops such as maize, sorghum, millet and rice (*Striga*) and legumes such as Faba beans and tomatoes (*Orobanche*). The seeds of these parasitic weeds germinate upon contact with a chemical stimulant present in the root exudates of the host plant [19]. The synthesis of the germination stimulants is an interesting challenge. The influence of the stereochemical structure of these stimulants on their bioactivity is of fundamental importance. Two natural stimulants that

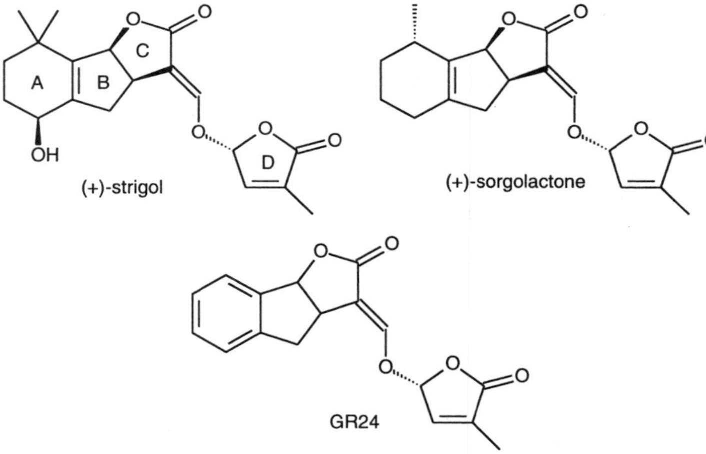

(+)-strigol

(+)-sorgolactone

GR24

Scheme 18

have been isolated from root exudates of host plants are strigol and sorgolactone (Scheme 18). The total synthesis of these compounds and their analogs (e.g. GR24) can be accomplished by coupling of the racemic ABC fragment with an enantiopure latent D-ring, whereby the final step the stereogenic auxiliary element is removed [20]

(Scheme 19). We used the Diels-Alder cycloadduct **6** of cyclopentadiene and citraconic anhydride as the principal starting material for this approach, as is shown in Scheme 20. After reduction to the racemic lactol **7** in a regioselective manner, this lactol was subjected to Lipase PS-mediated acetylation using vinyl acetate as the acyl donor. [21]

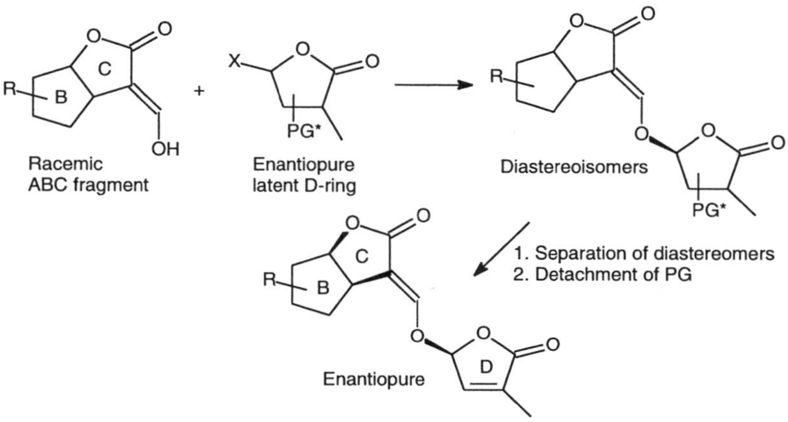

Scheme 19

This resulted in lactol **7** with high enantiopurity and the corresponding acetate 7-OAc (ee 87%). Remarkably, this acetate had the *endo*-configuration, which must be formed during the enzymatic acetylation with a concurrent epimarization at C_5 in **7**. Apparently, the *endo*-OH fits much better in the enzyme pocket than the *exo*-isomer. In spite of the fact that the *endo*-OH is present in a low not measurable concentration, this isomer reacts preferentially, whereby the dynamic equilibrium between these two alcohol isomers is eventually completely shifted to the *endo*-OH. The lactol **7** was then coupled to an ABC fragment, e.g. that of the stimulant GR7 (this contains no A-ring). The thus obtained diastereomers were separated and then subjected to cycloreversion by heating in *o*-dichlorobenzene at 180°C to give the GR7 stimulant with the natural configuration and its isomer, both in enantiopure form. [21] Using this strategy also the remaining two isomers of GR7 could be synthesized starting from 7-OAc. [21] This strategic use of enzymes in the synthesis of all stereoisomers of various germination agents, including GR24 and sorgolactone [22], has been successfully accomplished.

The concept of transient chirality and the strategic use of enzymes are playing a key role in the synthesis of these germination stimulants in a stereocontrolled manner.

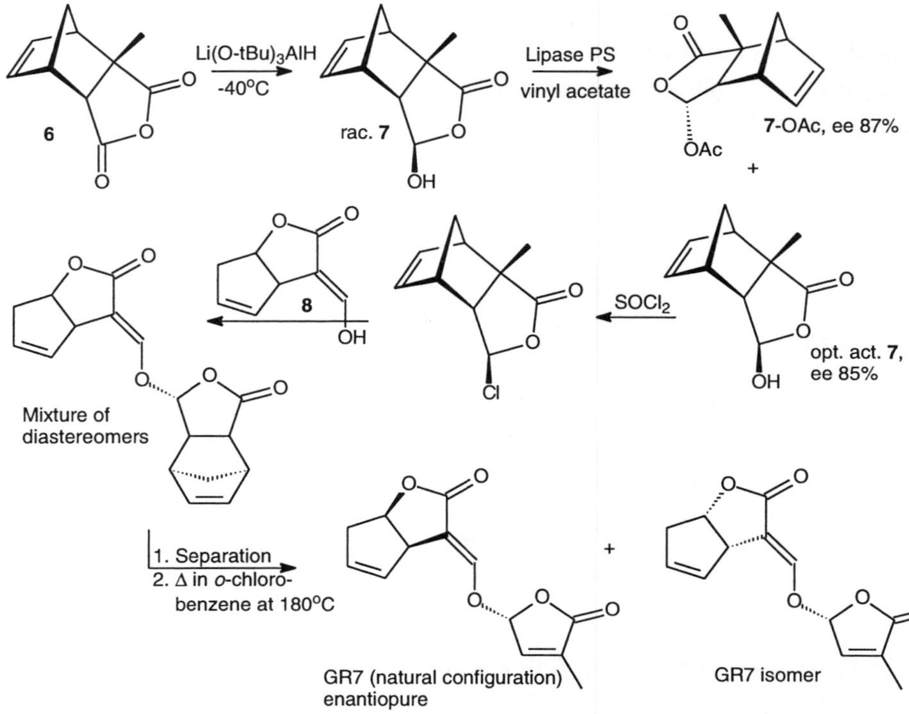

Scheme 20

5. Concluding Remark

Enzymes are excellent tools in organic synthesis; the wealth of possibilities arising from incorporation of an enzymatic reaction in a synthetic sequence has not been fully discovered yet. In the coming years many developments in molecular synthesis using enzymes in key operations are to be expected.

6. Acknowledgement

The authors thank their co-workers for their enthusiastic and skilled contribution to the work described in this chapter.

7.   References

1.   Wong, C.H. and Whitesides, G.M. (1994) Enzymes in synthetic organic chemistry, Pergamon Press (Tetrahedron Series vol 12); Faber, K. (1995) Biotransformations in organic chemistry, Springer-Verlag, Berlin (2nd Ed.).

2.   Gastel, F.J.C. van, Klunder, A.J.H. and Zwanenburg, B. (1988) unpublished results. See also: Gastel, F.J.C. van (1992) Enzymatic optical resolution of norbornene-type carboxylic esters and their application in synthesis, *Thesis*, University of Nymegen, The Netherlands.

3.   Dumont, M., Klunder, A.J.H. and Zwanenburg, B. (1984) unpublished results.

4.   Smeets, F.L.M. and Zwanenburg, B. (1981) unpublished results. See also: Smeets, F.L.M. (1981) Synthese van pyrenophorine, *Thesis*, University of Nymegen, The Netherlands. For colletallol: Legters, J., Thijs, L. and Zwanenburg, B. (1985) unpublished results.

5.   Gastel, F.J.C. van, Klunder, A.J.H. and Zwanenburg, B. (1992) unpublished results. See also ref. 2.

6.   Klunder, A.J.H., Bos, W., Verlaak, J.M.J. and Zwanenburg, B. (1981) A facile synthesis of functionalized cyclopentadienone epoxides by flash vacuum thermolytic cycloreversion of tricyclodecenones, *Tetrahedron Lett.*, 4553-4556.

7.   Klunder, A.J.H., Bos, W. and Zwanenburg, B. (1981) An efficient stereospecific total synthesis of (±)-terrein, *Tetrahedron Lett.*, 4557-4560.

8.   Klunder, A.J.H., Zhu, J. and Zwanenburg, B. (1999) The concept of transient chirality in the stereoselective synthesis of functionalized cycloalkenes applying the retro Diels-Alder methodology, *Chem. Rev.*, **99**, 1163-1190.

9.   Zwanenburg, B., Zhu, J. and Klunder, A.J.H. (1997) Cyclopentenoid natural products in *New Horizon in Organic Synthesis*, Eds. Nair, V. and Kumar, S., New Age International Publ., New Delhi, India.

10. Klunder, A.J.H., Valk, W.C.G.M. de, Verlaak, J.M.J., Schellekens, J.W.M., Noordik, J.H., Parthasarathi, V. and Zwanenburg, B. (1985) Nucleophilic eliminative ring fission of bridgehead substituted 1,3-bishomocubyl acetates, *Tetrahedron*, **41**, 963-973.

11. Klunder, A.J.H., Huizinga, W.B., Hulshof, A.J.M. and Zwanenburg, B. (1986) Enzymatic optical resolution and absolute configuration of tricyclo[5.2.1.0^{2,6}]decadienones, *Tetrahedron Lett.*, **27**, 2543-2546.

12. Klunder, A.J.H., Huizinga, W.B., Sessink, P.J.M. and Zwanenburg, B. (1987) Enzymatic optical resolution and flash vacuum thermolysis in concert for the synthesis of optically active cyclo-pentenones, *Tetrahedron Lett.*, **28**, 357-360.

13. Zhu, J., Yang, J.-Y., Klunder, A.J.H., Liu, Z.-Y. and Zwanenburg, B. (1995) A stereo- and enantioselective approach to clavulones from tricyclodecadienone using flash vacuum thermolysis, *Tetrahedron*, **51**, 5847-5870.

14. Takano, S., Inomata, K. And Ogasawara, K. (1989) Enantioconvergent route to α-cuparenone from dicyclopentadiene, *J. Chem. Soc., Chem. Commun.*, 271.

15. Liu, Z.-Y., He, L. and Zheng, H. (1993) Highly enantioselective synthesis of (+) and (-) endo-tricyclo[5.2.1.0^{2,6}]deca-4,8-dien-3-one by enzyme catalysed acetylation, *Tetrahedron Asymm.*, **4**, 2277.

16. Bakkeren, F.J.A.D., Ramesh, N.G., Groot, D. De, Klunder, A.J.H. and Zwanenburg, B. (1996) Asymmetric desymmetrization of a pseudo-meso-*endo*-tricyclo[5.2.1.0^{2,6}]deca-4,8-dien-3-one by chiral amines, *Tetrahedron Lett.*, **37**, 8003-8006.

17. Mohr, P., Waespe-Šarcevic, N., Tamm, C., Gawronska, K. and Gawronski, J.K. (1983) *Helv. Chim. Acta*, **66**, 2501.

18. Klunder, A.J.H., Gastel, F.J.C. van, and Zwanenburg, B. (1988) Structural requirements in the enzymatic optical resolution of bicyclic esters using Pig Liver Esterase, *Tetrahedron Lett.*, **29**, 2697-2700. Gastel, F.J.C. van, Klunder, A.J.H., and Zwanenburg, B. (1991) Enzymatic optical resolution of norbornene carboxylic esters using Pig Liver Esterase, *Recl. Trav. Chim. Pays-Bas*, **110**, 175-184.

19. Wigchert, S.C.M. and Zwanenburg, B. (1999) A critical account on the interception of *Striga* seed germination, *J. Agric. Food. Chem.*, **47**, 1320-1325 and literature cited therein.

20. Thuring, J.W.J.F., Nefkens, G.H.L., Schaafstra, R. and Zwanenburg, B. (1995) Asymmetric synthesis of a D-ring synthon for strigol analogues and its application to the synthesis of all four diastereoisomers of germination stimulant GR7, *Tetrahedron*, **51**, 5047-5056.

21. Thuring, J.W.J.F., Klunder, A.J.H., Nefkens, G.H.L., Wegman, M.A. and Zwanenburg, B. (1996) Enzymatic kinetic resolution of 5-hydroxy-4-oxa-*endo*- tricyclo[5.2.1.0^{2,6}]deca-4,8-dien-3-ones: a useful approach to D-ring synthons for strigol analogues with remarkable stereoselectivity, *J. Org. Chem.*, **61**, 6931-6935.

22. Sugimoto, Y., Wigchert, S.C.M., Thuring, J.W.J.F. and Zwanenburg, B. (1998) Synthesis of all eight stereoisomers of the germination stimulant sorgolactone, *J. Org. Chem.*, **63**, 1259-1267 and references cited therein.

GREEN SOLUTIONS FOR CHEMICAL CHALLENGES
Biocatalysis in the Synthesis of Semi-Synthetic Antibiotics

ALLE BRUGGINK

a) DSM Research, PO Box 18, 6160 MD Geleen, The Netherlands
b) University of Nymegen, Department of Organic Chemistry,
Toernooiveld 1, 6525 ED Nymegen, The Netherlands

Abstract: The impact of biocatalysis on the fine chemical industry is described, using the manufacturing processes for various penicillines and cephalosporins as examples. Reduction of the number of process steps and the amount of waste generated per kg of end product are major achievements. Perspectives towards a more sustainable fine chemical industry are discussed, whereby integration of biosynthesis and organic synthesis is expected to be a key development. Dedicated downstream processing, including molecular recognition techniques, and integrated (bio-)process technology will be needed to complement this.

1. Introduction

The introduction of biocatalysis in the synthesis of industrial chemicals, in particular fine chemicals, can be seen as a first step in integration of organic synthesis and biosynthesis. The onset of this development is known to be due to the need to replace traditional stoichiometric processes by catalytic processes thus improving the product/waste ratio's[1]. The cumbersome translation of petrochemical catalysis to catalysis during the synthesis of the more complex fine chemical molecules has prompted the fast acceptance of biocatalysis and biotransformations. Although (asymmetric) chemical catalysis allowing reactions to occur at ambient temperatures is developing fast, biocatalysis is in the lead from an industrial point of view. A development from single and relatively simple enzyme-catalyzed conversions to more complex biotransformations employing a number of enzymes, including cofactors,

449

B. Zwanenburg et al. (eds.), Enzymes in Action, 449–458.
© 2000 *Kluwer Academic Publishers. Printed in the Netherlands.*

effecting multi-step "one-pot" processes is well underway. An integration of organic synthesis and fermentations might be the end result, indeed green chemistry.

2.   Semi-synthetic antibiotics

The industrial manufacture of semi-synthetic antibiotics, i.e. penicillines and cephalosporins, with a history of over 30 years, is an outstanding example of applied biocatalysis. The successful shortening of the synthesis of Cefalexin from 10 to 6 steps serves as a first illustration (scheme 1)[2].

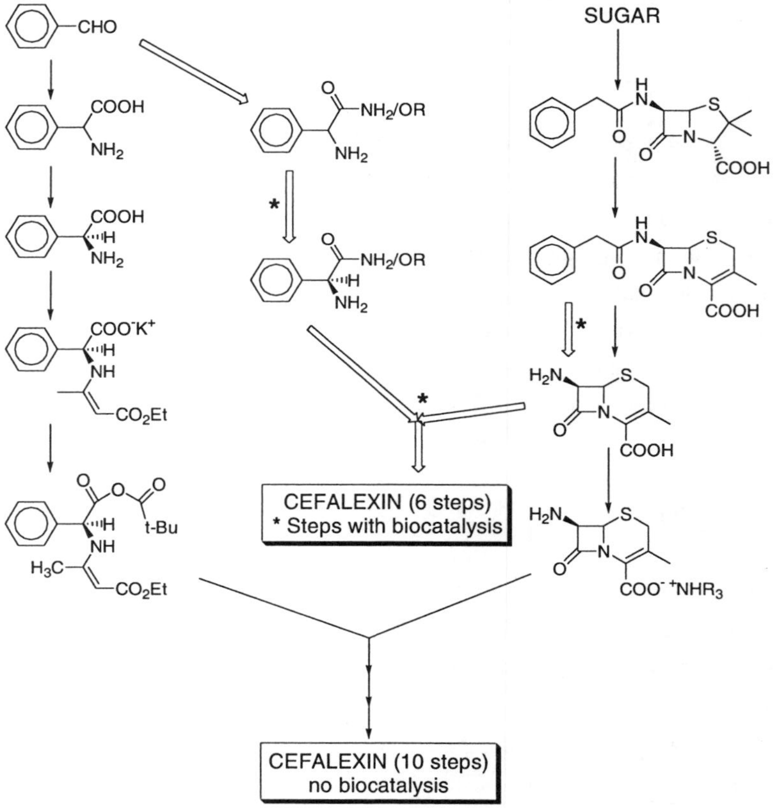

Scheme 1: Traditional (->) and modern (=>) biocatalytic process for Cefalexin.

The crucial final step, in which the cephalosporin nucleus 7-ADCA is coupled with phenylglycine, is one of the first industrial examples of biocatalytic synthesis. So far, enzymes were mainly employed in a hydrolytic mode as shown in the deacylation to give 7-ADCA and the kinetic resolution of the dl-phenylglycine derivatives. Similar processes were developed for all semi-synthetic antibiotics derived from phenylglycine and p-hydroxy-phenylglycine (scheme 2)[3].

A common drawback in these processes is the undesired enzymatic hydrolysis of the side chain derivative to give free, insoluble, amino acid[4]. Together with unfavorable equilibrium conditions in the coupling reaction this can give rise to laborious down stream processing, hampering economic use. From an environmental point of view great advantages have been gained by eliminating halogenated solvents as well as several harsh reagents and preventing the formation of large amounts of inorganic salts. Expressed as kg of waste/kg of product a reduction of 30/1 tot 5/1 has been achieved[1].

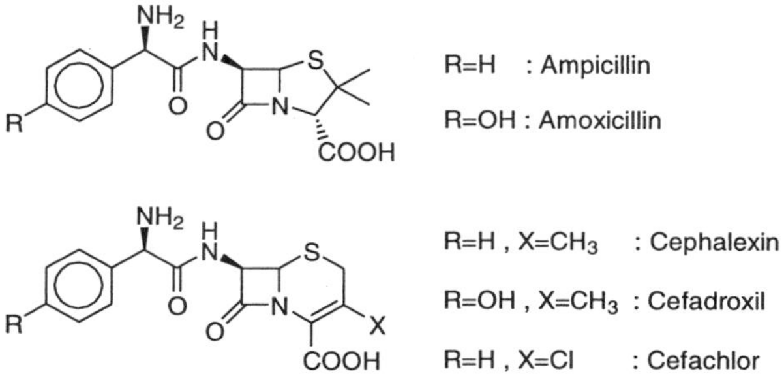

R=H   : Ampicillin
R=OH : Amoxicillin

R=H , X=CH$_3$: Cephalexin
R=OH , X=CH$_3$: Cefadroxil
R=H , X=Cl   : Cefachlor

Scheme 2: Penicillins and Cephalosporins for which enzymatic coupling processes have been developed.

3.   Ongoing greening

As penicillins and cephalosporins are expected to continue to be prominent antibacterials for another 10-20 years, (despite many resistance problems) research is continued towards further simplification and efficiency of the manufacturing process. Research at several Dutch universities in collaboration with DSM Life Science Products aims at improving enzymatic processes, development of new biocatalysts, as well as further integration of chemical synthesis and biocatalysis, alternative process technologies and efficient down stream processings. A number of approaches and results are presented below.

3.1. FERMENTATION OF 7-ADCA

A major step forward in further shortening of the industrial synthesis for cephalosporins has already been made. The multistep chemical conversion of Pen. G to 7-ADCA (see Scheme 1) has recently been replaced by a two-step biosynthesis. The 7-N-adipoyl-ADCA can now be obtained directly through fermentation with a modified *Penicilium chrysogenum*, followed by a simple enzymatic removal of the amino substituent[5].

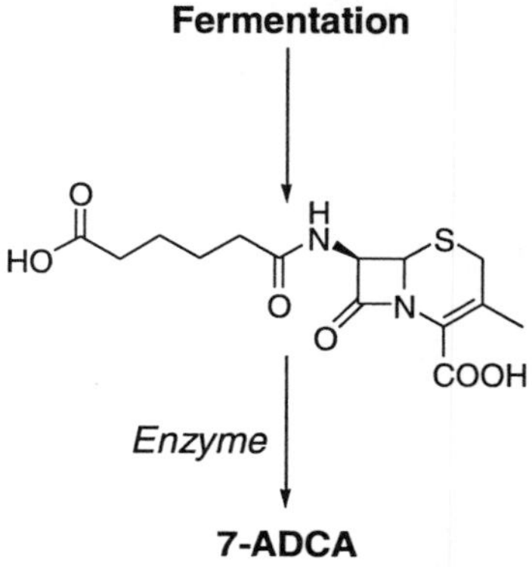

Scheme 3: Biosynthesis of 7-ADCA.

Again, a great number of organic reagents (silylating agents, phosphor halides, pyridine, DMF) and halogenated solvents have been replaced by biocatalysts in aqueous medium.

3.2. THERMODYNAMIC COUPLING

An early goal in our research program was the thermodynamically controlled direct coupling of the free side chain amino acid with the underivatized nucleus 7-ADCA (for cephalosporins; see Scheme 4) or 6-APA (for penicillines).

Some literature was available dealing with enzyme catalyzed equilibria between free nucleus and side chain[6]. These are, however, all based on hydrolysis experiments. A French patent claiming straightforward coupling of 6-APA with a simple phenylglycine salt in water and catalyzed by *Pen. G. Acylase* was proven to be false[7]. Also a Spanish patent application contains unproven claims about coupling with amino-acid side chains[8]. Research in collaboration with the Universities of Wageningen and Delft has shown that thermodynamic coupling can be done, provided the side chain does not contain an α-amino substituent[9-11]. Obviously, the zwitter ionic character of α-amino acids constitutes an energy minimum bringing them out of reach for activation towards coupling by commonly used Penicillin acylases. Replacing water by polar, hydrophilic solvents such as glycols and glymes some coupling activity could be detected. Conditions, however, are rather remote from industrial relevance.

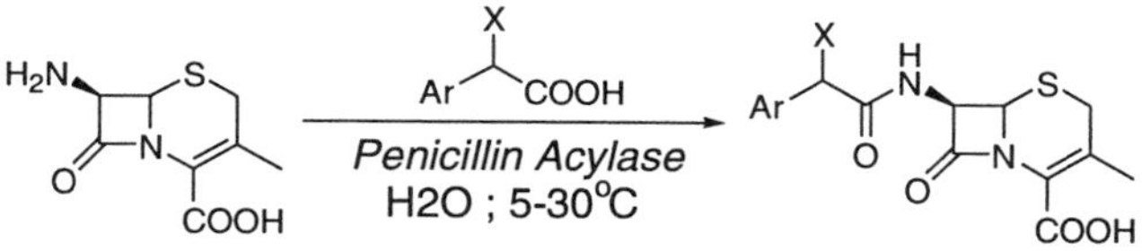

Ar	X	Coupling %	pH
Ph	H	28	6.0
3-Indole	H	4	5.0
Ph	Br	27	6.0
Ph	OH	9	5.0
4-HO-Ph	OH	6	6.0
Ph	NH_2	-	4-7
4-HO-Ph	NH_2	-	4-7

Scheme 4: Thermodynamic coupling towards β-lactam antibiotics.

3.3. PRODUCT-SPECIFIC COMPLEXATIONS

Specific complexation of cephalosporins was discovered by Eli Lilly and applied in enzymatic coupling to Cefalexin by NOVO, employing β-naphthol as complexing agent[12,13]. Further development is performed at DSM in collaboration with the University of Nijmegen. Surprisingly, β-naphthol does not interfere with the enzymatic coupling, although the reaction slows down. β-Naphthol complexation of Cefalexine brings the coupling equilibrium fully to synthesis. Undesired hydrolysis of the side chain precursor, however, still takes place resulting once again in complicated work-up procedures. Also, efficient removal of β-naphthol is of critical importance in order to meet all quality requirements of the bulk medicinal end product.

In order to find more environmentally compatible alternatives for β-naphthol, crystal structures of the β-naphthol complexes were determined and several other aromatics were tested in complex formation [14]. A range of complexing agents was found, as shown in scheme 6.

Although various crystal structure types are found, the common feature is a cage formed by four cephalosporin molecules. The cage is mostly filled with two host molecules and a varying number of water molecules to reach maximum crystal lattice stability. The flexibility of the cage combined with the usage of water as cement allows for the large number of hosts that can be accommodated. The results are currently used in a predictive model towards further product-specific complexation and product isolation[15].

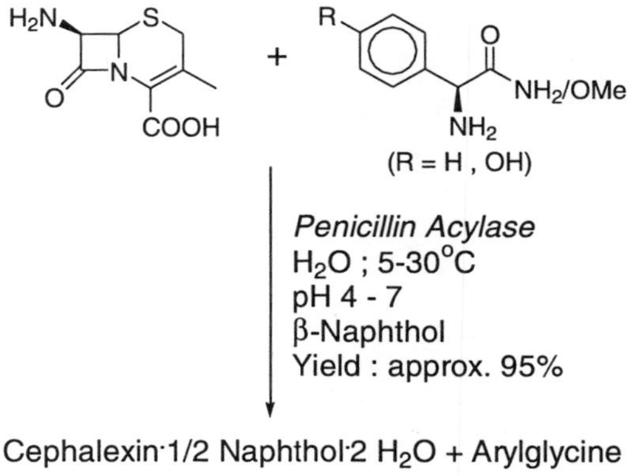
H2N
S
N
O
COOH
+
R
O
NH2/OMe
NH2
(R = H , OH)
Penicillin Acylase
H2O ; 5-30°C
pH 4 - 7
β-Naphthol
Yield : approx. 95%
Cephalexin·1/2 Naphthol·2 H2O + Arylglycine

Scheme 5: Complexation of Cephalosporins with β-naphthol..

Complexing agent	Cephradine	Cephalexin	Cefaclor	Cefadroxil
1. β-naphthol	A	A	A	B
2. α-naphthol	A	A	A	B
3. quinolin	A	A	A	*
4. naphthalene	A	A		*
5. 1,4-dihydroxynaphthalene	A	A		*
6. 1,5-dihydroxynaphthalene	A	A		*
7. 2,3-dihydroxynaphthalene	A	A		*
8. 1,6-dihydroxynaphthalene	A	A		B
9. 2,6-dihydroxynaphthalene	A	A		B
10. 2,7-dihydroxynaphthalene	A	A		B
11. 1-acetonaphtone	A			*
12. 2-acetonaphtone	A	A		*
13. 1-chloronaphthalene	A			*
14. coumarin	A			*
15. 8-hydroxyquinolin	A	A		*
16. 1,2,3,4-tetrahydro-1-naphthol	A			*
17. 1,5-dihydroxy-1,2,3,4-tetrahydronaphthalene	A			*
18. indole	A	A		*
19. indene	A			*
20. 2,2'-bipyridyl	A	A	A	*

Scheme 6: Complexation of Cephalosporins with aromatics.

4.   Outlook

The fast acceptance of biocatalysis by (fine-)chemical industry will continue to trigger a great deal of research in the areas of organic chemistry, bio-synthesis and process technology. At the same time much additional fundamental insight, i.e. in enzyme action, molecular biology of micro-organisms and biocatalyst formulation, is required to allow further industrial exploitation. A few challenges, both from a scientific and an applied point of view are shown below.

4.1.   CASCADE CATALYSIS

Even today's organic syntheses are still mainly governed by a step by step approach; bond cleavage and bond making are done one by one. Lack of selectivity and/or incompatible reaction conditions are the underlying causes. The high selectivity of enzymes all at similar conditions, i.e. aqueous systems, allows in principle the use of several biocatalysts in one reactor system (batch reactor, series of columns, etc.). A promising example is shown in Scheme 7 employing three enzymes and a consecutive substitution in one pot to give Cefazolin[14].

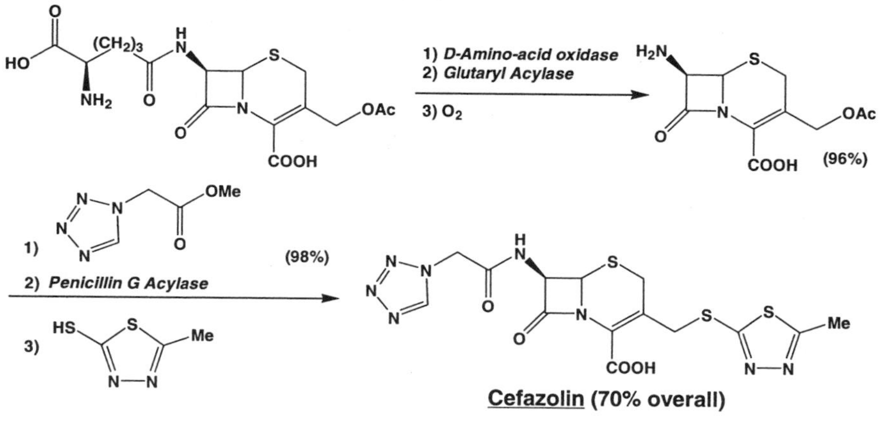

Scheme 7: Cascade catalysis in Cefazolin synthesis.

A challenging extension would be the introduction of those enzymes in the micro-organism employed in the fermentation of the starting material Cephalosporin C and thus allowing direct fermentation. Similar approaches can be envisaged for other penicillines and cephalosporines as outlined in Scheme 8.

456

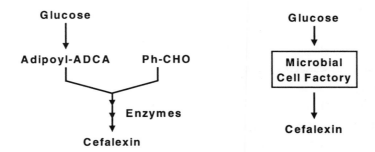

Scheme 8: Cascade catalysis and direct fermentation of Cefalexin.

Many problems at the molecular biology level, however have to be solved, before industrial application will be feasible. Transport mechanisms in micro-organisms and interaction of primary and secondary metabolism are just a few.

4.2.   BIOCATALYSTS

So far, industrially applied biocatalysts mainly serve hydrolytic reactions. In this paper some examples towards synthesis use are shown. Application of multi-enzyme systems, including cofactors, is still in its infancy and mainly investigated at academic institutions. Also the number of industrially available enzymes, ca. 350, is still rather limited compared to the vast natural reserves[17]. Several research lines need to be elaborated before biocatalysis and biotransformations will reach maturity:
- high speed screening techniques for multitudes of enzymes (natural or genetically engineered);
- reliable and speedy methods for the preparation of batches of enzymes;
- fast techniques for determining enzyme and active site structures as well as interactions with substrates;
- rational formulation methods, i.e. immobilization of enzymes to stable and robust industrial biocatalysts.

In particular dedicated biocatalyst formulation will be very useful. It is well known that immobilization of enzymes can greatly influence selectivity, efficiency and stability[18,19]. In fact, well chosen enzyme formulations may increase the present number of 350 industrially available enzymes to a few thousands biocatalysts. At the same time, closer collaboration between organic chemists and molecular biologists will lead to new bio-inspired catalyst systems.

4.3.   A DILEMMA?

The requirement for high speed techniques shown in the previous section is felt most profoundly in the development of fine chemicals for pharmaceuticals. Time to market is here the key success factor. At the same time several quality requirements in drug manufacturing allow only very limited changes in process routes once the first representative grams or kilograms have been produced. Extensive route scouting, careful process optimalisation and scale-up studies are severely hampered by these contradicting demands. Fast screenings techniques and automated process optimizations will be needed to solve this dilemma. Ideally these techniques are of such a level of reliability and reproducibility that time consuming scale-up studies and pilot-plant programs can be eliminated all together. Given the limited range of conditions in biotransformations when compared to chemical conversions, biocatalytic processes will stand a better chance to prevent the dilemma between development speed and process sustainability.

5.   Acknowledgement

This report is based on R&D efforts by many colleagues at DSM Research, DSM Anti Infectives, DSM Fine Chemicals and the Universities of Delft, Wageningen, Nymegen and Groningen. Financial support by the Ministry of Economic Affairs is kindly acknowledged.

6.   References

1.   Bruggink, A. (1998) Growth and efficiency in the (fine)chemical industry, *Chimica Oggi* **16**, 9/10, 44-47.
2.   Bruggink, A. (1996) Biocatalysis and process integration in the synthesis of semi-synthetic antibiotics, Chimia **50**, 431-432.
3.   Several patents have been applied for and published by DSM Life Science products, i.e. WO 99/20786 (1999), WO 98/56946 (1998), WO 98/56945 (1998), WO 99/15531 (1999), WO 95 /03420 (1995), WO 97/04086 (1997), WO 93/12250 (1993), WO 95/34675 (1995), WO 96/30376 (1996), WO 96/23796 (1996), WO 96/23797 (1996), WO 96/02663 (1996), WO 96/23897 (1996).
4.   Bruggink, A., Roos, E.C. and de Vroom, E. (1998) Penicillin Acylase in the industrial production of β-lactam antibiotics, *Organic Process Research & Development* **2**, 128-133.
5.   See patent applications by Gist-brocades WO 96/38580 (1996) and WO 97/20053 (1997).

6. Blinkovsky, A.M. and Markaryan, A.N. (1993) Synthesis of β-lactam antibiotics containing α-aminophenylacetyl group in the acyl moiety, *Enzyme Microbiol. Technol.* **15**, 965-973.

7. French Patent (1968), 2014689.

8. WO 91/09136 (1991).

9. Schroën, C.G.P.H., Nierstrasz, V.A., Kroon, P.J., Bosma, R., Janssen, A.E.M., Beeftink, H.H. and Tramper, J. (1999) Thermodynamically controlled synthesis of β-lactam antibiotics, *Enzyme Microb. Technol.* **24**, 498-506.

10. Nierstrasz, V.A., Schroën, C.G.P.H., Bosma, R., Kroon, P.J., Beeftink, H.H., Janssen, A.E.M., and Tramper, J. (1999) Thermodynamically controlled synthesis of Cefamandole, *Biocatalysis and Biotransformation* **17**, 209-223.

11. Diender, M.B., Straathof, A.J.J., Van der Wielen, L.A.M., Ras, C. and Heijnen, J.J. (1998) Feasibility of the thermodynamically controlled synthesis of amoxicillin, *J. Molecular Catalysis B: Enzymatic* **5**, 249-253.

12. Patents to Eli Lilly, WO 94/18209 (1994) and EP 637585 (1995).

13. See patent WO 93/12250 (1993), assigned to NOVO-Nordisk, transferred to DSM Life Science Products.

14. Kemperman, G.J., de Gelder, R., Dommerholt, F.J., Raemakers-Franken, P.C., Klunder, A.J.H. and Zwanenburg, B. (1999) Clathrate type complexation of cephalosporins with β-naphthol, *Chemistry*, 2163-2168.

15. Kemperman, G.J., de Gelder, R. Dommerholt, F.J., Raemakers-Franken, P.C., Klunder, A.J.H. and Zwanenburg, B. (1999) Design of inclusion compounds of cephalosporin antibiotics, submitted for publication.

16. Fernandez-Lafuente, R., Guisan, J.M., Pregnolato, M. and Terreni, M. (1997) Chemoenzymic one-pot synthesis of cefazolin, *Tetrahedron Lett.* **38**, 4693-4696.

17. See for example "Biocatalysts for Industry", Róche Diagnostics GmbH, Mannheim, Germany (1999).

18. Hartmeier, W. (1986) *Immobilisierte Biokatalysatoren*, Springer Verlag, Berlin.

19. Mohy Eldin, M.S., Schroën, C.G.P.H., Janssen, A.E.M., Mita, D.G. and Tramper, J. (1999) Immobilization of penicillin acylase onto a chemically grafted nylon matrix, *J. Molecular Catalysis B: Enzymatic*, submitted for publication